Archives of Toxicology • Supplement 20

Springer

Berlin
Heidelberg
New York
Barcelona
Budapest
Hong Kong
London
Milan
Paris
Santa Clara
Singapore
Tokyo

Diversification in Toxicology - Man and Environment

Proceedings of the
1997 EUROTOX Congress
Meeting Held in Århus, Denmark,
June 25–28, 1997

Edited by the Publication Commitee
J. P. Seiler, J. L. Autrup and H. Autrup

With 71 Figures and 58 Tables

Editors

DR. JÜRG P. SEILER
Intercantonal Office for the Control of Medicines (IOCM)
Erlachstrasse 8
CH-3000 Bern 9
Switzerland

PROF. DR. HERMAN AUTRUP
University of Åarhus
Steno Institute of Public Health
Universitetsparken, Bldg. 180
DK-8000 Åarhus C
Denmark

DR. JUDITH L. AUTRUP
Krathusvej 2A
DK-8240 Risskov
Denmark

ISBN 3-540-63660-9 Springer-Verlag Berlin Heidelberg New York

Cataloging-in-Publication Data applied for

Die Deutsche Bibliothek – CIP-Einheitsaufnahme

Diversification in toxicology: man and environment; proceedings of the 1997 EUROTOX Congress Meeting held in Århus, Denmark, June 25–28, 1997 – Berlin; Heidelberg; New York; Barcelona; Budapest; Hong Kong; London; Milan; Paris; Santa Clara; Singapore; Tokyo: Springer 1998
(Archives of toxicology: Supplement 20)
ISBN 3-540-63660-9
Früher Schriftenreihe
Fortlaufende Beil. zu Archives of toxicology
ISSN 0171-9750

Printed in Germany

Typesetting: camera ready by editors
Cover Design: design & production, Heidelberg

SPIN: 10654607 27/3136 – 5 4 3 2 1 0 – Printed on acid free paper

Contents

The Gerhard Zbinden Memorial Lecture

Plenary Lecture

Animals as Models in Toxicology: Animal Welfare and Ethics

Ecotoxicological Risk Assessment

Effects of Environmental Chemicals on Reproduction

Biomarkers of Exposure

Current Molecular Approaches in Toxicology

Chemoprevention

Regulatory Immunotoxicology: The Scientist's Point of View

Novel Issues of Risk Assessment of Chemical Carcinogens

Complex Mixtures

Receptor Mediated Toxic Responses

New Frontiers in Human and Ecological Toxicology: Determining Genetic vs. Environmental Causes of Susceptibility

Extrahepatic Metabolism in Target Organ Toxicity

The Young Scientist Poster Award

The Gerhard Zbinden Memorial Lecture

EUROTOX has instituted this lecture to honor the memory of Gerhard Zbinden (1924 - 1993), EUROTOX Honorary Member and recipient of the EUROTOX Merit Award. The Gerhard Zbinden Memorial Lecture aims at recognising scientific excellence in the area of drug and chemical safety. The lecture is held at the annual EUROTOX Congress by a scientist chosen for his/her outstanding research contributions to the science of toxicology. The lecture is sponsored by the Chemical Industries Basel (KGF: F. Hoffmann-La Roche AG, Lonza AG, Novartis AG), Switzerland.

Molecular Basis of Halothane Hepatitis

Josef Gut
Novartis Pharma, DMPK/Biotransformation, K-136.1.19, CH-4002 Basel

Introduction: Drug-Induced Hepatic Failure, Idiosyncratic Drug Reactions and the Hapten Theory

The formation of covalent adducts to cellular macromolecules, including proteins, phospholipids, and DNA or RNA, is associated with the exposure of humans and animals (most probably to any given living organism) to a large number of xenobiotics, including drugs as well as occupational and environmental pollutants (Hinson and Roberts, 1992; Park and Kitteringham, 1990; Nelson and Pearson, 1990). Only few parent compounds form, based on their intrinsic chemical reactivity, such adducts spontaneously; the majority of compounds do form adducts to cellular macromolecules only after metabolic activation to reactive intermediates (Park and Kitteringham, 1990; vanWelie et al., 1992). From an immunological point of view, adducted cellular target molecules constitute "modified self" or even "non-self" structures, which might provoke immune responses against themselves (Allison 1989; deWeck 1983). In general, low molecular mass organic compounds (<1000 Da), such as most drugs, are thought to be non-immunogenic *per se*. However, many of these compounds (coined haptens) may become immunogenic, when covalently linked to a macromolecular carrier such as a protein. Once formed, drug-carrier conjugates might act as immunogens and elicit immune responses at the humoral level, at the cellular level, or at both levels (Allison 1989; deWeck 1983). These immune responses might be directed against at least three different types of antigenic determinants. First, haptenic epitopes may include the derivative of the xenobiotic (i.e. hapten) bound to the carrier molecule. Second, new antigenic determinants (NAD) may comprise newly created linear or conformational structures on the carrier molecule elicited upon binding of the hapten to the carrier molecule. Last, cryptic autoantigenic determinants of the carrier molecule, which normally are seen as self or are ignored, could bypass the mechanisms of immunological self-tolerance as a consequence of hapten binding (Allison, 1989; Blooksma and Schuurman, 1989; Roitt et al., 1985). An immunological basis for the development, in susceptible individuals, of a number of severe, sometimes life-threatening adverse effects after therapeutic,

---

Abbreviations:
CF_3CO-Lys, N6-trifluoroacetylated-L-lysine; PDC, pyruvate dehydrogenase complex; OGDC, 2-oxoglutarate dehydrogenase complex; BCOADC, branched chain 2-oxoacid dehydrogenase complex; RSA, rabbit serum albumin;

occupational, behavioral, and environmental exposure of humans to xenobiotics has been implicated in debilitating syndromes including drug-induced lupus erythematosus, anaphylaxis, serum sickness, dermatitis, glomerular nephritis, etc. (deWeck, 1983). Many of these syndromes are caused by idiosyncratic toxins.

Drug-induced hepatitis in chronic, intermittent, and fulminant forms is a prominent clinical manifestation of a severe adverse drug reaction, and a recent update (Zimmermann, 1990) indicates that the number of drugs which induce hepatitis is on the rise and includes compounds from a broad range of pharmacological actions such as anti-inflammatory agents (diclofenac, sulindac), diuretic agents (tienilic acid), cardiovascular agents (ajmaline), anti-hypertensive agents (dihydralazine), anticonvulsant agents (diphenylhydantoin, carbamazepine) and anesthetic agents (halothane, enflurane). Evidence for the involvement of the immune system in the development of drug-induced hepatic failure stems from the fact that autoantibodies and/or antibody/antigen complexes against cellular self macromolecules are frequently found in the sera of afflicted patients (Homberg et al. 1985; Beaune et al., 1987; Bourdi et al., 1990; Leeder et al., 1992; Pirmohamed et al., 1992). Multiple exposure to, and metabolism of, the respective drug seemed to be prerequisites for induction of the autoimmune-type hepatic reactions, and discontinuation of the drug resulted in both the remission of the clinical syndrome and the reduction in the titer of autoantibodies (Homberg et al., 1985; Bourdi et al., 1992). Moreover, re-challenge of patients with the offending drug may result in re-appearence of high-titer autoantibodies and the clinical syndrome, as nicely documented in a recent case report of diclofenac-induced hepatitis (Scully et al., 1993).

Halothane Hepatitis

Halothane was developed as a volatile anesthetic agent in the 1950's and its safety and efficacy rendered it the most widely used general anesthetic agent by the 1970's. Isolated reports on unexplained hepatitis following halothane exposure precipitated the retrospective National Halothane Study (National Halothane Study, 1966) which confirmed an association of exposure of humans with the occurrence of otherwise unexplained hepatitis. There has been considerable progress in the understanding of the biotransformation of halothane, in the mechanisms of formation of drug-carrier conjugates, in the identification of such conjugates, and in their implication in the development of halothane-induced hepatotoxicity.

Metabolism of Halothane

Biotransformation of halothane appears to be a prerequisite for halothane hepatotoxicity. About 20% of the administered dose is metabolized through oxidative and reductive pathways (Casorbi et al., 1970). There exists ample in vitro evidence that the oxidative pathway of halothane biotransformation is catalyzed by rodent and human cytochrome P450 2E1 (Gruenke et al., 1992; Waskell, 1994, Spracklin et.al, 1997), while human cytochromes P450 2A6 and 3A4 catalyze the reductive biotransformation of halothane (Spracklin et. al.,

1996). For the sake of this report, the reductive pathways of halothane biotransformation are not considered any further. Under oxidative conditions (i.e., O_2 partial tensions > ~50 μM), P450 2E1-dependent monooxygenation of halothane ($CF_3CHClBr$) gives rise to a highly unstable initial geminal halohydrin ($CF_3COHClBr$) which spontaneously degenerates to afford the corresponding trifluoroacyl chloride (CF_3COCl). Hydrolysis of the trifluoroacyl chloride yields trifluoroacetic acid (CF_3COOH), a major urinary metabolite of halothane. However, the trifluoroacyl chloride does also react with nucleophilic centers available at the site of its generation, giving rise to adducts to cellular macromolecules, including trifluoroacetylated proteins (i.e., CF_3CO-proteins[1]) and phospholipids (Cohen et al., 1975; Müller and Stier, 1982; Trudell et al., 1991). In liver proteins, N^6-trifluoroacetyl-L-lysine (CF_3CO-Lys) was identified by ^{19}F-NMR spectroscopy as the major adduct formed under these conditions (Harris et al., 1991).

Features of Halothane Hepatitis: Immunological Basis

Halothane hepatitis (for review see Neuberger, 1989) is defined as the development of otherwise unexplained liver damage in patients with previously normal liver functions. Halothane hepatitis is characterized by severe disturbance of liver functions and associated fulminant hepatic failure. The incidence was estimated to be in the order of 1 in 35'000 individuals exposed to the agent for the first time (National Halothane Study, 1966); in patients receiving halothane on multiple occasions within a 1-month period, the frequency was increased to ~ 1 in 3'700. Some characteristic features of halothane hepatitis include a heightened female-to-male ratio (~1.6:1), an exposure of individuals to halothane more than once (~75% of cases), a high mortality rate, the occurrence of liver-kidney microsomal autoantibodies, and an increased incidence of circulating immune complexes (Neuberger and Williams, 1984; Neuberger, 1989). The latter findings implicate an immunological basis for the etiology of halothane hepatitis. In sera of afflicted patients but not in halothane-exposed control individuals, antibodies exist that are specifically directed against CF_3CO-proteins (coined neo-antigens) present in rat and human liver microsomal membrane preparations of rats and humans previously exposed to halothane (Kenna et al., 1987b; Kenna et al., 1988b). The adducts present on CF_3CO-proteins (i.e., CF_3CO-Lys (Harris et al., 1991)) are elicited after exposure of experimental animals (or humans) to halothane through covalent interaction of the halothane-derived trifluoroacylchloride intermediate with lysine moieties present in target proteins. The formation of CF_3CO-proteins after exposure of experimental animals to halothane is not restricted to the liver. Discrete CF_3CO-proteins have been identified by immunoblotting in the kidney (Huwyler et al., 1992), the heart (Huwyler and Gut, 1992), and the testis (Kenna et al., 1992) of rats and the lungs and the respiratory and olfactory epithelia of mice (Heijink et al.,

---

[1] The term "CF_3CO-protein" refers to a trifluoroacetylated protein with no reference made to the identity or function of that particular protein, except where explicitly indicated.

1993). Generally, however, the density of CF_3CO-proteins in the extrahepatic tissues is much lower than in the liver (i.e., kidney ~5%, heart ~0.5%, when compared to the liver; Huwyler et al., 1992; Huwyler and Gut, 1992). In the rat liver, CF_3CO-proteins are located not only in hepatocytes, but also in Kupffer cells, which partially process CF_3CO-proteins (Christen et al., 1991a); given the ability of Kupffer cells for antigen presentation (Rubinstein et al., 1987), this finding might become important in terms of presentation of CF_3CO-proteins (or fragments thereof) to the immune system in a competent form. Moreover, CF_3CO-proteins have also been found on the surface of rat and rabbit hepatocytes (Vergani et al., 1980; Satoh et al., 1985a).

Identification of CF_3CO-proteins

The exposure of humans or experimental animals to halothane results in the generation of a large number of CF_3CO-proteins of distinct molecular masses within liver microsomal membranes (Kenna et al., 1988a; Kenna et al., 1988b; Pohl et al., 1989; Christen et al., 1991b; Gut et al., 1992); these CF_3CO-proteins are readily recognized by anti-CF_3CO antibodies[2] on immunoblots. CF_3CO-proteins are not persistent; ten days after exposure of rats to halothane they are no longer detectable (Pohl et al., 1989; Christen et al., 1991b); similarly, in patients with halothane hepatitis, biopsied at days 6 or 7 after exposure to halothane, CF_3CO-proteins were no longer detectable (Gut et al., 1992). Sera of patients with halothane hepatitis seem to express a certain preference for the recognition of a few distinct CF_3CO-proteins with apparent molecular masses of 100 kDa, 76 kDa, 59 kDa, 57 kDa, and 54 kDa, present in liver microsomes of halothane-exposed rats (Kenna et al., 1988a; Pohl et al., 1989; Pohl, 1990). Based on their reactivity with patients' sera, several CF_3CO-proteins were isolated from liver microsomal membranes of rats previously exposed to halothane. Purified proteins were then subjected to NH-terminal sequence analysis, sequence analysis of tryptically generated internal peptides, and molecular cloning of the corresponding cDNA's. Using these strategies, microsomal protein disulfide isomerase (Martin et al., 1993a), microsomal carboxylesterase (Satoh et al., 1989), the stress protein ERp99 (Thomassen et al., 1989) – which is believed to be identical to the glucose-regulated protein GRP 94 and endoplasmin (Mazzarella and Green, 1987) –, the stress protein ER p72 (Pumford et al., 1993), the glucose regulated protein GRP 78 (BiP, a molecular chaperone (Davila et al., 1992)), and the Ca^{2+}-binding protein calreticulin (Butler et al., 1992) have been identified as CF_3CO-proteins (i.e., neo-antigens). The identity and function of a further isolated neoantigen of 58 kDa remains to be determined (Martin et al., 1991; Martin et al, 1993b). Recently, cytochrome P450 2E1 was identified both as an CF_3CO-protein and a presumed hepatocellular surface autoantigen associated with halothane hepatitits (Eliasson and Kenna, 1996; Bourdi et al., 1996). The availability of isolated CF_3CO-proteins and their corresponding native, not trifluoroacetylated forms, allowed the characterization of the reactivity of

[2] The term "anti-CF_3CO antibody" refers to the monospecific antibody obtained in this laboratory (Christen et al., 1991a) from an anti-CF_3CO antiserum through affinity-purification on a CF_3CO-Lys affinity matrix.

patients sera towards these neoantigens, and recent studies (Smith et al., 1993) confirmed that patients' sera contain autoantibodies to the native, recombinantly expressed, non-trifluoroacetylated form of human carboxylesterase. Similarly, protein disulfide isomerase (Martin et al., 1993a) and the neoantigen of 58 kDa (Martin et al., 1993b), again both in trifluoroacetylated and native form were recognized by some patients' sera.

Molecular Mimicry of CF_3CO-Proteins

Most of the research efforts by several groups are focused on the identification of CF_3CO-proteins as (the) ultimate neoantigen(s) (i.e., candidate immunogen(s)) which might be responsible for the development of halothane-induced hepatitis. Current evidence suggests that a wide variety of CF_3CO-proteins is formed in all individuals (or experimental animals) exposed to halothane. An immunological reaction towards a discrete number of these CF_3CO-proteins occurs in the small subset of susceptible individuals only. Obviously, in contrast to these susceptible individuals, CF_3CO-proteins are immunologically tolerated by the vast majority of exposed individuals. One of several possibilities for the lack of immunological responsiveness towards CF_3CO-proteins in healthy individuals might be the existence of natural immunological tolerance (Allison, 1989; Nossal, 1991) towards CF_3CO-proteins. A mechanisms for such tolerance to develop might involve the existence of a repertoire of constitutively expressed self-determinants which bear, due to molecular mimicry (Fujinami 1991; Dyrberg 1992), a very close structural resemblance, if not identity, to determinants present on CF_3CO-proteins (i.e., CF_3CO-Lys). Based on such molecular mimicry, tolerance towards CF_3CO-proteins could develop in individuals incidentally through the normal processes of thymic selective elimination (clonal deletion), the development of clonal ignorance, and/or by thymic and peripheral silencing (clonal anergy) (Kappler et al., 1987; Kisielow et. al., 1988; Kappler et al., 1988; Ramsdell and Folkwes, 1990; Sprent et al., 1990; Ohashi et al., 1991) of those cells in the maturing repertoires of T-cells and B-cells which are directed towards those self-determinants which sufficiently closely mimic corresponding determinants present on CF_3CO-proteins.

Molecular Mimicry of CF_3CO-Proteins by Constitutive Proteins

If crossreactive determinants that molecularly mimic determinants on CF_3CO-proteins (i.e., CF_3CO-Lys) would occur in the repertoire of self, one should be able, based on immunochemical crossreactivity, to detect such structures upon suitable screening with the help of an antibody which is highly specific towards CF_3CO-proteins. We had previously generated a monospecific antibody, coined anti-CF_3CO antibody, through affinity purification on a CF_3CO-Lys affinity matrix of an anti-CF_3CO antiserum, raised in rabbits towards CF_3CO-RSA (Christen et al., 1991b). In fact, this anti-CF_3CO antibody recognized two crossreactive proteins of 52 kDa and 64 kDa on immunoblots of both liver tissue obtained from human individuals who were never before exposed to halothane and also in the liver as well as in several other tissues (i.e., kidney, heart, thymus, spleen, and skeletal muscle) of unexposed rats (Christen et al., 1991b). In antibody exchange experiments, anti-CF_3CO antibodies, pre-adsorbed to the crossreactive proteins of 52 kDa and of 64 kDa present in human liver, spontaneously exchanged to target CF_3CO-proteins present in liver tissue of rats exposed to halothane (Christen et al., 1991b) or in the liver biopsy of a human individual who died shortly after halothane anesthesia (Gut et al., 1992). Moreover, in both types of experiments the recognition by anti-CF_3CO antibody of the proteins of 52 kDa and 64 kDa was abolished in presence of CF_3CO-Lys, suggesting that the recognized epitopes of proteins of 52 kDa and 64 kDa exhibit immunochemical properties very similar, if not identical, to epitopes present on CF_3CO-proteins.

Identification of the Protein of 64 kDa

By immunoaffinity chromatography on an anti-CF_3CO antibody affinity matrix, we have isolated the crossreactive protein of 64 kDa from solubilized rat heart subcellular fractions. The protein was identified as the dihydrolipoamide acetyltransferase subunit (E2 subunit) of the mitochondrial pyruvate dehydrogenase complex (PDC) by comparing amino acid sequences, obtained by microsequencing from internal tryptic peptides of the protein, with those available from the EMBL and SwissProt databases (Christen et al., 1993). The E2 subunit of the PDC is a highly structured molecule comprising a number of functional and structural domains. In particular, the E2 subunits of mammalian pyruvate dehydrogenase complexes contain two lipoyl-binding domains in which the prosthetic group lipoic acid is covalently linked to the N6-amino group of the central lysine residue of a consensus motif ETDKA found in human, rat, mouse, and bovine E2 subunit proteins (Yeaman, 1986; Perham, 1991). While the exact epitope(s) on the E2 subunit protein, reacting with anti-CF_3CO antibody, was (were) not identified at the time, the lipoyl-binding domain(s) were likely candidates in that free lipoic acid[3] inhibited, in both competitive immunoblotting and immunoprecipitation experiments, the recognition by anti-CF_3CO antibody of the E2 subunit of the PDC and of CF_3CO-proteins in a manner very similar to CF_3CO-Lys (Christen et al., 1993). In additional experiments, it was shown that in fact the presence of lipoic acid was responsible for the recognition of the E2 subunit of the PDC by anti-CF_3CO antibody. Thus, upon recombinant expression in E. coli of a subgene encoding the inner lipoyl domain of the human E2 subunit of the PDC, a lipoylated form (Llip) and an unlipoylated form (Ulip) were expressed and could be separated from each other by non-denaturing gel electrophoresis (Quinn et al., 1993). Upon analysis on immunoblots, Llip only was recognized by anti-CF_3CO antibody while no reactivity was evident with Ulip (Christen et al., 1994).

Structural Basis for Molecular Mimicry of CF_3CO-Lys by Lipoic Acid

In order to further elucidate the structural basis of molecular mimicry of CF_3CO-Lys by lipoic acid, we performed, by using the affinity purified E2 subunit of the

[3] Except where noted, (6RS)-lipoic acid was used throughout our experiments.

PDC as target antigen and anti-CF_3CO antibody as a molecular probe, competitive immunoblotting experiments in which the interaction between the antigen and the antibody was competed for by the presence of increasing concentrations of CF_3CO-Lys, lipoic acid (both the oxidized and reduced forms), Lys(Ac), octanoic acid, L-Lysine and trifluoroacetic acid. Moreover, we synthesized the lipoylated peptide $ETDK_{(lipoyl)}ATIG$, which comprises the consensus sequence ETDKA (Yeaman, 1986; Perham, 1991) and its non-lipoylated counterpart IETDKATIGFE and used them in competitive immunoblotting experiments also. In these experiments, the apparent half-maximal inhibitory constants (IC50) obtained reflected the relative structural relatedness of the tested compounds towards the primary antigenic target recognized by anti-CF_3CO antibody. Based on the IC_{50}-values obtained, it became evident that lipoic acid in its oxidized form (IC_{50} = 0.047 mM) is very closely related in overall structure to CF_3CO-Lys (IC50 = 0.004 mM). Furthermore, it appeared to be irrelevant whether lipoic acid was in free form or in covalent linkage to the N^6-amino group of the central lysine of the peptide $ETDK_{(lipoyl)}ATIG$ (IC_{50} = 0.033 mM). The collective evidence from these data suggested that in fact free lipoic acid is the minimal, yet sufficient structure to molecularly mimic CF_3CO-Lys.

We have used a molecular modelling approach (Christen et al., 1994) to rationalize at the level of molecular structures the phenomenon of mimicry of CF_3CO-Lys by lipoic acid. Thus, using Nemesis V2.0 on the Macintosh platform, we calculated Connolly dot surfaces of energy-minimized conformations of CF_3CO-Lys and of the (6S)- and the (6R)-enantiomers of lipoic acid. These calculations suggested the occurrence of energetically favorable conformations of the three compounds with a high degree of structural relatedness, which was revealed by the calculated union surface obtained after overlaying CF_3CO-Lys and (6S)-lipoic acid (for details, see Christen et al., 1994). Similar results were obtained with (6R)-lipoic acid also. However, molecular shape recognition by anti-CF_3CO antibody appears not to be the sole determinant for molecular mimicry of CF_3CO-Lys by lipoic acid for two reasons. First, only very minor deviations from the union surface of CF_3CO-Lys and (6S)-lipoic acid were observable in the union surfaces obtained after overlaying of maximally energy-minimized conformations of CF_3CO-Lys and Lys(Ac) or octanoic acid. Second, despite these minor deviations in surface structures, considerable increases in apparent IC_{50}-values were observed in competitive immunoblotting experiments when Lys(Ac) (IC_{50} = 0.63 mM) or octanoic acid (IC_{50} = 13.0 mM) were compared with lipoic acid (IC_{50} = 0.047 mM) or CF_3CO-Lys (IC_{50} = 0.004). Taken together, these data suggested that other factors such as the localization and/or number of critical sites (i.e., atoms) capable of forming hydrogen bonds or of undergoing hydrophobic interactions within the binding pocket of the anti-CF_3CO antibody are important parameters for molecular mimicry to occur.

Identification of Aadditional Lipoylated Proteins Crossreactive with Aanti-CF_3CO Antibody

The pyruvate dehydrogenase complexes of a number of species comprise three constituent subunit enzymes, namely the pyruvate decarboxylase (E1 subunit), the dihydrolipoamide acetyltransferase (E2 subunit), and the dihydrolipoamide dehydrogenase (E3 subunit) (Yeaman, 1986; Perham 1991). The E2 subunit protein is highly structured and comprises, at least in humans and rats, two lipoyl-binding domains, followed by a E1/E3-binding domain (peripheral subunit binding domain), which in turn is followed by the E2 subunit binding/catalytic domain (acyl transferase and inner core domain) which in mammals is responsible for the association of 60 copies of the E2 subunit to form the icosahedral core of the PDC; the distinct domains of the E2 subunit are linked to each other by segments (hinge regions) rich in alanine and proline which have a high degree of conformational flexibility (Perham, 1991; Gershwin and Mackay, 1991; Mattevi et al., 1992; Matuda et al., 1992). Interestingly, the E2 subunit of the PDC is structurally very closely related to the E2 subunits of other oxoacid dehydrogenase complexes, namely the oxoglutarate dehydrogenase complex (OGDC) and the branched chain oxoacid dehydrogenase complex (BCOADC) as well as to protein X, a constituent of the PDC (Coppel et al., 1988; Matuda et al., 1992; Yeaman, 1986; Perham 1991). A striking similarity among these E2 subunit proteins is the presence of one (OGDC, BCOADC, protein X) or two (PDC} lipoyl binding domains to which the prosthetic group lipoic acid is covalently attached to the N^6-amino group of a critical lysine residue (Yeaman 1986; Behal et al., 1989; Reed and Hackert, 1990; Perham, 1991). Consequently, we have identified the E2 subunits of the OGDC and the BCOADC as well as protein X as proteins immunochemically crossreactive with anti-CF_3CO antibody; both E2 subunits as well as protein X behave, with respect to all immunochemical and structural parameters examined, in the same way as the E2 subunit of PDC (Frey et al., 1995).

Lipoylated Proteins as Autoantigens Associated with Halothane Hepatitis

In the light of molecular mimicry of CF_3CO-Lys by lipoic acid, those sera of patients with halothane hepatitis which are reactive with CF_3CO-proteins should, conceptionally, contain (a) discrete population(s) of autoantibodies which also recognize the E2 subunits of the PDC, the OGDC, and the BCOADC, as well as protein X, because of the presence of the prosthetic group "lipoic acid" in these proteins. In fact, in antibody exchange experiments, a discrete population of autoantibodies in patients' sera was identified that could not discriminate between CF_3CO-RSA and the E2 subunits of the PDC or the OGDC and whose reactivity with either target antigen was completely abolished by the presence of CF_3CO-Lys or lipoic acid, but not by L-lysine throughout the antibody transfer step (Christen et al., 1994; Frey et al., 1995).

Up to 82% of those patients' sera which recognized the model antigen CF_3CO-RSA on immunoblots were found to also recognize the E2 subunit of the PDC as well as protein X, the additional subunit constituent of the PDC. Interestingly, in very recent experiments, a very low level of naturally occurring autoantibodies towards the E2 subunit of PDC was detected in ~20% of human individuals in the general population. In contrast, an extremely high titer of anti-E2 subunit autoantibodies was detected in one human individual who is highly sensitized to passive occupational exposure to halothane in an operating theater (J. Gut, unpublished data).

Expression of the E2 Subunit of the PDC and the Proteins of 52 kDa in Patients with Halothane Hepatitis

A wide interindividual variability was noted in the amounts of the proteins of 52 kDa (comprising the E2 subunit of the OGDC, the BCOADC, and protein X) and of the E2 subunit of the PDC (formerly referred to as protein of 64 kDa) recognized by anti-CF_3CO antibody on immunoblots of liver homogenates obtained from 19 control individuals who had not been exposed to halothane or metabolites thereof. On immunoblots of liver biopsies obtained *post mortem* (between day 6 to day 33 after exposure to halothane) or day 54 (diagnostically, from one patient) from patients afflicted with halothane hepatitis, unusually low amounts of both the proteins of 52 kDa and of the E2 subunit of the PDC were recognized by anti-CF_3CO antibody in 5 out of 7 patients (Gut et al., 1992). In the liver biopsies of the two remaining patients afflicted with halothane hepatitis, the amounts of the proteins of 52 kDa and of the E2 subunit of the PDC were well within the range of immunorecognizable protein observed in control individuals. Overall, in the human population, there appears to exist an appreciable variability in the immunodetectable (by anti-CF_3CO antibody) amounts of the proteins of 52 kDa and of the E2 subunit of the PDC. Notably, a fraction of patients with halothane hepatitis clusters at the very low end of the wide range of the amounts of the proteins of 52 kDa and of the E2 subunit of the PDC detectable in the liver of human individuals. The present data might indirectly suggest that the aberrant expression at low levels of the proteins of 52 kDa and the E2 subunit of the PDC in humans could be associated with the susceptibility of certain individuals for the development of halothane hepatitis.

Modulation of Autoantibodies Towards the Murine E2 Subunit of PDC in B6.CAST Mice upon Exposure to Halothane

If the concept of molecular mimicry between lipoic acid and CF_3CO-Lys is valid, one would expect that an immune response elicited through exposure of experimental animals to halothane should concomitantly also elicit an immune response towards lipoylated autoantigens. In support of this concept (Frey N., and Gut, J., manuscript in preparation) the exposure of presumably immune-compromised B6.CAST mice to halothane resulted in the modulation of the preexisting anti-E2 subunit autoantibody levels. Thus, when B6.CAST mice, which exhibited the highest pre-exposure level of natural anti-E2 autoantibodies of the five murine strains tested, were exposed to single doses of halothane on repeated (3, between-treatment intervals of 4 weeks) occasions, in 3 out of 8 animals preimmunized with human E2 subunit protein and CF_3CO-MSA (murine serum albumin) the level of anti-E2 subunit autoantibodies was elevated 6 to 9-fold. Two mice were partial responders, while the remaining mice and the mice in two control groups did not respond to halothane exposure.

Perspectives

The presence of constitutively expressed epitopes in the "self" repertoire of human individuals which structurally and immunochemically mimic CF_3CO-

motifs indirectly suggests that an immunological reaction towards CF_3CO-proteins involves the breakdown of tolerance towards the lipoyl domains. The exact mechanisms by which tolerance to lipoylated domains is overcome remain to be elucidated both in humans and in experimental animals. In humans the identification of genetic susceptibility factors will become a major goal and experiments to elucidate aberrancies in the genes coding for the E2 subunits of PDC, OGDC, and BCOADC, and of protein X have been initiated. However, aberrancies in genes coding for proteins with major roles in the regulation of immune responses or genes coding for CF_3CO-protein might be of similar importance and are likely to contribute to the overall susceptibility of afflicted individuals for the development of halothane hepatitis.

Acknowledgements

The author is indebted to his former coworkers U. Christen, N. Frey, J. Huwyler, V. Koch, and D. Stoffler, at the Biocenter of the University, Basel, Switzerland, and wishes to thank J.G. Kenna, Dept. of Pharmacology, St. Mary's Hospital Medical School, Imperial College of Science, Technology and Medicine, Norfolk Place, London, U.K., A.J. Gandolfi, Dept. of Anesthesiology, University of Arizona Health Science Center, Tuscon, Arizona, USA, L. Ranek, Medical Dept. A, Rijgshospitalet, Copenhagen, Denmark, S.J. Yeaman, Dept. of Biochemistry and Genetics, Medical School, University of Newcastle upon Tyne, Newcastle upon Tyne, U.K., and Y. Shimomura, Dept. of Bioscience, Nagoya Institute of Technology, Gokiso, Showa-Ku, Nagoya, Japan, for their contributions of antisera, tissue biopsies, and isolated autoantigens.

References

Allison, AC (1989) Theories of self tolerance and autoimmunity In: Kammüller, ME, Bloksma, N, and Seinen, W (Eds), Autoimmunity and Toxicology. Elsevier, Amsterdam, pp 67-104

Beaune, Ph, Dansette, PM, Mansuy, D, Kiffel, L, Amar, C, Leroux, JP, and Homberg, J-C (1987) Human anti-endoplasmic reticulum autoantibodies appearing in a drug-induced hepatitis are directed against a human liver cytocrome P450 that hydroxylates the drug. Proc Natl Acad SciUSA 84, 551-555

Behal, RH, Browning, KS, Hall, TB, Reed, LR (1989) Cloning and nucleotide sequence of the gene for protein X from Saccharomyces cerevisiae. Proc Natl Acad Sci USA 86, 8732-8736

Bourdi, M, Gautier, J-C, Mircheva, J, Larrey, D, Guillouzo,A, Andre, C, Belloc, C, and Beaune, Ph H (1992) Anti-liver microsomes autoantibodies and dihydralazine-induced hepatitis: specificity of autoantibodies and inductive capacity of the drug. Mol Pharmacol 42, 280-285

Bourdi, M, Larrey, D, Nataf, J, Bernuau, J, Pessayre, D, Iwasaki, M, Guengerich, FP, and Beaune, Ph H (1990) Anti-liver endoplasmic reticulum autoantibodies are directed against human cytochrome P450IA2 A specific marker of dihydralazine-induced hepatitis. J Clin Invest 85, 1967-1973

Bourdi, M, Chen, W, Peter, RM, Martin, JL, Buters, JT, Nelson, SD, and Pohl, LR (1996) Human cytochrome p450 2E1 is a major autoantigen associated with halothane hepatitis. ChemRes Toxicol 9, 1159-1166

Butler, LE, Thomassen, D, Martin, JL, Martin, BM, Kenna, JG, & Pohl, LR (1992) The calcium-binding protein calreticulin is covalently modified in rat liver by a reactive metabolite of the inhalation anesthetic halothane. Chem Res Toxicol 5, 406-410

Casorbi, HF, Blake, DA, and Helrich, M (1970) Differences in biotransformation of halothane in man. Anesthesiology 32, 119-123

Christen, U, Jenö, P, and Gut, J (1993) Halothane metabolism: The dihydrolipoamide acetyltransferase subunit of the pyruvate dehydrogenase complex molecularly mimics trifluoroacetyl-protein adducts. Biochemistry 32, 1492-1499

Christen, U, Bürgin, M, and Gut, J (1991b) Halothane metabolism: Immunochemical Evidence for molecular mimicry of trifluoroacetylated liver protein adducts by constitutive polypeptides. Mol Pharmacol 40, 390-400

Christen, U, Bürgin, M, and Gut, J (1991a) Halothane meatbolism: Kupffer cells carry and partially process trifluoroacetylated protein adducts. Biochem Biophys Res Commun 175, 256-262

Christen, U, Quinn, J, Yeaman, SJ, Kenna, GJ, Clarke, JB, Gandolfi, JA, and Gut, J (1994) Identification of the dihydrolipoamide acetyltransferase subunit of the human pyruvate dehydrogenase complex as an autoantigen in halothane hepatitis: molecular mimicry of trifluoroacetyl-lysine by lipoic acid. Eur J Biochem 223, 1035-1047

Cohen, EN, Trudell, JR, Edmunds, NN, and Watson, E (1975) Urinary metabolites of halothane in man. Anesthesiology, 43, 392-401

Coppel, RL, McNeilage, LJ, Surh, CD, Van de Water, J, Spithill, TW, Wittingham, S, and Gershwin, ME (1988) Primary structure of the human M2 mitochondrial autoantigen primary biliary cirrhosis: Dihydrolipoamide acetyltransferase. Proc Natl Acad Sci USA 85, 7317-7321

Davila, JC, Martin, BM, and Pohl, LR (1992) Patients with halothane hepatitis have serum antibodies directed against glucose regulated stress protein GRP78/BiB. Toxicologist 12, 255

deWeck, AL (1983) Immunopathological mechanisms and clinical aspects of allergic reactions to drugs In: deWeck, AL, and Bungaard, H (Eds), Allergic Reactions to Drugs. Springer, Berlin, pp 75-133

Dyrbeg,T (1992) Molecular mimicry in autoimmunity In: Talal, N (Ed), Molecular Autoimmunity. Academic Press, London, pp 197-204

Eliasson, E, and Kenna, JG (1996) Cytochrome P450 2E1 is a cell surface autoantigen in halothane hepatitis. Mol Pharmacol 50, 573-582

Frey, N, Christen, U, Jenö, P, Yeaman, SJ, Shimomura, Y, Kenna, JG, Gandolfi, AJ, Ranek, L, and Gut, J (1995) The lipoic acid containing components of the 2-oxoacid dehydrogenase complexes mimic trifluoroacetylated proteins and are autoantigens associated with halothane hepatitis. Chem Res Toxicol 8, 736-746

Fujinami, RS (1991) Molecular mimicry In: Rose, NR, and Mackay, IR (Eds), The Autoimmune Diseases II. Academic Press, San Diego, pp 153-169

Gershwin, ME, and Mackay, IR (1991) Primary biliary cirrhosis: paradigm or paradox for autoimmunity. Gastroenterology 100, 822-833

Gorsky, LD, Koop, DR, and Coon, MJ (1984) On the stoichiometry of the oxidase and monooxygenase reactions catalyzed by liver microsomal cytochrome P450. J Biol Chem 259, 6812-6817

Gut, J, Christen, U, Huwyler, J, Bürgin, M, and Kenna, JG (1992) Molecular mimickry of trifluoroacetylated human liver protein adducts by constitutive proteins and immunochemical evidence for its impairment in halothane hepatitis. Eur J Biochem 210, 569-576

Harris, JW, Pohl, LR, Martin, JL, and Anders, MW (1991) Tissue acylation by the chlorofluorocarbon substitute 2,2-chloro-1,1,1-triflouroethane. Proc Natl Acad Sci USA 88, 1407-1410

Heijink, E, DeMatteis, F, Gibbs, AH, Davies, A, and White, INH (1993) Metabolic activation of halothane to neoantigens in C57B1/10 mice: immunochemical studies. Eur J Pharmacol 248, 15-25

Hinson, J, and Roberts, DW (1992) Role of covalent and noncovalent interactions in cell toxicity: effects on proteins. Ann Rev Pharmacol Tox 32, 471-510

Homberg, J-C, Abuaf, N, Helmy-Kahlil, S, Biour, M, Poupon R, Islam, S, Darnis, F, Levy, VG, Opolon, P, Beaugrand, M, Toulet, J, Dana, G, and Benhamou, J-P (1985) Drug-induced hepatitis associated with anticytoplasmic organelle autoantibodies. Hepatology 5, 722-727

Huwyler, J, Aeschlimann, D, Christen, U, and Gut, J (1992) The kidney as a novel target tissue for protein adduct formation associated with metabolism of halothane and the canditate chlorofluorocarbon replacement 2,2-dichloro-1,1,1-trifluoroethane. Eur J Biochem 207, 229-238

Huwyler, J, and Gut, J (1992) Exposure to the chlorofluorocarbon substitute 2,2,-dichloro-1,1,1-trifluoroethane and the anesthetic agent halothane is associated with transient protein adduct formation in the heart. Biochem Biophys Res Commun 184, 1344-1349

Kappler, JW, Roehm, N, & Marrack, PC (1987) T-cell tolerance by clonal elimination in the thymus. Cell 49, 273-280

Kappler, JW, Staerz, U, White, J, & Marrack , PC (1988) Self-tolerance eliminates T cells specific for Mls-modified products of the major histocompatibility complex. Nature 332, 35-40

Kaufman, DL, Clare-Salzler, M, Tian, J, Forsthuber, T, Ting, GSP, Robinson, P, Atkinson, MA, Sercarz, EE, Tobin, AJ, and Lehman, PV (1993) Spontaneous loss of T-cell tolerance to glutamic acid decarboxylase in murine insulin-dependent diabetes. Nature 366, 69-72

Kenna, JG, Martin, JL, and Pohl, LR (1992) The topography of trifluoroacetylated protein antigens in liver microsomal fractions from halothane treated rats. Biochem Pharmacol 44, 621-629

Kenna, JG, Satoh, H, Christ, H, and Pohl, L (1988b) Metabolic basis for a drug hypersensitivity: antibodies in sera from patients with halothane hepatitis recognize liver neoantigens that contain the trifluoroacetyl group derived from halothane. J Pharmacol Exp Ther 245, 1103-1109

Kenna, JG, Neuberger, J, and Williams, R (1987) Identification by immunoblotting of three halothane-induced liver microsomal polypeptide antigens recognized by antibodies in sera from patients with halothane-associated heaptitis. J Pharmacol Exp Ther 242, 733-740

Kenna, JG, Neuberger, J, and Williams, R (1988a) Evidence for the expression in human liver of halothane-induced neoantigens recognized by antibodies in sera from patients with halothane hepatitis. Hepatology 8, 1635-1641

Kisielow, P, Blüthmann, H, Staerz, UD, Steinmetz, M, & von Boehmer, H (1988) Tolerance in T-cell receptor transgenic mice involves deletion of nonmature CD4+8+ thymocytes. Nature 333, 742-746

Leeder, JS, Riley, RJ, Cook, VA, and Spielberg, SP (1992) Human anti-cytochrome P450 antibodies in aromatic anticonvulsant-induced hypersensitivity reactions. J Pharmacol Exp Ther263, 360-367

Martin, JL, Pumford, NR, LaRosa, AC, Martin, BM, Gonzaga, HM, Beaven, MA, and Pohl, LR (1991) A metabolite of halothane covalently binds to an endoplasmic reticulum protein that is highly momologous to phosphatidylinositol-specific phospholipase Ca but has no activity. Biochem Biophys Res Commun 178, 679-685

Martin, JL, Kenna, JG, Martin, BM, Thomassen, D, Reed, G, and Pohl, LR (1993a) Halothane hepatitis patients have serum antibodies that react with protein disulfide isomerase. Hepatology 18, 858-863

Martin, JL, Reed, GF, and Pohl, LR (1993b) Association of anti-58 kDa endoplasmic reticulum antibodies with halothane hepatitis. Biochem Pharmacol 46, 1247-1250

Mattevi, A, Oblomova, G, Schulze, E, Kalk, KH, Westphal, AH, DeKok, A, and Hol, WGJ(1992) Atomic structure of the cubic core of the pyruvate dehydrogenase multienzyme complex. Science 255, 1544-1550

Matuda, S, Nakano, K, Ohta, S, Shimura, M, Yamanaka, T, Nakagawa, S, Titani, K, and Miyata,T (1992) Molecular cloning of dihydrolipoamide acetyltransferase of the rat pyruvate dehydrogenase complex: sequence comparison and evolutionary relationship to other dihydrolipoamide acetyltransferases. Biochim Biophys Acta 1131, 114-118

Mazzarella, RA, and Green, M (1987) ERp99, an abundant, conserved glycoprotein of the endoplasmic reticulum, is homologous to the 90-kDa heat shock protein (hsp90) and 94-kDa glucose-regulated protein (GRP94). J Biol Chem 262, 8875-8883

Müller, R, and Stier, A (1982) Modification of liver microsomal lipids by halothane metabolites: a multi-nuclear NMR spectroscopic study. Naunyn-Schmiedebergs Arch Pharmacol 321, 234-237

National Halothane Study (1966) Summary of the National Halothane Study: possible association of halothane anesthesia and postoperative hepatic necrosis. J Am Med Ass 197, 121-134

Nelson, SD, and Pearson, PG (1990) Covalent and noncovalent interactions in acute lethal cell injury by chemicals. Ann Rev Pharmacol Tox 30, 169-195

Neuberger, J (1989) Halothane hepatitis - an example of possibly immune-mediated hepatotoxicity In: Kammüller, ME, Bloksma, N, and Seinen, W (Eds), Autoimmunity and Toxicology. Elsevier, Amsterdam, pp37-65

Neuberger, J, and Williams, R (1984) Halothane anaesthesia and liver damage. Br Med J 289, 1136-1139

Nossal, GJV (1991) Auroimmunity and self-tolerance In: Rose, NR, and Mackay, IR (Eds), The Autoimmune Diseases II. Academic Press, San Diego, pp 27-44

Ohashi, PS, Oehen, S, Buerki, K, Pircher, H, Ohashi, CT, Odermatt, B, Malissen, B, Zinkernagel, RM, & Hengartner, H (1991) Ablation of "tolerance" and induction of diabetes by virus infection in viral antigen transgenic mice. Cell 65, 305-317

Park, KB, and Kitteringham, NR (1990) Drug-protein conjugation and its immunochemical consequences. Drug Metab Rev 22, 87-144

Perham, RN (1991) Domains, motifs, and linkers in 2-oxo acid dehydrogenase multienzyme complexes: A paradigm in the design of a multifunctional protein. Biochemistry 30, 8501-8512

Pirmohamed, M, Kitteringham, NR, Breckenridge, AM, and Park, BK (1992) Detection of an autoantibody directed against human liver microsomal protein in a patient with carbamazepine hypersensitivity. Br J Clin Pharmacol 33, 183-186

Pohl, LR (1990) Drug-induced allergic hepatitis, Sem Liver Dis 10, 305-315

Pohl, LR, Kenna, JG, Satoh, H, Christ, DD, and Martin, JL (1989) Neoantigens associated with halothane hepatitis. Drug Metab Rev 20, 203-217

Pumford, NR, Martin, BM, Thomassen, D, Burris, JA, Kenna, JG, Martin, JL, & Pohl, LR (1993) Serum antibodies from halothane hepatitis patients react with the endoplasmic reticulum protein ERp72. Chem Res Toxicol 6, 609-615

Quinn, J, Diamond, AG, Palmer, JM, Bassendine, MF, James, OFW, and Yeaman, SJ (1993) Lipoylated and unlipoylated domains of human PDC-E2 as autoantigens in primary biliary cirrhosis: Significance of lipoate attachment. Hepatology 18, 1384-1391

Ramsdell, F, & Folkwes, BJ (1990) Clonal deletion versus clonal anergy: The role of the thymus in inducing self tolerance. Science 248, 1342-1348

Reed, LJ, and Hackert, ML (1990) Structure-function relationships in dihydrolipoamide acyltransferases. J Biol Chem 265, 8971-8974

Roitt, IM, Brostoff, J, and Male, D (1985) Autoimmunity and autoimmune disease In: Immunology. CV Mosby, St Louis, pp 231-2312

Rubinstein, D, Roska, AK, and Lipsky, PE (1987) Antigen presentation by liver sinusoidal lining cells after antigen presentation in vivo. J Immunol 138, 1377-1382

Satoh, H, Martin, BM, Schulik, AH, Christ, DD, Kenna, JG, & Pohl, LR (1989) Human anti-endoplasmic reticulum antibodies in sera from patients with halothane-induced hepatitis are directed against a trifluoroacetylated carboxylesterase. Proc Natl Acad Sci USA 86, 322-326

Satoh, H, Gillette, JR, Davies, HW, Schulik, RD, and Pohl, LR (1985) Immunochemical evidence of trifluoroacetylated cytochrome P450 in the liver of halothane treated rats. Mol Pharmacol 28:568-474:

Scully, LJ, Clarke, D, and Barr, RJ (1993) Diclofenac induced hepatitis 3 cases with features of autoimmune chronic active hepatitis. Digest Dis Sci 18, 744-751

Smith, CCM, Kenna, JG, Harrison, DJ, Tew, D, and Wolf, CR (1993) Autoantibodies to hepatic microsomal carboxylesterase in halothane hepatitis. Lancet 342, 155-157

Spracklin, DK, Thummel, KE, and Kharasch, ED (1996) Human reductive halothane metabolism in vitro is catalyzed by P450 2A6 and 3A4. Drug Metab Dispos 24, 976-983

Spracklin, DK, Hankins, DC, Fisher, JM, Thummel, KE and Kharasch, ED (1997) Cytochrome P450 2E1 is the principal catalyst of human oxidative halothane metabolism in vitro. J Pharmacol Exp Ther 281, 400-411

Sprent, J, Gao, E-K, and Webb, SR (1990) T cell reactivity to MHC molecules: Immunity versus tolerance. Science 248, 1357-1363

Thomassen, D, Martin, BM, Martin, JL, Pumford, NR, & Pohl, LR (1989) The role of a stress protein in the development of a drug-induced allergic response. Eur J Pharmacol 183, 1138-1139

Trudell, JR, Ardies, CM, and Anderson, WR (1991) Antibodies rised against trifluoracetyl-protein adducts bind to N-trifluoroacetyl-phosphatidylethanolamine in hexagonal phase phospholipid micelles. J Pharmacol Ther Exp 257, 657-662

vanWelie, RTH, vanDijk, RGJ, and Vermeulen, NPE (1992) Mercapturic acid, protein adducts, and DNA adducts as biomarkers of electrophilic chemicals. Crit Rev Toxicol 22, 271-306

Vergani, D, Mieli-Vergani, G, Alberti, A, Neuberger, J, Eddleston, ALWR, Davis, M, and Williams, R (1980) Antibodies to the surface of halothane altered rabbit hepatocytes in patients with halothane associated hepatitis. New Eng J Med 303, 66-71

Wark, H, Earl, J, Chau, DD, Overton, J, and Cheung, HTA (1990) A urinary cycteine-halothane metabolite: validation and measurement in children. Br J Anaesth 64, 469-473

Waskell, L (1994) Metabolism of volatile anesthetics In: Rice, SA, and Fish, KJ (Eds), Anesthetic Toxicity. Raven Press, New York, pp 49-72

Yeaman, SJ (1986) The mammalian 2-oxoacid dehydrogenases: a complex family. TIBS 11, 293-296

Zimmermann, HJ (1990) Update on hepatotoxicity due to classes of drugs in common clinical use: non-steroidal drugs, anti-inflammatory drugs, antibiotics, anti-hypertensives, and cardiac and psychotropic agents. Sem Liver Dis 10, 322-338

Plenary Lecture

Receptor -Mediated Toxicity

Jan-Åke Gustafsson, George Kuiper, Eva Enmark, Eckhardt Treuter and Joseph Rafter.
Departments of Medical Nutrition and Biosciences, NOVUM, Karolinska Institute, S-141 86 Huddinge, Sweden

Introduction

The nuclear receptor superfamily, which includes receptors for steroid hormones, thyroid hormone, vitamin D, retinoic acid, peroxisome proliferators and ecdysone, consists of a surprisingly large number of genes (Tsai and O'Malley, 1994); a large number of genes having extensive sequence homology to the nuclear receptor family, but for which no ligands have yet been identified, are furthermore also known (Enmark and Gustafsson, 1996). It remains possible that, as with vitamin A and vitamin D, the natural ligands for these receptors are components of the diet. Equally likely is the possibility that, like the peroxisome proliferator activated receptor (PPAR), environmental contaminants or xenobiotics can activate some of these receptors and can thus interfere with normal physiological functions in the body. This lecture focuses on two members on the nuclear receptor super gene family, the newly discovered estrogen receptor, ERβ, and the PPAR to illustrate how these receptors could be involved in receptor mediated toxicity.

Estrogen is a modulator of cellular growth and differentiation. Its major targets are the mammary gland, uterus, vagina, ovary, testis, epididymis, prostate, bone, cardiovascular system and brain (Clark et al, 1992; Turner et al, 1994; Farhat et al, 1996). Estrogen mediates its functions through specific intracellular receptors that act as hormone-dependent transcriptional regulators (Beato et al, 1995). As is the case with other signalling molecules in the body which stimulate growth, estrogen is involved in carcinogenesis. The biological importance of estrogen is evident from the number of disease states associated with altered production of estrogen or abnormalities in the estrogen receptor. Osteoporosis, breast cancer, endometrial cancer and prostate hypertrophy are examples of disorders in which estrogen receptors are involved. Reflecting a medical need, several pharmaceuticals that regulate receptor function have been developed and are currently used in the clinic. Clinical experience with these compounds has revealed several unwanted side-effects and development of resistance. Our recent discovery (Kuiper et al, 1996) of a novel estrogen receptor, ERβ, reveals that the mechanism of the biological action of estrogen is more complex than previously thought. It also provides a unique opportunity for the development of improved modulators of estrogen action and the identification of new targets for estrogens.

Estrogen Receptor β

The long held dogma of the presence of a single estrogen receptor (ER) gene was difficult to reconcile with the striking differences in action of synthetic estrogens and antiestrogens with respect to responses in different target cells and tissues (Katzenellenbogen et al, 1996). We have recently cloned a novel estrogen receptor (ERβ) from a rat prostate cDNA library (Kuiper et al, 1996). As with other nuclear receptors ERβ is composed of three major functional domains (Farhat et al, 1996). The N-terminal A/B domain contains a transactivation function. The C-region contains two type-II zinc fingers, which are involved in specific DNA-binding and receptor dimerization. The ligand-binding domain is important for ligand binding and also for receptor dimerization, nuclear localization and interactions with transcriptional co-activators and co-repressors (Kuiper et al, 1996). ERβ is highly homologous to the well studied estrogen receptor protein (now called ERα), particularly in the DNA-binding domain (95% amino acid identity) and in the ligand-binding domain (55% amino acid identity). Saturation ligand binding experiments reveal a high affinity and specific binding of estradiol by ERβ. The ERβ protein is able to stimulate the transcription of an estrogen receptor target gene in an estradiol dependent manner. Some synthetic and naturally occurring ligands have different relative affinities for ERβ vs ERα, although many ligands bind with similar affinity to both subtypes (Kuiper et al, 1997). This suggests that alterations in ligand structure could produce agents that show preferential selectivity for ERβ vs ERα and *vice versa*.

In the rat, distribution and/or relative levels of ERα and ERβ expression vary in different tissues. There is moderate to high expression of ERα in the uterus, testis, pituitary, ovary, kidney, epididymis and adrenal. ERβ on the other hand, is predominant in the prostate, ovary, lung, bladder, brain and epididymis (Kuiper et al, 1997). The rat ovary expresses both receptors although ERβ predominates (Byers et al, 1997). In the rodent brain ERβ constitutes a significant fraction of total ER mRNA, and examination of ERβ mRNA expression by *in situ* hybridisation shows expression within various regions of the hypothalamus (Shugrue et al, 1996).

The mouse (Tremblay et al, 1997) and human homologues (Mosselman et al, 1996) of rat ERβ have been cloned and high expression in ovary and testis (human), among others, was found.

It is an interesting possibility that ERα and ERβ differentially regulate the expression of certain estrogen target genes in a tissue specific manner (Katzenellenbogen and Korach, 1997). Various mechanisms have been suggested as explanations for the striking cell- and promotor-specific effects of antiestrogens, all on the basis of the assumption that only a single ER gene existed (Kuiper and Gustafsson, in press). Different ratios of ERα and ERβ proteins in different cells, could provide a hitherto unrecognised explanation for the mechanism of tissue and cell-type specific actions of antiestrogens.

Estrogen and Breast Cancer

Most human breast cancers, at least initially, are hormone-dependent and they undergo regression when deprived of these supporting hormones. Patients whose breast tumors lack significant amounts of ER rarely respond to endocrine ablation or treatment with antiestrogens, whereas most patients with ER-containing cancers benefit from such treatment (Osborne et al, 1996; Jensen, 1995). Immunochemical determination of ER in tumor biopsies has become a routine clinical procedure on which the choice for therapy is based. A high proportion of patients with ER positive tumors, however, do not respond to endocrine therapy, or eventually become resistant to the hormone antagonist, tamoxifen. Recent findings indicate that, in many breast cancers, at least part of the ER is aberrant (dominant negative or dominant positive ER forms). The contribution of these aberrant ERs to hormone unresponsiveness is still not clear. The currently available immunochemical procedures for ER measurements are based on ERα protein and most likely do not detect ERβ protein. Investigation of the possible presence of ERβ protein in breast tumor biopsies and correlation with hormone responsiveness could prove to be very informative and important.

Environmental Estrogenic Compounds

There is increasing concern over the effect of various chemicals released into the environment on the reproduction of humans and animals (Nimrod and Benson, 1996; Gimeno et al, 1996; Simons, 1996; Arnold et al, 1996; Jensen et al, 1995). Male offspring born to mothers who were given diethylstilbestrol (DES), a very potent synthetic estrogen, have an increased incidence of undescended testes, urogenital abnormalities and semen abnormalities, compared with those from mothers who did not take DES (Nimrod and Benson, 1996; Jensen et al, 1995). The similarities between the observations made in male DES offspring and abnormalities being observed in the general population in recent years, has led to the hypothesis that one potential cause of the rise in male reproductive tract problems might be inappropiate exposure to estrogens or suspected environmental estrogenic chemicals (from pesticides, components of plastics, detergents, cosmetics, etc.) during fetal and/or neonatal life (Nimrod and Benson, 1996; Gimeno et al, 1996; Simons, 1996; Arnold et al, 1996; Jensen et al, 1995). Fetal or neonatal exposure to low doses of estrogenic chemicals has also been shown to lead to permanent enlargement of the prostate in adult life, possibly resulting in a higher risk of developing severe forms of benign prostatic hyperplasia (Saal et al, 1997). The expression of ERβ in the male urogenital system indicates that this receptor is of importance with regard to the described effects. Suspected environmental estrogenic compounds must be evaluated for their interaction with ERβ in comparison to ERα by ligand competition and transactivation assay systems. In addition, the expression of ERβ and ERα during embryonal development (fetal testis, urogenital sinus, Wolffian duct) must be investigated in detail.

Phytoestrogens

There are two additional classes of environmental compounds which are receiving extensive attention because of their influence on estrogen action. The first are the so-called phytoestrogens. These are nonsteroidal plant compounds of diverse structures (for instance genistein, daidzein, coumestrol) that produce estrogenic responses (Saal et al, 1997, and references therein). They are found in many fruits, vegetables and grain, and are relatively weak estrogens. When ingested in large amounts, however, they can have profound biological effects. It is thought that the presence of phytoestrogens in the human diet may be responsible for developmental changes, reduced fertility, reduced severity of menopausal symptoms and increased or decreased risks of hormonally related cancers, especially breast, endometrial and prostate cancer. Again , it will be necessary to compare the relative affinities of the various phytoestrogens for ERα and ERβ. Available data indicate that phytoestrogens bind to ERβ with about ten times as high affinity as to ERα.

Role of PPAR in signalling Pathways of Potential Importance in Colon Carcinogenesis

The cause of colorectal cancer is now widely accepted to be the accumulation of mutations in specific genes controlling cell division, apoptosis and DNA repair (Kinzler and Vogelstein, 1996). There is also a wealth of evidence that dietary factors, especially fat and fiber, influence the development of this disease. However, there has been little understanding of how these dietary and genetic factors interact.

In the normal colon, the epithelial cells arise from stem cells, they differentiate, and then they undergo apoptosis, a process which several groups have shown to be decreased in tumorigenesis (Kinzler and Vogelstein, 1996). Indeed, effects of colonic luminal components on apoptosis in colonic epithelial cells has recently become an area of interest in studying mechanisms underlying the dietary etiology of colon cancer (Hague et al, 1993; Hague et al, 1995). An important insight into this process came from Tsujii and DuBois (1995), who showed that overexpression of cyclooxygenase-2 (COX-2) in a rat intestinal cell line prevented apoptosis, and that apoptosis was restored by treatment with a cyclooxygenase inhibitor. This strongly suggested that a product of COX-2 biased the cell's fate away from apoptosis, and other experiments implicated an increase in bcl-2 as the mechanism. The signalling route from prostaglandins to apoptosis is not known; it could be through prostaglandin receptors, which are members of the G protein-coupled family of receptors. However, it could also be through the peroxisome proliferator-activated receptors (PPARs). This small sibship (three members: PPARα, PPARδ, PPARγ) in the steroid receptor pedigree has been shown to control the differentiation from preadipocytes to adipocytes in response to prostaglandins (mainly PPARγ and α). Thus, it is an intriguing hypothesis that PPARs could be downstream of COX-2 in colon carcinogenesis, an area that we feel deserves further scrutiny.

We recently demonstrated expression of PPAR in mouse colonic and small intestinal mucosa by western blot analysis and immunohistochemistry, indicating a higher expression level in the differentiated colonic epithelial cells facing the intestinal lumen (Mansen et al, 1996). Quantification of PPAR mRNA by ribonuclease protection assay revealed relatively high expression of PPARγ and PPARδ in the colon as compared to the small intestine. In contrast, PPARα expression was higher in the small intestine as compared to the colon.

Interestingly, several observations link APC to the ligand-activated PPAR:
i) The gene for secretory phospholipase A2 (sPLA2) has been suggested as one candidate gene responsible for genetic variation of cancer development in the APC background (MacPhee et al, 1995). The sPLA2 enzyme is involved in the production of arachidonic acid. Arachidonic acid is known to activate PPAR *in vivo* and is the common precursor for the suggested endogenous ligands for all three PPAR subtypes: leukotrienes (PPARα), prostaglandins (PPARγ) and prostacyclins (PPARδ).
ii) The fact that COX-2 inhibitors or disruption of the COX-2 gene reduces tumor incidence (Oshima et al, 1996), suggests a further link between PPAR activation and tumor development in the APC background, since the COX-2 enzyme is required for the production of prostaglandin ligands for PPAR.
iii) A two-hybrid screening of an activation domain-tagged cDNA library identified human β-catenin as a PPARα-interacting protein (Treuter et al, 1997). Preliminary experiments have shown, that β-catenin interacts directly with the LBD/AF-2 of the PPAR subtypes α and γ *in vivo* and *in vitro*. Importantly, ligands for PPARγ, such as the thiazolidinedione BRL 49635, enhance the *in vitro* interaction, suggesting that β-catenin might form a complex with the ligand-activated PPAR. We noticed that β-catenin shares a conserved peptide motif LxxLLL, referred to as the NR-box, together with other AF-2 cofactors, such as RIP140 and TIF-1. The motif is present in more than 50 % of all PPAR-interacting proteins and might bind to the conserved part of the AF-2 interaction domain.

β-Catenin and its *Drosophila* homologue Armadillo are components of the wnt/wingless signalling pathway, important for many key developmental processes, such as embryonic pattern formation in *Drosophila* and mouse (Sanson et al, 1996). The pathway links APC to PPAR, since β-catenin interacts directly with APC. Furthermore, it has been observed that cancer cells with mutant APC contain abnormally high levels of intracellular β-catenin. This suggests, that APC regulates β-catenin signalling and that dysfunction of this pathway contributes to carcinogenesis. Other evidence also implicates a direct role for β-catenin in cell proliferation, since an N-terminally deleted β-catenin (lacking the APC-interaction domain) transforms cells. In summary, a mutated

APC protein would perhaps interfere with β-catenin signalling to the nucleus and could involve ligand-activated PPAR as one target protein. A putative function for β-catenin through direct interaction with DNA-bound transcription factors has been suggested recently based on the finding that it might modulate the activity of the LEF transcription factor (Behrens et al, 1996).

Many questions remain to be answered. It is not clear, how mutated APC interferes with catenin signalling. Both positive and negative models have been proposed. It is also unknown, whether the observed interaction between catenin and PPAR influences the function of PPAR as a transcription factor. If so, PPAR could regulate unidentified target genes, the products of which might be critically involved in colon carcinogenesis, apoptosis etc. Thus, a differential display or subtraction screening is required to identifiy such PPAR regulated target genes. Again, PPAR could act positively or negatively on the expression of these genes. It is also possible that PPAR interferes with the binding of catenin to another as yet unidentified target protein. However, the ligand-dependency of the interaction, the almost exclusively nuclear localization of PPAR and the obvious link between the generation of putative PPAR-ligands and colon cancer all favour a direct involvement of PPAR as a transcription factor. Much more information is needed, however, about PPAR function, the nature of its ligands, its interaction with β-catenin as well as its downstream protein targets (coactivators, corepressors).

References

Arnold, S, Klotz, DM, Collins, BM, Vonier, PM, Guilette, LJ and McLachlan, JA (1996) Synergistic activation of estrogen receptor with combinations of environmental chemicals. Science 272, 1489-1491.

Beato, M., Herrlich, P. and Schutz, G. (1995) Steroid hormone receptors: many actors in search of a plot. Cell, 83 , 851-857.

Behrens J, et al. (1996) Functional interaction of β-catenin with the transcription factor LEF. Nature, 382: 638-642.

Byers, M, Kuiper, GGJM, Gustafsson, J-Å and Park-Sarge, O-K (1997) Estrogen receptor β mRNA expression in rat ovary: down-regulation by gonadotropins. Molecular Endocrinology 11, 172-182.

Clark, JH, Shrader, WT and O'Malley, BW (1992) Mechanisms of action of steroid hormones, in: Textbook of Endocrinology (Wilson, J.D. and Foster, D.W. Eds.) pp. 35-90, W.B. Saunders Company, New York.

Enmark, E and Gustafsson, J-Å (1996) Orphan nuclear receptors - the first eight years. Mol. Endocrinol. 10, 1293-1307.

Farhat, MY, Lavigne, MC and Ramwell, PW (1996) The vascular protective effects of estrogen. FASEB J. 10, 615-624.

Gimeno, S., Gerritsen, A., Bowmer, T and Komen, H. (1996) Feminization of male carp. Nature 384, 221-222.

Hague A, Manning AM, Hanlon KA, et al. (1993) Sodium butyrate induces apoptosis in human colonic tumour cell lines in a p53-independent pathway - implications for the possible role of dietary fiber in the prevention of large-bowel cancer. Int J Cancer, 55: 498-505.

Hague A, Elder DJ, Hicks DJ, Paraskeva C. (1995) Apoptosis in colorectal tumour cells: induction by the short chain fatty acids butyrate, propionate and acetate and by the bile salt deoxycholate. Int J Cancer, 60: 400-406.

Jensen, EV (1995) Steroid hormones, receptors and antagonists. Ann. N.Y. Acad. Sci. 761, 1-17.

Jensen, TK, Toppari, J, Keiding, N and Skakkebaek, NE (1995) Do environmental estrogens contribute to the decline in male reproductive health? Clinical Chemistry 41, 1896-1901.

Katzenellenbogen, B and Korach, K (1997) Editorial: A new actor in the estrogen receptor drama-enter ERβ. Endocrinology 138, 861-862.

Katzenellenbogen, J, O'Malley, BW and Katzenellenbogen, BS (1996) Tripartite steroid hormone receptor pharmacology: interaction with multiple effector sites as a basis for the cell- and promoter-specific action of these hormones. Molecular Endocrinology 10, 119-131.

Kinzler KW, Vogelstein B. (1996)Lessons from hereditary colorectal cancer. Cell, 87: 159-170.

Kuiper, GGJM, Enmark, E, Pelto-Huikko, M, Nilsson, S and Gustafsson, J-Å (1996) Cloning of a novel estrogen receptor expressed in rat prostate and ovary. Proc. Natl. Acad. Sci. USA 93, 5925-5930.

Kuiper, GGJM, Carlsson, B, Grandien, K, Enmark, E, Häggblad, J, Nilsson, S and Gustafsson, J-Å (1997) Comparison of the ligand binding specificity and transcript tissue distribution of estrogen receptors α and β. Endocrinology 138, 863-870.

Kuiper, GGJM and Gustafsson, J-Å (in press) The novel estrogen receptor β subtype: potential role in the cell- and promoter specific actions of estrogens and antiestrogens. FEBS Letters.

MacPhee M, et al. (1995) The secretory phospholipase A2 gene is a candidate for the Mom1 locus, a major modifier of Apc^{min}-induced intestinal neoplasia. Cell, 81: 957-966.

Mansen A, Guardiola-Diaz H, Rafter J, et al. (1996) Expression of the peroxisome proliferator-activated receptor (PPAR) in the mouse colonic mucosa. Biochem Biophys Res Commun, 222: 844-851.

Mosselman, S, Polman, J and Dijkema, R (1996) ERβ : identification and characterization of a novel human estrogen receptor. FEBS Letters 392, 49-53.

Nimrod, AC and Benson, WH (1996) Environmental estrogenic effects of alkylphenol ethoxylates. Critical Reviews in Toxicology 26, 335-364.

Osborne, CK, Elledge, RM and Fuqua, SA (1996) Estrogen receptors in breast cancer therapy. Scientific American, pp.32-41.

Oshima M, Dinchuk JE, Kargman, SL, et al. (1996) Suppression of intestinal polyposis in APC delta716 knockout mice by inhibition of prostaglandin endoperoxide synthase-2 (COX-2). Cell, 87: 803-809

Saal, FS, Timms, B, Montano, MM, Palanza, P, Thayer, KA, Nagel, SC, Dhar, MD, Ganjam, VK, Parmigiani, S and Welshons, WV (1997) Prostate enlargement in mice due to fetal exposure to low doses of estradiol or diethylstilbestrol and opposite effects at high doses. Proc. Natl. Acad. Sci. USA 94, 2056-2061.

Sanson B, et al. (1996) Uncoupling cadherin-based adhesion from wingless signalling in *Drosophila*. Nature, 383: 627-630.

Simons, SS (1996) Environmental estrogens : can two 'alrights' make a wrong? Science 272, 1451.

Shughrue, PJ, Komm, B and Merchenthaler, I (1996) The distribution of estrogen receptor-β mRNA in the rat hypothalamus. Steroids 61, 678-681.

Tremblay, GB, Tremblay, A, Copeland, NG, Gilbert, DJ, Jenkins, NA, Labrie, F and Giguere, V (1997) Cloning, chromosomal localization, and functional analysis of the murine estrogen receptor β. Molecular Endocrinology 11, 353-365.

Treuter E, et al. (manuscript submitted, 1997) Identification and functional properties of nuclear factor RIP140 as a cofactor for PPAR..

Tsai, MJ and O'Malley, BW (1994) Molecular mechanisms of action of steroid/thyroid receptor superfamily members. Annu. Rev. Biochem. 63, 451-486.

Tsujii M and DuBois RN. (1995) Alterations in cellular adhesion and apoptosis in epithelial cells overexpressing prostaglandin endoperoxide synthase 2. Cell, 83: 493-501.

Turner, RT, Riggs, BL and Spelsberg, TC (1994) Skeletal effects of estrogens. Endocrine Reviews 15, 275-300.

Animals as Models in Toxicology: Animal Welfare and Ethics

(Chair: M. Jackson, United Kingdom, and J. Ladewig, Denmark)

Animal Burdens versus Human Benefits – How Should the Ethical Limits be Drawn for Use of Animals as Models in Toxicology?

Peter Sandøe[1] and Ove Svendsen[2]

[1] Bioethics Research Group, University of Copenhagen, Department of Education, Philosophy and Rhetorics, Njalsgade 80, DK-2300 Copenhagen S, Denmark

[2] Department of Pharmacology and Pathobiology, The Royal Veterinary and Agricultural University, Bülowsvej 13, DK-1870 Frederiksberg C, Denmark

A View on Animal Ethics

Before making any specific recommendations regarding the ethical limits for the use of animals as models in toxicology we want to make explicit the general ethical view on the basis of which these recommendations are made. This view will be referred to as the *animal welfare view*. It may be presented by way of contrast with two other views, the *self-interest view* and the *animal rights view*. These two views are in most respects opposed to one another, but they have in common a tendency to give very simple answers to questions about the ethical limits which should be placed on man's use of animals, including laboratory animals.

According to the self-interest view there is no ethical limit to the ways in which we may use animals, and according the animal rights view all animal experimentation should be banned.

One philosopher (Narveson 1983) who has defended the self-interest view writes:

> A major feature of this view of morality is that it explains why we have it and who is party to it. We have it for reasons of long-run self-interest, and parties to it include all and only those who have both of the following characteristics: (1) they stand to gain by subscribing to it, at least in the long run, compared with not doing so, and (2) they are capable of entering into (and keeping) an agreement. Those not capable of it obviously cannot be parties to it, and among those capable of it, there is no reason for them to enter into it if there is nothing to gain for them from it, no matter how much the others might benefit. Given these requirements, it will be clear why animals do not have rights. For there are evident shortcomings on both scores.

Underlying the self-interest view is egoism. When one is required to show consideration for others it is really for one's own sake. By respecting the rules of morality one contributes to the maintenance of a society whose existence is essential to one's own welfare.

On this view there is a clear, morally relevant difference between one's relation to other human beings and one's relation to animals. Every person is dependent on the respect and co-operation of other people. If you treat your fellow human beings badly, they will also treat you badly, whereas the animal community will not strike back if, for example, some of its members are used in painful experiments. From an egoistic point of view one need only treat the animals well enough for them to be fit for one's own purposes.

The self-interest view accords with certain aspects of morality prevalent in our society. Thus it serves to explain why legislation, allegedly for the protection of animals, usually protects those animals most which matter most to us – for example dogs and cats enjoy a much better protection than mice and rats.

However, we disagree with the self-interest view because it claims that causing animals to suffer, even for a trivial reason, or for no reason whatsoever, is morally unproblematic as long as the well-being of human beings (and in particular, that of the perpetrator) is not reduced by it. We think that it is, in itself, morally problematic to cause innocent creatures to suffer, be they human or animal – and we believe that most people share this intuition with us. Therefore we reject the claim of the self-interest view that animal ethics (as distinct from animal-involving human ethics) is a non-issue. Animals matter morally, in and of themselves, and a plausible ethical view must make room for that.

According to the second view, the animal rights view, there is no trouble finding this room and animals are bearers, not only of intrinsic moral significance, but rights. This view claims that all higher animals (including all vertebrates) must be treated with respect. No such animal may be used as a mere means for human purposes. Thus it is morally unacceptable to use animals in research. The animal rights view is expressed in the following way by its main proponent (Regan, 1985):

> In the case of the use of animals in science, the rights view is categorically abolitionist. Lab animals are not our tasters; we are not their kings. Because these animals are treated routinely, systematically as if their value were reducible to their usefulness to others, they are routinely, systematically treated with lack of respect, and thus are their rights routinely, systematically violated. This is just as true when they are used in trivial, duplicative, unnecessary or unwise research as it is when they are used in studies that hold out real promise for human benefits.

This view has the advantage of setting a very clear limit to the use of animals for research and for most other purposes – not least animal production.

However we want to reject the animal rights view for two reasons: First it does not allow for any sort of trade-off between animal burdens and human benefits. Secondly, it has a too broad a notion of what matters in the way we treat animals.

As regards the first of these reasons we find it implausible that animal research can never be justified by reference to human benefits, even when the benefits are vital and the harm imposed on the animals is only slight. There must be some middle position between the liberal position endorsed by the self-interest view and the abolitionist position endorsed by the animal rights view.

Concerning our second reason for rejecting the animal rights view we find it implausible that the mere fact that animals are used as means for our purposes should matter morally. What matters in our view is how *the animals* are affected. The fact that an animal is used as a means for our purposes does not by itself imply that, from the perspective of the animal, something bad has happened. Thus, animal experiments may be conducted in an environment which fully caters for the needs of the animals and in such a way that no suffering or discomfort is inflicted on the animals.

The moral issue according to the view we adopt, the animal welfare view, is not whether animals are used as means for our purposes, but whether their welfare is adversely affected by the way we treat them. On this view animal experimentation is morally problematic when and only when the animals suffer or otherwise experience reduced welfare.

According to the welfare view a morally problematic experiment may nonetheless be morally acceptable if the following two conditions are fulfilled: 1) important human benefits are at stake, and 2) it is impossible to get the same information in a way which is less burdensome to the animals.

Ethical assessment of animal experiments on the welfare view is very much a matter of weighing human benefits against the burdens put on the animals. However, in our view an absolute limit should be imposed upon the sort of burden that may be put on the animals. Animals should under no circumstances be allowed to experience strong pain and other forms of intense suffering.

To claim on the one hand that human benefits *and* animal burdens should be weighed against each other and on the other hand that there is an absolute limit to the burden that may be put on the animals may look like an unstable hybrid. It might seem more logical to claim that any burden put on the animals can be justified if the expected benefits are great enough.

We reject this claim for the following reason: In dealing with our fellow human beings we accept in many areas of life that burdens are put on persons for the benefit of others. For example, we accept that people living where a new road is to be built may be forced to move house for the benefit of those who are going to use the road. We also endorse clinical trials conducted on human patients for the benefit of future patients. However, it is generally thought that there is a limit to the sort of burden that may be put on one person for the benefit of others - no matter how great that benefits might be. And this is quite analogous to the restriction that we propose should be placed on burdensome use of laboratory animals.

Of course, situations can be imagined where the benefit at stake is the avoidance of a disaster. If the disaster and the chance of avoiding it were huge enough many people might think that we should no longer observe ordinary limits on the sorts of burden that may be put on others (be they human beings or animals). However, there may still be good moral reasons for upholding the rule that we should not be allowed to impose intense suffering on the animals.

Even if an absolute limit to the burdens we may impose on the animals is adopted, researchers will nearly always be able to design their experiments so that the relevant human benefits are derived without causing animals intense pain or other kinds of severe suffering.

Drawing the Ethical Limits

In the light of the animal welfare view it is possible to set out a strategy for establishing the ethical limits to the use of animals as models in toxicology. This strategy may be formulated by means of the following three questions:

1) will the animals experience reduced welfare?

2) are important human benefits at stake?

3) is it possible to get similar information in a way that is less burdensome to the animals?

In relation to the first question an experiment may be categorized as either *unproblematic*, *problematic* or *unacceptable*. If an experiment falls in either the unproblematic or the unacceptable category then no more questions are necessary. For those experiments categorized as problematic the second question is relevant. If the answer to the second question is positive one can move on to the third question. The experiment will be considered ethically acceptable if the answer to that question is negative.

We shall try, in what follows, to explain the three steps of the strategy in more detail:

Regarding question 1): Over the last two decades the study of animal welfare has grown into a scientific discipline of its own. Methods have been developed which can provide information about how well or badly animals fare. However, the study of animal welfare also gives rise to a number of methodological controversies. One such controversy concerns the definition of "animal welfare".

Duncan (1996) claims that animal welfare should be defined in terms of feelings, where a feeling in turn is defined as 'a specific activity in a sensory system of which an animal is aware'. The key notion here is 'awareness' - only states of which an animal is aware contribute, positively or negatively, to its welfare. Another author (Broom, 1996) claims that animal welfare should be defined in broader terms than this. Broom does not deny that feelings are relevant to welfare. He merely wishes to include, in the definition of welfare, the requirement that the welfare of an animal is impaired if the animal is not able to cope successfully with its environment.

There is no simple way to decide who is right and who is wrong where this definition is concerned, for a definition is not a statement of fact which can be tested empirically. One way to make progress might be to ask why in the first place we are interested in animal welfare.

This question has in our view been convincingly answered by Dawkins (1990):

> Animal welfare involves the subjective feelings of animals. The growing concern for animals in laboratories, farms, and zoos is not just concern about their physical health, important though that is. Nor is it just to ensure that animals function properly, like well-maintained machines, desirable though that may be. Rather, it is a concern that some of the ways in which humans treat other animals cause mental suffering and that these animals may experience 'pain,' 'boredom,' 'frustration,' 'hunger,' and other unpleasant states perhaps not totally unlike those we experience.

The point here is not that the mentioned concerns should blindly be accepted. Rather, it is that for the study of animal welfare to be relevant to the worries which initially prompted interest in the issue, animal welfare must be defined in terms of subjective feelings. And when it comes to the use of animals as models in toxicological research the prospect of the animals experiencing pain or other kinds of suffering is certainly the main cause for concern.

It should be noticed that the occurrence of unpleasant mental states does not by itself imply that there is suffering. Such occurrences are an unavoidable part of a normal animal life. Pain and other unpleasant mental states serve as signals which under normal circumstances point the animal in a useful direction and thereby ensure the satisfaction of its biological needs. Often such negative experiences are compensated for by corresponding positive experiences. This may, for example, be the case with a hungry animal that finds food. In such a case the animal is not suffering. Unpleasant states make up a welfare problem only if they are not compensated for by corresponding positive feelings, or of long duration, or frequent. For example, there is a welfare problem if a hungry animal is not being fed, or if an animal which is strongly motivated to root or scratch in the ground is kept in an environment where it has no opportunity to exercise this kind of behaviour.

As regards the measurement of animal welfare it is important to be aware that mental states cannot be measured directly. All that can be measured are physiological, behavioural, pathological and other similar objective parameters which may serve as evidence for the occurrence of the relevant feelings in the animals.

We shall not attempt here to give a detailed account of the various parameters which may serve as indicators of reduced (or enhanced) welfare. Overviews may be found in Dawkins, (1980) , Rushen and de Passille (1992), and Sandøe et al. (1996). We just want to point out a distinction between two kinds of parameters: Those which indicate that there is an actual welfare problem and those which indicate that the welfare of the animals is at risk. For example, certain abnormal behavioural patterns, like stereotypies and self-mutilation, are good indicators of an actual welfare problem, whereas the lack of normal immune function is an example of a piece of information from which it is only possible to infer that the welfare of the animals is at risk.

According to the animal welfare view the expected level of pain, suffering or discomfort in the animals must be assessed and classified according to the above mentioned trio of categories: unproblematic, problematic and unacceptable. If the expected level of pain, suffering and discomfort is only trivial then the experiment is unproblematic; and if, on the other hand, strong pain or other intense suffering is to be expected, then the experiment is clearly unacceptable.

Of course, there is no exact way of measuring the relevant level of pain, suffering or discomfort. This is so for two reasons. First, the words "trivial", "strong" and "intense" are relative terms. Secondly, assessing pain, suffering and discomfort is a matter of interpretation where there is room for individual judgement and consequent variation. However, a practice may still develop within which those responsible for the welfare of the animals are able to draw the relevant limits in a relatively uniform manner.

Problematic experiments may turn out to be either acceptable or unacceptable according to the answers given to question 2 and question 3.

Regarding question 2): Toxicological research and testing as well as other research and testing done on animals is typically justified by reference to important human benefits - for example, the provision of a foundation for risk evaluation and management. Therefore it is crucial to ask whether important human benefits really are at stake.

An obvious case for saying that important human benefits are *not* at stake is when no new information is to be expected from an experiment. This is so if either the experiment is badly designed and therefore does not yield any information at all (Rodd 1990) or the information yielded by the experiment is not new. Of course, experiments have to be repeated to confirm the results, but there may also be experiments which deserve to be regarded as needless repetitions.

Repetitions of dubious value may occur on a considerable scale within toxicological testing when different regulatory agencies require the same kind of test conducted according to different guidelines. For example Koëter (1991) describes the situation for skin irritation in the following way:

> Disturbing differences are noticeable between the various guidelines: numbers are either 3 (EC and OECD) or 6 (FDA, EPA, MAFF). Further, the duration of exposure as well as the observation time-points after exposure are different in the FDA guideline from all others. Where the OECD, EC and EPA guideline cover both irritation and corrosion, the FDA-guideline only covers irritation. For the detection of skin corrosion a separate test guideline has been described in the Code of Federal Regulations (1981), which allows classification of corrosive substances according to a system applied by the US Department of Transportation. This corrosivity test is quite comparable with the OECD and EG method for the detection of skin irritation/corrosion. However, still 6 animals are required instead of 3. As a result of these differences in skin irritation guidelines, studies are now unnecessarily being repeated; the US regulatory agencies do not accept studies with exposures of less than 24

> hours, whereas for instance, there are various examples of EG countries that do not accept studies with observation times that are different from the EC protocol. Further harmonisation is therefore urgently needed.

To avoid dubious repetitions of toxicological experiments all major regulatory authorities should therefore have the same guidelines. Achieving this is the aim of the International Conference for Harmonisation (ICH) which was set up in 1992 as a global initiative. Although ICH has made significant progress in harmonisation of guidelines, there is still work to be done before toxicological guidelines are rendered uniform.

What, then, about *basic research*? Does it satisfy the requirement that animal experiments should only be allowed when important human benefits are at stake? One defining feature of basic research is precisely, that it cannot be foreseen what findings will emerge from it. It may be said that basic research is directed at new knowledge and that increased knowledge is always good. However, this argument has been questioned by a British psychologist (Boyle, 1976):

> Enormous suffering is imposed on animals to establish theoretical viewpoints (possibly more in other countries than here, but the argument is universal). This is because we hold the view that the advancement of knowledge is an absolute good. I really believe that this view needs to be challenged.

As should be clear from what has been said above, we think that no experiments should be allowed where "enormous suffering" is imposed on the animals. We also agree that the advancement of knowledge for its own sake is not an absolute good. However, the justification for doing basic research is not just that it advances knowledge. Rather it is that this kind of research in the long run will turn out to be most useful. In many areas of applied research basic knowledge is needed.

This is also the case with toxicology. In many areas we lack knowledge of the mechanisms of toxicity; and therefore the results that we get from toxicological testing on animals may be difficult to interpret and utilise in scientifically based risk management. The point has been dealt with by Roberfroid (1991):

> Currently used experimental models for collecting data do not, of course, yield the truth about human risk and it is the lack of fundamental biological knowledge that is the major bar to this truth. ...
>
> Without any doubt, the second half of the XXth century will remain in history as the era of the explosion of knowledge in biological sciences. Highly sophisticated techniques have been developed which have allowed immense steps forward in the understanding of biology. Yet toxicology as a science ignores most of the mechanisms of the effects it observes. As compared to biology as well as to sister discipline pharmacology, toxicology as a science is still in its infancy. The future of toxicology requires a

rapid and immense effort to integrate these techniques with the objective to do more basic research. It is the only way to develop better and more predictive tests the rationale of which will be the cellular and molecular mechanisms of the adverse effects which need to be identified.

When it comes to basic toxicological research it can be safely assumed that important human benefits are certainly at stake. Incidentally there is good reason to believe that a better understanding of the mechanisms of toxicity will encourage the development of alternatives to the use of live animals as models in toxicology. Thus lighter burdens will be put on laboratory animals in the future.

Between the extremes of experiments either giving no new information whatsoever or having the prospects of providing valuable basic knowledge a line has to be drawn between experiments which promise to provide information that is important enough to justify the use of live animals and those which don't. Of course, there are no sharp and clear criteria for drawing this line. However, the mere fact that the issue is raised may have a significant effect on those who take the decisions on use of animals in toxicological experiments.

It is important to be aware that these decisions are taken at several levels. Thus the authorities which issue guidelines, the companies who do the experiments, the researchers and technicians who conduct the experiments, and the regulatory authorities which recognise the experiments all share responsibility for the ethical quality of toxicological research and testing. The discussion about whether the benefits are important enough to justify animal experimentation should take place at all levels – including the editorial boards of scientific journals (Svendsen et.al. 1997). It is also important at every level to consider whether the same results could be achieved in a way causing the animals to suffer less. This is where the third question comes in.

Regarding question 3): There are several ways to reduce the burden put on animals without foregoing the human benefits. Significant reduction of potential suffering can be obtained by refining the use of the animals (Smith and Boyd 1991). Thus it is essential in toxicity experiments and toxicity testing to stop the studies at a stage where mild signs of toxicity have been observed. This will still fulfil the purpose of the study while meeting the requirements imposed by the regulatory authorities. From a scientific point of view there is no need to continue studies to a stage where severe signs of toxicity appear. This aspect of the refinement of toxicity testing may reduce the potential for animal suffering significantly, although it obviously demands somewhat greater observational prowess.

Stepwise toxicity testing, including pilot studies employing small numbers of animals, may reduce the risk of suffering even more. In types of toxicity testing with more unpleasant end-points, such as eye or skin irritation/corrosion, the stepwise procedure may be a significant refinement – particularly where it includes an evaluation of physico-chemical properties of the relevant compound, *in vitro* testing and inclusion of only few animals in the subsequent initial testing. If signs of toxicity are observed in these few animals, further testing seems redundant. For the sake of proper safety assessment of no apparent risk, further confirmatory testing on the animals has to occur when no signs of toxicity have been observed in the first few animals included. But in these cases the risk that the animals will suffer is very small. Stepwise procedures can also be introduced in tests for systemic toxicity, thereby reducing total suffering.

A general and sometimes neglected aspect of the refinement of animal experimentation is the training and general attitude towards animals of the staff involved. Proper training is a critical issue, but it needs to be regularly reviewed. Where training was received years ago, for example, recent experience and practice of procedures is essential. Updating in training may be achieved using dead or anaesthetised animals. This can serve to refine the subsequent study(ies).

The significance of the attitude of people involved in animal research is often overlooked as a welfare issue. The majority of people actively involved with animal experimentation have a positive and professional attitude towards the welfare of the animals. Reporting to the responsible scientists by those who actually handle and observe the animals is very important, not least in toxicity testing. Professionalism, here, should not be interpreted as neglect of animal welfare, but rather as requiring an essential understanding of the needs of the animals. Procedures carried out by properly trained people with a solid understanding of the relevant study ought to present fewer problems than they do when carried out by researchers with less training or a weaker understanding of the purpose of the study. It goes without saying that the most important factor in refinement and reduction of suffering in toxicity testing is the human factor: *Thinking and considerate scientists and technicians make all the difference.*

References

Boyle, D.G. (1976) Animals as experimental subjects (letter). Bull. Br. Psychol. Soc. 29: 312.

Broom DM (1996) Animal welfare defined in terms of attempts to cope with the environment. Acta Agric, Scand. Sect. A, Animal Sci., Suppl. 27: 22-28.

Code of Federal Regulations (1981) Method of testing corrosion to skin. 49 CFR 100 to 199, part 173, Appendix A.

Dawkins MS (1980) Animal Suffering: the Science of Animal Welfare. Chapman & Hall, London.

Dawkins MS (1990) From an animal's point of view: Motivation, fitness, and animal welfare. Behav. and Brain Sci. 13: 1-61.

Duncan IJH (1996) Animal welfare defined in terms of feelings, Acta Agric, Scand. Sect. A, Animal Sci., Suppl. 27: 29-35.

Koëter HBWM (1991) Current guidelines and regulations in toxicological research. In: Hendriksen, CFM & Koëter HBWM (eds.) Animals in biomedical research. Elsevier, Amsterdam, pp 17-34.

Narveson J (1983) Animal rights revisited. In: Miller HB & Williams WH, Ethics and Animals, The Humana Press, Clifton NJ, 1983, pp 56-58.

Regan T (1985) The Case for animal rights. In: Singer P (ed.) In Defense of Animals, Basil Blackwell, Oxford, pp 21-25.

Roberfroid MB (1991) Long term policy in toxicology. In: Hendriksen CFM & Koëter HBWM (eds.) Animals in biomedical research. Elsevier, Amsterdam, pp. 35-48.
Rodd R (1990) Biology, Ethics and Animals, Clarendon Press, Oxford
Rushen J, de Passillé. AMB (1992) The scientific assessment of the impact of housing on animal welfare: A critical review, Can. J. Anim. Sci. 72: 721-743.
Sandøe P, Giersing MH, Jeppesen, LL (1996) Concluding remarks and perspectives. Acta Agric, Scand. Sect. A, Animal Sci., Suppl. 27: 109-115.
Smith JA, Boyd KM (eds) (1991) Lives in the balance, The ethics of using animals in biomedical research, Oxford University Press, Oxford.
Svendsen O, Thorn N, Sandøe P (1997) Laboratory animal science, Welfare and ethics in pharmacology and toxicology, Pharmacology & Toxicology, 80: 3-5

Behavior of Laboratory Animals under Unnatural Conditions

Jan Ladewig
Department of Animal Science and Animal Health, The Royal Veterinary and Agricultural University, Bülowsvej 13, DK-1870 Frederiksberg C

Abstract

Domestic animals are animals whose living conditions and reproduction, among other things, are controlled by man. As such, the current discussion about the welfare of domestic animals is similar for farm, companion, laboratory, and zoo animals. It concerns identification of the behavioral and physiological needs of the animals and development of living conditions that enable them to satisfy these needs. The paper describes two approaches that have been used in behavior biology to identify such needs. One approach is the measurement of stress responses that may be activated when an animal's needs are not fulfilled. The other approach is the use of operant conditioning techniques to establish demand functions by which the motivation of an animal to perform a specific behavior is measured. It is concluded that, since welfare is characterized by the absence of a number of factors, such as stress, pain, fear, disease, hunger etc., many types of measurements must be applied to ensure optimal welfare.

Introduction

Concern about the welfare of domestic animals has been the incentive for much research on animal behavior in the last two decades. Whereas this research in the beginning was focused primarily on housing conditions of farm animals, the interest has expanded into areas such as housing and management under intensive versus extensive conditions. In addition, not only farm animals are studied, but all species that are under the direct influence of man, i.e. also companion animals, zoo and laboratory animals.

For laboratory animals, the concern about their welfare is not new. In fact, the anti-vivisection movement is considerably older than the anti-agribusiness movement. What is new in the concern is the attempt to improve the living conditions of laboratory animals in order to meet their behavioral needs. In principle, such attempts are not different from similar attempts within agriculture and consist of identifying the behavioral needs of the animals and developing systems in which these needs can be satisfied. Obviously, the first part of this process, to identify the behavioral needs of animals, is by far the most difficult task.

Basically, there are two approaches to analyze behavioral needs. One is to look for signs of stress when an animal is not able to satisfy its behavioral needs. The other approach is to measure how motivated an animal is to be able to perform various species specific behaviors.

Signs of Stress

When an animal is restricted in its behavioral repertoire to some extent, it will show a wide variety of responses. It is customary to divide these responses into behavioral and physiological responses, where the physiological responses may consist of stress responses per se and their influence on the metabolic, the reproductive, and the immunological functions of the organism. The result of these events is that the symptoms of stress may consist of many different types which means that it can be necessary to look at several or all aspects to identify that something is wrong. For instance, if an animal develops stereotypic behavior, it can be concluded with a reasonable degree of certainty that the animal is kept under conditions that do not fulfill its behavioral needs. The opposite conclusion, however, that animals that do not develop stereotypic behavior are kept satisfactorily, is not possible. To make such a conclusion, it is necessary also to analyze for stress reactions, reduced growth and fertility, or increased disease rate.

For many years we have argued that it is necessary to include physiological measure ments in behavioral research in order to get objective information about how the animals feel (Ladewig, 1994). Unfortunately, however, inclusion of physiological data does not always increase the understanding, but rather adds to the confusion. The main reason for this situation is not that physiological stress indicators are too insensitive or too erratic in their response, but rather is due to the complexity of the operating mechanisms and to their sensitivity to various intervening factors. As in other "real life" situations, many of these intervening factors cannot be controlled, causing a considerable amount of variation in the results and making comparison between studies, or replicates of studies, difficult (Rushen, 1991).

When we consider a phenomenon such as stress, the biggest problem is that we tend to treat it as a well-defined state that can be diagnosed on the basis of one or a few behavioral and physiological reactions. That this is not so is best illustrated by the following definition on stress (Veith-Flanigan & Sandman, 1985): Stress encompasses all extraindividual events capable of eliciting a broad spectrum of intraindividual responses after being mediated by a complex filter, termed individual differences. Apart from helping us to realize the complexity of stress, this definition makes it possible to analyze the complexity at three different levels: at the level of the stressors, the individual differences and at the level of the stress reactions.

Stressors are usually categorized according to their qualitative (e.g. physical, chemical, etc) and to their quantitative (e.g. acute or chronic) characteristics. There are, however, several important factors that we need to be aware of, before such a categorization is of any practical use. Firstly, as demonstrated by Mason (1974), many stressors only result in stress reactions when they exert their effect at the cognitive level, i.e. when the animal becomes emotionally upset. Secondly, as pointed out by Burchfield (1979), because of this emotional involvement, it is questionable whether chronic stress (i.e. a stressor that exerts its effect at a constant level over a longer period of time) really exists. Rather, the chronic stress is a series of acute type stressors that are repeated with varying intervals. In addition, many stress situations consist of several or many stressors, of which only some may be of importance for the animal. For instance, a sheep subjected to electroshocks may respond differently, if it is separated from its group members, than if it is kept in the group.

Individual differences affect the stress responses either due to genetic differences or due to differences in the earlier experience. In this connection it is important to realize that humans play an important role in the ontogeny of the domestic animals, a role that can directly affect their stress responses. From research, particularly on companion animals, but also on farm animals, it is known that, if the animal has a certain relationship to its caretaker, his or her presence during a stress situation can alleviate the stress response (Korff & Ladewig, 1995).

The type of stress reaction that an animal may show depends primarily upon the coping pattern of the animal, as shown by Henry and Stephens (1977). Thus, an animal subjected to the threat of loss of control will respond primarily by behavioral activation, i.e. by fighting or fleeing. This response is supported physiologically by activation of the amygdala of the limbic system, the sympathetic nervous system and adrenaline secretion. In contrast, an animal that has lost control over the situation and is aware of this loss will respond by behavioral depression, a response that is supported by activation of the hippocampus of the limbic system and of the hypothalamo-pituitary adrenocortical system.

Taking these precautions into consideration, it is possible to evaluate the living conditions of domestic animals and to change them, such that the activation of the behavioral and physiological stress responses are minimized. Similarly, the various handling procedures to which we subject the animals can be altered to obtain the same result.

Measurement of Behavioral Needs by Demand Function

The quantitative measurement of animals' behavioral needs is based on a principle established in economics (Lea, 1978). Economists have observed that consumption of a commodity depends on its price. For some commodities, consumption remains relatively constant when their price increases and these commodities are said to have an inelastic demand. An example of a commodity with an inelastic demand is gasoline, for which the consumer does not have an alternative to switch to, when gasoline prices increase. For other commodities, consumption decreases when their price increases, and these commodities are said to have an elastic demand. Examples of commodities with elastic demands are wine or eggs, for which the consumer can switch to alternative items when their prices increase. If the relationship between consumption and price increase is described by a curve on log-log coordinates, the slope of this demand function indicates the elasticity of the commodity.

In behavioral biology it is possible to use the same principle as in economics by using the operant conditioning technique. The price increases are simulated by requiring that a test animal performs a certain behavioral response (such as operating a manipulandum) an increasing number of times (also referred to as work-load) in order to obtain a commodity in the form of a reinforcement or reward. This reinforcement can be access to some factor in the environment which allows the animal to perform a certain behavior, e.g. food - eating behavior, nest material - nest building behavior, sand - dust bathing, etc.

For instance, a pig is trained to push a plate with its nose one time to obtain a small (and constant) amount of food. The ratio of responses to reinforcement is fixed on 1:1 and called Fixed Ratio (FR)1. As soon as it has learned the connection between its behavioral response and the reinforcement, the pig is only fed during the tests and it is tested every day. Its daily food consumption is measured. When a relatively constant level of consumption is reached, the test conditions are changed in that two nose plate responses are necessary per reinforcement (FR2). After stabilization of food intake, the number of nose plate responses is increased gradually over the following days or weeks (FR5, FR10, up to FR80 responses per food reinforcement). At each level of work-load, food intake is plotted and the demand curve and its slope calculated. In order to ensure replicability of the result, some or all of the different levels of work-load can be repeated, or the different levels can be tested in a random fashion. If food is important for the pig (which it is in case it is only fed during tests), its consumption at a high work-load will be almost as high as at a low work-load. The demand curve will have a small slope.

In a different test series (with the same or other experimental pigs), each pig is trained to push the nose plate to get a limited (and constant) amount of social contact with a known neighbor pig. Outside the test situation the test pig is kept socially isolated. Again the pig is tested for number of reinforcements at the different work-loads, a demand curve is plotted and a slope calculated. If social contact is less important than food, the pig will obtain fewer contacts at a high work-load than at a low work-load, and the demand curve will have a larger slope (Matthews & Ladewig, 1994). Using the same basic principle we have tested pigs' demand for various bedding materials (straw, saw dust, wood shavings, and sand), for straw in a rack and for locomotion (Ladewig & Matthews, 1996).

Substitutability of Different Environmental Factors

The demand for a commodity depends not only upon need or motivation and price or work-load, but also upon the availability of alternatives that can substitute the commodity to some extent. In order to be able to take this substitutability into consideration and to analyse the effect of confounding factors quantitatively, we have changed the method in the following way. Instead of testing the demand for related commodities in two separate experiments, they are tested simultaneously. If, for instance, we want to analyze to what degree straw can be substituted by saw dust as a bedding material, the test pig can operate one nose plate to get access to saw dust and another to get access to straw. The price or work-load for one commodity is kept constant throughout the experiment (e.g. FR20) and the price for the other commodity is varied as before. Ideally, two experiments should be conducted, in that the constant and varied price should be alternated between the two commodities. If one commodity is strongly preferred over the other, the animal will accept a relatively higher price for the preferred item compared to the less preferred item. On the other hand, the more one commodity substitutes the other, the closer to the constant price should the switch-over from one item to the other be. The slope of the items with the varied price indicates to what degree they are substituted by the items with the constant price.

Using this approach it is possible to rank the various species specific behaviors of an animal (e.g. feeding, nest building, social behavior and locomotion) according to their importance to the animal and to develop living conditions in which the performance of important behaviors is possible.

Conclusion

Welfare is a state that is characterized by the absence of pain, discomfort, fear, boredom, hunger, and numerous other factors, just like health is a state that is characterized by the absence of infectious agents, malnutrition, toxins and numerous other factors. And just as it is not possible to diagnose health, but only the absence of disease A, B, C etc., so it is not possible to diagnose welfare but only the absence of the many factors that may reduce welfare. Combined with studies that show the importance of the different species specific behaviors, it is possible to develop living conditions and management programs for domestic animals that ensure the fulfillment of important behavioral and physiological needs of the animals.

References

Burchfield SR (1979) The stress response: a new perspective. Psychosom Med 41:661-672

Henry JP, Stephens PM (1977) Stress, health, and the social environment. A sociobiologic approach to medicine. Topics in Environmental Physiology and Medicine. Springer-Verlag, New York

Korff J, Ladewig J (1995) Analyse der sozialen Beziehung zwischen Mensch und Nutztier im Hinblick auf einer Optimierung dieser Interaktion. KTBL-Schrift 359, Darmstadt.

Ladewig J (1994) Stress. In: Döcke F (ed) Veterinärmedizinische Endokrinologie. Gustav Fischer Verlag, Jena, Stuttgart, pp 379-398

Ladewig J, Matthews LR (1996) The role of operant conditioning in animal welfare research. Acta Agric Scand Sect A Anim Sci Suppl 27:64-68

Lea SEG (1978) The psychology of economics and demand. Psychol Bull 85:441-466

Mason JW (1974) Specificity in the organization of neuroendocrine response profiles. In: Seeman P, Brown G (eds) Frontiers in Neurology and Neuroscience Research. Univ. of Toronto, Toronto 1974, pp 68-80

Matthews LR, Ladewig J (1994) Environmental requirements of pigs measured by behavioural demand functions. Anim Behav 47:713-719

Rushen J (1991) Problems associated with the interpretation of physiological data in the assessment of animal welfare. Appl Anim Behav Sci 28:381-386

Veith-Flanigan J, Sandman CA (1985) Neuroendocrine relationships with stress. In: Burchfield SR (ed) Stress. Psychological and Physiological Interactions. Hemisphere, Washington, D.C. pp 129-161

The Impact of Stress and Discomfort on Experimental Outcome

Wolf H. Weihe, Ganghoferstr.6, D-83098 Brannenburg, Germany

Summary

Stress refers to a physiological and emotional state of man and higher animals in which the autonomic regulation is overstrained and temporarily disturbed under the impact of conflicting stimuli. Stress activates, invigorates, acts life-sustaining, and initiates and drives adaptive changes towards improved fitness. While the positive action is commonly underestimated, much attention is given to the discomfort and the strain of efforts required during coping. The label of stress as being bad and the core of suffering has been applied with particular empathy to laboratory animals, for they are kept in captivity and are exposed to experimental procedures. The husbandry conditions to which the animals are adapted are commonly standardized. This applies to procedures for subacute and chronic toxicity testing. Acute toxicity tests are the classical example of stress research in which the demands on the organism exceed the limits of its regulative capacity. Stressors are: the test compound, the procedure proper and preceeding treatment of the animal. The experimental stress contributes to model the real situation. The weighting between the stressors may modify the outcome of the test.

Introduction

Stress is a familiar word, readily used in colloquial language to refer to unpleasant feelings in states of bereavement, desperation, weakness, fatigue, exhaustion, associated with extraordinary physical and psychic demands or on involvement in conflict situations. A remark "I am under stress" or a judgement "this animal is under stress" derives from an awareness or passion of the individual that there is something which is irritating, disturbing or distracting the balance of the internal state. This something is only vaguely recognized, cannot be quantified, not even identified. It is called nonspecific, because it is noticed to be complex, something of which, for the time being, no precise diagnosis can be made. The word stress primarily sets the stage for an alarming situation, claiming sustained attention that the control over the course of life got out of hands and an effort for its correction is instantly needed.

It is fashionable to refer to physical, preferably emotional stress in daily life and it is expected of individuals to express or extend on the basis of personal experience with stress a compassionate attitude towards all living beings. On account of the experiences with disturbances of one's own personal condition it is concluded by analogy that these must be comparably bad for all those creatures which are kept and bred under the control of man for the maintenance, improvement or safety of human health and life in general. Of particular interest in this context are laboratory animals because they are purposedly used to investigate normal and disturbed processes in health and disease of which stress is a case in point. Regarding their function to serve as models for all kinds of exploratory investigations, not performable in humans, stress of laboratory animals deserves particular attention. Models are incontestable as scientific instruments; they provide circuitous routes for advances in biomedical research. Working in experimental research with animals needs to consider both how representative they are as models for real and realistic conditions in life and how apt the conditions of the experiment are for the animal. The animal should be maintained free of, or shielded from, additional and possibly interacting influences (i.e. those in excess of the demands brought about by the procedure proper) which may harm it beyond the unavoidable and thus may obscure results. Such influences are readily assigned to stress. Manser (1992) stated: "Stressful situations in the laboratory may arise either because of the way the animal is kept or because of the experimental procedures to which the animal is subjected". There is continuous concern that stress can lead to physiological and behavioral changes which may render experimental results invalid.

The Stress Concept

Behind today's worldwide lay application of the word "stress" to depict almost any disbalancing state in humans as well as in animals lies a scientific concept. The term "stress" was first used by Hans Selye in 1936 to describe a general biological phenomenon. He introduced the term and concept in 1950 in the monograph "Stress. A treatise based on the concepts of the general-adaptation-syndrome and the diseases of adaptation" (Selye, 1950). Selye had demonstrated that "the organism responds in a stereotypical manner to a variety of widely different factors such as infections, intoxications, trauma, nervous strain, heat, cold, muscular fatigue, or irradiation. The specific actions of these agents are quite different. It was concluded that the stereotypical response, which is superimposed upon all specific effects, represents the somatic manifestations of nonspecific stress itself" (Selye and Heuser, 1956, p.26). What is nonspecific stress? Selye defined stress as "the sum of all nonspecific changes caused by function or damage". He described this common response pattern as a typical and almost independent humoral stimulus-response tract within the physiological regulation network of the organism. The tract consists of the hypothalamus-pituitary-adrenal cortex axis (HPA) which he considered to be the central mechanism of the General-Adaptation-Syndrome (GAS) involving several hormonal negative feedback mechanisms. The time course of changes within the GAS from the first stimulation to the end was divided into three stages: (1) the alarm-reaction, (2) the stage of resistance at which he claimed adaptation to be optimal, and (3) the stage of exhaustion which led over into diseases of adaptation or death.

As the organism reacted to the impact of different stimuli with the general pattern of the GAS and not, as would be anticipated, with a specific stimulus-resonse reaction, Selye introduced the word "stressor", a noncommittal reference to a stimulus that can be described but is not clearly specified. Often a stressor is not primarily observed but inferred from, or looked for, when stress is diagnosed from an activation of the HPA. In 1956 Selye defined: stress is the state manifested by a specific syndrome which consists of all the nonspecifically induced changes within the biological system, a stressor is a cause of stress. With reference to laboratory animals Manser (1992, p.2) declares: "A state of stress occurs when an animal encounters adverse physical or emotional conditions which cause a disturbance of its normal physiological and mental equilibrium".

Selye developed the stress theory on results from experiments with rats. A state of stress was diagnosed when the level of blood corticosteroids was increased in live animals, and the adrenals were enlarged and the thymus disintegrated at autopsy. Selye applied extreme procedures. He wanted to see effects and these effects should be as uniform as possible. To arrive at these changes he augmented the strength of plain damaging stimuli. By biostatistical standards, which were increasingly applied in the Fifties, the results should be highly significant, the variances of means small. The peculiarity of these procedures, overloaded to the extreme, was never clearly addressed in conjunction with the GAS. Considering strength of stressors and survival time Selye's experiments can be compared with acute toxicity tests; they ended with death of the animal within hours or 1-2 days. For example, immobilization of the animal in dorsal position on a board as the stressor led to spontaneous death within a few hours. Considering the short duration of test experiments terminated by death within shorter time than one day the description of the central phase as adaptation was misleading. The GAS as derived from his experiments is just the opposite of adaptation (Fig.1). Adaptation requires time for the autonomic regulation to arrange responses; they are always specifically and selectively related to the properties of the stimulus. Adaptive processes lead to modifications of the physiological functions as compared to the previous state of the organism before the response to the stimulus. Those relief changes can only develop if the strength of the stimulus remains well within the limits of the adaptive capacities of the organism. Selye did not respect this. His standard experiments to elucidate stress were designed deliberately in excess of the animal's adaptive capacities as indicated by their fatal ending. Under the overwhelming, incapacitating strength of the stressors the organism was not given a chance to initiate adaptive regulation. Instead, after the autonomic regulation was alarmed, it was raised to the uttermost activation of resistance and remained on this level as the stressor continued to affect it. Such demanding activities of the HPA for defence can be maintained only for a limited time before breaking down. Resistance was the appropriate term for this phase of the GAS. To describe the overstretched resistance as "optimal adaptation" was a false

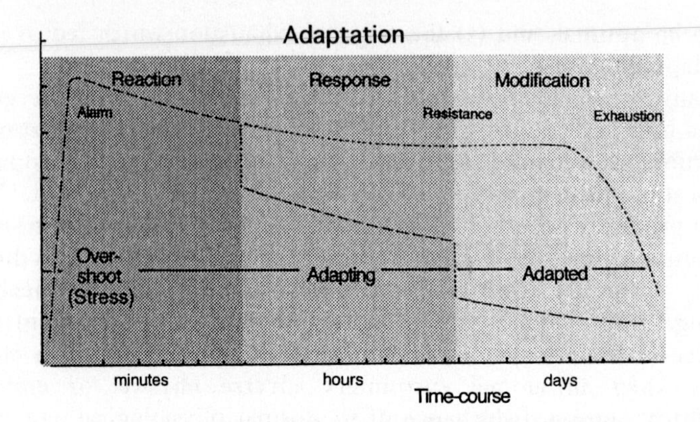

Fig 1. Time course of three phases of adaptation processes from left to right (dashed line): (1) the stimulus elicits a reaction; (2) reaction leads to response of adaptation; (3) response ends when adapted with modification: adaptive regulation is set to a lower level. Increasing strength of stressor is indicated on the ordinate. In the GAS (dotted line) the reaction phase is similar to the alarm-reaction. The response phase is a general resistance with maximum effort of regulation. With adaptation the activity of regulation proceeds downwards, with resistance in the GAS it proceeds at high level up the point of breakdown from exhaustion.

explanation and led to much confusion and misunderstanding. Selye could have avoided this if he had based stress on a General-Resistance-Syndrome instead of the GAS. Though resistance and adaptation represent different activities and directions of physiological regulation, Selye treated them like synonyms. Adaptation is the opposite of resistance, it regulates the demand on physiological systems down to maintain its efficiency. Selye also objected to associate the alarm-reaction of the GAS with Cannon's emergency-reaction based on the catecholaminergic neurovegetative system.

The Stress System

The GAS concept has been modified since it was first propagated. Interesting enough, the key word "adaptation" is not listed in the index of Selye's last cumulated literature review published in 1971 under the title "Hormones and resistance". What used to be the stress theory is now known as the stress system (Chrousos et at, 1993). The HPA has been replaced by the hypothalamic-corticotropin-releasing hormone (CRH) system. The original observation is proven that there are short-loop feedbacks of hypothalamic releasing or inhibiting hormones; involved are ACTH, ADH (antidiuretic hormone), β-endorphin, α-MSH (melanocyte-stimulating hormone). Peripherally, other hormones are included, such as LH (luteinizing hormone), FSH (follicle stimulating hormone), GH (growth hormone) and TSH (thyroid-stimulating hormone) (Chrousos and Gold 1992). The two principal reaction chains, the CRH hormonal chain and the sympathetic-adrenal medullary system (SAMS) neural chain are combined into one system. The stress system unites the concept of Selye with that of Cannon. It represents an essential move towards an integration of all systems participating in autonomic regulation. The incorporation of the sympathetic nervous system into the stress system was facilitated by a change from extreme to moderate, less consuming experimental procedures. By reducing the strength of stressors or stimuli to tolerable levels the adaptive capacities of the animals could be studied. In as much as the strain in these experiments is maintained below exhausting levels the organism has a chance to activate specific adaptive responses for strain relief. It is now generally accepted that the "nonspecific" changes, indicated by the activation of the CHR and SAM systems, appear in conjunction with the action of specific stimuli not only under exceedingly strong but also with all degrees of moderate condition. As soon as their strength is reduced following the initial arousal via the general pattern of release of CRH and noradrenaline from cells of the CNS there is time and chance for the organism to respond with adjustment or activation of specific adaptive regulation processes. These processes are programmed to search and aim towards physiological and morphological modification for attenuation of the stress or for stress relief.

The release of numerous hormones, their humoral circulation and increased urinary excretion during arousal of the stress system has led to call them "stress hormones". It is claimed that the stress system is activated when the values of one or more hormones are increased. In that case the subject is under stress, and with reference to laboratory animals, it is claimed that they suffer.

Stress and Homeostasis

Stress is defined as a state of the organism at which the autonomic and conscious regulation are aroused to a level that threatens or violates homeostasis (Kopin 1993). But what is homeostasis? Homeostasis – the term was introduced by Cannon in 1928 – refers to the maintenance of an integrated balance of internal functions. It is regarded as a state of harmony among balances of the body's physiological functions on which its viability, activity, and physical and mental efficiency depend, thrive, and from which comfort and well-being result. Cannon observed that passions such as hunger, thirst, fear and rage elicit an "emergency-reaction" through stimulation of the sympathetic autonomic nervous system with the discharge of adrenaline. The autonomic nervous system plays a key role in activating and organizing all related physiological functions to regulate any minor or major disturbed balance from internal strain back to a balanced state of harmony among them. Cannon as well as Selye found that strong stimuli aggravate the internal strain to levels high enough to activate physiological mechanisms beyond daily or common variation ranges of regulation. This is what is meant by saying stress is irritated or threatened homeostasis.

Stress and Adaptation

The word "stress" fills a gap in communication to refer to the awareness of a person of internal strain and conflict. The complex response patterns to strain or complex stress are collectively described as coping. Coping is guided by homeostasis. There is a strong genetic disposition for coping potentials to overcome stress. Jacob et al (1986) observed in mice that the amplitude and the mechanism of forming analgesia during stress varied between strains. Coping is an integral part of adaptation. Adaptation means up- and down-regulation of functions and activities for relief of the strain of stress to economize regulation and to attain fitness. Adaptation processes proceed with and without major stress as part of regulation in response to changing conditions, a continuous succession as long as the organism is alive. The processes to accommodate stress and regain comfort and efficiency proceed parallel on the level of the autonomic nervous system and the level of the conscious mind. Man has great advantages for coping because of the active engagement of his mind in decision making. This is not so for animals in the laboratory; they cannot differentiate feelings or reflect in conflict situations. The diagnosis of stress in animals can be made only on the basis of a small number of physiological factors and a narrow repertoire of behaviors. The gravity of stress cannot be measured in quantitative terms because it is an ongoing process to which, even if a stressor is properly defined, a variety of uncertainties contribute, such as genetic disposition, state of health, preceeding adaptive modifications. In most cases the gravity of stress can only be assumed and judged by analogy with reference to signs of coping.

Stress and Suffering

Manser (1992) claims that stress is invariably bad for laboratory animals and objects to divide stress into eustress and distress. She does not accept the concept of over-stress, neutral stress and under-stress, based on adaptive conditioning of animals with regard to measures of husbandry. According to Manser every experiment involves stress: the animal is seen like a solid block which resists to deformation when hit upon. This view of quantifyable resistance is in accord with the description of stress in technical sciences. Stress may occur due to both the conditions of husbandry and of experimental procedures. " A tremendous amount of suffering has been inflicted upon animals which have been subject of stress research" and " there is no place for research involving the application of severe stressors". This may be correct for research for stress's sake. Indeed, research into extreme stress appears superfluous today, because the phenomenon "stress" is well known. However, further observations have been made such as the occurrence of analgesia during stress, and new questions have arisen such as the mobilization of opoids during stress leading to analgesia (Curzon et al 1986; Grosman 1986; Mayer 1989). The discussion continues as to whether the aspects of pain during stress should be studied (King 1986). Prospective pharmaceuticals have been, and still need to be, tested in animals in order to find out whether and what sort of damage they may cause. Bioassaying is not reasearch on stress but on the effectiveness of a chosen compound for which the chances of causing stress and lasting damage need to be known for future prevention. Testing of new compounds profits from the research on stress. It provides a lead for specific investigations into the mechanisms of action of the compound. Obstruction for the animal may be involved, but must not necessarily follow. The unproven "suffering" accompanying bioassaying has to be weighted against the general and expected practical importance of the compound and the need to find out about its toxic action. In a review on the "Assessment of stress in laboratory animals" Manser (1992) argues that the measurement of physiological and behavioral changes caused by stress "provide a means of recognising and assessing suffering in animals". After checking a large array of physiological, hormonal, and behavioral indicators of stress she concluded that the information based on these parameters is not reliable. This is not surprising because in sub-extreme stress there are always adaptive processes regulating against the disturbed homeostasis. However, if even the intensity of stress cannot be diagnosed, how can it then be justified to extrapolate from a general reaction to claim suffering? Because of the many uncertainties Manser recommends further evaluation and validation of biochemical as well as behavioral stress indicators such as the discomfort index. Thus, even among animal protectionists a continuation of research on stress is demanded. The contradictions over the need and justification of research with animals dwindle as soon as the parties realize uncertainties over details. While animal protectionists are never short of arguments to point to the differences between animals and man, with regard to the stress system they accept, and insist on, unreserved comparability.

Stress versus Standardization

Stress due to measures of husbandry should not be a problem in toxicity testing. Husbandry conditions and health control are standardized, approved and further refined. Agreement on many issues has led to recommendations and testing guidelines, such as the "OECD Guidelines for the Testing of Chemicals". It should never be forgotten that they represent a compromise among multiple suggestions derived from prevailing climatic conditions, technical, organisational, and financial facilities in the industrialized countries. Almost all toxicity research is carried out in those countries where the respective potentials exist. The recommended or requested conditions are not necessarily optimal but they are practical, and animals have little difficulty to adapt to them. It cannot be emphasized enough that there are no optimal conditions in husbandry. Recently Poole (1997) argued that "happy animals make good science". Happiness is a unique human phenomenon. Poole upratedselected behaviors of animals in the laboratory to represent happiness. A happy animal is "one which is alert and busy, is able to rest in a relaxed manner, is confident, does not display fear towards trivial non-threatening stimuli and does not show abnormal behavior."

These are signs of undisturbedness. How can one produce happy animals considering genetic dispositions and environmental potentials including the personnel? Can an animal claim desires of fulfilment and rights to be happy in utalitarian thinking? Will solely "happy" animals secure research? Is there agreement over happiness as it is in the case of health, and how will I feel chosing a happy in preference of a simply healthy animal for a toxicity test? It is generally agreed that healthy and emotionally undisturbed animals will make good research. Emotional undisturbedness, however, is not automatically happiness. The assignment of happiness to animals may stem from a misinterpretation of the World Health Organization's definition of health as "a state of complete physical, mental , and social wellbeing and not merely the absence of disease". The common interpretation of health as absence of disease is idealistically extended to psychological and social dimensions. After much practical experience with this definition it has been realized that a state of complete physical, mental, and social wellbeing is closer to happiness than to health (Saracci 1997). Health and happiness are distinct experiences and their relationships are neither fixed nor constant. Common existential problems, as Saracci points out, involving emotions, passions, personal values, can make one's days less than happy or even uncomfortable, but are not reducible to health problems. "The distinction between health and happiness is crucially relevant in terms of rights, in particular positive rights or entitlements, requiring societal actions..". This would involve new aspects in the discussion on animal rights. Happiness is strictly subjective. Quoting from Saracci: "Failing to distinguish health from happiness has main consequences. Firstly, any disturbance to happiness, however minimal, may come to be seen as a health problem. Secondly, because the quest for happiness is essentially boundless, the quest for health also becomes boundless. Thirdly, annexing happiness to health and regarding health as a universal positive right introduces an underlying prescriptive view of happiness in society. Trying to guarantee the unattainable, i.e. happiness, will inevitably subtract resources".

There has never been a problem that absence of disease and careful adherence to the requirements of husbandry for the maintenance of health will provide unstressed animals fit to tolerate a fair amount of stress without conflicting with experimental outcome. Doubts are always left. One concerns the perturbation of animals after transport. Van Ruiven et al (1996) compiled reports on time needed for adaptation after moving animals between rooms and after transport. Depending on circumstances and on the kind of transport the minimum length of time required for adaptation varied from none to 7-8 weeks; in most investigations 7 days appeared to be sufficient for adaptation. One wonders whether such statements are realistic and of any practical value. In this case it might be useful to take man as model of the animal. Would it be necessary to stay in bed for one or more days, up to a week, after a trip by taxi from the airport to the hotel, or after a drive home from the office by car, in order to bring corticosterone or adrenaline levels down? They vary widely during the day and are never stable anyway. Elevated levels of these so-called stress hormones are not reliable parameters for estimating whether the power of stress has been strong enough to conflict with coping capacity, involving obstruction (Manser 1992). Transport as a novel or unfamiliar time restricted stimulus may, and often will, cause a transitory irritation. But this is a reaction which will hardly be that serious to require a recovery period of 1 to 7 weeks. Recovery from irregularities is not identical with adaptation, because it ends in a return to the previous state before exposure; no particular modification in the physiology of the animal will follow apart from a new experience. Elsewhere I have discussed that the variation of the milieu factors needs to be considered in context with the standardization schemes (Weihe 1994); transport of animals is one of them. Animals will adapt with down-regulation to within narrow limits rigidly controlled conditions and with up-regulation of their physiological functions to moderate variation under liberalized conditions. The former mode of adaptive modification will make the animals susceptible to general and abnormal variations, the latter mode will render them resistant; they will hardly react because they are familiarized to variations of factors as a common feature. Overreactions in down-regulated animals because they were pushed to adapt to the prescribed narrowly controlled husbandry cannot be put on the same level with stress. They should be seen to indicate that these animals are artificially sensibilized in life of under-stress under a restricted standardization scheme in which every form of variation is excluded. Stress is not absolute to the stressor but relative to the conditioning of the animal to which it is fitted through adaptation.

A comprehensive account of defined conditions for care and husbandry of animals is one of the oldest principles in toxicological bioassaying in order to ensure comparability and reproducibility of results. The principle is observed like a dogma according to which every source of variation should be excluded through standardization of all major factors on which the conditions are based. The classical bioassay was developed by insisting that distinct criteria should be fulfilled for the performance of the procedure, such as animal species, strain, sex, age, composition of food and nutritional regime, number of cage mates, type of cage, cage bedding, cleaning, handling, microclimatic and further environmental conditions like noise level, air purity, air admixtures. A standardization scheme is arranged around the choice of those conditions which appear to require a minimum of effort for the animal's physiological regulation. This is said to be optimal. Optimal cannot be absolute; it needs to be seen as relative to the species and strain, the condition of the animal, the method, and the purpose of the experiment. Every animal can readily adapt within the genetically determined capacities depending on the species and strain without major effort to other measures of standardization if specific susceptibilities are requested. The standardization frame is not a permanent fixture. On the contrary it should be open to modifications if there are reasons such as the suitability of the model. For example, to raise the susceptibility of the animal for a toxicity test of a compound with possible side-effects on thermoregulation it would be more reasonable to keep it at lower or higher temperatures than of comfort for the resting animal. The animal serves as a model but to serve its function the model proper has to be modeled in such a way that it can serve its purpose best.

Stress in Toxicology

The compound chosen to be tested in large and small dosages for toxicity fits with the classical conception of a stressor, and the effects follow the activation of resistance in accordance with the former GAS (Fig.1). Bioassaying of high dosages of a compound for acute poisoning will undoubtedly be accompanied by high power stress; testing low dosages may, and if, will elicit only mild stress reactions not more than a transitory irritation or perturbation of the physiological regulation from which the animal can quickly recover without adverse consequences or shortcomings for its comfort. Needless to say, the release and extent of toxicity stress depends on the dosage of the compound.

There is a real problem incorporated in screening of a new compound in animals for evaluation of unwanted versus wanted effects. Every new compound is selected on account of the likelihood to display a specific action at a target side where it can accumulate to improve or relieve diseased functions and facilitate adaptation of the organism. The pharmacological action of the compound as specific stimulus follows the adaptation curve in Fig.1. Simultaneously it is widely distributed over the body and may invade to, or be deposited at, other sites which makes it an unspecific toxic stimulus. The toxicological effect follows the resistance curve in Fig.1. The specific effect of the compound is wanted, the unspecific action not. The pharmacological and toxicological actions of a compound diverge, the former providing relief, the latter bringing about enhancement of burden. The unspecific action will cause stress, and the more toxic the compound turns out to be, the more the organimsm will be stressed. However, the interest of the scientist is not in the initiation of stress but in the damage caused in the body. The stress is incidental in conjunction with damage summarized as side-effects. The animal is valued to be representative of man and not solely a model for selected functions. If so, is the animal at the time of the test representative of a healthy or diseased human subject? The first requirement for preparing the animal for the test is that it should be healthy without the burden from stress of disease and emotional disturbance. This ensures a quality standard, but not representativeness. To achieve the latter the inherited properties of the animal and man should be closely similar and the features of the state in the healthy and diseased humans should be established in the animal in advance of the testing. The differences in genetic make-up are beyond control, they have to be accepted as they come. With respect to the possible responses regarding the test situation there are striking differences in two features between the human subject and the animal: the attitude towards the method of testing and the anticipation about the action of the compound. The method of administration is of little importance for the human subject but of great influence for the reaction of the animal, especially if it has never experienced such measures before. Anticipations relating to the effects of the compound do not occur in the animal, while they can be very strong in human subjects, particularly in patients with hopes for curative effects. Animals are entirely naive about the effects of the compound while humans are fully aware of it and conditioned to accept or object to it. Consequently stress in animals will arise from the manipulations in conjunction with the administration of the compound and from its unspecific toxic actions that follow. The latter will be largely alike in man and animal, the former will be largely different. The stress deriving from manipulation is accessible to control by the scientist in charge of the test. He can direct and greatly reduce strain by familiarizing the animal with the procedure in order to eliminate fear and panic through habituation. Even if the animal is alarmed through fear about a sudden interference with its life this impact is a single, short lasting event unless other stressors are introduced in the procedure. In the hands of an experienced scientist stress for the animal caused by the initiation of the test with the administration of the compound can be held at a low level if not be entirely avoided. There must be major mistakes in manipulating unprepared naive animals to raise stress to the level of suffering from pain. Considering the scientist's responsibility for the welfare and fate of animals used for assaying toxicity he is obliged to make every effort to reduce the chances for manipulation-stress to occur. By adhering to this well-reasoned argument he will succeed in both directions: the reliability of results from assaying will be made safe and the stress forced on the animal kept low. This will be to the advantage of both drug testing on the one side and the welfare of the purpose-needed animal on the other.

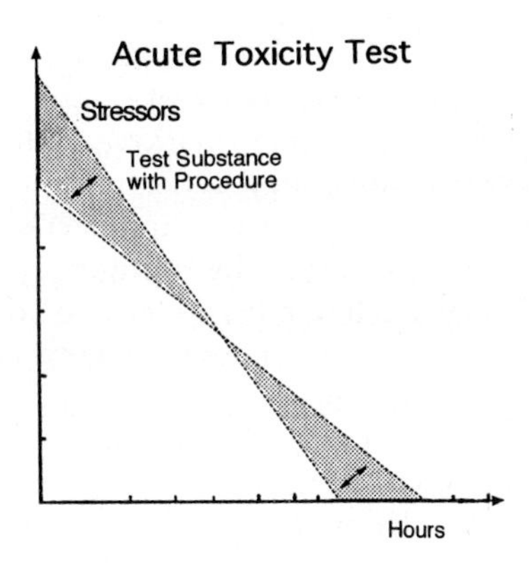

Fig. 2. In acute toxicity tests the stress caused by the procedure contributes to the stress caused by the tested compound. This may enhance and alter poisoning processes within some degrees of freedom (shaded field, indicated by arrows) and accelerate or delay breakdown with death.

The weight given here to the manipulation of the animal as part of the procedure to be an independent stressor separated from the stress affected by the compound in the course of assaying it for toxicity allows an evaluation between the three classes of test: the acute, the subchronic, and the chronic toxicity test. The largest provocation of stress can be expected to occur in testing for acute toxicity. Because it is a single, short-lasting event, the animal is often

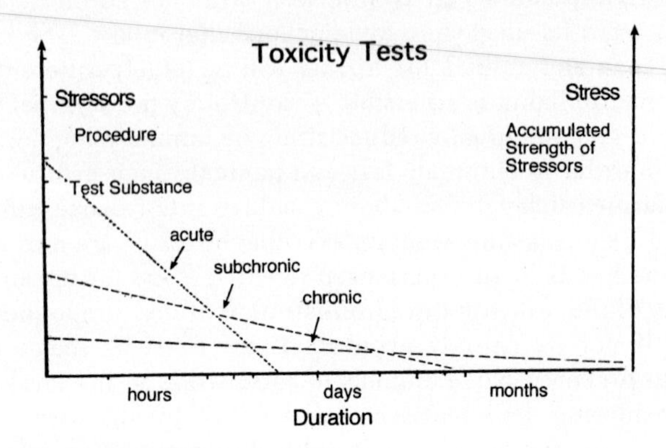

Fig. 3. The strength of the accumulated stressor action decreases when, for subtoxic tests, the dosages of the compound are reduced and the observation periods are prolonged, because the animal will habituate to the procedure. Minimum or no test-related stress is anticipated in long-term tests for chronic toxicity.

not prepared for the manipulation. Animals are stressed by events preceding the manipulation such as transport, grouping, marking, withdrawal of food and water. The effect of these stressors may accumulate or even-up with the manipulation stressor when the compund is administered. If they accumulate with the stress caused by the compound, the poisoning process may produce changes of symptoms and may accelerate the process to come to an earlier end (Fig.2). These stressors gradually loose power as they are attenuated when prolonged observation periods are planned: the manipulation measures are more moderate routine which will initiate habituation of the animal to the manipulation, and when the dosage is gradually lowered the compound will weaken as stressor in as much as habituation to the compound develops (Fig. 3). In chronic toxicity tests manipulation and dosage as stressors are not isolated anymore. With time alterations in the animal and variations in environmental conditions will add to the daily load from which no creature, neither man nor animals can be protected. As Selye stated "Stress is life and life is stress".

Conclusion

Assaying substances for specific pharmaco-biological action and toxicity can be affected, and results can be distorted, when the condition of the animal and the consequences of manipulation for the assaying are improperly respected which may give rise to elicit stress. The stress due to manipulation can be neutralized or enhanced by additional stress resulting from the toxicity of the compound. The experimental outcome, summing up comparability and applicability of test results, is chiefly affected, directed and controllable by the scientist. Whether the stress associated with bad management will enhance toxic effects of the compound or fuse without stress reinforcement is neither known nor predictable.

References

Chrousos GP, Gold PW (1992) The concepts of stress and stress system disorders. JAMA 267: 1244-1252.

Chrousos GP et al (Eds) (1993) Stress. Basic mechanisms and clinical implications. Ann NY Acad Sci 771: 1-775.

Curzon G, Hutson PH and others (1986) Characteristics of analgesias induced by brief and prolonged stress. Ann NY Acad Sci 467: 93-103.

Grossman A (1989) Opioids and stress: The role of ACTH and epinephrine. In: Neuropeptides and Stress. Taché Y, Morley JE, Brown MR (Eds), Springer-Verlag, Berlin.Heidelberg.New York, 313-324.

Jacob JJ, Nicola MA, MIchaud G, Vidal C, Prudhomme, N (1986) Genetic modulation of stress-induced analgesia in mice. Ann NY Acad Sci 467: 104-115.

King FA (1986) Philosophical and practical issues in animal research involving pain and stress. Ann NY Acad Sci 467: 405-409.

Koob GF, Tazi A, LeMoal M, Thatcher-Britton K (1989) Corticotropin-realeasing factor, stress and arousal. In: Neuropeptides and stress. Y Taché, JE Morley, MR Brown (Eds), Springer-Verlag, Berlin.Heidelberg.New York, 49-60.

Kopin IJ (1995) Definitions of stress and sympathetic neuronal responses. Ann. NY Acad Sci 771: 19-30.

Manser CE (1992) The assessment of stress in laboratory animals. RSPCA, Horsham, West Sussex.

Mayer DJ (1989) Stress, analgesia, and neuropeptides. In: Neuropeptides and stress. Y Taché, JE Morley, MR Brown (Eds), Springer-Verlag, Berlin.Heidelberg.New York, 276-296.

van Ruiven R, Meijer GW, van Zutphen LFM, Ritskes-Hoitinga J (1996) Adaptation period of laboratory animals after transport: a review. Scand J Lab Anim Sci 23:185-190.

Poole T (1997) Happy animals make good science. Lab Anim 31:116-124.

Saracci R (1997) The World Health Organization needs to reconsider its definition of health. Brit Med J 314: 1409-1410.

Selye H (1950) Stress. Acta, Montreal.

Selye H, Heuser G (ed.) (1956) Fifth Annual Report on Stress, 1955-56. MD Publications, New York.

Selye H (1956) The stress of life. McGraw-Hill, Toronto.

Selye H (1971) Homones and resistance. Part I and II. Springer-Verlag, Berlin, Heidelberg, New York.

Weihe WH (1993) Adaptation in animal husbandry and experiment. In: Welfare and Science. J. Bunyan (Ed.), Royal Society of Medicine Press, London, 294-299.

Priorities in the Development of Alternative Methodologies in the Pharmaceutical Industry

M R Jackson
Glaxo Wellcome Research and Development, Park Road, Ware, Hertfordshire SG12 0DP, UK.

Abstract

Promotion of animal welfare is an underlying and laudable goal for toxicologists and there is good reason to adopt practical, focused, investigative approaches towards this aim.

Pharmaceutical regulatory toxicology can be subdivided into the areas of systemic (target organ), reproductive, genetic and topical toxicology, as well as immunotoxicology and oncology. These areas can be assessed for prioritisation as to where adoption of measures to promote any or all of the 3Rs (reduction, replacement, refinement) would lead to the most tangible benefit for animals.These measures can range, for example, from replacement of animal experimentation with alternative *in vitro* techniques, to adoption of regulatory protocols that reduce the number of animals required.

This paper is confined to consideration of *in vitro* technology in terms of reducing/replacing laboratory animal use, and a suggested list of criteria for prioritisation is potential for:-

- Regulatory acceptability
- Reducing development cost
- Reducing animal numbers
- Promoting welfare aspects
- Elucidating toxic mechanisms
- Usefulness in compound selection
- Advancing the science of toxicology

Clear messages emerge from such an analysis which could influence prioritisation of the application of *in vitro* toxicology from the standpoints of animal welfare, feasibility and resources.

Introduction

The scope of animal welfare and ethics in regulatory toxicology testing is extremely broad. It is helpful within such a breadth to prioritise interest and activities that offer the greater opportunities for progress in these respects and to see what synergies emerge from consideration of other factors that are important in pharmaceutical development. One approach to this is to examine animal use in the various components of toxicity testing in the Pharmaceutical Industry in order to identify, from a number of perspectives, the best opportunities and benefits for reducing, refining and replacing the use of laboratory animals. This paper is confined to consideration of prioritisation in the search for *in vitro* methodologies that will contribute to the reduction and replacement of animal usage. The factors considered in sequence for the purpose of prioritisation are potential for: regulatory acceptability; reducing animal numbers; reducing development costs; promoting animal welfare; elucidation of toxic mechanisms; selection of compounds; advancing the science of toxicology. These factors are each applied in turn to the various components of preclinical safety assessment, i,e., acute, repeat dose for identification of target organ toxicity, oncogenicity, reproduction, genotoxicicity, irritancy and immunotoxicity studies. In association with the consequent prioritisation the feasibility of advancing animal welfare by adoption of *in vitro* methodology in each particular area is then examined.

Regulatory Acceptability

Regulatory agreements in partnership with Industry within the recent International Conference on Harmonisation (ICH) activities have made progress on refinement and reduction (Van Cauteren, H. 1996). This has been done by consensus on modification of protocols that affect the character and extent of testing. In contrast reducing laboratory animal use by adoption of alternatives has made little progress in terms of regulatory acceptability. Genotoxicology provides the sole example where *in vitro* tests for point mutation and clastogenicity allow early evaluation of a candidate medicine in man without the need for confirmatory data derived from use of animals (expected to be ratified by ICH 4). Nonetheless the role of *in vitro* technology in providing explanation of toxic mechanisms or information on cross-species extrapolation can be pivotal in providing both Industry and Regulatory Agencies with interpretative insights on safety issues.

Regulatory, and indeed Industrial acceptance of *in vitro* methodology as *replacements* will continue to be a major challenge for the foreseeable future. As elaborated later on in this paper, the basis for *in vitro* replacement will need to be grounded firmly on a fundamental understanding of the science in any particular aspect of toxicology.

Reducing Animal Numbers

The UK Government (Home Office) statistics for 1995 (Statistics of Scientific Procedures on Living Animals, Great Britain, 1995) reveal a decrease of 4.7% in the number of procedures (total 2.7m) on animals from the previous year; 80% of the total number of procedures employed rodents. Almost half of toxicology tests in 1995 were for applied studies in relation to human medicines. About one fifth of all procedures were performed to comply with legislation.

Within Glaxo Wellcome the proportionality between research (80% use) and development (20% use), most of the latter being toxicology, has not altered substantially over the past few years. (Table 1).

For the past 5 years there has been a steady reduction in animal numbers used within the company. Very little of this decrease has occurred in Development (mainly used in toxicology) in contrast to Research use which has incurred a 50% reduction over this period. A reason for this difference is that preclinical safety evaluation is regulatory driven and so, unless there are changes in regulatory requirements, the number of animals used for this purpose will be more or less static. The International Conference on Harmonisation process is undoubtedly having some influence on reducing animal use (Van Cauteren, H. 1996) as is the growing interest in biotechnological development of large molecules, this incurring less safety evaluation in animals since the amount of *in vivo* toxicology feasible with proteins is limited. However, the increasing emphasis on pharmacokinetics in safety evaluation is bringing a modest pressure in the opposite direction.

Table 1. Trends in Animal Usage 1990-1994 (GlaxoWellcome)

Activity	Development	Research
• Proportion of animal usage	20%	80%
• Change	Very Little	50% reduction
• Influences	*Regulatory Driven* (Safety)	*Discovery Driven* (Innovation)
	↓ International Harmonisation	↓ Acceptability of *In vitro* Techniques
	↓ Biotechnology Products	↑↓ Molecular Biology
	Products	
	↑ Pharmacokinetics - Toxicokinetics - Metabolic profiles	↓ High Throughput screens

↑ increased animal usage
↓ decreased animal usage

Research on the other hand is driven by innovation from the perspective of efficacy. In the absence of safety considerations at this stage, there is a more positive inclination towards the employment of *in vitro* techniques and particularly high throughput screens. The trend is not entirely straightforward since molecular biology can influence animal usage in two directions. Molecular biologists are primarily interested in pathways within cells and therefore work with *in vitro* systems. However, the increasing interest in genetic manipulation (transgenic animals) is tending to drive the number of animals both bred and used upwards (Statistics of Scientific Procedures on Living Animals, Great Britain 1995).

The various components of regulatory toxicology required for the development of a New Chemical Entity (NCE) into a chronically used medicine for a single route of administration are shown in Table 2 together with an approximation of the number of animals used. It is clear from the perspective of numbers of animals used, that the area of oncogenicity would give the greatest return from replacement by *in vitro* alternatives. By contrast, studies concerned with areas of immunotoxicology, irritancy, acute toxicity and *in vivo* genotoxicity use comparatively small numbers of animals and from this standpoint are less critical for reduction/replacement goals.

Table 2. Animal Usage and Cost in Regulatory Toxicology

Area	Number of animals used[1]	Cost (£)
Oncogenicity	1800	1,160,000
Genetic Toxicology	60	25.000
Acute Studies/ Dose Ranging	50	6,000
Reproductive Toxicology	390	200,000
Immunology	36	2,000
Repeat Dose Studies	810	700,000
Skin/Eye Irritancy	21	1,000

[1] Approximate number of animals used for development of an NCE for chronic use for single route of administration

Reducing Development Costs

Table 2 also shows approximate costs for each component of regulatory toxicology. Oncogenicity studies are the most expensive costing in excess of 1m sterling for preliminary and main studies. By comparison the costs for immunotoxicology and irritancy evaluation are vanishingly small. The promotion of *in vitro* toxicology on financial grounds would most favour resource application in the area of oncology - areas such as irritancy are of much lesser importance in this respect.

Promoting Welfare from the Perspective of the Individual Animal

The UK Government (Home Office) categorises the severity of toxicological procedures as shown in Table 3. The basis of this categorisation has never been explicitly stated, but is reasonably consistent with criteria published from another source (Baumans V, et.al). Most of the procedures fall within the category of potentially moderate severity, although very often in practice there are trends toward mild outcomes. For example, although teratology falls within the moderately severe category, overt maternal toxicity is deliberately avoided and the trend is therefore more towards mild than towards substantial. Similarly in oncogenicity studies encompassing lifespan exposure, undue toxicity is deliberately avoided by employing dose levels that will allow longevity and with it the possibility of tumour expression. *In vitro* replacement in the area of acute and dose ranging studies are obvious targets to gain maximum benefit for laboratory animals in terms of welfare of *individual* animals.

Table 3. Home Office Project Licence Categorisation for Regulatory Toxicology

Severity of Procedure (Potential)		
Mild	*Moderate*	*Substantial*
	Repeat Dose Reproduction Irritancy Genotoxicology Immunology	Acute Dose Ranging

Elucidating Toxic Mechanisms

There are numerous instances where *in vitro* toxicology has led to an understanding of a toxic mechanism and/or species extrapolation that has allowed progression of a useful medicine. On purely pragmatic grounds the later a significant toxicity problem is encountered in a development programme , the more potential significance it has for the company. By its nature and timing, therefore, oncology, accomplished by rodent bioassays, is of critical importance. These studies usually take a minimum of 2½ years to conduct and analyse; failure to resolve a significant safety issue at this late stage can lead to serious delays in registration or, worse, abandonment of the compound into which huge financial investment has already been put. On the grounds of understanding toxic mechanisms the most important target area for *in vitro* problem solving is oncology.

Compound Selection

Radical changes are occurring within the Pharmaceutical Industry in relation to approaches to the discovery of substances as potential medicines. These approaches are concerned broadly with three areas; vastly greater opportunities to identify small molecules as potential medicines through application of combinatorial chemistry; increased emphasis on identifying medicinal usefulness of macromolecules through biotechnology; application of genetics and an understanding the human genome in relation to disease.

Combinatorial chemistry is able to generate libraries of millions of compounds in a relatively short space of time and these can be evaluated at a molecular level in high throughput efficacy screens. This greatly increased productivity at the discovery phase inevitably presents significant challenges to exploratory development processes. The earliest possible selection of compounds that offers the best possibility for realisation as a medicine needs to involve toxicology as a critical component. In order to cope with the numbers of compounds already emerging from the combinatorial approach, 'early safety selection' can only be effected by reliance on high throughput cell or molecular screens. The use of traditional animal-based toxicity screens is wholly inappropriate for this phase, both from a logistic standpoint and from the ethical consideration of the inevitability of huge increases in animal use. Consequently the search for predictive high throughput *in vitro* safety screens is now being pursued earnestly throughout the Pharmaceutical Industry. This search is driven very much by a 'business need'. However, spin-offs in relation to developing alternative *in vitro* methodologies for use in later stages of development are likely to be considerable.

Advancing the Science of Toxicology

In vivo safety evaluation is based on the science of toxicology. Examples of where *in vitro* methodology has already played a central role in an understanding of mechanisms of toxicity are numerous and include interference with membrane function, interference with cellular energy production, binding of reactive moieties to macromolecules, the significance of calcium homeostasis and DNA damage to somatic cells. Replacement of traditional safety evaluation procedures will only come about because of a thorough understanding of mechanisms that predict for various toxic outcomes. *In vitro* methodology is bound to continue to be a very critical component to this understanding and on this basis must attract high priority for resources. Investment by Industry into fundamental academic research that is involved in the science of toxicology through *in vitro* technology is very important.

To summarise at this point, none of the individual areas of regulatory toxicology attracts first priority from replacement by *in vitro* when all of the above factors are taken into consideration. With this in mind a subjective model for overall prioritisation is suggested.

Prioritisation

A matrix can be constructed in which benefits derived from replacement of *in vivo* areas of regulatory toxicology by *in vitro* methodology are scored against the various considerations already outlined (Table 4). It is accepted that the tabular numerical data are based on no more than subjective opinion. Indeed, the reader may be interested in running through this exercise independently based on his/her own values and judgement The model may serve to stimulate a thoughtful approach as to where resources on the search for alternatives can be applied with greatest benefit.

Table 4. Weighted Benefits[1] for Development of *In vitro* Techniques in Various Toxicological Areas

Area / Target	Acute Studies	Repeat[2] Dose Studies	Onco-genicity	Repro-ductive Toxico-logy	Genetic Toxico-logy	Irritancy	Immu-nology
Promoting animal welfare	4	3	1	1	1	2	2
Reducing animal numbers	1	3	4	2	1	1	1
Financial Costs	1	3	4	2	1	0	1
Mechanism	0	4	4	2	2	0	1
Compound selection	1	3	4	4	4	2	1
Science	1	4	4	3	4	1	2
Total	8	20	21	14	13	6	8

[1] In this model weighted benefits have been given to each combination on a scale of 0-4, with zero representing no benefit and 4 representing significant benefit derived from adopting *in vitro* approaches, i.e., the higher the total score the greater the perceived overall benefit.

[2] For the purpose of identifying target organ toxicity

A number of conclusions can be drawn from Table 4, and the following serves to illustrate such interpretation.

The area of acute/dose ranging studies yields a low overall score of 8 for value of *in vitro* alternatives, a score that is attained by the following components:

For individual animals this type of study has the potential to lead to substantial pain and distress (although thorough monitoring procedures of the condition of the animals minimises this possibility) and therefore a high score is given for *in vitro* methodology within the Animal Welfare category of Table 4. However, low scores of zero or 1 are allocated to other considerations, i.e., with the abandonment of "exact" LD_{50} tests only small numbers of animals are used; acute tests are not expensive to run; they do not usually present mechanism problem-solving issues, nor contribute overmuch to decisions on compound selection. Finally since this type of evaluation is aimed almost exclusively at establishing safety margins in situations of overdosage, results rarely attract a need for scientific understanding that could be gleaned from *in vitro* methodology.

By contrast the number of animals used in oncogenicity testing attracts a high score of 4 for the potential benefits of *in vitro* replacement, but only a low score of 1 for animal welfare, since the animals do not usually suffer unduly in this type of study - longevity must be assured. On the other hand these studies are very expensive to conduct; compound selection to avoid oncogenic potential is very important; scientific comprehension of carcinogenesis will be fundamental to *in vitro* replacement in this area - these last three areas therefore each attract high scores of 3 or more to yield an overall score of 21.

For irritancy evaluation the overall score in Table 4 is low. The animal welfare aspect for the individual animal is an important consideration, as is the usefulness for selection of parenteral/topical formulations. The few animals used, the low cost of the investigation, rarely encountered mechanistic issues, and relatively low importance in terms of contribution to science all serve to outweigh these factors in overall prioritisation for resource application - hence an overall score of six.

The conclusion drawn from Table 4 is that *in vitro* methodology aimed at replacing the use of animals in oncogenicity and repeat dose studies (identifying target organ toxicity) should attract a high prioritisation in terms of resource application for the purposes of the Pharmaceutical Industry. As a contrasting example a need for irritancy evaluation by *in vitro* methodology commands a low priority for resource.

Practicability

The virtual model described above may serve a purpose for initiating a thoughtful process on resource prioritisation in the search for *in vitro* alternatives. However, such prioritisation must also be grounded firmly in feasibility. The practicability of adopting *in vitro* methodology as replacement depends on the status of the fundamental science in the particular area, and the future prospects for replacement on the likely rate of progress of scientific understanding. The initial component of this paper reflected on regulatory/industrial acceptability for adoption of replacement *in vitro* technology. This is inevitably a prime consideration and will vary from one component of safety evaluation to another; for example the consequences of misleading data from an *in vitro* irritancy evaluation are not at all the same as the consequences of making an incorrect interpretation in areas of teratogenicity or oncology. This type of risk management is important to how prioritisation and resource application are developed.

Table 5 contains a weighted matrix comparison that attempts to score the current practicability of adopting *in vitro* methodology (score weighting 0-4 as given for Table 4).

Table 5. Weighted Benefits for Feasibility for Adopting *in vitro* Techniques

Area / Purpose	Acute Studies	Repeat Dose Studies	Onco-genicity	Repro-ductive Toxico-logy	Genetic Toxico-logy	Irritancy	Immuno toxico-logy
Status of science	2	1	2	2	3	4	2
Regula-tory accepta-bility	2	0	1	0	2	3	2
Industry accepta-bility	2	0	2	0	2	4	2
Total	6	1	5	2	7	11	6

Although not attracting a high overall priority for resource application for pharmaceutical purposes (see Table 4) irritancy has been given considerable attention in the past. As a consequence practical and acceptable *in vitro* techniques are available that reduce considerably the need for animal testing. It is perhaps worthwhile reflecting on why *in vitro* irritancy evaluation has progressed substantially - is it perhaps on the grounds of practicality rather than prioritised need? Oncogenicity gets a good overall score (Table 5) because of the established relationship between genotoxicity and oncogenicity. Nonetheless regulatory agencies continue to require rodent bioassays. The requirement for two species is a current focus for debate (Van Cauteren H. 1996). Genetic toxicology attracts a comparatively high practicability score because of the acceptability of *in vitro* evaluation in this area. On the other hand, development of *in vitro* systems to predict target organ toxicity remain very problematical.

Table 6 illustrates an overall ranking for adoption of *in vitro* methodology derived from both benefit and feasibility considerations.

Table 6. Overall ranking for development/adoption of *in vitro* methodologies.

Overall Ranking	Component	Benefit ranking (from Table 5)	Feasibility ranking (From Table 6)
1	Oncogenicity	1	4
2	Genetic Toxicology	4	2
3	Irritancy	7	1
4	Repeat dose (target organ)	2	7
4	Reproductive toxicology	3	6
6	Acute studies	6	4
6	Immunotoxicology	6	4

Conclusion

A prime value of *in vitro* techniques is undoubted by virtue of an animal welfare perspective alone. Within the area of Pharmaceutical Regulatory Toxicology it is clear that employment of *in vitro* methodology in certain areas will generate greater benefit than in others. This paper concludes from an overall pharmaceutical perspective that the greatest need for *in vitro* toxicology is in the area of oncogenicity evaluation and that such a prioritisation does not conflict with intentions to advance animal welfare. Oncology retains the top ranking for attention even when the problematical feasibility of adoption of *in vitro* technology in this area is taken into account.

The role of *in vitro* techniques will increase in safety evaluation - indeed in early pharmaceutical development the rapid establishment of high throughput safety screens is crucial for selection purposes.

The regulatory acceptability of replacement *in vitro* methodology depends on thorough scientific understanding of particular toxic mechanisms.

References

Baumans V, Brain PF, Brugére H, Clausing P, Jeneskog T, Perretta G. (1994) Pain and Distress in Laboratory Animals and Lagomorphs : Report of the Federation of European Laboratory Animal Science Associations (FELASA) Working Group on Pain and Distress, Lab Animals 28:97-112.

Statistics of Scientific Procedures on Living Animals, Great Britain 1995 (1996) ISBN. 0.10.135162.3

Van Cauteren H (1996) Implementation and Impact of the ICH Safety Guidelines In: PF D'Arcy and WG Harron, (eds) Proceedings of the Third International Conference on Harmonisation,Yokohama, 1995. The Queens University of Belfast, pp 213-220.

Development and Strategic Use of Alternative Tests in Assessing the Hazard of Chemicals

Iain F. H. Purchase
Central Toxicology Laboratory, Zeneca Ltd., Alderley Park, Macclesfield, Cheshire SK10 4TJ, England.

There are scientific, ethical and financial reasons why toxicologists seek alternatives to the use of animals in toxicity testing. The scientific drive to develop alternatives is aimed at improving the process of hazard assessment which currently relies predominantly on the results of animal tests. Ethical concern for the welfare of animals is a potent driving force for the development of alternatives. However, the protection of human health also places strong ethical demands on the toxicologist to provide the best available hazard information and currently it is believed to be that based on the results of animal experiments. Alternative testing which is cheaper and carried out in a shorter time, provided that the concern for scientific and ethical standards is met, is beneficial both to industry and to the research community

The lead given by Russel and Burch in 1959 to biological science, with the publication of their book on 'The Principles of Humane Experimental Technique', has been followed in the field of toxicology, particularly in the last 20 years. This is seen in the development of scientific journals, such as Toxicology *In vitro*, and in the evolution of regulations mandating the use of alternatives when they are available (for example article 7.2 of the EU Council Directive 86/609/EEC states that 'An experiment shall not be performed if another scientifically satisfactory method of obtaining the result sought, not entailing the use of an animal, is reasonably and practicably available'. Provisions in the UK Animals (Scientific Procedures) Act of 1986 also require the use of alternatives whenever possible).

With these developments spanning many years, what stage have we reached in deploying alternatives in toxicity testing? This paper reviews the current use of alternatives and, in particular, analyses the experience of the development and use of the most widely used *in vitro* method, the Ames test.

Test Validation

It is generally accepted that the principal purpose of toxicity testing, namely the assessment of human hazard, requires that any new test must be validated to ensure that it is at least as good as the existing test methods. Recently, the validation process has been defined as consisting of 5 steps, namely

1) Test development,
2) Prevalidation,
3) Validation ,
4) Independent assessment, and
5) Progression towards regulatory acceptance (Balls et. al., 1995).

Qualitative validation: At is simplest level, tests aim to identify whether a chemical has or does not have certain properties (e.g. carcinogenicity or teratogenicity). The method of deciding whether a test has performed well is to test a large number of chemicals of known activity and compare the results of the two methods. The terms used to describe the outcome of validation are given in Table 1, using carcinogens as an example.

Table 1. Definition of terms used in validation (After Cooper et al., 1979)

Carcinogenicity classification based on existing (animal) methods	Alternative test result +	-
Carcinogen	a	b
Non-carcinogen	c	d

$$\text{Sensitivity} = \frac{\text{No. of carcinogens positive in alternative test}}{\text{Total number of carcinogen}} = \frac{a}{a+b}$$

$$\text{Specificity} = \frac{\text{No. non-carcinogens negative in alternative test}}{\text{Total number of non-carcinogens}} = \frac{d}{c+d}$$

$$\text{Predictive value} = \frac{\text{Number of positive results from carcinogens}}{\text{Total number of positives}} = \frac{a}{a+c}$$

$$\text{Prevalence} = \frac{\text{Number of carcinogens}}{\text{Total number of chemicals}} = \frac{a+b}{a+b+c+d}$$

Quantitative validation: Ultimately, hazard assessment requires information on the dose at which toxicological events occur so that an assessment of the risk of exposure to certain doses can be determined. In the case of the Ames test, there was a debate about the correlation between the potency of carcinogenicity results and the results of the Ames test (Ashby and Styles, 1978). By 1985 it was clear that the results of the Ames test could not be used for prediction of carcinogenic potency (Purchase, 1985) and there is no evidence to suggest that the Ames results are currently being used to assess potency.

Experience with the Ames Test

Retrospective validation: The initial report of the large scale validation of the Ames test was published in 1975 (McCann et al., 1975). Subsequently several other validations were carried out and reviewed (Purchase, 1982). More recently the results of the US National Toxicology Program's (NTP) carcinogen testing have been examined retrospectively with a view to establishing the correlation between mutagenicity and carcinogenicity (Zeiger, 1987; Tennant and Ashby, 1990). When a sample of these results are viewed together (Fig.1), it can be seen that the sensitivity and specificity figures have declined from about 90% to between 50 and 75%.

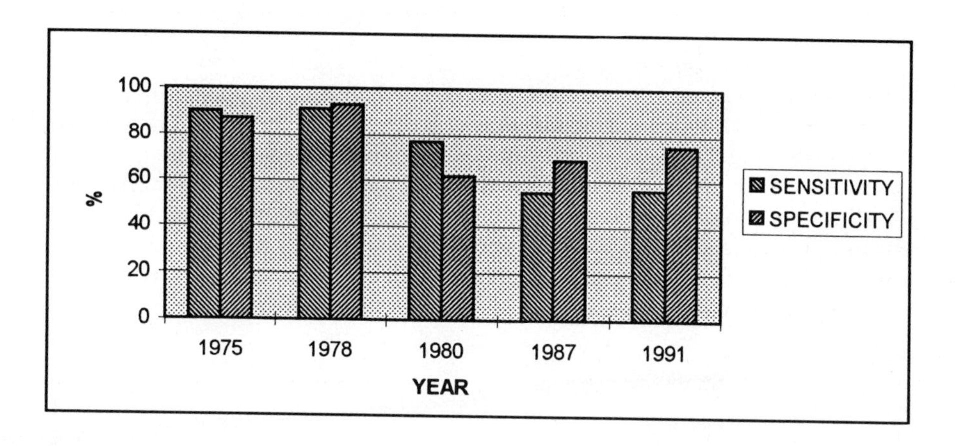

Fig. 1 Sensitivity and specificity results from the validation of the Ames *Salmonella* mutation assay against rodent carcinogenicity data.

What are the reasons for these changes in sensitivity and specificity values? The method used for testing chemicals in the Ames' test has remained relatively constant; the change has been in the chemicals tested. The use of the results of validation studies depends on several assumptions, namely

a) that the chemicals used in the validation are representative of those subsequently tested (prevalence is an important consideration in this respect, Table 2),

b) the alternative tests have the same degree of reproducibility in future use as they did in the validation study and

c) that the criteria for classifying chemicals as carcinogens remains unchanged (Purchase et al., 1978).

Subsequent analysis of the results of validation against the NTP results revealed that the Ames test results could predict carcinogenicity for certain classes of chemicals with much greater accuracy than others; indeed the sensitivity values are similar to those found in the early validation studies.

Table 2. The number of positive results and the predictive value of a predictive test used to screen 1000 chemicals. The figures refer to the number of positive results for each set of circumstances. Those in brackets give the predictive value (or accuracy of a positive result) for each set of circumstances.

Sensitivity and Specificity of test	Prevalence of carcinogens 0.001	0.01	0.05	0.5
95%	51 (9%)	59 (16%)	95 (50%)	500 (95%)
90%	101 (1%)	108 (8%)	140 (32%)	500 (90%)

Further, the prediction of carcinogenicity for chemicals which produce cancer in both sexes of both species is much greater than for single sex/species carcinogens (Fig. 2). The large number of single species carcinogens and chemicals classified as 'equivocal' by the NTP suggested that the animal tests have become more sensitive over this period. The recognition that chemicals can produce cancer by mechanisms which do not involve direct DNA damage, the non-genotoxic carcinogens, has occurred over the last 20 years, coinciding with the reduction in the specificity of the Ames' test prediction.

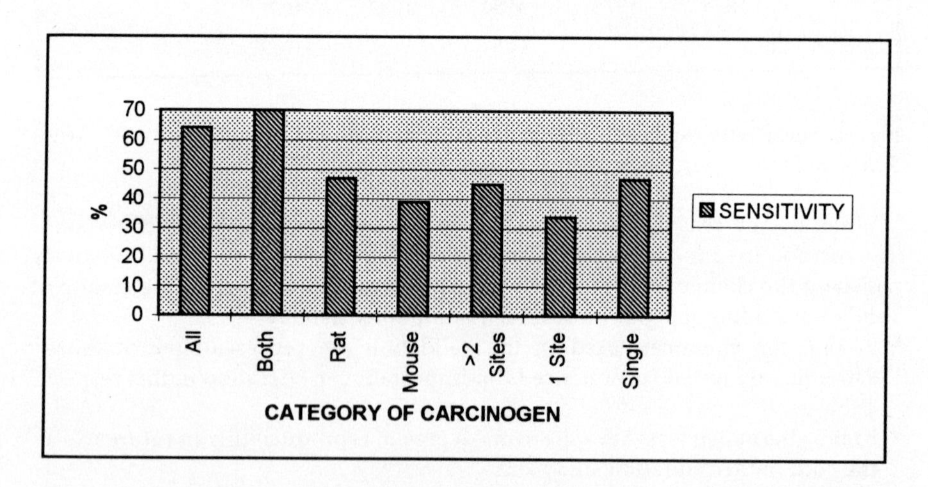

Fig. 2. Correlation values between Ames test results and results of NTP carcinogenicity studies by category.
All = All carcinogens; Both = carcinogenic to both rat and mouse; Rat = rat specific carcinogens; Mouse = mouse specific carcinogens; 2 Sites = carcinogenic to 1 species at 2 or more sites; 1 Site = carcinogenic in 1 species at 1 site; Single = carcinogenic to a single species/sex/site. (After Tennant and Ashby, 1991).

Prospective validation: When used as a toxicological test, the results of the Ames test must be considered in a prospective sense. For this purpose, a prospective study of the ability of various methods to predict the outcome of carcinogenicity studies on 44 chemicals was organised. A summary of the results, after 40 compounds had been tested (Parry, 1994), revealed that no method which relied solely on a single method (either computer based calculations or the results of the Ames test) predicted the carcinogenicity accurately. The use of a combination of information (structural alerts, Ames results, results of 90 day toxicity studies etc.) predicted the outcome in 30 (75%) of the chemicals (Fig 3.)

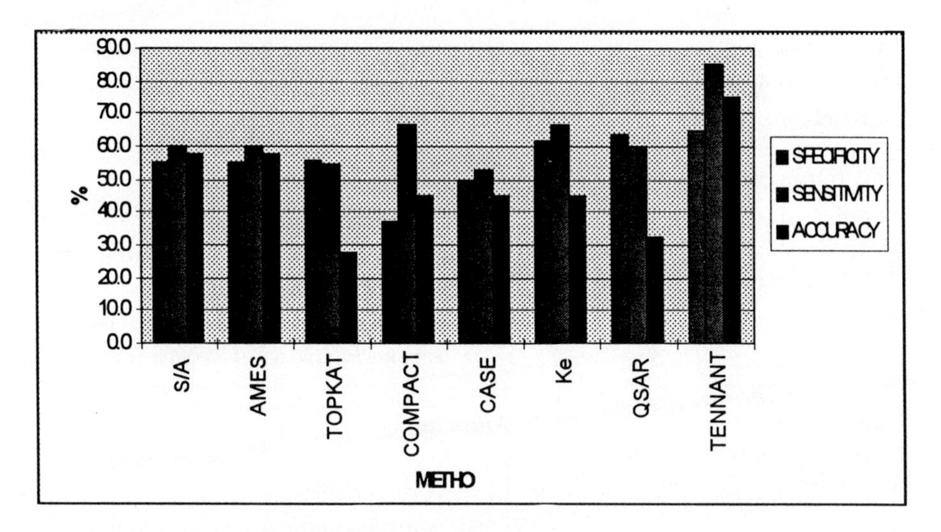

Fig. 3. Sensitivity and specificity figures for the various methods used to predict prospectively the carcinogenicity of 40 chemicals tested by the NTP (recalculated from Ashby and Tennant, 1994).

Principal uses: The results of the Ames test are rarely used alone to predict carcinogenicity, because there is usually much more information available about the chemistry, physical properties and other toxicity of a chemical.

The principal uses of the Ames test are:

- to assist in the selection of chemicals during development of pharmaceuticals, agrochemicals, household or industrial products.
- to provide data for regulatory submissions
- to identify potentially genotoxic chemicals in natural products, with a view to further testing to identify germ cell mutagens or carcinogens
- to identify the presence of genotoxic impurities in products
- in studies of the mechanism of action, including metabolism, of carcinogens

Regulatory requirements: The registration of virtually all chemicals requires information on mutagenicity for identifying chemicals that are either potential germ cell mutagens or potential carcinogens. The results of the Ames test are a general requirement. It is also a general requirement that additional information from *in vitro* and often *in vivo* studies are also required (Table 3). In each case there are requirements for *in vivo* mutation assays in the initial requirements or to confirm positive results. Only in the case of testing of new chemicals (e.g. the base set notification in the EU) is the requirement based on *in vitro* results only, and this is superseded by requirements for *in vivo* tests if the result is positive or when higher tonnages are reached.

Table 3. Examples of regulatory requirements for various types of chemicals.

CHEMICAL TYPE	TESTING BATTERY

EU Notification for industrial chemicals	• Ames
	• *In vitro* chromosome analysis
	• Positive results in the Ames test require
	    • *in vitro* gene mutation assay
	    • *in vivo* chromosomal damage assay
	    • second *in vivo* assay at high tonnage

USA Registration of agrochemicals	• Ames test • *In vitro* gene mutation assay (L5178Y) • *In vivo* cytogenetics • Positive results in the Ames test require • *In vivo* test for interaction with germ cell DNA
International Conference on Harmonisation guidelines for pharmaceuticals	• Ames test • *In vitro* mammalian test (L5178Y or IVC) • *In vivo* chromosomal damage assay

Reduction in animal usage: It is very difficult to provide a reliable estimate of the effect of the introduction of the Ames test on the use of animals. The simple paradigm that testing a chemical for potential carcinogenicity saves a standard carcinogen test using aver 400 animals is not a correct analysis. For any chemical which has widespread use and causes significant exposure, every regulatory authority requires a carcinogenicity study in at least one, and often two, species. Certainly, in the selection of chemicals in development, negative results from the Ames test lead to a higher probability that the candidate selected will have the correct toxicological profile for registration. Unfortunately, some chemicals which are not genotoxic on the basis of Ames and other tests, may turn out to be non-genotoxic carcinogens.

On the other hand, positive Ames results lead inevitably to rejection of the chemical or to further testing in animals to establish if the potential for mutagenicity is reflected *in vivo*. If *in vivo* results are negative, the results of the Ames test may be overruled. In this case it could be argued that the Ames result required the use of animals.

Lessons from the Use of the Ames Test

The Ames test has a number of attributes which make it an ideal case study for evaluating the use of *in vitro* methods in toxicology. It is a well-established method in use in a large number of laboratories around the world, it is based on a mechanistic understanding of the induction of mutations and their role in cancer induction, it has been used very effectively in mechanistic studies of carcinogenic chemicals and it has been part of the regulation of chemicals for nearly 20 years. The lessons relevant to the use on *in vitro* methods are:

- the results of validation studies may change with time because of:
    - changes in the classes of chemicals used
    - changes in the criteria used to classify chemical toxicity
    - changes in the prevalence of the chemicals tested
- validated tests are only likely to perform well prospectively if the chemicals used in the validation are representative of those tested prospectively, particularly with respect to mechanisms of toxicity (in the absence of knowledge of mechanisms of toxicity, similarity in respect of chemical class is the surrogate most frequently used)
- *in vitro* methods will have variable performance in prospective testing depending on the match between the mechanism in the *in vitro* method and the mechanism of toxic action of the chemical tested
- *in vitro* methods are unlikely to be of value in predicting the potency of systemic toxicity *in vivo*, because:
    - *in vitro* metabolic systems differ quantitatively and qualitatively from *in vivo* metabolism
    - there are usually pharmacodynamic differences between the cells *in vitro* and those which form the target *in vivo*
- even when a well validated *in vitro* method is used in regulatory testing, confirmation of the results in *in vivo* tests may be required
- it is difficult to assess the benefits, in terms of reduction of animal use, which results from the application of validated *in vitro* methods

Practical Strategies in the Development of Alternative Tests

Regulatory toxicology: The process of developing and validating novel test methods for regulatory purposes must reconcile the importance that regulatory guidelines give both to the protection of human health and the international harmonisation of test methods as well as to animal welfare. A recent analysis of the process suggests that from test development to international acceptance of a validated protocol will take on average about 10 years (Purchase, 1997).

Considerable progress has been made in developing *refinement alternatives* to the standard LD_{50} and various methods have been accepted internationally (by the OECD), including the fixed dose procedure and the up-and-down method. These methods can also be considered as *reduction alternatives.*

When it comes to replacement alternatives, progress has been much slower. It is anticipated that the number of animals required for the regulatory toxicology testing of a chemical which has widespread use will only be reduced by about 5% within the next 10 years (Purchase, 1997). The justification for this conclusion can be seen in Fig. 4.

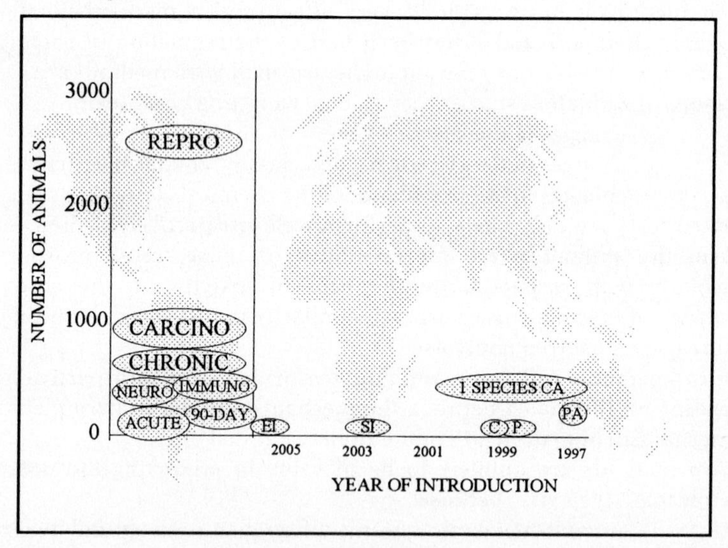

Fig. 4. A schematic representation of the progress towards the introduction of alternative toxicity tests into international regulatory guidelines. As it takes about 10 years to develop and validate a new test, the likely timescale for successfully validated tests to reach international acceptance can be plotted on a calendar.

The tests are: EI - Eye irritation: SI - Skin irritation: C - Corrosivity: P - Phototoxicity: PA - Percutaneous absorbtion: 1 SPECIES CA - Testing for carcinogenicity using only one species: NEURO - Neurotoxicity test: ACUTE - Acute toxicity test (oral): 90-DAY - 90 day toxicity test: IMMUNO - Immunotoxicity test: CHRONIC - Chronic toxicity test in dogs and rodents: CARCINO - Lifetime carcinogenicity test in two species (usually rat and mouse): REPRO - Multigeneration reproduction study and teratology test.

The other main area of development has been in the development of schemes for the stepwise testing of chemicals. For example, testing for skin and eye irritation may follow a sequence of tests, starting with an *in vitro* cytotoxicity test, followed by *in vitro* skin corrosivity test. If the physico-chemical properties or the results of these tests suggest that the chemical will be a severe irritant, it can be labelled as such. If there is an important reason why *in vivo* data are required, the initial test can be carried out on a single animal. Only if this is negative will further testing on animals be required; otherwise the chemical will be classified as an irritant or corrosive. These stepwise schemes are both *reduction and refinement alternatives.*

Product selection in development: The overall process of identifying chemical structures which may have properties of value in the pharmaceutical or agrochemical industries has been revolutionised in the last few years by the advent of High Throughput Screening. It is not uncommon for the primary screen in these industries to be capable of screening between 500,000 and 5 million chemicals per year. Once the chemical series has been demonstrated to have some efficacy, chemical synthetic effort is deployed to optimise the structure with respect to the property of interest. Many of the chemicals shown to have very potent activity in the *in vitro* screens are found not to be active *in vivo.*

These developments have provided toxicologists with particular challenges. Unlike the situation with tests for efficacy, those for toxicity may lead to dropping a chemical and the loss of potentially valuable product. The irony of the situation is that it is technically possible to mount high throughput screens for *in vitro* toxicology - cell cytotoxicity, differential killing of cell lines selected to demonstrate a particular mode of action or expression of particular gene products are all technically possible. The interpretation of the results as a contribution to improving product selection is the challenge which has yet to be overcome.

Animal tests, specifically designed to provide information useful in development decisions, remain the principal methods in product development.

Research: There is no doubt that *in vitro* methods are used extensively in toxicological research. They have the great advantage of providing precise information on particular biological mechanisms and, combined with *in vivo* results, provide the mainstay of research to investigate mechanisms of toxicity.

References

Ashby J and Styles JA (1978) Does carcinogenic potency correlate with mutagenic potency in the Ames assay? Nature 274:20-22

Ashby J and Tennant RW (1994) Prediction of rodent carcinogenicity for 44 chemicals: results. Mutagenesis 9:7-15

Cooper JA, Saracci, R and Cole P (1979) Describing the validity of carcinogen screening tests. Br J Cancer 39:87-89

Balls M, Blaauboer BJ, Fentem JH, Bruner L and 11 others. (1995) Practical aspects of the validation of toxicity test procedures: A report and recommendations of ECVAM workshop 5. ATLA 23: 129-147

McCann J, Choi E, Yamasaki E and Ames BN (1975) Detection of carcinogens as mutagens in the Salmonella/microsome test. Part1, Assay of 300 chemicals. Proc Natn Acad Sciences USA 72:5135

Purchase IFH (1982) An appraisal of predictive tests for carcinogenicity. Mutation Research 99:53-71

Purchase IFH (1985) A comparison of the potency of the mutagenic effect of chemicals in short-term tests with their carcinogenic effect in rodent carcinogenicity experiemnts. In: Vouk VB, Butler GC, Hoel DG and Peakell DB (eds) Methods for estimating risk of chemical injury: Human and non-human biota and ecosystems. SCOPE

Purchase IFH (1997) Prospects for reduction and replacement alternatives in regulatory toxicology. Toxicology *In Vitro*: In Press

Parry JM (1994) detecting and predicting the activity of rodent carcinogens. Mutagenesis 9:3-5

Purchase IFH, Longstaff E, Ashby J, Styles JA, Anderson D, Lefevre PA and Westwood FR (1978) An evaluation of 6 short-term tests for detecting organic chemical carcinogens. Brit J Cancer 37:873-959

Russel WMS and Burch RL (1959) The principles of humane experimental technique. London: Methuen

Tennant RW and Ashby J (1990) Classification according to chemical structure, mutagenicity to *Salmonella* and level of carcinogenicity of a further 39 chemicals tested for carcinogenicity by the U.S. National Toxicology Program. Mutation Research 257:209-227

Zeiger E (1987) Carcinogenicity of mutagens: predictive capability of the *Salmonella* mutagenesis assay for rodent carcinogenicity. Cancer Research 47:1287-1296

Ecotoxicological Risk Assessment

(Chair: P. Bjerregaard, Denmark)

Soil Ecotoxicological Risk Assessment: How to Find Avenues in a Pitch Dark Labyrinth

H. Eijsackers
Laboratory for Ecotoxicology, National Institute for Human Health and the Environment, P.O.B. 1, NL 3720 BA Bilthoven, The Netherlands

Introduction

When assessing ecological risks of chemical substances for soil ecosystems, one is faced with a threefold challenge:
- how to assess the impact of many compounds occurring simultaneously in variable compositions,
- how to encounter the considerable variance in environmental conditions, especially with repect to exposure and effect assessment.
- how to assess the impacts on a large variety of species which have a heterogeneous distribution over a variety of ecosystems,

To meet these challenges various instruments are described which have been developed and tested so far: laboratory test strategies and the validation in the field (experminents and samplings), and various models to be used to inter- and extrapolate results between different species, substances and environmental conditions. These different approaches are applied in the different forms of risk assessment: preventive labelling and ranking, site specific assessment of actual risk, and general forecasting of risks in the framework of envirornmental planning.

Inter- and Extrapolation Models on Compounds' Behaviour and Bioavailability

Quantitative Structure Activity Relationships (QSARs). In order to assess the impact of a huge variety of compounds present and emitted in our environment a number of interpolation techniques have been developed. With respect to the relation between the physico-chemical properties of chemicals and clearly defined processes, like biodegradation and lipophilic accumulation, chemists have developed the concept of Quantitative Structure Activity Relationships. These QSAR-approaches provide valuable tools to describe the physico-chemical behaviour and to categorize groups of chemicals according to their adsorption/availability, biodegradation, and bioaccumulation. This is applied more and more for ranking and labelling of chemicals, providing a first screening mechanism.

Next to organic compounds with a general narcotic mechanism of action for which this concept has been worked out extensively (Van Leeuwen and Hermens, 1995), also for inorganic compounds (heavy metals) some promising results have been obtained (Janssen et al., 1995 and Pretorius et al., 1995) in which the availability of heavy metals could be modelled taking into account the various adsorption mechanisms involved. For predictive purposes this approach will suffice.

However, we have to realize that in reality things are more complicated. Compounds are very seldom present singularly. Due to the composition of mineral ores, emissions of heavy metal smelters comprise various elements; in agriculture pesticides are used in various combinations and, together with manuring, diffuse pollution adds to this. So far within soil ecotoxicology combination toxicology has been studied with various heavy metals and some combinations of heavy metals and organic compounds. Outcomes vary, indicating both synergistic and antagonistic action (e.g. Posthuma et al. in press with earthworms and Doelman pers. comm. with microorganisms). Studies on the combined impact of contaminants with nutrients, and inorganic with organic compounds are even scarcer.

Next to the interference between various compounds we are dealing with dynamic instead of static situations. Regarding the ways in which chemical compounds are available under varying and changing environmental conditions, Salomons and Stigliani (1995) recently published a review on the biogeodynamics of pollutants in soil and sediments. It shows that variable environmental conditions like pH, redox-condition, salinity change the mobility of heavy metals and other major elements considerably. By selecting and defining a number of regularly occurring contamination patterns, the predictive static approach could be extended with a number of dynamic model scenarios to be used for auditive purposes or for forecasting in the context of environmental outlooks.

In addition to this environmental behaviour oriented approach we need a more exposure and effect driven classification. A method to classify compounds according to their acute toxicity (for aquatic species) and their chemical properties has been described by Verhaar et al. (1992). The first group, for which the present QSAR-approach for organic compounds is relevant, comprises groups of chemicals with non-specific generally acting mechanisms like narcotic acting chemicals. These chemicals do not show specific binding processes to biological receptors and are inert in this perspective. Other contaminants show more specific working mechanisms in addition to these general mechanisms which can therefore be defined as baseline toxicity. The second group shows, due to the polarity of the contaminants, more specific binding processes. The third group comprises contaminants with specific but nonselective working mechanisms like the organophosphate compounds. The fourth group includes those contaminants with very specific mechanisms, sometimes even constricted to specific species or species groups. For these specifically acting compounds further research is needed, especially for biologically highly active and specialized compounds like hormone-mimicking compounds. Basic testing is still the best approach in these cases. More advanced testing is needed to generate data for further development of QSARs for these more specifically acting compounds.

Further, specific uptake and internal storage mechanisms found in various soil animals may play a major role. Also here a QSAR-approach may provide some first general answers. Van Gestel and Ma (1991) already showed for a number of chlorobenzenes that there is a clear relation between lipophilicity and bioaccumulation in earthworms. Belfroid (1994) further extended this approach for uptake directly or via the food by earthworms, the last route being especially relevant with compounds of high lipophilicity (log K_{ow}> 5). Janssen et al. (1991) showed variable uptake rates and patterns for cadmium by various soil arthropods.

As a further extension a stepwise approach is suggested for heavy metals (Peijnenburg et al, 1997). This is based on the observation that both abiotic (physico-chemical) and biotic (uptake and relase processes) are for the greater part related to the dissolved fraction of heavy metals. Therefore it is possible to subdivide availability in three phases: a physico-chemically driven desorption process (environmental availability), a physiologically driven uptake process (environmental bioavailability) and a physiological/metabolic driven internal reallocation process (toxicological bioavailability); the last one leading to loading of a target organ, an accumulation organ, or release. (Fig. 1).

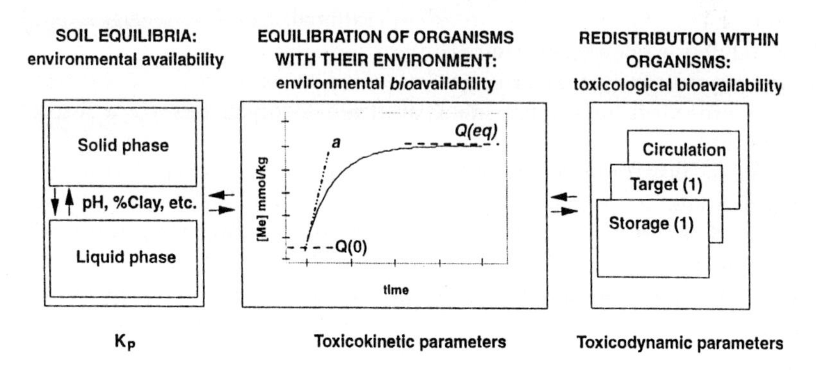

Fig. 1 Conceptual model for a three phase availability approach (acc. to Dickson et al. 1994)

The work on environmental availability by Janssen et al (1995) mentioned previously has been extended to bioavailability by Posthuma (1997) with earthworms. From the multiple regression formula describing the impact of various sorption factors it can be deduced that pH and Fe_{ox} are important, although soil factors like clay and organic matter (OM)play a more prominent role in BCF (Table 1).

Table 1: Multiple regression of environmental availability (k_p) and Bio Concentration Factors (BCF) in the earthworm *Eisenia fetida* of cadmium, copper, lead and zinc in relation to various soil factors in decreasing order of importance, according to the following general formula:
Log k_p respectively log BCF = a*pH($CaCl_2$) + b*log OM + c*log clay +d*log Fe_{ox} + e*log Al_{ox} +f* log DOC + g*log I + h (Janssen et al in press[a], Janssen et al, in press[b])

Log k_p

Cd	= 0.48*pH + 0.28
Cu	= 0.15*pH + 0.45*log Fe_{ox} - 0.71*log DOC + 1.33
Pb	= 0.24*pH + 0.40*log Fe_{ox} + 1.98
Zn	= 0.61*pH - 0.65

BCF

Cd	= -0.43*pH + 1.36*log clay - 1.39*log OM + 3.19
Cu	= -0.65*log Fe_{ox} - 0.38*log clay + 1.38
Pb	= -0.78*log clay - 0.45*log Fe_{ox} + 0.46
Zn	= -0.39*pH - 1.06*log Al_{ox} + 0.73*log clay + 3.04

In order to make this three-phase approach operational, six steps are suggested:

1. Analyse and define relevant exposure routes;
2. Develop a procedure to predict concentration in pore water;
3. Develop a procedure to predict metal available to organisms via a water exposure route;
4. Develop a data base for biological effects for species whose principal exposure is from pore water;
5. Develop a procedure to predict metal available to organisms via a soil/food exposure route;
6. Develop a data base for biological effects for species whose principal exposure is from soil.

How to Structure Inter-Species Sensitivity: Statistically, Functional Groups and/or Key-Species ?

Statistical interpolation. Inter-species sensitivity can be approached in a ecologically defined (indiactor or key species) or in an undefined way. This last approach has been used in the development of the HC5-derivation. For this methodology to derive hazardous concentrations the total sensitivity range of all species was derived from a limited number of test species. For this risk derivation method (developed by a series of scientists Kooijman, Van Straalen and Denneman, Wagner and Løkke, and in its final form written out by Aldenberg and Slob (1991)), it is assumed that:

- species are statistical entities and equally representative;
- all species are effectively protected, by ensuring that 95% of all species are not harmed at all, hence 5% may show a sign (even a slight) of adverse impacts;
- protection of 'all' species will also protect the ecosystems structure;
- protecting the ecosystem structure will protect the system functions

The limit of 5% potentially affected species was set by policy.

In order to further classify the differences in sensitivity between species in a non-predefined way the concept of Quantitative Species Sensitivity Relationships has been developed. Vaal and Hoekstra (1994) analyzed an extensive data-base of (aquatic) experiments with results of 35 compounds, tested on, depending on the compound, 12 to 62 different test species. Their general conclusion was that both with respect to acute and chronic effects the majority of the contaminants which have a general non-ionic narcotic working mechanisms exerted similar toxicity (Fig. 2). The sensitivity distribution curves of these compounds are narrow and have an almost identical position on the toxicity rank scale. Compounds with more specific working mechanisms have very different sensitivity distribution curves which have a different place on the toxicity rank scale, like for instance the OP-esters.

Similarly the toxicities of non-ionic narcotic compounds show a very good correlation with their K_{ow} -value. Relatively few chemical compounds show a different position, and act differently. In general practice these outliers are treated as experimental aberrations. Vaal and Hoekstra (pers. comm.) state that especially this non-general behaviour needs, however, further research on the reasons why they behave in a different way. As a next step Vaal and Hoekstra (1994) applied a simple model, including body weight, surface area (= uptake surface) and fat content of the test organisms, to describe the inter-species differences in sensitivity. For short term exposure to hydrophobic compounds the model proved to be reasonably successful, but application to a compound with a more specific mode of action asked for an extension of the model with, apparently, data on species metabolism.

It has to be noted explicitely that, given the present poor database of soil ecotoxicity tests, a similar approach with soil animals is far from being effectuated. However, when assuming that availability is governed by the same mechanisms in water, sediments and soils, i.e. the dissolved fraction according to the pore water partitioning theory, it can be stated that the results of Vaal and Hoekstra have a general validity.

To categorize groups of species on ecological premises, one can apply concepts like key species or indicator species. However, it is hard to specify what kind of 'lock' a specific key species is supposed to open, or particularly which compound or type of contamination is indicated. Hence, these kinds of concepts seem to overgeneralize the relations between specific ecosystem conditions, species communities or types of contamination. Moreover, in a great number of national environmental and nature policies very specific species are protected. A great number of these species can only be protected in a sustainable way by conserva-

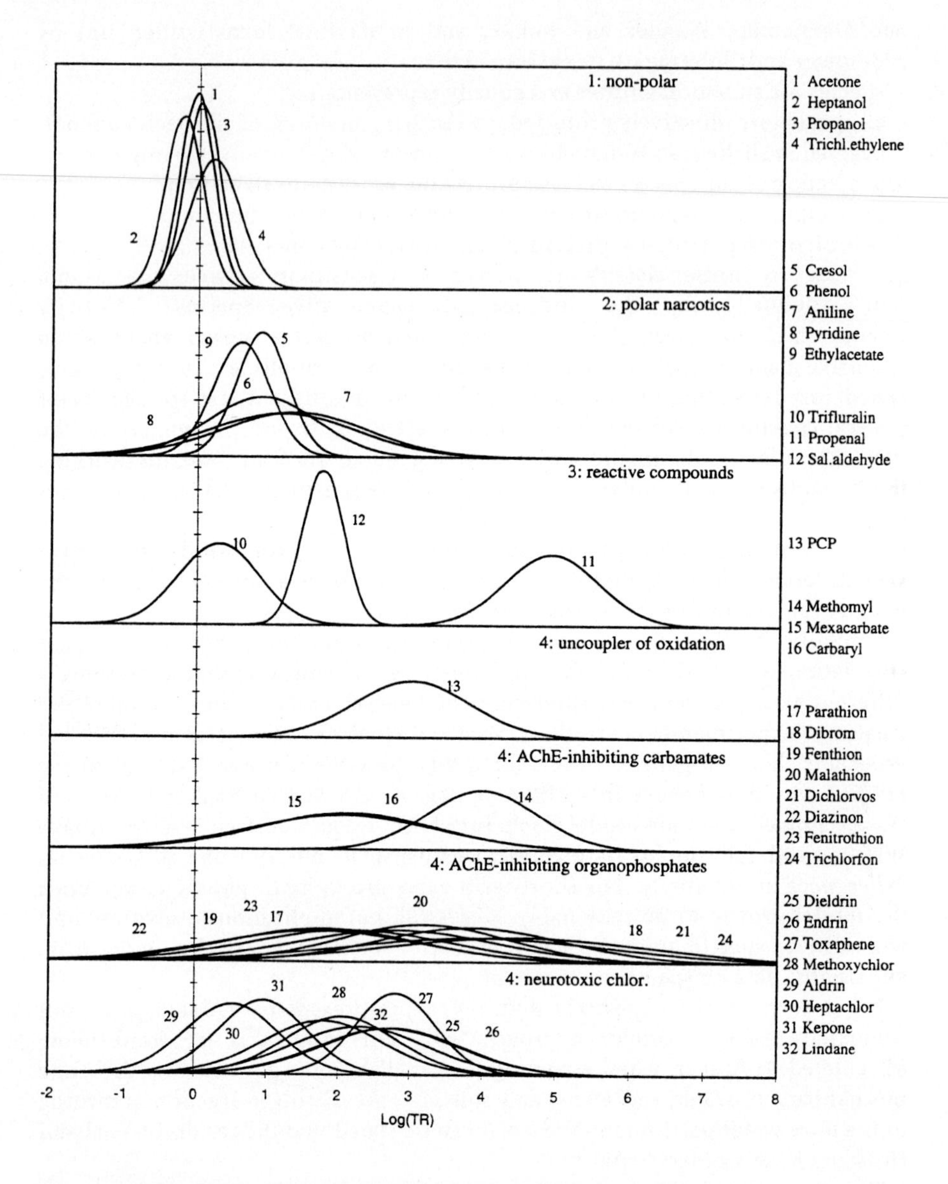

Fig. 2. Distribution curves of groups of aquatic tests species for various groups of compounds with a general (narcotic) of specific intoxication mechanism, arranged on a relative toxicity scale.

tion or the establishment of the proper, sometimes rather specific, environmental conditions. As their mere rareness makes it already unacceptable to manipulate them, it might be more useful to use common species which are nonetheless typical (= frequently occurring) for the ecosystems in focus as indication for the success of the management measures.

Along these lines Eijsackers and Løkke (1997) suggest a hierarchical set of species to establish:

1. specific (mostly rare) target species,
2. species which have a key function (whether or not as indicator) in the functioning of ecosystems and their interrelation with these target species,
3. specific system processes as related to these key function species and
4. the environmental conditions necessary for these processes, key functions and specific species.

So key and indicator species have to be combined with specific processes. An example of a terrestrial key species group are the earthworms, as they have a crucial role in burial and communition of plant litter, in creating and maintaining a loose soil structure and so stimulating general soil processes like litter degradation. An example of a key process is nitrification, a crucial step in the total N-cycle and carried out by a very limited number of soil micro-organisms.

This hierarchical set should not be defined for all kinds of soils, but should specify the specific target and key species, key processes and related environmental conditions of those situations which provide high risks.

How to adress ecological complexity. A basic question with the above described mathematical species-structure modelling (and in fact also with statistics-based interpolation systems which ask for 'representative' test species) is to what extent they are relevant for the function driven organization of ecological systems. Therefore other approaches may be applied like the concept of functional groups as developed for earthworms (Lavelle, 1983), springtails (Gisin, 1960) and soil macro-fauna (Faber, 1991). These are species which exert similar activities within the various soil ecosystems.

Examples of ecosystems built up as C- and N-driven food-webs of functionally related soil organism groups are given by De Ruiter, et al. (1995) for agro-ecosystems and Berg (1997) for forest ecosystems. This approach addresses interconnectivity, and combines structural with functional ecosystem features. They have shown that internal ecosystem stability to a great extent is governed by both top-down and bottom-up system relations between these functional groups.

These models do not comprise, however, the spatial heterogeneity of soils which, in combination with the temporal dynamics of these interacting processes, make soils highly complicated and complex to study and generalize. In the discussion about the functional meaning of biodiversity some ecologists state that a great number of species fulfill the same function and role (see the functional group modelling described above). Consequently soil ecologists like Swift and Anderson (1993) state that a great number of soil species is superfluous (redundant) and can be missed without impairing soil system functions. Beare et al (1995) claim, however, that the combination of species and specific niche makes soil species unique and not redundant.

It is interesting to note that the spatial heterogeneity, habitat diversity, and niche differentiation have not been interpreted as an ecological role in itself for a long period. Only recently Lawton and Jones (1994) called attention to the specific role of animals in creating and maintaining the physical structure of their environment. Especially in soil this ecosystem-engineering role is apparent. Earthworms exert a major impact on the physico-structural characteristics of a soil by influencing aeration, drainage and derived physico-chemical processes. In general the composition of the vegetation and litter quality together with the mineralogical soil characteristics provide the prime source of soil formation. But it is the composition of the soil fauna which defines the amount of bioturbation, changing for instance a clearly layered podzol-profile into a brown forest soil after plantation of oak in Denmark (Løkke et al. 1996). Consequently, soil biologists have defined a soil classification system according to the biological activity.

To what extent might this engineering function influence soil ecotoxicological risk assessment ? For this, we have to realize that testing is carried out mostly with nominal compound concentrations. Only very few examples studied ecotoxicity together with biodegradation, biotransformation and the derived ecotoxicity of metabolites. Doelman et al. (1992) studied fate and ecotoxicity of endosulphan and its metabolites on aquatic and soil animals in order to assess the impact of contaminated sediment *in situ* repectively after dredging and spreading on land. An at first sight erratic increase in toxicity after 30 days appeared to coincide with the formation of the metabolite endosulphan-ether. Ma et al (1995) studied the fate and effects of polycyclic aromatic hydrocarbons (PAHs) with and without earthworms. Earthworms improved the breakdown of PAHs presumably by improving soil aeration. PAHs are persistent in anaerobic consitions but readily broken down under aerobic conditions.

Because earthworms change environmental conditions and thus contaminant behaviour, they could be used in first succession of contaminated areas. In these areas contamination has lead to an eradication of soil life and a consequent deterioration of soil structure, thereby also changing environmental conditions to the worse. A literature survey showed that in most of these cases the same earthworm species appear as first invader (Eijsackers, 1997). A literature scan on the sensitivity of these species for contaminants showed that these species demonstrate a lower sensitivity than species which invade at later stages. In this way earthworms could act as real ecosystem-engineers fulfilling a major role in system restoration. Further field and laboratory research to confirm this assumption is presently under way.

Expressing the Potentially Affected Fraction of Species

In order to incorporate ecotoxicological impact data on environmental contaminants into environmental planning products some further activities are needed and methodes have to be developed to:

- assess the combined impact of various contaminants (1);
- measure the total contaminant load of environmental compartments (2);
- calculate the potential fraction of species affected by this total load (3);
- measure the actual fraction of species affected (4).

To this end various projects have been started at our institute of which some results on the points 2, 3 and 4 above are described briefly here.

To measure the total contaminant load, a 'total' extraction procedure has been developed called 'potential Toxicity ' or pT-extraction. For a mixture of generally occurring organic contaminants this procedure gave representative recoveries, albeit at a limited absolute level. To improve the recovery Supercritical Fluid Extraction is tested now, with positive results so far. Further activities will be concentrated on the application of this technique for sediment and soil substrates, and on the extension of the methodology to other contaminant groups.

The total fraction of potentially affected species (PAF) has been calculated for heavy metals (copper, zinc, cadmium and lead) and a selection of 20 pesticides. It is based on:

a. chemical monitoring results in the framework of the national soil monitoring network,
b. recent scientific views on (bio)availability modelling for heavy metals and pesticides, and
c. on modelling of combined effects of contaminants.

The total impact of individual heavy metals and compounds or various combinations thereof have been calculated, and expressed as the percentage of all species that are exposed to concentrations above the NOEC (No Observed Effect Concentration). This fraction is supposed to be negatively affected and, hence, to show toxic effects. For an example on cadmiun and soil animals, see Fig.3.

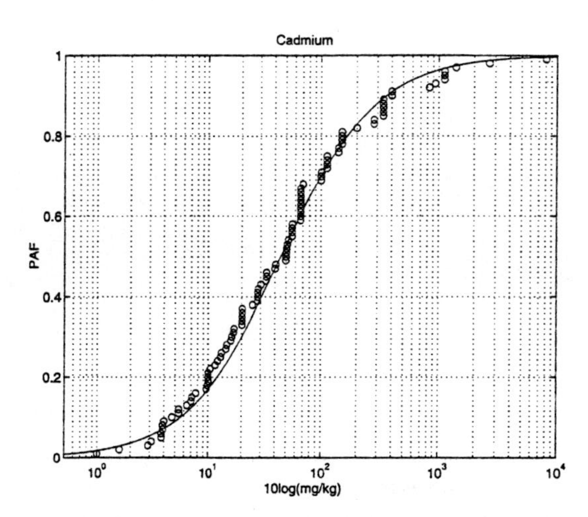

Fig. 3 Dose-PAF curve for cadmium and soil animals, as derived from various literature sources.

By combining the actual concentrations of contaminants on a national scale (though the calculations can be carried out at different spatial scales), it is possible to construct maps with this dose-PAF curve showing the fractions of affected species in various regions (fig. 4). These first results triggered a lot of discussion because in some regions quite a percentage of species is potentially affected, whereas in the actual situation one does not observe massive mortality.

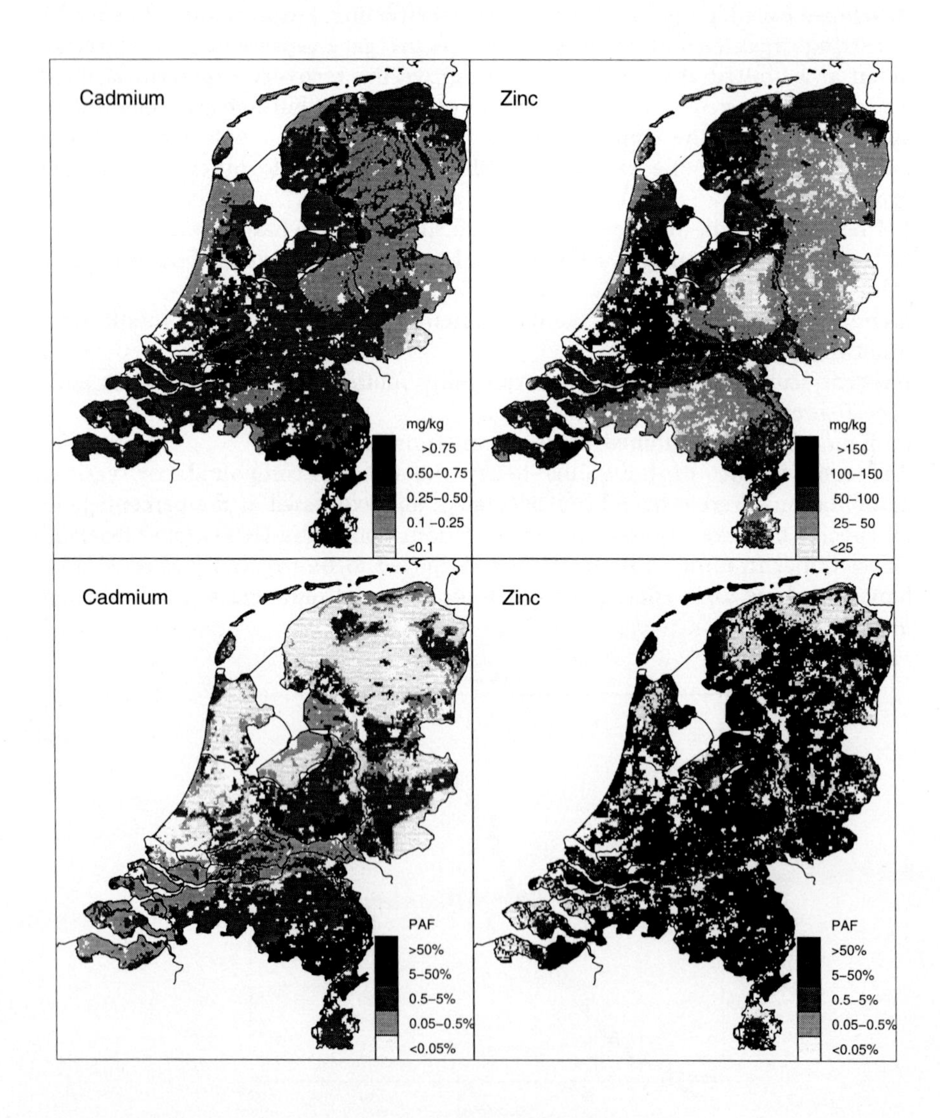

Fig. 4 Calculation of the Potentially Affected Fraction of Species in relation to the bioavailable amount of cadmium and zinc in the soil on a national (Dutch) scale.

Irrespective of possible flaws in the methodology, or refinements to be made, we have to be aware that the diffuse contamination of our environment during many centuries may have resulted already in a gradual deterioration of the total species diversity of these areas. Hence, these calculations of potentially affected species need field validation. Combination of PAF-calculations with the outcomes of a national survey on the species composition of soil nematodes (Klepper et al. 1997) revealed that nematodes do show detrimental impacts of the heavy metal concentrations.

Within the pT-project the toxic stress in our main river and estuarine systems is surveyed. Sixty liter samples of surface water are sampled regularly and extracted for organic contaminants. The extracted contaminants are subsequently tested with micro-toxicity tests, and calculations are made in order to interpret the outcome of both projects (Figure 5). Although the results still show considerable vaiability they are consistent with other monitoring results (de Zwart 1997).

Besides this assessment of total impact on species, there is a need in nature conservancy policy to assess impacts on target species. Therefore the consequences of these diffusely present contaminants have been modelled for a number of nature conservation 'target' species (species designated to get special protection due to their declining numbers, importance on a international scale and/or scarcity). The model calculations (Traas et al. 1996) have been made for mammals and birds applying various food chain models. Based on the available concentrations of inorganic (heavy metals like zinc, cadmium and copper) and organic (PCBs, PAHs) pollutants their uptake by plants and soil organisms and the subsequent transfer in aquatic and terrestrial food chains is modelled. Calculations have been made now for various target species, mammals and birds, both on a national and a regional scale, the latter a heavy metal contaminated area in the southern part of The Netherlands, de Kempen. A more specialized application dealt with the possible transfer of PCBs in the otter (*Lutra lutra*).

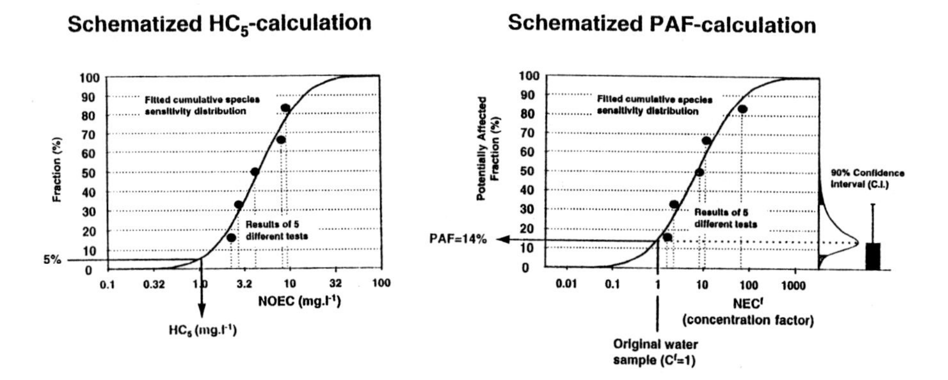

Fig. 5 Survey of toxic stress in the Dutch river and estuarine system as measured with pT-total extraction and microtox methodology

References

Aldenberg T, Slob W (1991) Confidence limits for hazardous concentrations based on logistically distributed NOEC toxicity data. RIVM-report 719102002, Bilthoven.

Beare MH, Coleman DC, Crossley jr DA, Hendrix PF and Odum EP (1995) A hierarchical approach to evaluating the significance of soil biodiversity to biogeochemcial cycling. In: Collins HP, Robertson GP and Klug MJ (Eds) The Significance and Regulation of Soil Biodiversity. Kluwer Acad. Publisher, Dordrecht: 5-22.

Belfroid AC (1994) Toxicokinetics of hydrophobic chemicals in earthworms. Ph.D. thesis. University Utrecht, 98 pp.

Berg M (1997) Decomposition, nutrient flow and food web dynamics in a stratified pine forest soil. Thesis Free University Amsterdam, 310 pp.

Brussaard L and Kooistra MJ (Eds.) (1993). Soil structure/soil biota interrelationships. International Workshop on Methods of Research on Soil Structure/Soil Biota Interrelationships, held at the International Agricultural Centre, Wageningen, The Netherlands, 24-28 November 1991. Elsevier Sci. Publ., Amsterdam.

Council of Europe (Steering Committee for the Conservation and Management of the Environmental and Natural Habitats) (1990). Feasibility study on possible national and/or European actions in the field of soil protection. Report presented by the Belgian delegation to the 6th European Ministerial Conference on the Environment. Council of Europe, Strasbourg.

Dickson KL, Giesy JP and Wolfe L (1994) Summary and conclusions. In Hamelink JL, Landrum PF, Bergman HL and Benson WH (Eds) Bioavailability. Physical, chemcial and biological interactions. CRC Press, Boca Raton, USA: 221-230.

Doelman P (199) Microbial degradation of HCH-isomeres in mineral soil and of endosulfan in organic soil in connection with soil and water quality: ecotoxicological research. CHO-TNO 23: 73-91.

Domsch KH, Jagnow G and Andersson TH (1983) An ecological concept fro the assessment of side-effects of agrochemicals on soil micro-organisms. Residue Reviews 86: 65-106.

Eijsackers H and Løkke H (1992) SERAS - Soil Ecotoxicity Risk Assessment System. A European Scientific Programme to Promote the Protection of the Health of the Soil Environment. Report from a Workshop, Silkeborg, Denmark 13-16 January 1992, National Environmental Research Institute, Silkeborg, 60 pp.

Eijsackers H (1997) Natuurbeheer voor én door Milieubeheer (The mutual Interaction of Nature management and Environmental management). Inaugural adress in Dutch. Free University Amsterdam, 13 decmber 1996. 21 pp.

Eijsackers H (submitted to Ambio) Soil quality assessment in an international perspective.

Faber JH (1991) Functional classification of soil fauna: a new approach. Oikos 62: 110-117.

Gestel CAM van, Ma W (1991) An approach to quantitative structure-activity relationships (QSARs) in earthworm toxicity studies. Chemosphere 21: 1023-1033.

Gisin H (1960) Collembolenfauna Europas. Museum d'Histoire Naturelle, Genève.

Janssen MPM, Bruins A, Vries TH de, Van Straalen, NM (1991) Comparison of cadmium kinetics in four soil arthropod species. Archives of Environmental Contamination and Toxicology 20: 305-312.

Janssen RPT, Peijnenburg WJGM, van den Hoop MAGT (in press a) Equilibrium partitioning of heavy metals in Dutch field soils. I. Relationships between metal partitioning coefficients and soil characteristics. Environmental Toxicology and Chemistry.

Janssen RPT, Posthuma L, Baerselman R, den Hollander HA, van Veen RPM and Peijnenburg WJGM (in press b) Equilibrium partitioning of heavy metals in Dutch field soils. II. Prediction of metal accumulation in earthworms. Environmental Toxicology and Chemistry.

Jones CG, Lawton JH and Shachak M (1994) Organisms as ecosystem engineers. Oikos 69: 373-386.

Lavelle P (1983) The structure of earthworm communities. In Satchell JE (Ed.) Earthworm ecology: from Darwin to vermiculture. Chapman and Hall, London: 449-466.

Leeuwen, CJ van, Hermens, JLM (eds) (1995) Risk Assessment of Chemicals: An Introduction. Kluwer Academic Publisher, Dordrecht, 374 pp.

Løkke H, Bak J, Falkengren-Gerup U, Finlay RD, Ilvesniemi H, Nygaard PH and Starr M (1996). Critical loads of acidic deposition for forest soils - is the current approach adequate. Ambio25: 510-516.

Ma WC, Immerzeel J and Bodt J (1995) Earthworms and Food Interaction on Bioaccumulation and Disappearance in Soil of Polycyclic Aromatic Hydrocarbons: Studies on Phenanthrene and Fluoranthene. Ecotoxicology and Environmental Safety 32: 226-232.

Peijnenburg WJGM, Posthuma L, Eijsackers H and Allen HE (in press) Implementation of Bioavailability for Policy and Environmental Management Purposes. Ecotoxicology and Environmental Safety.

Posthuma L, Baerselman R, van Veen RPM and Dirven - van Breemen EM (in press) Single and joint toxic effects of copper and zinc on reproduction of *Enchytraeus crypticus* in relation to sorption of metals in soils. Ecotoxicology and Environmental Safety.

Posthuma L and Notenboom J (1995) Effects of heavy metals in the oligochaetes *Eisenia andrei* and *Enchytraeus crypticus* in OECD-artifical soil and in soil from a field site polluted by zinc smelter activities. In: Notenboom, J and Posthuma, L (eds.). Progress report 1994 on the project 'Validation toxicity data and risk levels of soil contamination'. RIVM-report nr. 719102044, Bilthoven.

Posthuma L, Baerselman R, Zweers P, Janssen R, Peijnenburg W and Notenboom J (1997) Principal processes for Exposure Quantification of Worms. Introductory Paper Second International Workshop on Earthworm Ecotoxicology, Amsterdam, April 3-5 1997.

Pretorius PJ, Janssen RPT, Peijnenburg WJGM,van den Hoop, MAGT (1995) Chemical equilibrium modelling of metal partitioning in soils. In: Wilken R-D, Förstner U, Knöchel A (eds) Heavy Metals in the Environment Volume 2. CEP-Consultants, Edinburgh, 145-148.

Ruiter PC de, Neutel A-M and Moore JC (1994) Modelling food webs and nutrient cycling in agroecosystems. Trends Ecol. Evolution 9: 378-383.

Ruiter PC de, Neutel A-M and Moore JC (1995) Energetics, Patterns of Interaction Strengths, and Stablity in Real Ecosystems. Science 269: 1257-1260.

Rundgren S (1993) The Swedish soil research programme (MATS): aims, preliminary results, limitations and perspectives. In: Eijsackers H, Hamers T (eds) Integrated Soil and Sediment Research: A Basis for Proper Protection. Kluwer Scientific Publishers, Dordrecht: 289-294.

Salomons W, Stigliani WM (eds) (1995) Biogeodynamics of Pollutants in Soils and Sediments. Springer Verlag, Berlin, 352 pp.

Straalen NM van, Denneman CAJ (1989) Ecotoxicological evaluation of soil quality criteria. Ecotoxicology and Environmental Safety 18: 241-251.

Straalen NM van and Løkke H (1997) Ecological approaches in Soil Ecotoxicology. Chapman and Hall,354pp.

Swift MJ and Anderson JM (1993) Biodiversity and ecosystem function in agricultural systems. In: Schultze ED and Mooney HA (Eds) Biodiversity and ecosystem function. Springer, Berlin: 15-41.

Traas TP, Luttik R and Jongbloed RH (1996) A probabilistic model for deriving soil quality criteria based on secondary poisoning of toppredators. I: Model description and uncertainty analysis. Ecotox. Envrion. Saf. 34:264-278.

Vaal MA, Hoekstra JA (Subm). Species sensitivity in relation to pollutant classification. Subm. to Environmental Chemistry & Toxicology. (See also: Vaal MA, Van der Wal JT, Hoekstra JA (1994) Ordering aquatic species by their sensitivity to chemical compounds: a principal component analysis of acute toxicity data. Report 719102028 National Institute of Public Health and the Environment, Bilthoven The Netherlands; See also: Wal, J.T. van der , Vaal, M.A., Hoekstra, J.A., Hermens, J.L.M. (1995) Ordering chemical compounds by their chronic sensitivity to aquatic species. A principal component analysis. Report 719102041 National Institute of Public Health and the Environment, Bilthoven The Netherlands.)

Vaal MA and Hoekstra JA (1994) Modelling the sensitivity of aquatic organisms to toxicants, using simple biological and physico-chemcial factors. Report 719102034 National Institute of Public Health and the Environment, Bilthoven, The Netherlands;

Verhaar HJM, Van Leeuwen CAJ, Hermens JLM (1992) Classifying environmental pollutants. 1. Structure-activity relationships for prediction of aquatic toxicity. Environmental Contamination and Toxicology 53: 98-105.

Vries W de (1994) Soil response to acid deposition at different regional scales: Field and Laboratory data, critical loads and model predictions. Thesis Agricultural University Wageningen, 487 pp.

Wagner C, Løkke H (1991) Estimation of ecotoxicological protection levels from OECD toxicity data. Water Research 25: 1237

Zwart D de, van de Meent D, Roghair C, Struijs J, Sterkenburg A, Maas H and van Beusekom B (1997) Monitoring Toxicity of Surface Water. Poster Setac Conference, Amsterdam.

Monitoring and Risk Assessment for Endocrine Disruptors in the Aquatic Environment: A Biomarker Approach

P. Bjerregaard, B. Korsgaard, L. B. Christiansen, K. L. Pedersen, L. J. Christensen, S. N. Pedersen and P. Horn.
Institute of Biology, Odense University, Campusvej 55, DK-5230 Odense M, Denmark

Introduction

Evidence that a number of chemicals affect wildlife populations or individuals via interaction with endocrine systems has been increasing in recent years. Worldwide effects of tributyltin from antifouling paints on mollusc populations (Langston 1996; Oehlmann et al. 1996), effects of polychlorinated biphenyls on Baltic and Wadden Sea seals (Reijnders 1986; Brouwer et al. 1989), masculinisation of North American fish affected by pulp and paper mill effluents (Howell 1980; Munkittrick et al. 1991, 1992), feminisation of male fish in British rivers receiving effluents from waste water plants (Jobling and Sumpter 1993; Purdom et al. 1994; Harries et al. 1997), demasculinisation of alligators in Lake Apopka after a chemical spill (Guillette et al. 1994, 1995a,b, 1996) and effects on North American birds (Fry and Toone 1981; Fry 1995) are some of the most prominent and best documented examples, all attributed to chemicals exerting endocrine disrupting effects.

Not all of the mechanisms of action are fully understood, but endocrine disrupting chemicals may work at various biochemical levels, e.g. affecting the synthesis of hormones, interfering with hormone transporting proteins in the blood, affecting the metabolisation of hormones or by direct effects on cellular hormone receptors. In dogwhelks *Nucella lapillus* tributyltin inhibits the aromatase that converts testosterone to oestrogen thereby masculinising the females (Oehlmann et al. 1996). Metabolites of polychlorinated biphenyls interfere with thyroxin transporting proteins in the blood of seals (Brouwer et al. 1989). Chemicals that induce MFO-activity may indirectly lead to altered hormone levels by increasing the metabolisation of hormones (Arukwe and Goksøyr 1997). Alkylphenols react directly with the oestrogen receptor (White et al. 1994) which in turn may lead to feminisation of male organisms exposed (Gimeno et al. 1997; Gray and Metcalfe 1997).

Routine testing of chemicals according to accepted guidelines has failed to reveal these effects, which have been identified in the environment by chance rather than by environmental monitoring programmes. Methods for routine testing of new and existing chemicals (according to OECD or US-EPA guidelines) may be improved by incorporating end points specifically relevant to endocrine systems and reproduction (Benson et al. 1997).

In nature, carefully designed monitoring programmes that emphasize studies on reproductive and developmental effects (especially with regard to gonad histopathology, sex ratios etc.) may be developed, allowing us to improve our capabilities of identifying types of endocrine disrupting effects that we are not already aware of (Vethaak et al. 1997). The practical difficulties, however, of implementing such programmes should not be neglected.

A more pragmatic initial approach is to develop and validate biomarker methods that allow us to quantify and assess the extent of the effects already identified from endocrine disrupting chemicals. Among these effects, the oestrogenicity of some commonly used industrial chemicals has caused a greater public concern than any of the other examples mentioned above, probably because a role of oestrogenic chemicals in human reproductive disorders has been suggested but not yet elucidated (Sharpe and Skakkebæk 1993). Therefore, our laboratory has engaged into the development of methods for determination of fish and crustacean vitellogenin with the aim of being able to assess oestrogenic effects in laboratory tests as well as in the field.

Vitellogenin as a biomarker for oestrogenicity: The oestrogenic activity of chemicals can be assessed in egg producing organisms *in vivo* by utilizing the physiological processes involved in egg production - in fish termed vitellogenesis. During their reproductive period, female fish produce precursors of yolk protein (vitellogenin) in the liver. The production of vitellogenin is under oestrogenic control. Vitellogenin is liberated to the blood and subsequently taken up by the ovary. Male fish possess the gene for vitellogenin but due to the normally very low concentrations of oestrogen in the male, the gene is normally not expressed. Treatment of male fish with oestrogen initiates the same series of events as in the female fish, i.e. the male starts to produce vitellogenin and blood vitellogenin concentrations increase rapidly. Increases in the vitellogenin concentration of the blood by exposure of male fish to chemicals consequently indicate that the chemicals have oestrogen-like activity. The vitellogenin response is directly related to the binding of the chemicals to the oestrogen receptor, and is as such a specific biomarker for *in vivo* oestrogenic effect. The vitellogenin response can be used as a screening tool in both laboratory and field investigations. This *in vivo* method has the advantage compared to *in vitro* systems that effects of uptake mechanisms and metabolism of the chemicals are included in the test.

The concentrations of vitellogenin can be estimated indirectly by the large amount of calcium and alkali-labile phosphate associated with the protein (Korsgaard and Kjær 1974; Korsgaard and Petersen 1976; Korsgaard et al. 1986; Tinsley 1985; Povlsen et al. 1990; Korsgaard and Mommsen 1993), and, with a higher specificity and sensitivity, by immunological methods such as radio immuno assays [RIA] (Sumpter 1985; Jobling and Sumpter 1993) and enzyme linked immuno sorbent assays [ELISA] (Kishida and Specker 1993; Manãnos et al. 1994). Vitellogenin can also be demonstrated qualitatively and semi-quantitatively by polyacrylamide gel electrophoresis [PAGE].

Whereas the link between oestrogenic activity and vitellogenin production in fish is well established, the precise role of oestrogen in the hormonal regulation (and possibly in vitellogenin induction) in crustaceans is far from being understood. In many decapod crustaceans yolk protein is produced both in the ovaries and in the midgut gland from which it is transported to the former by the haemolymph (Quackenbush and Keeley 1988; Quackenbush 1989; Fainzilber et al. 1992).

The control of vitellogenin production in crustaceans is not yet understood but is believed to be influenced by the levels of several hormones including vitellogenesis inhibiting hormone (Meusy and Payen 1988), juvenile hormone (Laufer and Borst 1988), ecdysone (Hagedorn 1989), progesterone (Quinitio et al. 1994) and 17-β-estradiol (Couch and Lee 1987).

To promote a better understanding of the hormonal regulation of vitellogenin production in crustaceans and to be able to investigate the possible impact of endocrine disruptors in other taxonomic groups than vertebrates, ELISAs for crayfish (*Astacus astacus*) and shore crab (*Carcinus maenas*) vitellogenin are being developed in our laboratory.

Research avenues: The oestrogenic activity of certain alkylphenols has been conclusively demonstrated (Jobling and Sumpter 1993; White et al. 1994; Gimeno et al. 1997). However, little is known about dose-response relations and we have therefore initiated *in vivo* studies on e.g. nonylphenol in flounder.

Further studies in this area are currently facilitated by the development of ELISAs, e.g. against rainbow trout vitellogenin, which is also used in field studies, studies on pesticide interactions and in screening of xenooestrogens.

Recent investigations have demonstrated marked oestrogenicity of effluent from some British sewage treatment works (Harries et al. 1997) and we have initiated investigations of this issue in Denmark.

The finding by Arnold et al. (1995) that the weakly oestrogenic (*in vitro*) pesticides dieldrin and endosulfan showed strong synergy when given together has been challenged by other groups that have not been able to reproduce these results neither *in vitro* nor in *in vivo* (Shelby et al. 1996; Ashby et al. 1997). Since endosulfan has been reported to be anti-oestrogenic in catfish *Clarias bathrachus* (Chakravorty et al. 1992) it was interesting to test if endosulfan and dieldrin given alone and in combination could induce production of vitellogenin in rainbow trout.

Recent results obtained *in vitro* show that the branching of the alkylgroup in alkylphenols affects the oestrogenicity (Routledge and Sumpter 1997). This had also been seen in our initial *in vivo* screening studies and investigations that compare the potency of branched and linear akylphenols *in vivo* in rainbow trout are now being carried out.

Materials and Methods

Effect of technical nonylphenol on vitellogenin production in flounder: Five groups of 5 male flounders (*Platichthys flesus)* were injected intra peritoneally 4 times over a two week period with technical grade 4-nonylphenol at doses of 10, 50 100, 150 and 200 mg/ kg body weight/week. The chemical (from Fluka®) was dissolved in peanut oil. A control group received peanut oil only. Blood samples were drawn and analysed for alkali-labile protein phosphorous according to Korsgaard and Kjær (1974).

Development of ELISA for rainbow trout vitellogenin, Purification of vitellogenin and immunisation: Vitellogenin for the assay standards and affinity chromatography was purified from 17-β-estradiol-treated rainbow trout. The purification was a two step procedure using gel filtration (Sephacryl S300 HR) followed by anion exchange chromatography (DEAE Sephacel). Vitellogenin for immunisation of rabbits to raise polyclonal antibodies was purified by a modified version of the precipitation method described by Wiley et al. (1979). The specific antibodies against vitellogenin were purified from ammonium sulphate precipitated rabbit serum by affinity chromatography on a CNBr-activated Sepharose 4B column coupled with rainbow trout vitellogenin, and the specificity of the antibodies was tested by Western blotting after SDS-PAGE.

ELISA procedure: The ELISA is a direct sandwich ELISA, where: 1) microtiter plates are coated with specific antibodies against vitellogenin and blocked with BSA (3%), 2) vitellogenin is added as standard or samples, followed by 3) incubation with HRP-coupled antivitellogenin antibodies, and 4) the colour development after supplying the enzyme substrate OPD is monitored at 490/650 nm.

Oestrogenic effect of a Danish waste water effluent: Immature rainbow trout (230-320 g) were placed in a steel cage directly in the effluent from the sewage treatment work Ejby Mølle, which is processing waste water from approximately 200,000 inhabitants of Odense. A pre-exposure control group of 6 fish was sampled at the onset of the experiment. Groups of 5 fish were sampled after one, two, four and six weeks of exposure. A control group was kept in a laboratory tank supplied with tap water (ground water). The fish were starved during the entire experiment. Vitellogenin levels in the blood were determined.

Oestrogenic effect of endosulfan and dieldrin: Groups of 5 or 6 juvenile rainbow trout (each group kept in 55 l glass tanks) were acclimated for one week. On day 0, 6 and 12 in the exposure period, the fish received intraperitoneal injections of endosulfan (100% α-isomer) and dieldrin dissolved in peanut oil. The chemicals (both obtained from Riedel-de Haën®) were administered alone and in combination in the doses given in Table 1. Blood samples were drawn on day 18 in the exposure period and examined for vitellogenin by gelelectrophoresis and/or ELISA.

Effect of different isomers of alkylphenols: Groups of 8 juvenile rainbow trout (body weights between 103 and 168 g) were placed in 96 l steel tanks containing 80 l of aerated tap water. The fish were acclimated for one week before the beginning of the experiment. The fish were kept under a 12 h light :12 h dark photoperiod and they were not fed during acclimation and experimental periods. Two groups were intraperitoneally injected with 50 mg/kg body weight 4-n-octylphenol and 50 mg/kg 4-*tert*-octylphenol(4-(1,1,3,3-tetramethyl-butyl)phenol), respectively, on day 0 and 6 in the exposure period. Both chemicals were obtained from Aldrich and they were dissolved in peanut oil. One group receiving i.p. injections of peanut oil only, served as a control. Injection volumes were adjusted to 1 ml/kg. Blood samples were drawn on day 0 (before injection), 6 and 12 in the exposure period and the vitellogenin concentration was determined by ELISA.

Results

Effect of technical nonylphenol on vitellogenin levels in flounder: After 15 days' exposure to intraperitoneally administered technical nonylphenol, a dose dependent increase in vitellogenin levels was observed (Fig. 1).

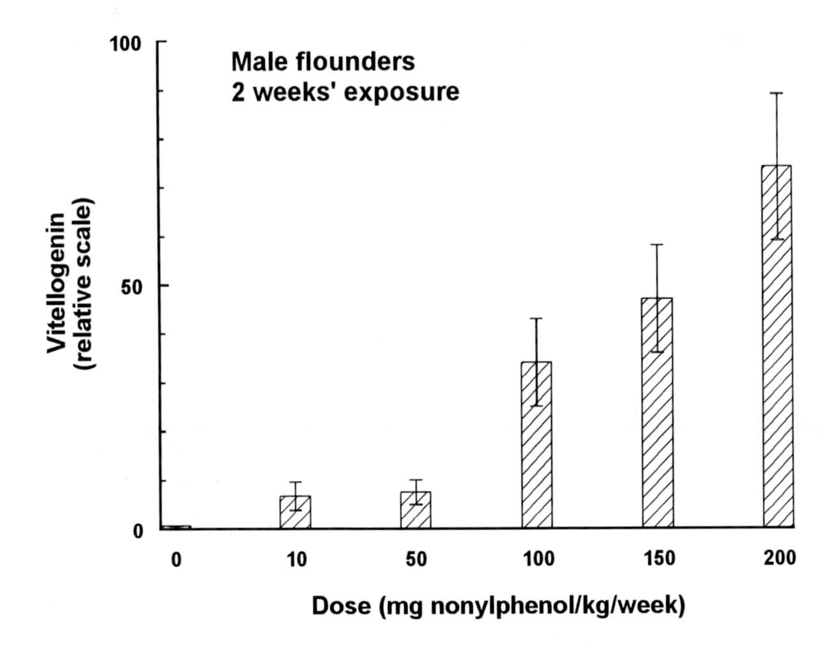

Fig. 1. Effect of technical nonylphenol on vitellogenin levels in male flounders receiving two i.p. injections per week. The vitellogenin concentrations are assessed by the concentration of alkali labile protein-phosphorus.

Development of ELISA for rainbow trout vitellogenin: The presence of vitellogenin in the plasma and the progress of its purification was determined by native- and SDS-PAGE (results not shown). A single band with a molecular weight of approximately 500 kD was observed after both gel filtration and ion chromatography. Western Blotting of plasma after SDS-PAGE was used to confirm the specificity of the affinity-purified antibodies. The antibodies did not recognize any proteins from male control plasma. ELISA standard curves fit a quadratic equation with correlation coefficients close to 1.00. The assay has a detection limit of 500 ng/ml plasma, since samples had to be diluted 1:100 to avoid matrix effects.

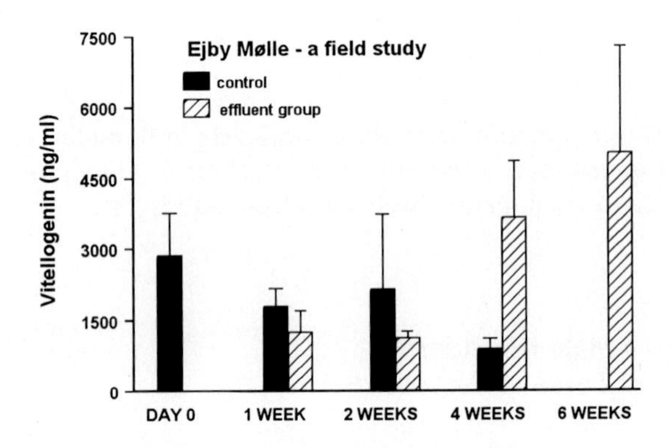

Fig. 2. Concentrations of vitellogenin in the blood of rainbow trout caged in a waste water effluent. Mean± SEM for the fish exceeding the detection limit (500 ng/ml) are given. After 6 weeks all values in the control group were below the detection limit.

Oestrogenic effect of a Danish waste water effluent: There was a gradual increase in the vitellogenin level with time in the effluent exposed group, although large interindividual variations were observed (Fig. 2). Concomitant with the rise in the exposed group, a decrease of the vitellogenin concentration in the control group was observed, possibly due to starvation. After six weeks all individuals in the control group were below the detection limit of 500 ng/ml, while the average vitellogenin level in the effluent exposed group was 5019 ± 2029 ng/ml.

Oestrogenic effect of endosulfan and dieldrin: Neither endosulfan nor dieldrin given alone at doses around 4 µmoles/kg showed oestrogenic effects in rainbow trout (Table 1). Combinations of the two compounds also failed to elicit oestrogenic responses.

Table 1. Doses of endosulfan and dieldrin injected into rainbow trout on day 0, 6 and 12 in the exposure period. Vitellogenin was measured on day 18.

Endosulfan µg (µmoles)/kg	Dieldrin µg (µmoles)/kg	Vitellogenin response
0 (0)	0 (0)	None
1600 (3.93)		None
	1600 (4.20)	None
826 (2.03)	774 (2.03)	None
1291 (3.17)	1209 (3.17)	None

Effect of different isomers of alkylphenols: Exposure to 4-n-octylphenol augments vitellogenin levels slightly, indicating a weak oestrogenic effect (Fig. 3). 4-*tert*-octylphenol augments vitellogenin levels by several orders of magnitude, indicating a much higher oestrogenic potency (Fig. 3).

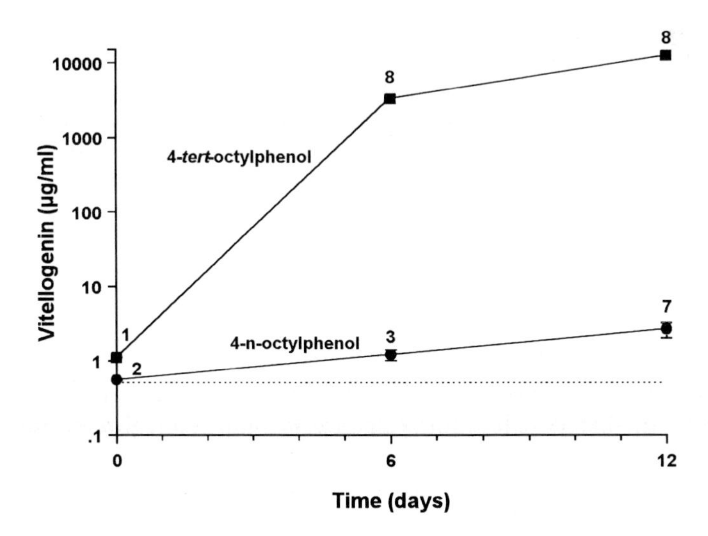

Fig. 3. Vitellogenin concentration in blood of rainbow trout injected 50 mg 4-*tert*-octylphenol (■) or 4-n-octylphenol (●) per kg on day 0 and 6. Mean±SEM for animals (n given for each point) exceeding the detection limit (dotted line) are given. There were 8 fish in each group.

Discussion

Although the detection limit in the present rainbow trout ELISA is not as low as that of a previously reported RIA (Sumpter 1985), it would seem to have picked up all the effluents with oestrogenic activity identified in a recent British field study (Harries et al. 1997).

Technical 4-nonylphenol is a mixture of many isomers, most of them having either a secondary or a tertiary α-carbon in the alkylgroup (Wheeler et al. 1997). It is therefore fully consistent with recent *in vitro* results obtained with alkylphenols of different structural features (Routledge & Sumpter 1997) that the technical mixture should elicit the response seen in flounders. The finding that 4-*tert*-octylphenol *in vivo* is several orders of magnitude more potent than 4-n-octylphenol also confirms the *in vitro* results of Routledge and Sumpter (1997).

A weak oestrogenic effect of the effluent from Ejby Mølle was found. The high initial levels found in the control group might be due to the influence of steroid-like compounds in commercial fish food which has been known to stimulate vitellogenin synthesis in the Siberian sturgeon (Pelissero et al. 1989). In an examination of five United Kingdom rivers induction of high levels of vitellogenin was found in caged rainbow trout in four of the rivers with increases as high as 1000-10.000 fold during a 3 week exposure period (Harries et al. 1997). Since the effluent tested in the presented study is led out from the sewage treatment plant of Odense, the third biggest town in Denmark, the contamination of the Danish rivers with oestrogenic compounds might not be as profound as the one observed in England. At the doses and routes of administration tested, endosulfan and dieldrin (alone or in combination) showed no oestrogenic effect *in vivo* in rainbow trout. Likewise, *in vivo* uterotrophic assays failed to identify oestrogenic effects of endosulfan [mouse (Shelby et al. 1996); rat (Ashby et al. 1997)] and dieldrin [rat (Ashby et al. 1997)]. Other *in vitro* systems (Shelby et al. 1996; Ashby et al. 1997) than the one used by Arnold et al. (1995) have also failed to show the oestrogenic effects of dieldrin and endosulfan as well as the marked synergistic effect of the combination of the two compounds shown by Arnold et al. (1995).

Risk assessment: Induction of vitellogenin production in male fish has been shown to be a valid biomarker for oestrogenic activity. It is, however, important to link the biomarker (vitellogenin) response to more ecologically important parameters, such as semen quality, testicular development, histological changes and offspring sex ratio and development.

Although these links between biomarker responses and parameters, relevant to population dynamics, are important to be established, specific biomarker responses, once they are validated, could be considered sufficient in environmental risk assessment and subsequent regulatory measures (e.g.. in approval of new chemicals and in monitoring effluents).

As an example of this approach, the Danish government requires that the phase-out of alkylphenols is completed by the year 2000. This regulation is based on the inherent properties of the chemicals (i.e. their oestrogenicity and their general toxicity) rather than a formal risk assessment in which the expected environmental concentrations of the chemicals are related to no observable effect levels.

Acknowledgements

This research is supported by grants from 'The Danish Interministerial Research Programme on Pesticides' and 'The Danish Environmental Research Programme'.

References

Arnold SF, Klotz DM, Collins BM, Vonier PM, Guillette LJ Jr, McLachlan JA (1996) Synergistic activation of estrogen receptor with combinations of environmental chemicals. Science 272:1489-1492

Arukwe A, Goksøyr A (1997) Comparative sensitivity of fish vitellogenin and zona radiata (eggshell) protein to environmental estrogens using immunoassays. In: Development of ecotoxicity and toxicity testing of chemicals. Nordic Council of Ministers. TemaNord 1997:524, pp. 46-48

Ashby J, Lefevre PA, Odum J, Harris CA, Routledge EJ, Sumpter JP (1997) Synergy between synthetic oestrogens ? Nature 385:494

Benson WH, Tyler C, Brugger KE, Daston G, Fry M, Gahr M, Gimeno S, Kolossa M, Länge R, Matthiessen P, van der Kraak G (1997) Strategies and approaches to *in vivo* screening and testing in identifying the hazards of endocrine modulating chemicals to wildlife. In: Tattersfield L, Matthiessen P, Campbell P, Länge R (eds) SETAC-Europe/OECD/EC expert workshop on endocrine modulators and wildlife - Assessment and testing (EMWAT). In press

Brouwer A, Reijnders PHJ, Koeman JH (1989) Polychlorinated biphenyl (PCB)-contaminated fish induces vitamin A and thyroid deficiency in the common seal (*Phoca vitulina*). Aquat Toxicol 15:99-106

Chakravorty S, Lal B, Singh TB (1992) Effect of endosulfan (thiodan) on vitellogenesis and its modulation by different hormones in the vitellogenic catfish *Clarias batrachus*. Toxicology 75:191-198

Couch EF, Lee JW (1987) Changes in estradiol and progesterone immunoreactivity in tissues of the lobster *Homarus americanus* with developing and immature ovaries. Comp Biochem Physiol 87A: 765 - 770

Fainzilber M, Tom M, Shafir S, Applebaum SW, Lubzens E (1992) Is there extraovarian synthesis of vitellogenin in penaid shrimps? Biol Bull 183: 233 - 241

Fry DM, Toone CK (1981) DDT-feminization of gull embryos. Science 231:919-924

Fry M (1995) Reproductive effects in birds exposed to pesticides and industrial chemicals. Environ Health Perspect 103:165-171

Gray MA, Metcalfe CD (1997) Induction of testis-ova in japanese medaka (*Oryzias latipes*) exposed to p-nonylphenol. Environ Toxicol Chem 16:1082-1086

Gimeno S, Gerritsen A, Bowmer T, Komen H (1996) Feminization of male carp. Nature 384:221-222

Guillette JL Jr, Pickford DB, Crain DA, Rooney AA, Percival HF (1996) Reduction in penis size and plasma testosterone concentrations in juvenile alligators living in a contaminated environment. Gen Comp Endocrinol 101:32-34

Guillette LJ Jr, Crain DA, Rooney AA, Pickford DB (1995a) Organization versus activation:The role of endocrine-disruption caontaminants (EDCs) during embryonic development in wildlife. Environ Health Perspect 103:157-164

Guillette LJ Jr, Gross TS, Gross DA, Rooney AA, Percival HF (1995b) Gonadal steroidogenesis *in vitro* from juvenile alligators obtained from contaminated or control lakes. Environ Health Perspect 103:31-36

Guillette LJ Jr, Gross TS, Masson GR, Matter JM, Percival HF, Woodward AR (1994) Developmental abnormalities of the gonad and abnormal sex hormone concentrations in juvenile alligators from contaminated and control lakes in Florida. Environ Health Perspect 102:680-688

Hagedorn HH (1989) In : Koolmann J (ed) Ecdysone. Thieme, Stuttgart, pp. 279 - 289

Harries JE, Sheahan DA, Jobling S, Matthiessen P, Neall P, Sumpter JP, Taylor T, Zaman N (1997) Estrogenic activity in five United Kingdom rivers detected by measurement of vitellogenesis in caged male trout. Environ Toxicol Chem 16: 534-542

Howell WM, Black DA, Bortone SA (1980) Abnormal expression of secondary sex characteristics in a population of mosquitofish, *Gambusia affinis holbrooki*: Evidence for environmentally induced masculinization. Copeia 4:676-681

Jobling S, Sumpter JP (1993) Detergent components in sewage effluent are weakly oestrogenic to fish: An *in vitro* study using rainbow trout (*Oncorhynchus mykiss*) hepatocytes. Aquat Toxicol 27: 361-372

Kishida M, Specker JL (1993) Vitellogenin in tilapia (Oreochromis mossambicus): Induction of two forms by estradiol, quantification in plasma and characterization in oocyte extract. Fish Physiol Biochem 12: 171-182

Korsgaard B, Kjær K (1974) Seasonal and hormonally induced changes in the serum level of the precursor protein vitellogenin in relation to ovarian vitellogenic growth in the toad *Bufo bufo bufo* (L.). Gen Comp Endocrinol 22: 261-267

Korsgaard B, Petersen I (1976) Natural occurrence and experimental induction by estradiol of a lipophosphoprotein vitellogenin in the flounder *Platichthys flesus*. Comp Biochem Physiol 55: 315-321

Korsgaard B, Mommsen TP, Saunders R (1986) The effect of temperature on the vitellogenic response in Atlantic salmon post-smolts. Gen Comp Endocrinol 62: 193-201

Korsgaard B, Mommsen TP (1993) Gluconeogenesis in hepatocytes of immature rainbow trout (*Oncorhynchus mykiss*): Control by estradiol. Gen Comp Endocrinol 89:17-27

Langston WJ (1996) Recent developments in TBT ecotoxicology. Toxicol Ecotoxicol News 3:179-187

Laufer H, Borst DW (1988): In : Laufer H, Downer RGH (eds) Invertebrate Endocrinology. Vol. 2., AR Liss, New York, pp 305 - 313

Manãnós E, Núñez J, Zanuy S, Carrillo M, Le Menn F (1994) Sea bass (*Dicentrarchus labrax* L.) Vitellogenin. II - Validation of an enzyme-linked immunosorbent assay (ELISA). Comp Biochem Physiol 107B: 217-223

Meusy JJ, Payen G (1988) Female reproduction in malacostracan crustacea. Zool Sci 5: 217 - 265

Munkittrick KR, Portt CB, Van der Kraak GJ, Smith IR, Rokosh DA (1991) Impact of bleached kraft mill effluent on population characteristics, liver MFO activity, and serum steroid levels of a Lake Superior white sucker (*Catostomus commersoni*) population. Can J Fish Aquat Sci 48:1371-1380

Munkittrick KR, Van der Kraak GJ, McMaster ME, Portt CB (1992) Response of hepatic MFO activity and plasma sex steroids to secondary treatment of bleached kraft pulp mill effluent and mill shutdown. Environ Toxicol Chem 11:1427-1439

Oehlmann J, Stroben E, Schulte-Oehlmann U, Bauer B, Fioroni P, Markert B (1996) Tributyltin biomonitoring using prosobranchs as sentiel organisms. Fresenius J Anal Chem 354:540-545

Pelissero C, Cuisset B, Le Menn F (1989) The influence of sex steroids in commercial fish meals and fish diets on plasma concentrations of estrogens and vitellogenin in cultured Siberian sturgeon *Acipencer baeri*. Aquat Living Resour 2: 161-168

Purdom CE, Hardiman PA, Bye VJ, Eno NC, Tyler CR, Sumpter JP (1994) Estrogenic effects of effluents from sewage treatment works. Chem Ecol 8: 275-285

Povlsen AF, Korsgaard B, Bjerregaard P (1990) The effect of cadmium on vitellogenin metabolism in estradiol-induced flounder *Platichthys flesus* males and females. Aquat Toxicol 17: 253-262

Quinitio ET, Hara A, Yamauchi K, Nakao S (1994) Changes in the steroid hormone and vitellogenin levels during the gametogenic cycle of the giant tiger shrimp, *Penaeus monodon*. Comp Biochem Physiol 109C: 21-26

Quackenbush LS (1989) Vitellogenesis in the shrimp *Penaeus vannamei*: *in vitro* studies of the isolated hepatopancreas and ovary. Comp Biochem Physiol 94B: 253 - 261

Quackenbush LS, Keeley LL (1988) Regulation of vitellogenesis in the fiddler crab *Uca puligator*. Biol Bull 173: 321 - 331

Reijnders PHJ (1986) Reproductive failure in common seals feeding on fish from polluted coastal waters. Nature 324:456-457

Routledge EJ, Sumpter JP (1997) Structural features of alkylphenolic chemicals associated with estrogenic activity. J Biol Chem 272: 3280-3288

Sharpe RM, Skakkebæk NE (1993) Are oestrogens involved in falling sperm counts and disorders of the male reproductive tract ? The Lancet 341:1392-1395

Shelby MD, Newbold RR, Tully DB, Chae K, Davis VL (1996) Assessing environmental chemicals for estrogenicity using a combination of *In Vitro* and *In Vivo* assays. Environ Health Perspect 104:1296-1300

Sumpter JP (1985) The purification, radioimmunoassay and plasma levels of vitellogenin from the rainbow trout, *Salmo gairdneri*. In: Lofts B, Holmes WH (eds) Trends in Comparative Endocrinology. Hong Kong University Press, pp 355-357.

Tinsley D (1985) A comparison of plasma levels of phosphoprotein, total protein and total calcium as indirect indices of exogenous vitellogenesis in the crusian carp, *Carassius carassius* (L.). Comp Biochem Physiol 80B: 913-916

Vethaak D, Jobling S, Waldock M, Bjerregaard P, Dickerson R, Giesy J, Grothe D, Karbe L, Munkittrick K, Schlumpf M, Sumpter J (1997) Approaches for the conduct of field surveys and toxicity identification and evaluation in identifying the hazards of endocrine modulating chemicals to wildlife. In: Tattersfield L, Matthiessen P, Campbell P, Länge R (eds) SETAC-Europe/OECD/EC expert workshop on endocrine modulators and wildlife - Assessment and testing (EMWAT). In press

Wheeler TF, Heim JR, Latorre MR, Janes AB (1997) Mass spectral characterisation of p-nonylphenol isomers using high resolution capillary GC MS. J Chromatogr Sci 35:19-30

White R, Jobling S, Hoare SA, Sumpter JP, Parker MG (1994) Environmentally persistent alkylphenolic compounds are estrogenic. Endocrinology 135: 175-182

Wiley SH, Opreske L, Wallace RA (1979) New methods for purification of vertebrate vitellogenin. Anal Biochem 97: 145-152

Effects of Environmental Chemicals on Reproduction

(Chair: I. Brandt, Sweden, and J.P. Bonde, Denmark)

Developmental and Reproductive Toxicity of Persistent Environmental Pollutants

Ingvar Brandt, Cecilia Berg, Krister Halldin, and Björn Brunström
Dept of Environmental Toxicology, Uppsala University, Norbyvägen 18A, S-752 36 Uppsala, Sweden

Introduction

A multitude of persistent environmental pollutants are known to bioaccumulate, and to biomagnify in food chains. Adverse biological effects caused by such chemicals may occur in heavily exposed animals at high trophic levels such as fish-eating mammals and birds. Different types of toxicity of persistent chlorinated environmental pollutants have been documented in mammals, birds, reptiles and fish, both in the field and in the laboratory. Since exposure to foreign chemicals in the environment generally is complex, it may be difficult to establish which individual compounds are responsible for the effects observed, and what mechanisms of action are in operation.

In recent years, there has been intense discussion in the scientific community and the society whether chemicals in the environment can act as xenohormones and give rise to adverse health effects in humans and wildlife, particularly following exposure during early life stages. The so called "estrogen hypothesis" has gained support from the fact that medication with the potent synthetic estrogen diethylstilboestrol (DES) during pregnancy was associated with reproductive organ changes in boys and girls born by medicated mothers (reviewed by Palmlund et al. 1995). Other arguments for this hypothesis are reports on increased incidences of testicular and breast cancer, decreased sperm counts, and on reproductive toxicity caused by persistent environmental pollutants in wildlife. While there is evidence to support that developmental and reproductive toxicity in certain wildlife populations may involve pollutant-induced hormonal disturbances, the hypothesized role of environmental endocrine disruptors in the above human health disorders seems to lack scientific support at present. The concern for endocrine disruptors has drawn attention to developmental toxicology as a rapidly developing field of research. Evidently, there is an urgent need to develop test methods and procedures for human and environmental risk assessment of hormonally active chemicals.

Definition of an Endocrine Disruptor

There are numerous ways by which chemicals can interact with the hormonal systems and cause reproductive and developmental toxicity. Some hitherto recognized mechanisms of action involve estrogen and androgen receptor binding (agonism and/or antagonism), inhibition of steroid hormone synthesis, interaction with thyroxin transport proteins and toxicity in hormone-producing tissues such as the adrenal cortex and gonads. So far, binding of phenolic compounds to the estrogen receptor (ER) and stimulation of ER-regulated gene transcription is a prevalent mechanism of action, and a fairly large number of both synthetical and natural chemicals are weak estrogens *in vitro*. To enable a rapid testing of chemicals, specific *in vitro* test methods to measure receptor-regulated gene transcription, hormone synthesis etc. need to be developed. Such *in vitro* systems are already useful tools to examine mechanisms of action and to select chemicals for further toxicity testing *in vivo*. The *in vitro* systems will not, however, account for the situation *in vivo*, where bioavailability, rates of metabolism and transformation to active metabolites, excretion or accumulation in target or non-target tissues will also influence the toxic potency.

"An endocrine disruptor is an exogenous substance that causes adverse health effects in an intact organism, or its progeny, consequent to changes in endocrine function". This *in vivo*-oriented definition, endorsed by a recent major European workshop in Weybridge (1996), reflects the awareness that "an endocrine disruptor could only be adequately defined using an *in vivo* test model where a functional endocrine system was present and the full interplay between normal physiological and biochemical processes could occur" (OECD Draft Detailed Review Paper: Appraisal of test methods for sex-hormone disrupting chemicals). To date, there are several examples of persistent environmental pollutants such as chlorinated dioxins, dibenzofurans, biphenyls, DDT and their persistent metabolites, which are known or suspected to cause reproductive and developmental toxicity in the laboratory or in the field. So far, only a few of these compounds have been shown to fulfill the Weybridge definition of an endocrine disruptor.

Many Persistent Environmental Pollutants are Active as Metabolites

The concept of persistency implies that a persistent environmental pollutant is more or less resistant to physico-chemical and biological degradation in the environment. Hence, it would be expected that compounds which are biomagnified in the food chain should not exert toxicity following metabolic transformation to reactive or stable metabolites. This is true for a large group of chlorinated dioxins and dibenzofurans, coplanar PCBs, and chlorinated naphthalenes, which bind to the Ah-receptor. Many Ah-receptor ligands are potent teratogens and reproductive toxicants in sensitive species, and there is evidence for reproductive toxicity in certain populations of mammals, birds and fish. Recent studies have shown that 2,3,7,8-tetrachlorodibenzo-p-dioxin (TCDD) may cause demasculinization of *in utero*-exposed male rats following administration of doses as low as 64-160 ng/kg b.w. to the pregnant dams (Mably et al. 1992 a), even though fertility was not significantly reduced at these low doses (Mably et al. 1992b). Moreover, Ah-receptor agonists may act as functional antiestrogens through a mechanism that excludes a primary binding to the estrogen receptor. Research during the last few years has shown, however, that PCBs and DDT are biotransformed to lipophilic metabolites, e. g. hydroxy-derivatives and methyl sulphones, which may be found in human and wildlife blood and other tissues. Consequently, a large number of hydroxy-PCBs are present in blood of Swedish human subjects as well as in Baltic grey seals and Aroclor 1254-treated rats (Bergman et al. 1994). These types of metabolites are particularly interesting because of their structural similarity to thyroxin (T4) and their ability to bind to the T4 transporter transthyretin, a T4 binding protein expressed in the liver and the *plexus choroideus* of rodents and other species. Considering the critical role of thyroid hormones in the developing brain, the observed interaction of hydroxy-PCBs with the T4 binding to transthyretin is of particular interest because of neuro-behavioural effects observed in neonatally PCB-exposed mice and rats (Eriksson & Fredriksson 1996). Some T4-like hydroxy-PCBs formed after administration of the parent PCB to pregnant mice are enriched in the blood and tissues of late gestational fetuses (Darnerud et al. 1986). Among the methylsulphones, the DDT metabolite 3-methylsulphonyl-DDE is a highly potent adrenocorticolytic agent following cell-specific metabolic activation in the *zona fasciculata* cells in mice and some other species (see below). In addition, DDD and DDE are stable DDT metabolites with a capacity to induce toxicity in the endocrine system.

DDT as a Multimechanistic Model Compound for Endocrine Disruption and Reproductive Toxicity

The classical pollutant DDT provides a unique example of a reproductive toxicant acting by several distinct mechanisms of action. This mechanistic diversity results from the fact that DDT is metabolised to biologically active metabolites which are present in different combinations in human milk and in tissues of wild mammals, birds, reptiles, and fish. Moreover, there are major species differences in the response to the various DDT metabolites. DDT may therefore serve as an interesting model compound, demonstrating the biological complexity of endocrine disruption and reproductive toxicity.

The idea that persistent environmental pollutants may induce reproductive toxicity following interaction with the hormonal system is at least 30 years old. Based on the structural similarity of DDT and the potent synthetic estrogen diethylstilboestrol (DES), Bitman & Cecil (1970) examined the estrogenic activity of 52 related compounds. Using an uterotrophic assay in the juvenile rat, they found that several compounds with a para-position unoccupied or occupied with a hydroxy or methoxy group were weak estrogens (o,p'-DDT, o,p'-DDE, methoxychlor, bisphenol A, two hydroxy-biphenyls, some low-chlorinated technical PCB-mixtures), whereas DDT-compounds with chlorines in both para-positions lacked estrogenic activity (p,p'-DDT, p,p'-DDE). These authors also suggested that phenolic metabolites are the active estrogens derived from o,p'-

substituted DDT analogs. It has later been shown that the insecticide methoxychlor is an active estrogen following demethylation to the corresponding mono- and bis-phenol derivatives. Over the years, a number of toxic effects relating to the estrogenic activity of DDT, PCB, and other persistent environmental pollutants have been published, e.g. prolonged estrous cycles, decreased prostatic and seminal vesicle weights in mice etc. Some DDT-induced effects will be discussed below.

Table 1 Some hormone-related toxicities induced by DDT and its persistent metabolites. Notably, there are major differences in sensitivity between different species of both mammals and birds.

Compound	Mechanism	Effects	Vertebrate class/species	Author
o,p'-DDT	ER agonist	Feminization Uterine growth	Birds Rats	Fry & Toone 1981 Bitman & Cecil 1970
p,p'-DDE	AR antagonist	Demasculinization	Rats	Kelce et al. 1995
	Prostaglandin synthetase inhibition	Egg shell thinning	Birds	Ratcliffe 1970 Lundholm 1997
o,p'-DDD/ p,p'-DDD	P450-catalysed covalent binding	Adrenocortical necrosis, atrophy	Humans Dogs Mink	Martz & Straw 1980 Martz & Straw 1980 Brandt et al 1992
MeSO2-DDE	P450-catalysed covalent binding	Adrenocortical necrosis, Decreased corticosterone production	Mouse, Birds (Humans)*	Lund et al. 1988 Brandt et al. 1992 Jõnsson & Lund1994 Lund & Lund 1995

* Only *in vitro*-data available

p,p'-DDE as an Antiandrogenic Developmental Toxicant: p,p'-DDE is the mostdominating persistent environmental pollutant present in human milk. Using a competitive binding assay with (^{3}H)R1881 in rat ventral prostate cytosol, Kelce et al. (1995) recently reported that p,p'-DDE binds with high affinity to the androgen receptor (AR) (K_i about 3.5 μM). Using transiently transfected monkey kidney CV1 cells, p,p'-DDE was further found to inhibit AR transcriptional activity with an efficacy about equal to that of the potent antiandrogen hydroxyflutamide. Quite surprisingly, p,p'-DDE was thus shown to be a potent AR antagonist *in vitro*. In order to determine the activity *in vivo*, fetal, pubertal and adult male rats were exposed to repeated doses of p,p'-DDE. Following treatment of pregnant rats with a total dose corresponding to about 500 mg/kg of p,p'-DDE, reduced anogenital distance, and retained thoracic nipples were observed in postnatal male pups. Treatment of prepubertal rats (days 21-57) with daily doses of 100 mg/kg delayed the onset of puberty by 5 days without affecting the serum testosterone levels. Finally, treatment of adult rats with a total dose of 800 mg/kg significantly reduced seminal vesicle and prostate weights without changing testosterone levels. These data show that p,p'-DDE is an active androgen receptor antagonist *in vivo*, even though demasculinization was demonstrated at high doses with regard to the long biological half-life of p,p'-DDE. Hence, this study does not seem to confirm that p,p'-DDE is as potent an AR antagonist *in vivo* as it is *in vitro*.

o,p'-DDT as an Estrogenic Developmental Toxicant: There is evidence showing that birds are sensitive to endocrine disrupting chemicals, which may give rise to changed sex organ development and impaired sexual behaviour. Unlike the situation in mammals, the females represent the heterogametic sex among birds, and estrogens are of crucial importance for sex differentiation in this class of vertebrates. Several bird species are highly sensitive to estrogens such as DES and o,p'-DDT. These compounds may induce feminization of the testicle (ovotestis) and the tubular reproductive system, and changed sexual behaviour in males.

The first report linking experimental DDT-induced feminization of male bird embryos with suspected DDT-induced effects in the environment was presented by Fry & Toone (1981) and Fry et al. (1987). Resulting from a large DDT spill contaminating the marine ecosystem offshore southern California, breeding failure due to eggshell thinning was observed in brown pelicans and double-crested cormorants. In western gulls, however, poor breeding success seemed to result from reduced numbers of adult males in the nesting area, where a skewed sex ratio and female-female pairing were observed. Experimental studies to examine possible developmental effects of different DDT derivatives were performed, using gull eggs collected at uncontaminated sites. Embryonated eggs were injected either with o,p'-DDT, p,p'-DDT, p,p'-DDE, methoxychlor or estradiol (positive control) and incubated. Examination at hatching showed that doses of o,p'-DDT and methoxychlor as low as 2-5 ppm (whole egg) resulted in feminized male embryos similar to those given estradiol (0.5-2 ppm). The most sensitive indicator of feminization was localization of primordial germ cells (PGC) in a thickened ovary-like cortex in the left testis. Development of both left and right oviducts were also observed at doses of 5 ppm and higher. It should be noted that also p,p'-DDE caused localization of primordial germ cells in cortical tissue of male embryos at higher doses (20-100 ppm), while p,p'-DDT was inactive. Since p,p'-DDE was recently reported to act as an antiandrogen in male rats (Kelce et al. 1995), an antiandrogenic component could also be possible in the feminization of male gull embryos. Notably, p,p'-DDE is most often the dominating DDT substance found in bird eggs .

Feminized sex organs have been confirmed also in quail, chicken and cormorant embryos exposed to o,p'-DDT (Bryan et al. 1989; Berg et al. unpublished). Moreover, Bryan et al. (1989) observed reproductive behavioural effects of o,p'-DDT in male quail, a species suitable for reproductive behavioural studies in birds.

p,p'-DDE as a Toxicant in the Shell Gland in Birds: DDE-induced reproductive failure due to egg shell thinning represents one of the most serious and well-documented effects of persistent environmental pollutants in a number of bird species (Ratcliffe 1967; 1970). A series of studies on the mechanism of action of p,p'-DDE-induced egg shell thinning in ducks have been published by Lundholm (reviewed by Lundholm 1997). Based on these and other studies, he concludes that egg shell thinning is a specific effect caused by p,p'-DDE at doses which do not give rise to general toxicity in the birds. No similar effect was seen with p,p'-DDT, o,p'-DDT or o,p'-DDE at comparable doses.

Based on studies *in vivo* and *in vitro* using homogenates from the duck shell gland, Lundholm presented evidence that p,p'-DDE-induced egg shell thinning involves inhibition of prostaglandin synthetase resulting in a decreased calcium transport, and that the level of prostaglandin E_2 in the shell gland was reduced following *in vivo* treatment. Similar effects could be produced with the cyclooxygenase inhibitor indomethacin. No prostaglandin synthetase inhibition or decreased prostaglandin E_2 level in the shell gland occurred in the domestic fowl, a species which was insensitive to DDE-induced egg shell thinning (Lundholm 1997). The relation between egg shell thinning and sex hormonal action seems unclear at present, although an interaction of p,p'-DDE with the progesterone receptor may be possible.

$MeSO_2$-DDE and DDD isomers as toxicants in the adrenal cortex: The persistent DDT-metabolite 3-methylsulphonyl-DDE ($MeSO_2$-DDE) was originally identified in blubber of Baltic grey seal, a species suffering from adrenocortical hyperplasia and a disease state resembling Cushing's disease (Bergman & Ohlsson 1986). Similarly to the $MeSO_2$-PCBs, $MeSO_2$-DDE seems to be formed in a pathway involving enterohepatic circulation and sequential metabolism of glutathione conjugates in the liver and intestinal microflora. Once formed, $MeSO_2$ -DDE is further metabolised by the mitochondrial cytochrome P450 form P450 11_β (P450c11) to a reactive intermediate that binds covalently to cellular constituents at its site of formation (Lund et al. 1988; Brandt et al. 1992; Lund & Lund 1995). Since P450 11_β is specifically expressed in the adrenal *zona fasciculata*, the covalent metabolite binding is confined to the glucocorticosteroid-producing cells at this site. Despite a very high rate of metabolism of $MeSO_2$-DDE in this restricted population of target cells, it has a long half-life in the body. Moreover, in mice a single dose as low as 3 mg/kg results in mitochondrial destruction in the adrenal *zona fasciculata*. At higher doses, complete necrosis of the *zona fasciculata* cells may occur. $MeSO_2$-DDE is readily transferred over the placenta to fetuses, where covalent metabolite binding and mitochondrial destruction is observed as early as the fetal adrenal cortex can be observed (Jönsson et al. 1995). $MeSO_2$-DDE is also efficiently transferred via mother's milk to suckling pups, which attain higher levels of bound adducts in their adrenals than do the dams. Consequently, $MeSO_2$-DDE is a potent transplacental and transmammary adrenal toxicant in mice. Decreased corticosterone levels in plasma in suckling pups were observed following administration of $MeSO_2$-DDE to the lactating dam (Jönson 1994).

The DDT metabolite o,p'-DDD was early associated to adrenocortical atrophy in dogs (Nelson & Woodard 1989; Martz & Straw 1980), and o,p'-DDD is currently used as a cytostatic for the treatment of adrenocortical carcinoma in humans. With regard to adrenocortical toxicity and/or P450-dependent covalent binding, there are remarkable species differences both for $MeSO_2$-DDE, o,p'-DDD and p,p'-DDD (Brandt et al. 1992). Interestingly, however, these persistent environmental pollutants (which can be indentified in human milk), are metabolised by cytochrome P450 to reactive, tissue-binding metabolites in human adrenal cortex homogenate (Jönsson & Lund, 1994). Considering the high toxic potency of the methyl sulphone as compared to DDD, and its cell-specific metabolic activation by adrenal cytochrome P450 11_β, $MeSO_2$-DDE should be further examined as a possible replacement for o,p'-DDD in the treatment of adrenocortical carcinoma in humans. Also, the role of these adrenocorticolytic DDT-metabolites in the etiology of adrenocortical hyperplasia in Baltic seals requires further investigation.

Concluding Remarks

The information available demonstrates a complex interaction of DDT and its persistent metabolites with the endocrine and/or reproductive system in different species. The effects discussed include both agonistic and antagonistic interactions with sex hormone receptors, targeted toxic effects on the glucocorticoid synthesis in the fetal, postnatal, and adult adrenal cortex and on the prostaglandin synthesis in the eggshell gland. Both teratogenic lesions resulting from disturbances in the developing embryo/fetus and toxic lesions resulting from postnatal exposure have been observed. Taken together, these diverse effects of DDT and its persistent metabolites demonstrate that DDT may serve as a multimechanistic model compound for endocrine disruption and reproductive toxicity.

References

Bergman Å, Klasson-Wehler, Kuroki H (1994) Selective retention of hydroxylated PCB metabolites in blood. Environ Health Perspect 102: 464-469.

Bergman A, Olsson M. (1986) Pathology of Baltic grey seal and ringed seal females with special reference to adrenocortical hyperplasia: Is environmental pollution the cause of a widely distributed disease syndrome ? Finn Game Res 44: 47-62.

Bitman, J., Cecil H. C. (1970) Estrogenic activity of DDT analogs and polychlorinated biphenyls. J Agr Food Chem 18: 1108-1112.

Brandt I, Jönsson CJ, Lund BO (1992) Comparative studies on adrenal corticolytic DDT-metabolites. Ambio 21: 602-605.
Bryan TE, Gildersleeve RP, Wiard RP (1989) Exposure of Japanese quail embryos to o,p'-DDT has long-term effects on reproduction behaviours, hematology and feather morphology. Teratology 39: 525-535.
Darnerud PO, Brandt I, Klasson-Wehler E, Bergman Å, d'Argy R, Sperber GO (1986) 3,3',4,4'-Tetrachlorobiphenyl in pregnant mice: enrichment of phenol and methyl sulphone metabolites in late gestational foetuses Xenobiotica 16: 295-306.
Eriksson P, Fredriksson A (1996) Developmental neurotoxixity of four ortho-substituted polychlorinated biphenyls in the neonatal mouse. Environm. Toxicol. Pharmacol. 1: 155-165.
Fry DM, Toone CK (1981) DDT-induced feminization of gull embryos. Science 213: 922-924.
Fry DM, Toone CK, Speich SM, Peard RJ (1987) Sex ratio skew and breeding patterns of gulls: demographic and toxicological considerations. Studies Avian Biol 10: 26-43.
Jönsson J, Rodriguez-Martinez H, Brandt I (1995) Transplacental toxicity of 3-methylsulphonyl-DDE in the developing adrenal cortex of mice. Reproduct Toxicol 9: 257-264.
Jönsson CJ, Lund BO (1994) *In vitro* bioactivation of the environmental pollutant 3-methylsulphonyl-2,2-bis(4-chlorophenyl)-1,1-dichloroethene in the human adrenal gland. Toxicol Lett 71: 169-175.
Jönsson CJ (1994) Decreased plasma corticosterone levels in suckling mice following injection of the adrenal toxicant, $MeSO_2$-DDE to the lactating dam. Pharmacol Toxicol 74: 58-60.
Lund BO, Bergman Å, Brandt I (1988) Metabolic activation and toxicity of a DDT-metabolite, 3-methylsulphonyl-DDE, in the adrenal *zona fasciculata* in mice Chem-Biol Interactions 65: 25-40.
Lund BO, Lund J (1995) Novel involvement of a mitochondrial steroid hydroxylase (P450c11) in xenobiotic metabolism. J Biol Chem 270: 20895-20897.
Lundholm CE (1997) DDE-induced eggshell thinning in birds: Effects of p,p´-DDE on the calcium and prostaglandin metabolism of the eggshell gland. Comp. Biochem. Physiol. in press.
Mably TA, Moore RW, Peterson RE (1992a) *In utero* and lactational exposure of male rats to 2,3,7,8-tetrachlorodibenzo-p-dioxin. I: Effects on androgenic status. Toxicol Appl Pharmacol 114: 97-107.
Mably TA, Bjerke DL, Moore RW, Gendron-Fitzpatrick A, Peterson RE (1992b) *In utero* and lactational exposure of male rats to 2,3,7,8-tetrachlorodibenzo-p-dioxin. II: Effects on spermatogenesis and reproductive capability. Toxicol Appl Pharmacol 114: 118-126.
Martz F, Straw JA (1980) Metabolism and covalent binding of 1-(o-chlorophenyl)-1-(p-chlorophenyl)-2,2-dichloethane (o,p'-DDD). Correlation between adrenocorticolytic activity and metabolic activation by adrenocortical mitochondria. Drug Metab Dispos 8:127-130.
Nelson AA, Woodard G (1949) Severe adrenal cortical atrophy (cytotoxic) and hepatic damage produced in dogs by feeding 2,2-bis(parachlorophenyl)-1,1-dichloroethane (DDD or TDE) Arch Pathol 48: 387-394.
Norén K, Lundén Å, Pettersson E, Bergman Å (1996) Methylsulfonyl metabolites of PCBs and DDE in human milk in Sweden 1972-1992. Environ Health Perspect 104: 766-773.
Palmlund I, Apfel R, Buitendijk S, Cabau A, Forsberg JG (1993) Effects of diethylstilbestrol (DES) medication during pregnancy. J Psychosom Obstet Gynaecol 14: 71-89.

Ratcliffe DA (1967) Decrease in eggshell weight in certain birds of prey. Nature 215: 208-210.
Ratchliff DA (1970) Changes attributable to pesticides in egg breakage frequency and egg shell thickness in some British birds. J Appl Ecol 7: 68-115.
Weybridge (1996) European workshop on the impact of endocrine disruptors on human health and wildlife. Weybridge, U.K., 2-4 December 1996, Report of proceedings.

Effects of the In Vitro Chemical Environment During Early Embryogenesis on Subsequent Development

Donald Rieger
Animal Biotechnology Embryo Laboratory, Dept. of Biomedical Sciences, University of Guelph, Guelph, Ont., Canada N1G 2W1

Abstract

The development of the preimplantation embryo seems morphologically very simple, and embryologists previously assumed that an embryo that developed to the blastocyst stage was fully capable of normal development after transfer to the uterus of a recipient female. This complacency was disturbed by reports that exposure of early embryos to mutagens such as methylnitrosourea led to fetal abnormalities, decreased birth rates, and decreased life-span. Even more disturbing are recent reports that culture of early embryos in supposedly benign conditions can adversely affect their subsequent development. Techniques have been developed for the production of cattle and sheep embryos by in-vitro fertilization and by cloning. Such embryos must be cultured for several days before they can be transferred, and, in some cases, this has been related to abortion, very high birthweight, physical abnormalities and peri-natal mortality of the calves and lambs. This syndrome may result from an unbalanced development of the trophoblast relative to the inner-cell mass, possibly related to the presence of serum, glucose, or ammonium in the culture medium. An analogous phenomenon has been observed in human in-vitro fertilization where babies from single pregnancies have below-normal birth-weight. There is also evidence to suggest that the in-vitro environment of the gametes before fertilization can affect subsequent embryonal and fetal development. Exposure of mouse oocytes to vitrification solutions has been shown to lead to fetal malformations, and treatment of bull sperm with glutathione improves early embryo development. The common thread in these diverse observations is that development can be affected by events that occur long before any defect is apparent. Consequently, the production of a morphologically normal embryo is no guarantee that fetal development and post-natal life will be normal. This is of immediate concern in human reproductive medicine due to the increasing use of sperm injection for fertilization, and the emergence of in-vitro oocyte maturation. Further development and application of reproductive techniques would benefit from a toxicological evaluation of risk factors and exposure limits.

Introduction

Figure 1 shows the development of the early cattle embryo over the first 9 days after fertilization of the oocyte, and this pattern is generally representative of all eutherian mammals. This early development appears simple but includes a number of significant events including initiation and continuation of cleavage, activation of the embryonic genome, maturation of the mitochondria, compaction and formation of tight junctions at the morula stage, formation of the blastocoelic cavity, initial cell differentiation, and hatching from the zona pellucida. Activation of the embryonic genome generally occurs between the 4- and 16-cell stages. Before this, the structures and functional control of the embryo are inherited from the oocyte, arising from transcription and translation of the maternal genome. After the transition control of cell form and functions are controlled by the embryonic genome. Differentiation at the blastocyst stage yields the inner cell mass, which will form the embryo proper, and the trophoblast cells, which will form the embryonic contribution to the placenta. (see Betteridge and Fléchon, 1988)

The modern history of the manipulation of reproduction can be traced to the experiment of Walter Heape, who transfered embryos from one rabbit to another, over 100 years ago (Biggers, 1991). The practical application of embryo transfer in domestic animals began in the 1970's and has increased to over 400,000 reported transfers of cattle embryos world-wide in 1995 (Thibier, 1996). For the most part, embryos are collected from superovulated donor animals at the morula to blastocyst stage and evaluated for morphology before being transfered to the recipients. Morphological quality is reasonably well related to the viability of populations of embryos, (Shea, 1981; Hasler et al., 1983) but is notoriously unreliable for predicting the developmental potential of any given individual embryo (Rieger, 1984; Butler and Biggers, 1989). The corollary is that the developmental potential of early embryos may be disturbed without having any effects on their morphology.

This was dramatically demonstrated by Iannaccone (1984) who showed that the 50% lethal dose for immediate death of mouse blastocysts after a 1 h exposure to methylnitrosourea (MNU) *in vitro* was 4200 µg/ml. In contrast, for blastocysts exposed to MNU *in vitro* and then transfered to recipient females, the EC_{50} for implantation was 160 µg/ml, and the EC_{50} for live birth was 4.7 ng/ml. The author interpreted this hierarchy of sensitivity of different aspects of the development of the mouse blastocyst to MNU to suggest that "processes with increasing levels of integration and sophistication are increasingly sensitive to chemical disruption at an early stage of development." Most remarkably, the pups that developed from blastocysts exposed to MNU had a 58% crude mortality rate before one year of age, compared with 22% for the controls. It was similarly shown that fetal malformations can be induced by pre-implantation exposure to ionizing radiation, medroxyprogesterone acetate, nickel chloride and ethylene oxide. The mechanism(s) by which pre-implantation exposure to these agents can affect later development are unknown, but may involve the induction of non-lethal mutations, or d isruption of the early programming of

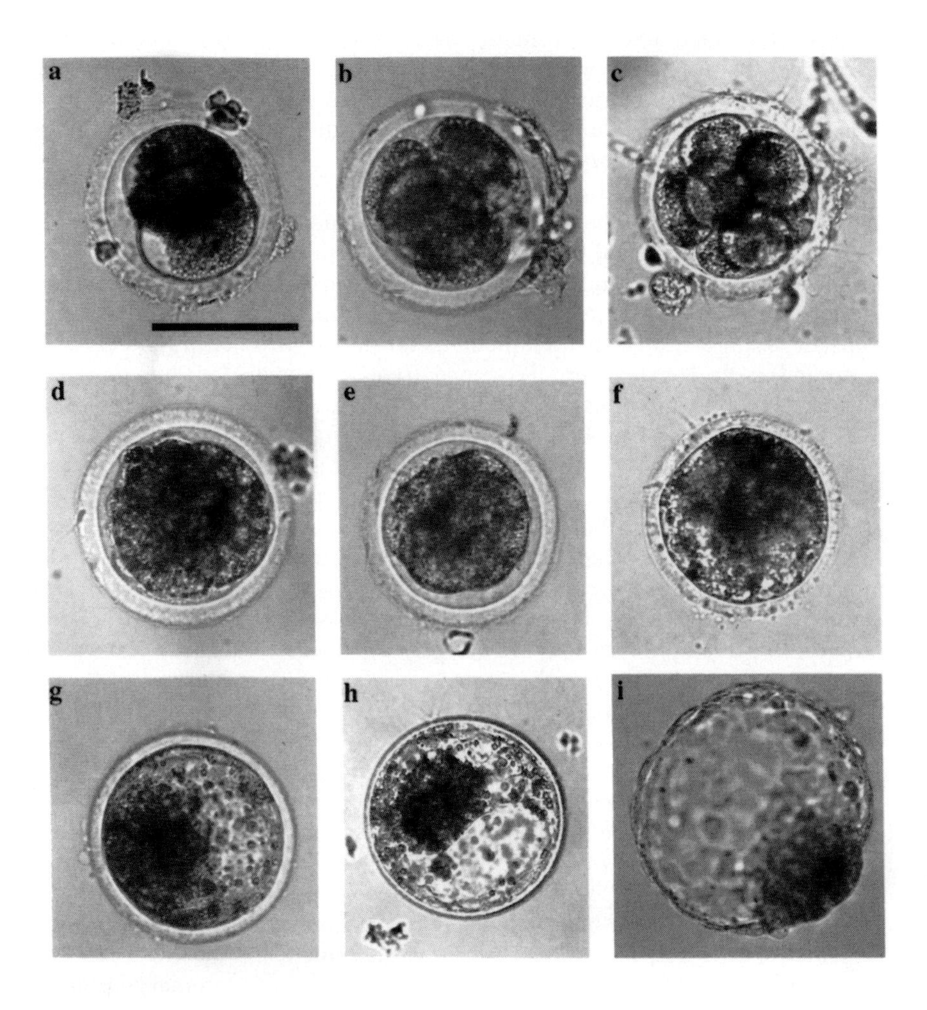

Fig. 1. Early development of the cattle embryo. a Zygote, b 2-cell stage, c 4-cell, d 8-cell, e 16-cell, f compacted mourla, g early blastocyst, h blastocyst, i expanded blastocyst, j hatching blastocyst, k hatched blastocyst. Ruler in frame a indicates 100 µm. (Photo: Esther Semple)

gene expression (Kimmel et al., 1993).

These reports were of significant interest to developmental biologists because their implications for the fundamental understanding of the control of embryonic, and later development. However, they appeared to be of limited concern for the applied aspects of embryo biology, because embryos would never be exposed to such agents in culture. For the most part, embryos of domestic animals were cultured in media and conditions that were well established as favorable for somatic cells, or in some cases, that had been specifically developed to support early embryo development. The supposedly benign nature of such culture conditions was called into question by two major advances in the reproductive technology of domestic animals: *in vitro* fertilization (IVF) and embryo cloning.

Brackett et al.(1982) produced the first calf by in-vitro fertilization of an oocyte that had matured within the follicle of the cow, similar to the technique used by Steptoe and Edwards (1978) to produce the first IVF human baby. This is still the normal approach in human IVF, but it requires intensive monitoring of follicular development and hormone levels and is thus impractical for routine use in domestic animals. Consequently, methods for the in-vitro maturation (IVM) of sheep and cattle oocytes were developed (see Greve and Madison, 1991), and IVM is now used for oocytes collected from live animals and from ovaries obtained at slaughter (see Betteridge and Rieger, 1993). Human IVF embryos can be, and normally are, introduced in the woman's uterus at very early cleavage stages (Gerrity, 1992), whereas cattle embryos must be at the morula or later stage to survive in the uterus of the cow (Newcomb and Rowson, 1975). Consequently, methods were developed for the culture of IVF cattle (and other domestic animal) embryos to the morula or blastocyst stage. Initially, the oviducts of intermediate recipients (rabbits or sheep) were used for this purpose, but this approach has been largely superseded by in-vitro co-culture with oviductal epithelial, or other somatic cells. Efforts continue to develop completely defined culture conditions that are free of other cells, serum or bovine serum albumin with mixed success (see Betteridge and Rieger, 1993; Hasler et al., 1995).

Willadsen (1986) produced the first mammalian clones, groups of genetically identical sheep, and later, of cattle (Willadsen, 1989) by fusing single blastomeres from cleavage stage embryos to enucleated oocytes. The resultant embryos are cultured through several cleavages and the process repeated, or, ultimately, cultured to the morula or blastocyst stage for implantation into recipients (see Betteridge and Rieger, 1993). More recently, the same technique has been used to produce clones from an embryonic cell line and adult mammary epithelial cells (Wilmut et al., 1997). Such clones offer significant advantages for the genetic improvement of livestock (Smith, 1989), and consequently attempts were made to commercialize the technique (see Barnes et al., 1991; Betteridge and Rieger, 1993). Perhaps the most significant outcome of these commercial ventures was a major problem they encountered.

The Large Calf/Lamb Syndrome

Willadsen et al. (1991) reported a high incidence of congenital deformities and very high birthweights among cloned calves. Similar observations were reported by other investigators (Bondioli, 1992; Garry et al., 1996). These observations led to the demise of cloning as a commercial enterprise because the birthing problems associated with large calves are severely disadvantageous to both the animals and the livestock producers. Walker et al. (1992) showed that in-vitro culture of early sheep embryos was associated with increased mean gestation length, birth weight, and perinatal mortality of the lambs, suggesting that the problem originated from the in-vitro culture associated with embryo cloning, rather than the mechanical manipulations. A recent meta-analysis of the results of 28 studies of IVF and cloned calves indicated that IVF calves were as, or more, likely as cloned calves to have extended gestations, high birth weights, and high rates of dystocia and perinatal mortality (Kruip and den Daas, 1997). In a practical sense, this is even more significant because there is an increasing commercial interest in producing calves by IVF (see Betteridge and Rieger, 1993) and even relatively infrequent occurrences of large calves and the problems associated with them could severely damage the image of the usefulness of the technique.

There are only a few reports that directly demonstrate that specific components of embryo culture can induce later developmental defects. Thompson et al. (1995) showed that the inclusion of 20% human serum into sheep embryo culture medium caused the accumulation of lipid within the embryonic cells, and resulted in significant increases in the gestation length and birthweight of the lambs. Amino acids, particularly glutamine, are generally favorable for the in-vitro development of mouse embryos, but the spontaneous breakdown of amino acids over extended culture periods produces significant concentrations of ammonium (Gardner and Lane, 1993). Exposure of mouse embryos to these concentrations of ammonium for 69 h caused dose-dependent increases in fetal retardation and exencephaly in the pups after transfer to recipient females (Lane and Gardner, 1994). At the blastocyst stage and later, high glucose concentrations *in vitro* or *in vivo* cause retarded development or fetal abnormalities in the mouse, rat and human (see De Hertogh et al., 1991). It is perhaps most significant that high glucose concentrations cause a relatively larger loss of cells from the inner cell mass than from the trophoblast of rat blastocysts (De Hertogh et al., 1991; De Hertogh et al., 1992). Glucose, oxygen concentrations greater than 5%, and reactive oxygen species inhibit or delay the development of early embryos *in vitro* (see Rieger, 1992), while catalase, superoxide dismutase and glutathione peroxidase protect Day 9 rat embryos against the teratogenic effects of high glucose (Eriksson and Borg, 1991). It may be that the early inhibitory effects and the later teratogenic effects of glucose share a common etiology, mediated by reactive oxygen species.

The conditions to which oocytes and sperm are exposed before fertilization can also affect embryonic and later development. Rose and Bavister (1992) matured cattle oocytes in 7 different media and then fertilized them and cultured the embryos in a common medium. They found significant differences in cleavage rates, and development to morula and blastocyst stage due to maturation medium treatment. Lonergan et al. (1996) showed a similar effect of EGF during maturation of cattle oocytes on subsequent cleavage and development to the blastocyst stage. Exposure of mouse oocytes to vitrification chemicals has been shown to result in fetal malformations (Kola et al., 1988). The cleavage rate and development of cattle embryos to the blastocyst stage were significantly improved by treatment of sperm with glutathione (Earl et al., 1997) or by the addition of casein phosphopeptides to the fertilization medium (Nagai et al., 1996).

The mechanisms by which IVM, sperm preparation, or embryo culture can affect subsequent development are unknown. Walker et al. (1996) speculated that the *in vitro* culture of the early embryo may affect subsequent development by influencing the imprinting of genes essential to fetal development, or by disturbing extragenetic cytoplasmic factors or conditions. Large offspring can also result from transfer of early sheep embryos into an asynchronous recipient (Wilmut and Sales, 1981) or from progesterone treatment during early pregnancy (Kleemann et al., 1994). Culture of early embryos or early progesterone treatment of ewes leads to excessive growth of the trophoectoderm and, subsequently, increased placental weight (Walker et al., 1996), leading these authors to suggest that asynchrony between the development of the early embryo and the physiological state of the recipient uterus may be a common intermediary event in the large calf/lamb syndrome.

Implications for Human Medicine

There is evidence to suggest that human fetal development can also be affected by IVF procedures. Wang et al. (1994) found that the incidence of babies that were small or very small for their gestational age was significantly greater in singleton IVF pregnancies than in the normal obstetric population. Circulating maternal concentrations of hCG are significantly lower in IVF pregnancies than in spontaneous pregnancies (Johnson et al., 1994), indicating that the development and function of the trophectoderm is disturbed in IVF pregnancies. As for domestic animals and other species, human fetal development may be affected by the conditions to which the oocyte is exposed before fertilization. Gregory et al. (1994) reported that the proliferative capacity of the cumulus-corona cells which surround the oocyte was unrelated to fertilization rate or embryo morphology, but was directly related to the incidence of clinical pregnancy.

It is by no means clear that the developmental abnormalities in domestic animals and humans following in-vitro embryo culture have a common etiology. However, the observations in domestic animals must be cautionary for the current practice of, and especially for new developments in, assisted human reproductive medicine (see Seamark and Robinson, 1995). This is particularly true for techniques such as intra-cytoplasmic sperm injection which bypasses the normal control mechanisms of fertilization, in-vitro oocyte maturation, oocyte cryopreservation, and extended embryo culture. M□n□zo and Dale (1995) have said that "... many, if not all of the assisted human reproduction technologies have been applied enthusiastically with little concern for basic research." This is most certainly unacceptable, and domestic animals such as sheep and cattle are very valuable models for such research. Moreover, toxicological perspectives of risk factors and exposure limits should be integrated into the developmental studies of these techniques, at the earliest stage possible.

References

Barnes, FL, Looney, CR and Westhusin, ME (1991) Embryo cloning in cattle: the current state of technology. Embr Trans 6 1-5.

Betteridge, KJ and Fléchon, J-E (1988) The anatomy and physiology of pre-attachment bovine embryos. Theriogenology 29 155-187.

Betteridge, KJ and Rieger, D (1993) Embryo transfer and related techniques in domestic animals, and their implications for human medicine. Hum Reprod 8 147-167.

Biggers, JD (1991) Walter Heape, FRS: a pioneer in reproductive biology. Centenary of his embryo transfer experiments. J Reprod Fertil 93 173-186.

Bondioli, KR (1992) Commercial cloning of cattle by nuclear transfer. In: Symposium on Cloning Mammals by Nuclear Transplantation., pp. 35-38, Fort Collins, CO.

Brackett, BG, Bousquet, D, Boice, ML, Donawick, WJ, Evans, JF and Dressel, MA (1982) Normal development following *in vitro* fertilization in the cow. Biol Reprod 27 147-158.

Butler, JE and Biggers, JD (1989) Assessing the viability of preimplantation embryos *in vitro*. Theriogenology 31 115-126.

De Hertogh, R, Vanderheyden, I, Pampfer, S, Robin, D and Delcourt, J (1992) Maternal insulin treatment improves pre-implantation embryo development in diabetic rats. Diabetologia 35 406-408.

De Hertogh, R, Vanderheyden, I, Pampfer, S, Robin, D, Dufrasne, E and Delcourt, J (1991) Stimulatory and inhibitory effects of glucose and insulin on rat blastocyst development *in vitro*. Diabetes 40 641-647.

Earl, CR, Kelly, J, Rowe, J and Armstrong, DT (1997) Glutathione treatment of bovine sperm enhances *in vitro* blastocyst production rates. Theriogenology 47 255 (Abstract).

Eriksson, UJ and Borg, LAH (1991) Protection by free oxygen radical scavenging enzymes against glucose-induced embryonic malformations *in vitro*. Diabetologia 34 325-331.

Gardner, DK and Lane, M (1993) Amino acids and ammonium regulate mouse embryo development in culture. Biol Reprod 48 377-385.

Garry, FF, Adams, R, McCann, JP and Odde, KG (1996) Postnatal characteristics of calves produced by nuclear transfer cloning. Theriogenology 45 141-152.

Gerrity, M (1992) Determinants of human embryo quality following *in vitro* fertilization. Theriogenology 37 147-160.

Gregory, L, Booth, AD, Wells, C and Walker, SM (1994) A study of the cumulus corona cell complex in in-vitro fertilization and embryo transfer; a prognostic indicator of the failure of implantation Hum Reprod 9 1308-1317.

Greve, T and Madison, V (1991) *In vitro* fertilization in cattle: a review. Reprod Nutr Dev 31 147-157.

Hasler, JF, Henderson, WB, Hurtgen, PJ, Jin, ZQ, McCauley, AD, Mower, SA, Neely, B, Shuey, LS, Stokes, JE and Trimmer, SA (1995) Production, freezing and transfer of bovine IVF embryos and subsequent salving results. Theriogenology 43 141-152.

Hasler, JF, McCauley, AD, Schermerhorn, EC and Foote, RH (1983) Superovulatory responses of Holstein cows. Theriogenology 19 83-99.

Iannaccone, PM (1984) Long-term effects of exposure to methynitrosourea on blastocysts following transfer to surrogate female mice. Canc Res 44 2785-2789.

Johnson, MR, Abbas, AA, Irvine, R, Riddle, AF, Normantaylor, JQ, Grudzinskas, JG, Collins, WP and Nicolaides, KH (1994) Regulation of corpus luteum function. Hum Reprod 9 41-48.

Kimmel, CA, Generoso, WM, Thomas, RD and Bakshi, KS (1993) A new frontier in understanding the mechanisms of developmental abnormalities Toxicol Appl Pharmacol 119 159-165.

Kleemann, DO, Walker, SK and Seamark, RF (1994) Enhanced fetal growth in sheep administered progesterone during the first three days of pregnancy. J Reprod Fertil 102 411-417.

Kola, I, Kirby, C, Shaw, J, Davey, A and Trounson, A (1988) Vitrification of mouse oocytes results in aneuploid zygotes and malformed fetuses. Teratology 38 467-474.

Kruip, TAM and den Daas, JHG (1997) *In vitro* produced and cloned embryos: effects on pregnancy, parturition and offspring. Theriogenology 47 43-52.

Lane, M and Gardner, DK (1994) Amino acids increase post-implantation development of cultured mouse embryos while ammonium induces exencephaly and fetal retardation.. J Reprod Fert 102 305-312.

Lonergan, P, Carolan, C, Van Langendonckt, A, Donnay, I, Khatir, H and Mermillod, P (1996) Role of epidermal growth factor in bovine oocyte maturation and preimplantation embryo development *in vitro*. Biol Reprod 54 1420-1429.

M□n□zo, Y and Dale, B (1995) Paternal contribution to successful embryogenesis. Hum Reprod 10 1326-1328.

Nagai, T, Hori, N, Abe, S and Hirayama, M (1996) Effect of casein phosphopeptides on fertilization *in vitro* of bovine oocytes matured in culture. Anim Sci Technol 67 1037-1042.

Newcomb, R and Rowson, LEA (1975) Conception rate after uterine transfer of cow eggs, in relation to synchronization of oestrus and age of eggs. J Reprod Fert 43 539-541.

Rieger, D (1984) The measurement of metabolic activity as an approach to evaluating viability and diagnosing sex in early embryos. Theriogenology 21 138-149.

Rieger, D (1992) Relationships between energy metabolism and development of early mammalian embryos. Theriogenology 37 75-93.

Rose, TA and Bavister, BD (1992) Effect of oocyte maturation medium on *in vitro* development of *in vitro* fertilized bovine embryos. Molec Reprod Dev 31 72-77.

Seamark, RF and Robinson, JS (1995) Potential health hazards of assisted human reproduction. Potential health problems stemming from assisted reproduction programmes. Hum Reprod 10 1321-1322.

Shea, BF (1981) Evaluating the bovine embryo. Theriogenology 15 31-42.

Smith, C (1989) Cloning and genetic improvement of beef cattle. Anim Prod 49 49-62.

Steptoe, PC and Edwards, RG (1978) Birth after the reimplantation of a human embryo. Lancet 2 366.

Thibier, M (1996) The 1995 statistics on the world embryo transfer industry. Intl Emb Trans Soc Newslett 14 27-30.

Thompson, JG, Gardner, DK, Pugh, PA, McMillan, WH and Tervit, HR (1995) Lamb birth weight is affected by culture system utilized during *in vitro* pre-elongation development of ovine embryos. Biol Reprod 53 1385-1391.

Walker, SK, Hartwich, KM and Seamark, RF (1996) The production of unusually large offspring following embryo manipulation: concepts and challenges. Theriogenology 45 111-120.

Walker, SK, Heard, TM and Seamark, RF (1992) *In vitro* culture of sheep embryos without co-culture: successes and perspectives. Theriogenology 37 111-126.

Wang, JX, Clark, AM, Kirby, CA, Philipson, G, Petrucco, O, Anderson, G and Matthews, CD (1994) The obstetric outcome of singleton pregnancies following in-vitro fertilization/gamete intra-Fallopian transfer. Hum Reprod 9 141-146.

Willadsen, SM (1986) Nuclear transplantation in sheep. Nature 320 63-65.

Willadsen, SM (1989) Cloning of sheep and cow embryos. Genome 31 956-962.

Willadsen, SM, Jansen, RE, McAlister, RJ, Shea, BF, Hamilton, G and McDermand, D (1991) The viability of late morulae and blastocysts produced by nuclear transplantation in cattle. Theriogenology 35 161-170.

Wilmut, I and Sales, DI (1981) Effect of an asynchronous environment on embryonic development in the sheep.. J Reprod Fertil 61 179-184.

Wilmut, I, Schnieke, AE, McWhir, J, Kind, AJ and Campbell, KHS (1997) Viable offspring derived from fetal and adult mammalian cells. Nature 385 810-813.

Influence of Hormones and Hormone Antagonists on Sexual Differentiation of the Brain

Klaus D. Döhler
Haemopep Pharma GmbH, Feodor-Lynen-Str. 5, D-30625 Hannover, Germany

Sexual Differentiation: Environmental Influences

The question about which factors determine the fate of a developing fetus to become either male or female has occupied many previous cultures and scientists. The "*thermal hypothesis*", put forth by the ancient Greek philosopher and scientist Empedokles of Akras (about 460 BC), claimed that *temperature* was an important factor in sex determination (Plato, translated by Jowett 1953). Conception in a hot uterus would produce a male, in a cold uterus a female. Aristotle of Stagirus (384 to 322 BC) was convinced that sheep and goats would produce male offspring when warm winds were blowing from the south during copulation, but female offspring when cold winds were blowing from the north (Aristotle, translated by Cresswell 1862).

The "*thermal hypothesis*" of Empedokles may actually not be that far off after all. It has been shown that frog larvae develop a male phenotype when raised at elevated water temperature, at low temperature they develop into females (Piquet 1930). In some species of lizards breeding of the eggs at temperatures below 26°C will prime the embryos for female development, whereas at temperatures above 26°C the embryos will develop into males (Short 1982). In two species of turtles, *Emys orbicularis* and *Testudo graeca* the temperature effect on sexual differentiation is reversed. Male development is induced during breeding at temperatures below 28°C and female development is induced during breeding at above 32°C (Pieau 1975).

Another environmental influence which may affect sexual differentiation is the concentration of potassium and calcium ions. Three- to four-fold elevation of calcium ions in the water will stimulate the larvae of the toad *Discoglossus pictus* to develop into females. Five- to six-fold elevation of calcium ions will stimulate the same larvae to develop into males (Stolkovski and Bellec 1960).

In higher vertebrae, like birds and mammals, sexual differentiation is under *genetic* and *hormonal* control (see below). As we know from a great number of wildlife observations (see Colborn and Clement 1992; Colborn et al. 1996) natural or man-made compounds *in the environment* may interfere with the normal endogenous endocrine milieu of developing animal and human embryos and may, thus, exert deleterious effects on the differentiation of reproductive structures and functions. These endocrine disruptors in the environment render the developing organisms permanently infertile and may, thus, endanger whole species in the affected areas to become extinct.

Sexual Differentiation: How the Dinosaurs Became Extinct

Scientists have long been searching for clues to explain the sudden extinction of the dinosaurs 65 million years ago. The primary event, leading to their extinction, apparently was the collision of the earth with a huge meteorite. This collision created the Chicxulub-crater in the Gulf of Mexico and caused huge amounts of soil and dust to be tossed into the athmosphere. Scientists agree that, due to the darkening of the atmosphere, the average temperature on earth dropped by several degrees. Scientists speculate that the collision with the meteorite and the subsequent decrease in environmental temperature resulted in a shortage of food supply for the dinosaurs, which led to their individual deaths and eventually resulted in the extinction of most dinosaur species.

This hypothesis is, however, burdened with serious flaws. The probability is most unlikely, that a minor change in temperature would have killed every dinosaur on earth, irrespective of the geographical territory or the ecological niche they were living in. A change in environmental temperature might have influenced the speed of metabolism and a shortage of food might have starved and killed many dinosaurs. These events, however, provide no explanation for the extinction of every individual species, large or small, plant-eaters or meat-eaters, adapted to warm or to cold climates, living at sea level or at high altitude, living in Central America, in China or anywhere else on this planet. Moreover, this hypothesis does not explain, why predominantly the robust dinosaurs were affected by this environmental catastrophe, but not the filigree birds. Many bird species, like poultry, parrots, sea-gulls, starlings, hawks and cranes have developed long before the disaster happened 65 million years ago and their species still exist today.

There is a simple explanation for the extinction of the dinosaurs: environmental changes in temperature altered the sex ratio during sexual differentiation. In contrast to birds, which breed their eggs by body contact, dinosaurs left the breeding of their eggs to the environment as reptiles and amphibians do. We know that sexual differentiation of many reptiles and amphibians is determined by the environmental temperature at which the eggs are bred. The alteration in environmental temperature 65 million years ago lasted long enough to cover one generation of reproductive cycles. The eggs of the dinosaurs were incubated at reduced temperature, generating offspring of only one sex. It only took one generation of reproductive cycles and all living dinosaurs in this world belonged to the same sex. Since one sex by itself is unable to reproduce, the extinction of the dinosaurs became irrevocable. Thus, an environmental catastrophe may cause the extinction of a species without necessarily forcing a single individual to die of unnatural causes.

Sexual Differentiation: Hormonal Influences

In mammals sexual differentiation is independent of environmental temperature. Differentiation of the mammalian gonads is under chromosomal control and differentiation of other sexual structures, like the reproductive tract, the external genitalia and the brain, is controlled by an imprinting action of hormones during fetal or neonatal life.

The sertoli cells in the mammalian testes produce a locally acting substance, the *Müllerian inhibiting factor*. This factor causes regression of the female embryonic reproductive tract, the Müllerian ducts. Without the priming action of this factor, which is not existing in females, the Müllerian ducts will develop into fallopian tubes, uterus and upper vagina (for review see Döhler and New 1989).

The Leydig cells in the testes produce *testosterone*, a male sex hormone (androgen). Under the priming influence of testosterone the male embryonic reproductive tract, the Wolffian ducts, will develop into vasa deferentia, seminal vesicles and epididymis. Differentiation of the external male sex organs (penis, scrotum and urogenital sinus) occurs under the priming influence of another androgen, *5α-dihydrotestosterone* (DHT). In the target cells this androgen is converted from testosterone by the enzyme 5α-reductase. In cases where DHT is absent during the sensitive developmental phase — this is normally the case in females, but also in males with deficiency in 5α-reductase — the external genitalia will develop into female direction (vulva and lower vagina). Female type of genital development also occurs when the genital tissue is non-responsive to androgens, due to an androgen receptor defect, as it is the case in the syndrome of testicular feminization (for review see Döhler and New 1989).

In several species of fish and amphibians it was also shown that hormones may influence sexual differentiation. When female larvae of the medaka fish *Oryzias latipes* are treated with androgens, they will develop into fully reproductive males. When male larvae of *Oryzias latipes* are treated with estrogens, they will develop into fully reproductive females. Sex reversal was also observed in several amphibian species when the larvae were raised in water, which contained estrogenic or androgenic hormones respectively (for review see Döhler 1986).

Sexual Differentiation of the Brain

The most obvious functional differences between male and female animals are those involved in reproductive physiology and reproductive behavior. The best-studied animal model in this respect is the rat. In the female rat, rising plasma titers of estrogens trigger a cyclic neural stimulus which activates the release of gonadotropin-releasing hormone(s) (GnRH) from the hypothalamus (positive feedback). GnRH, in turn, stimulates the release of luteinizing hormone (LH) and follicle stimulating hormone (FSH) from the pituitary gland. The gonadotropins FSH and LH stimulate follicular maturation in the ovaries and trigger ovulation. In the male rat, rising plasma titers either of estrogens or of androgens are unable to stimulate the release of GnRH.

The neural substrate which controls GnRH release has developed differently in males and females. The neural substrate which controls sexual behavior has also developed along different lines in males and females. Under the influence of estrogens and progesterone adult female rats will respond to the mounting attempts of a sexually active male by an arching of the back, the so-called lordosis reflex. Adult male rats will hardly show any lordosis behavior, even if given the same hormone treatment. Under the influence of testosterone, adult male rats will show vigorous mounting, intromission and ejaculatory behavior towards a receptive female, whereas female rats will show little or no such responses when so treated with testosterone.

Differentiation of Sexually Dimorphic Brain Functions: In 1936 Pfeiffer presented evidence that there is a critical period during early postnatal development of the rat, during which differentiation of the pattern of anterior pituitary hormone secretion can be influenced permanently by testicular hormone action. Pfeiffer removed the testes of newborn male rats and replaced them with ovaries when the animals were adult. These male animals showed the female capacity to form corpora lutea in the grafted ovarian tissue. Newborn females on the other hand, which were implanted with testes from littermate males, were unable to show estrous cycles or to form corpora lutea in their ovaries when adult.

Present knowledge of hormonal influences on the development of sexually dimorphic brain structures and functions is based on a great number of studies, most of which have been carried out in the last 30 to 40 years. The individual contributions to the field of sexual brain differentiation have been discussed in several excellent reviews (Goy and McEwen 1980; Dörner 1981; Gorski 1987; Döhler 1991). In summary, there is a sensitive developmental period during which sexual differentiation of neural substrates proceeds irreversibly under the influence of gonadal hormones. In the rat, this period starts a few days before birth and ends approximately 10 days after birth.

Female rats, treated during this sensitive period with androgenic or estrogenic hormones, will permanently lose the capacity to ovulate and the capacity to show female lordosis behavior. The permanent loss of female characteristics is termed "*defeminization*". Instead, female rats which are treated postnatally with androgens will develop the capacity to show the complete masculine sexual behavior pattern following administration of testosterone in adulthood. The acquisition of male characteristics is termed "*masculinization*".

If castrated perinatally, *male rats* become unable to display male sexual behavior patterns after treatment with testosterone in adulthood. The permanent loss of male characteristics is termed "*demasculinization*". Instead, perinatally castrated male rats will develop the capacity to show lordosis behavior and — if implanted with ovaries from female littermates — they will show the capacity to ovulate. The acquisition of female characteristics is termed "*feminization*".

Not only androgens and estrogens influence sexual differentiation of brain functions, when applied during the sensitive differentiation period, but estrogen antagonists and androgen antagonists as well. Female rats, treated perinatally with an estrogen antagonist will permanently lose the capacity to ovulate and the capacity to show female lordosis behavior (Döhler et al. 1993). Perinatal treatment of male rats with an androgen antagonist (Neumann and Elger 1966) or with an estrogen antagonist interferes with differentiation of male sexual behavior patterns (Döhler et al. 1993).

These studies indicate that androgens and/or estrogens, whether released by the testes or applied exogenously during the perinatal period, will permanently defeminize and masculinize neural substrates which control sexually dimorphic brain functions. The molecular mechanisms of steroid-induced structural and functional organization of the brain are still unknown. Evidence is accumulating that sex steroids act on the developing brain and promote growth of responsive neurons (Toran-Allerand 1984). Steroids may also influence neurotransmitter metabolism, neuronal conductivity and synaptic connectivity of developing neurons, which may lead to permanent changes in synaptic transmission and overall neuronal activity.

In a series of experiments we treated newborn rats for several days with compounds which stimulated or inhibited the alpha- and beta-adrenergic, the serotoninergic, or the cholinergic system (Jarzab and Döhler 1984; Sickmöller et al. 1985; Jarzab et al. 1987; 1989; 1990). Postnatal application of compounds which stimulate or inhibit adrenergic activity affected the neural control of gonadotropin secretion and differentiation of behavior patterns.

Postnatal stimulation of serotonin synthesis by L-tryptophane inhibited the expression of lordosis behavior in female and in androgenized female rats in adulthood. Postnatal treatment with L-tryptophan also inhibited the expression of male mounting and intromission behavior in androgenized female rats after substitution with testosterone propionate in adulthood.

Cholinergic stimulation postnatally increased the capacity for male sexual behavior in male rats, inhibited differentiation of lordosis behavior in female rats, but attenuated the defeminizing activity of postnatal treatment with testosterone on differentiation of lordosis behavior. Whereas stimulation of muscarinic and nicotinic receptors postnatally had no influence on the LH-release response in adulthood inhibition of nicotinic receptors by mecamylamine postnatally resulted in permanent hypersensitivity of the LH-surge mechanism in adulthood, which may be an indication for a longer than normal reproductive period of cyclicity.

Differentiation of Sexually Dimorphic Brain Structures: The first discovery of a gross sexual dimorphism of brain structure was made by Nottebohm and Arnold (1976) on two species of song birds. During a reinvestigation of the male and female rat brain Gorski et al. (1978) observed a striking sexual dimorphism in gross morphology of the medial preoptic area (Fig. 1). The volume of an intensely staining area, called the sexually dimorphic nucleus of the preoptic area (SDN-POA), is several times larger in adult male rats than in females.

Analogous gross sexually dimorphic structures have subsequently been identified in a variety of other species such as the gerbil, guinea pig, ferret, quail, and also in the human (for review see Döhler 1991).

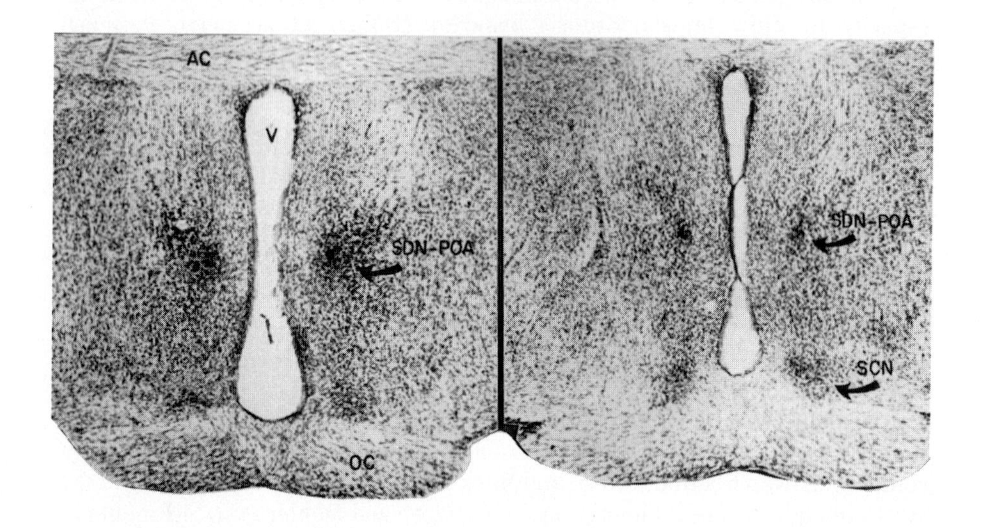

Fig. 1: Representative coronal sections through the sexually dimorphic nucleus of the preoptic area (SDN-POA) in a normal adult male rat (left) and in a normal adult female rat (right). AC, anterior commissure; OC, optic chiasma; SCN, suprachiasmatic nucleus; V, third ventricle

A series of studies has been performed in order to test the perinatal influence of hormones and neurotransmitters on development and differentiation of the SDN-POA. Neonatal castration of male rats (Gorski et al. 1978) reduced the volume of the SDN-POA permanently (Fig. 2). Reimplantation of a testis or treatment with a single injection of testosterone propionate (TP) one day or up to 4 days after neonatal castration restored SDN-POA volume in male rats to normal (Fig. 2). Treatment of female rats with a single injection of TP postnatally (Gorski et al. 1978) increased SDN-POA volume significantly; however, the volume of the SDN-POA in these animals was still significantly smaller than that of normal male rats (Fig. 2). Only the extended pre- and postnatal treatment of female rats with TP (Dšhler et al. 1984) resulted in SDN-POA differentiation equivalent to that of normal males (Fig. 2). The treatment of male rats pre- and postnatally with TP did not increase the size of their SDN-POA above normal (Fig. 2).

Although pre- and postnatal treatment of rats with TP was shown to substitute fully for testicular activities in stimulating SDN-POA development, the prime candidates for the control of SDN-POA differentiation do not seem to be androgens as such, but rather estrogens. This conclusion is supported by several observations:

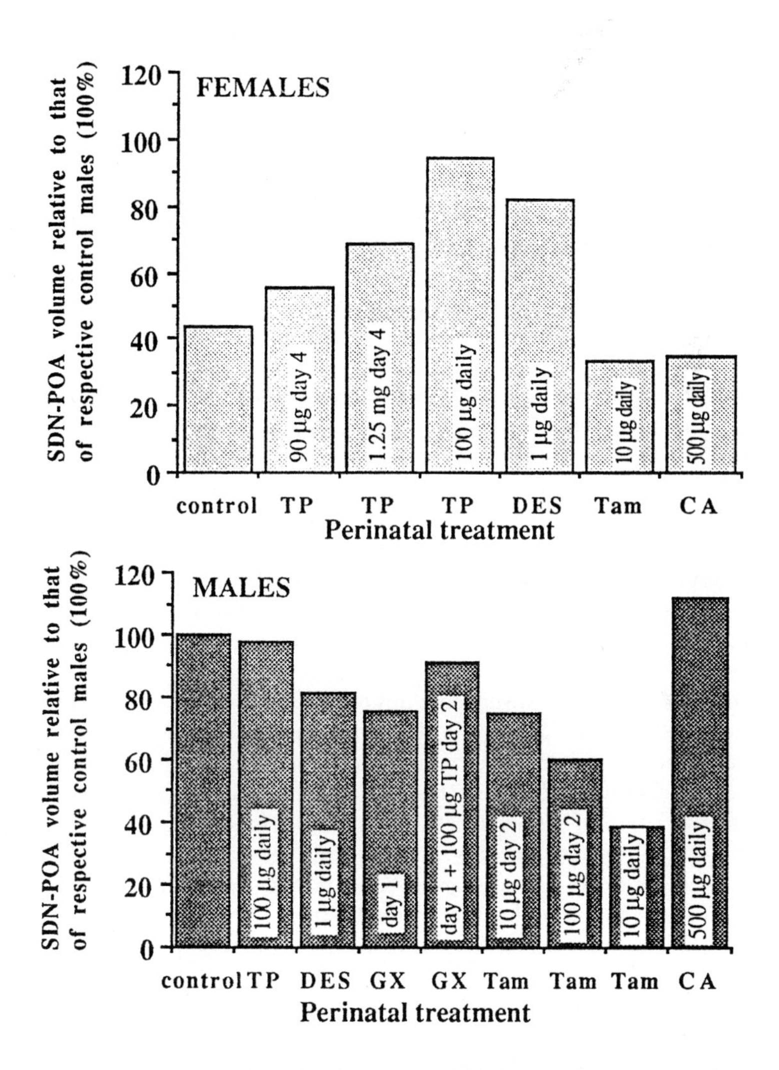

Fig. 2: Perinatal hormonal influence on differentiation of the sexually dimorphic nucleus of the preoptic area (SDN-POA) in female (top) and male (bottom) rats. Female rats received either 90 µg or 1.25 mg testosterone propionate (TP) on day 4 after birth, or daily treatment with TP, diethylstilbestrol (DES), the estrogen antagonist tamoxifen (Tam), or the androgen antagonist cyproterone acetate (CA) from day 16 of fetal life until day 10 after birth. Two groups of male rats were gonadectomized (GX) on the day of birth, one group received 100 µg TP one day after GX. Two groups of males received either 10 µg or 100 µg Tam on day 2 after birth. Four groups of male rats received daily treatment with TP, DES, Tam, or CA respectively from day 16 of fetal life until day 10 after birth. SDN-POA volume is indicated in percent as compared to SDN-POA volume of normal adult male rats (100 percent). (Data from Döhler et al, 1984; 1986; Gorski et al, 1978; Jacobson et al., 1981)

a. Female rats, which had been treated pre- and postnatally with the synthetic estrogen diethylstilbestrol (Döhler et al. 1984), developed a significantly enlarged SDN-POA which was similar in volume to that of control males (Fig. 2). This observation indicates that estrogens can stimulate SDN-POA development directly.

b. Male rats, treated pre- and postnatally with the androgen antagonist cyproterone acetate (Döhler et al. 1986), developed female genitalia, but the volume of their SDN-POA was not reduced (Fig. 2).

c. Male rats, treated pre- and postnatally with the estrogen antagonist tamoxifen (Döhler et al. 1986), developed male genitalia, but the volume of their SDN-POA was significantly reduced and was similar to that of control female rats (Fig. 2).

In the adult organism tamoxifen is known to bind to intracellular estrogen receptors and to prevent estrogen uptake, as it inhibits cytosol receptor replenishment (Jordan et al. 1977). Tamoxifen may act similarly in the developing organism. After aromatization of testicular androgens into estrogens, tamoxifen may interfere with estrogen uptake into cell nuclei of the SDN-POA by occupying intracellular estrogen receptors. The inhibitory effect of pre- and postnatal tamoxifen on growth and differentiation of the SDN-POA in male rats indicates that structural differentiation of the male rat brain may be dependent on aromatization of testicular androgens into estrogens and the subsequent interaction of these estrogens with the nuclear material.

The SDN-POA is sexually dimorphic not only in terms of its volume but also in terms of neurochemicals present in the cell-bodies of neurons comprising this nucleus and in the fibers innervating the nucleus and its vicinity. These differences are established during the critical pre- and postnatal period when sexual differentiation of the SDN-POA takes place. Thus, it seems that during the perinatal period gonadal steroids act not only as differentiation signals for morphological and functional parameters of the brain, but also influence neurotransmitter activity in the developing brain.

The mechanisms of steroid-neurotransmitter interaction in the developing brain are not well investigated. Testosterone has been described to influence neurotransmitter and neuropeptide content in the brain (for review see Döhler, 1991). Compounds which change neurotransmitter activity have been reported to influence postnatal sexual differentiation of the brain and to interfere with the effects of steroids during this process (for review see Döhler, 1991).

In a series of studies we (Jarzab and Döhler 1984; Jarzab et al. 1987; 1989; 1990) demonstrated that postnatal alteration of serotoninergic or adrenergic neurotransmission has profound effects on development and differentiation of the SDN-POA. Both, postnatal stimulation and inhibition of serotonin synthesis, provided a stimulus for SDN-POA morphogenesis in female rats. Postnatal treatment with TP potentiated the stimulatory effect of the serotonin precursor L-tryptophan (Jarzab and Döhler, 1984).

Adrenergic effects on differentiation of the SDN-POA are mediated mainly by alpha$_2$- and beta$_2$-adrenergic receptors (Jarzab et al. 1990). The alpha$_2$-receptor agonist clonidine was shown to augment the stimulatory effect of TP on SDN-

POA differentiation in female rats and the beta2-receptor agonist salbutamol causes the volume of the SDN-POA to increase in both female and male rats. The observed effects are independent from postnatal levels of circulating testosterone, as judged in male rats on day 3 of life (Jarzab et al. 1990).

Summary and Conclusion

In summary, a number of studies have shown that not only estrogenic and androgenic steroids and their antagonists influence sexual differentiation of the mammalian brain but also drugs which stimulate or inhibit the adrenergic, the serotoninergic, or the cholinergic system in the developing brain.

The present knowledge on the possible participation of neurotransmitter systems in sexual differentiation of the brain and their mode of interaction in this process perinatally with gonadal steroids is still rather limited. Sexual differentiation of the central nervous system is a complex integrated process, which relies on proper chronological and quantitative interactions of various endocrine and neuroendocrine mediators. Any disturbance of this delicate endogenous hormonal balance during ontogenetic development, e.g. by means of environmental influences, can result in permanent manifestation of anatomic and functional sexual deviations.

A large number of man-made chemicals that have been released into the environment have the potential to disrupt the endocrine system of animals and humans. They do so because they mimick the effects of natural hormones or neurotransmitters by recognizing their binding sites, or they antagonize the effects of endogenous hormones or neurotransmitters by blocking their interaction with their physiological binding sites. Interaction of environmental endocrine disruptors with animals or humans during ontogeny may have deleterious effects on the differentiation of reproductive structures and functions, rendering the individuals in question permanently incapable to reproduce and, thus, endangering survival of the species.

References

Aristotle (1862) In: Cresswell R (translated) History of animals, Book VI, Chapter XIX/2, HG Bohn, London, pp 165

Colborn T, Clement C (1992) Chemically-induced alterations in sexual and functional development: the wildlife/human connection, Princeton Scientific Publishing Co, Princeton, New Jersey

Colborn T, Dumanoski D, Myers JP (1996) Our stolen future, Dutton/Penguin, New York

Dōhler KD (1986) The special case of hormonal imprinting, the neonatal influences of sex. Experientia 42: 759-769

Dōhler KD (1991) The pre- and postnatal influence of hormones and neurotransmitters on sexual differentiation of the mammalian hypothalamus. Int Rev Cytology 131: 1-57

Döhler KD, Coquelin A, Davis F, Hines M, Shryne JE, Gorski, RA (1984) Pre- and postnatal influence of testosterone propionate and diethylstilbestrol on differentiation of the sexually dimorphic nucleus of the preoptic area in male and female rats. Brain Research 302: 291-295

Döhler KD, Coquelin A, Davis F, Hines M, Shryne JE, Sickmšller PM, Jarzab B, Gorski RA (1986) Pre- and postnatal influence of an estrogen antagonist and an androgen antagonist on differentiation of the sexually dimorphic nucleus of the preoptic area in male and female rats. Neuroendocrinology 42: 443-448

Döhler KD, Ganzemüller C, Veit C (1993) The development of sex differences and similarities in brain anatomy, physiology and behavior is under complex hormonal control. In: Haug M, Whalen RE, Aron C, Olsen KL (eds) The Development of sex differences and similarities in behavior. NATO ASI Series, Kluwer Academic Publ, Dordrecht/Boston/London, pp 341-361

Döhler KD, New MI (1989) Sexualentwicklung. In: Hesch RD (ed) Endokrino-logie, Urban & Schwarzenberg, München-Wien-Baltimore, pp 501-512

Dörner G (1981) Sexual differentiation of the brain. Vitam Horm 38: 325-381

Gorski RA (1987) Sex differences in the rodent brain: their nature and origin. In: Reinisch JM, Rosenblum LA, Sanders SA (eds) Masculinity/Feminity, basic perspectives, Oxford University Press, New York, pp 37-67

Gorski RA, Gordon JH, Shryne JE, Southam AM (1978) Evidence for a morphological sex difference within the medial preoptic area of the rat brain. Brain Research 148: 333-346

Goy RW, McEwen, BS (1980) Sexual differentiation of the brain, The MIT Press, Cambridge, Massachusetts

Jacobson CD, Csernus YJ, Shryne JE, Gorski RA (1981) The influence of gona-dectomy, androgen exposure or a gonadal graft in the neonatal rat on the volume of the sexually dimorphic nucleus of the preoptic area. J Neurosci 10: 1142-1147

Jarzab B, Döhler KD (1984) Serotoninergic influences on sexual differentiation of the rat brain. In: De Vries GJ, De Bruin JPC, Uylings HBM, Corner MA (eds) Progress in brain research, Elsevier Science Publ, Amsterdam, pp 119-126

Jarzab B, Sickmöller PM, Geerlings H, Dähler KD (1987) Postnatal treatment of rats with adrenergic receptor agonists or antagonists influences differentiation of sexual behavior. Horm Behav 21: 478-492

Jarzab B, Gubala E, Achtelik W, Lindner G, Pogorzelska E, Döhler KD (1989) Postnatal treatment of rats with beta-adrenergic agonists or antagonists influences differentiation of sexual brain functions. Exp Clin Endocrinol 94: 61-72

Jarzab B, Kaminski M, Gubala E, Achtelik W, Wagiel J, Döhler KD (1990) Postnatal treatment of rats with the $beta_2$-adrenergic agonist salbutamol influences the volume of the sexually dimorphic nucleus in the preoptic area. Brain Research 516: 257-262

Jordan VC, Dix CJ, Rowsby L, Prestwich G (1977) Studies on the mechanism of action of the nonsteroidal antiestrogen tamoxifen in the rat. Mol Cell Endocrinol 7: 177-192

Neumann F, Elger W (1966) Permanent changes in gonadal function and sexual behavior as a result of early feminization of male rats by treatment with an antiandrogenic steroid. Endokrinologie 50: 209-224

Nottebohm F, Arnold AP (1976) Sexual dimorphism in vocal control areas of the songbird brain. Science 194: 211-213

Pfeiffer CA (1936) Sexual differences of the hypophysis and their determination by the gonads. Am J Anat 58: 195-226

Pieau C (1975) Temperature and sex differentiation in embryos of two chalonians, *Emys orbicularis* L and *Testudo graeca* L. In: Reinboth R (ed) Intersexuality in the animal kingdom, Springer, New York, pp 332-339

Piquet J (1930) Détermination de sexe chez les batraciens en fonction de la température. Rev Suisse Zool 37: 173-281

Plato (1953) Symposium. In: Jowett B (translated) The dialogues of Plato, Vol I, 4th ed, 189d-190a, Clarendon Press, Oxford, pp 521

Short RV (1982) Sex determination and differentiation. In: Austin CR, Short RV (eds) Reproduction in animals, Vol 2, Cambridge Univerity Press, Cambridge, pp 70-113

Sickmöller PM, Jarzab B, Döhler KD (1985) Cholinergic influence on sexual differentiation of the LH release mechanism in rats. Acta Endocrinol 108, Suppl 267: 110-111

Stolkovski J, Bellec A (1960) Influence du rapport potassium/calcium du milieu d'élevage sur la distribution des sexes chez *Discoglosus pictur* (Otth). CR Acad Sci Paris 251: 1669-1671

Toran-Allerand CD (1984) On the genesis of sexual differentiation of the central nervous system: Morphogenetic consequences of steroidal exposure and possible role of alpha-fetoprotein. In: De Vries GJ, De Bruin JPC, Uylings HBM, Corner MA (eds) Progress in brain research, Elsevier Science Publ, Amsterdam, pp 63-98

Reproductive Effects from Oestrogen Activity in Polluted Water

John P. Sumpter
Department of Biology and Biochemistry, Brunel University, Uxbridge, Middlesex UB8 3PH, U.K.

Introduction

A new area of ecotoxicology, now usually called endocrine disruption, has arisen in the last few years. However, despite the present topicality of this issue, some of the best documented examples of endocrine disruption were reported a decade or more ago (see, for example, Fry, 1995). The issue is concerned with the effects of chemicals that mimic endogenous hormones on the physiology of exposed wildlife and humans. As many of these xenohormones mimic steroid hormones, especially oestrogens and androgens, most of the reported effects have involved effects on the reproductive system of exposed organisms. Many of these reported effects on wildlife have concerned aquatic, rather than terrestrial, organisms (even the well-documented effects on birds are primarily concerned with water birds, which feed predominantly on fish); the reproductive abnormalities seen in alligators living in some lakes in Florida (Guillette et al, 1994), and the oestrogenic effects on fish reported in British rivers (this example is discussed in detail below) provide good examples of the type of effects observed in aquatic organisms. This predominance of effects on aquatic organisms could reflect an unconscious bias of the interests of research scientists (are there more wildlife biologists interested in aquatic, rather than terrestrial, animals?), but is perhaps more likely a consequence of the fact that the aquatic environment is the ultimate "sink" for the intentional or unintentional disposal of much waste. Thus, this brief review is focused exclusively on the aquatic environment, and particularly on the effects on fish. However, I have attempted to emphasize the general nature of the phenomena illustrated by studies on endocrine disruption in fish, because they apply to most, if not all, examples of endocrine disruption in all wildlife.

Chemicals Present in the Aquatic Environment

An obvious, but very important, point to make is that an extremely large number of both man-made and natural chemicals are present in the aquatic environment. The realisation that around 70,000 man-made chemicals are in regular use, and another 2000 new ones are introduced each year, serves to illustrate this point. Many of these chemicals will undergo degradation, especially biodegradation, in the aquatic environment, leading to many more, often unidentified, chemicals. Further, many man-made chemicals are not homogeneous, but instead consist of a variety of isomers and oligomers; thus, what is often considered a single chemical is actually very many (and, as far as biological effects are concerned, these apparently subtle structural differences are extremely important in determining the type of activity and potency). A single example, nonylphenol, will serve to demonstrate these points. Nonylphenol is a fairly ubiquitous aquatic pollutant (see, for example, Naylor et al, 1992; Blackburn and Waldock, 1995), which originates primarily from the biodegradation of one group of non-ionic surfactants, the alkylphenol polyethoxylates, of which nonylphenol polyethoxylate is by far the major one. Nonylphenol polyethoxylate degrades in the aquatic environment via shortening of the ethoxylate chain, leading to a variety of short-chain ethoxylates, which can then be carboxylated, producing nonylphenol carboxylates. All of these intermediates can degrade to nonylphenol. Thus, a single chemical (actually usually a mixture of ethoxylates of somewhat different chain length) produces a fairly large number of different degradation products, all of which will be present at any one time. Further, the nonylphenol used in the manufacture of nonylphenol polyethoxylates is itself not homogeneous, but a mixture of many isomers (Wheeler et al, 1997) and thus each degradation product (for example, NP1EO) will exist in many isomeric forms. All this would be unimportant if all chemicals within a "family" had similar biological activities, but this is not so. Instead, quite the reverse is true, and apparently small differences in structure can have profound biological consequences (see, for example, Kelce et al, 1995 and Routledge and Sumpter, 1997).

It is equally important to realize that very many natural chemicals will be present in the aquatic environment. Much less emphasis seems to have been placed on these chemicals than on the man-made ones, presumably because they have always been present, and hence aquatic organisms are likely to have adapted to their presence, whereas man-made chemicals are a very recent addition to the aquatic environment. Nevertheless, they should not be overlooked when attempting to identify the cause of a particular biological phenomenon.

This very wide variety of chemicals can enter the aquatic environment in many ways. Some, particularly many natural ones, will originate from the decomposition of decaying animal and plant material. Man-made chemicals enter from a variety of sources, including effluents from sewage-treatment works (STWs), leachate from landfill sites, agricultural run-off (especially pesticides and herbicides), road run-off, and direct industrial effluents (e.g. pulp and paper mill effluent). Some of these inputs can be very significant; for example, effluent from STWs often contributes 50% of the flow of lowland rivers in England (and many other European countries), a figure that can rise to over 90% during periods of low rainfall.

Each of these various sources of chemical input into the aquatic environment is very variable in composition. For example, the composition of industrial inputs will depend upon which industries contributed to the input. Likewise,

STW effluent varies in composition depending upon both the inputs into the treatment works (which is usually of both domestic and industrial origin) and the type and degree of treatment that the material receives before it is released as effluent.

The general message is that a very large number of both natural and man-made chemicals is present in many aquatic environments. No two situations are alike; a river in lowland England will be receiving primarily domestic effluent, but a lot of it, whereas a river in northern Europe or Canada might be receiving primarily pulp and paper mill effluent - these two effluents being very different in composition. Thus, it should come as no surprise that quite different biological phenomena have been reported in different aquatic situations.

Endocrine Activity of Chemicals Present in the Aquatic Environment

An increasing number of chemicals known to be present in the aquatic environment are now known to have endocrine activity; that is, they can mimic the activities of endogenous hormones (i.e. act as agonists), or they can antagonize endogenous hormones. It is important to realize, however, that in nearly all cases this endocrine activity of chemicals has usually been demonstrated only in the laboratory (rather than the "real" world), and then often only in *in vitro* , rather than *in vivo*, assays. Thus, although it is now fairly well established that some individual chemicals, or groups of chemicals, possess endocrine activity, it has very rarely been demonstrated that these chemicals are present in the aquatic environment at concentrations high enough to cause the effects which can be demonstrated in the laboratory under very controlled (and possibly environmentally unrealistic) conditions. This is an extremely important point, to which I will return later in this paper.

There are three structurally-distinct groups of hormones: the protein/peptide hormones, the amines, and the steroid hormones. To date, all the environmental chemicals reported to possess endocrine activity have mimicked hormones within the steroid group; no chemicals have yet been reported which mimic protein/ peptide or amine hormones. Within the steroid hormone group, most of the focus has centred on chemicals which possess oestrogenic activity (so-called xenoestrogens), but chemicals with androgenic, anti-androgenic, and progestogenic activity have been reported (see, for example, Kelce et al, 1995; Tran et al, 1996). It is also likely that environmental chemicals capable of acting as thyroid hormone agonists and/or antagonists exist. The xenoestrogens have attracted the most attention, partly because many of the best-documented effects in wildlife appear to be due to "feminization" of the males, and partly because more, and better validated, laboratory techniques are available to investigate oestrogens than are available to study androgens, progesterones, or thyroid hormones. However, a number of anti-androgenic environmental chemicals have been identified which, by de-masculinizing exposed organisms, can produce effects very similar to those observed in "feminized" organisms.

Wildlife is very rarely (if ever) exposed to single chemicals, but instead is usually exposed to complex mixtures. These mixtures might contain many chemicals with different endocrine-disrupting activities, which makes interpretation of the observed biological effects very difficult. For example, if a fish has an ovo-testis, is it a male that has been partially "feminized", or a female that has been "masculinized"? Further, if it is a genetic male that has been partially "feminized", (and this was due to chemicals in the environment), was it "feminized" by exposure to oestrogenic chemicals, or "de-masculinized" by exposure to anti-androgenic chemicals?

The number of environmental chemicals shown to possess endocrine activity has increased rapidly in the last few years, as interest in the field of endocrine disruption has grown. Groups of chemicals demonstrated to possess endocrine-disrupting activity include the organochlorine pesticides (e.g. DDT and its metabolites, methoxychlor, lindane and kepone), polychlorinated biphenyls (PCBs: arochlor is a good example), dioxins (e.g. TCDD), alkylphenolic chemicals (e.g. nonylphenol), biphenolic chemicals (e.g. Bisphenol-A), some phthalates, some fungicides (e.g. Vinclozolin), and some organotin compounds (e.g. tributyl tin, or TBT). Besides these hormone (or anti-hormone) mimics, both natural (e.g. oestradiol, oestrone) and synthetic (e.g. ethinyl oestradiol) oestrogens have been reported to be present in the aquatic environment. Other natural oestrogens, derived from both plants (so-called phytoestrogens) and fungi (mycoestrogens) will be present, particularly in some effluents, such as those from pulp and paper mills. Natural sterols that have masculinizing effects are also likely to be present; they may well be responsible for the masculinized fish reported in creeks downstream of paper mills discharging bleached kraft-mill effluent (this work is summarized in Bortone and Davies, 1994). The existing data on the types of endocrine activities, and potencies, of these chemicals, which has primarily been derived from *in vitro* laboratory studies, suggest that most of those that mimic or antagonize the actions of steroid hormones are only weakly active. Their potencies are usually three or more orders of magnitude less than those of the natural hormones; for example, nonylphenol is about one ten-thousandth of the potency of 17β-oestradiol, and butyl benzyl phthalate is one millionth of the potency of 17β-oestradiol. Generally, therefore, most endocrine mimics have the potential to cause endocrine disruption in wildlife only if they circulate in the environment at high concentrations, or they are very persistent, or they bioaccumulate, or because they are constantly entering the aquatic environment. We still know very little about the levels of exposure of aquatic organisms to many of these endocrine-disrupting chemicals, and hence it is very difficult to extrapolate from effects observed in laboratory experiments, where the exposure is defined, to the field.

Oestrogenic Effects of Sewage Treatment Works Effluents

Despite the finding of hermaphrodite fish in the settlement lagoons of two STWs in the U.K. nearly 20 years ago, only recently was it demonstrated that effluent from STWs was oestrogenic to fish. The hypothesis that treated domestic and/or industrial sewage effluents may have endocrine-disrupting effects, and more specifically might contain oestrogen-mimicking chemicals at high enough concentrations to elicit biological effects, was tested by Purdom et al (1994). They placed caged adult rainbow trout (*Oncorhynchus mykiss*) in undiluted STW effluent at various sites in England, and measured induction of vitellogenin (VTG) after 3 weeks' exposure. VTG is the main yolk precursor in oviparous vertebrates; it is synthesized in the liver under the control of oestrogens, primarily oestradiol, originating in the ovary, secreted into the blood, and sequestered by the growing oocytes (Tyler and Sumpter, 1996). It can also be induced by exogenous oestrogen, or oestrogen mimics; thus, measurement of plasma VTG concentrations provides a very sensitive and specific indicator of oestrogen exposure (Sumpter and Jobling, 1995). Purdom et al (1994) reported that all effluents tested were strongly oestrogenic to caged rainbow trout and also to caged carp (*Cyprinus carpio*), although limited results were reported for the latter species.

As STW effluent can contribute very significantly to the flow of many lowland rivers in Britain and other countries (especially densely populated ones), it was possible that receiving waters downstream of the entry of effluents might also show oestrogenic activity. An initial investigation on the River Lea, near London, which receives effluent from five large STWs, was conducted by placing caged fish along the entire length of the river. Highest responses were observed immediately downstream of each STW discharge, and these then diminished fairly rapidly with distance downstream, although significant effects were seen up to 4.5 km downstream of discharges (Harries et al, 1996). Essentially similar results, demonstrating elevations in plasma vitellogenin concentrations in male carp (*Cyprinus carpio*) residing in the vicinity of a STW in the United States, accompanied by reduced serum testosterone concentrations, have since been reported (Folmar et al, 1996).

Further field investigations, similar in design to those conducted on the River Lea, were carried out on five other English rivers. Although the vitellogenin responses were of variable magnitude (probably because the oestrogenic "strengths" of the STW effluents varied), in four out of five cases the effluent, and a variable length stretch downstream of where the effluent entered, were oestrogenic (Harries et al, 1997). Very pronounced responses were seen on one river, which receives effluent from wool-scouring mills as well as domestic effluent. At all stations on a 5 km stretch of the river below the disharge, plasma vitellogenin concentrations were essentially maximal (Harries et al, 1997). Simultaneous measurements of liver size and testis size showed that liver weights were elevated with respect to control over the entire stretch of river surveyed, and testis weights were all significantly depressed.

When these field studies were conducted, the causative agents in the STW effluents were unknown. Informed guesswork had suggested that the synthetic oestrogen 17β-ethinyloestradiol, which is a component of most contraceptive pills, was a likely candidate (Purdom et al, 1994), although other possibilities, such as natural oestrogens and alkylphenolic chemicals (Purdom et al, 1994;

Harries et al, 1997) were also considered. Due to the fact that STW effluents are extremely complex mixtures of chemicals (see above), it was decided to use a toxicity-based fractionation approach to try and identify the oestrogenic chemical, or chemicals in STW effluent. This approach utilizes chemical fractionation of the effluent, combined with a bioassay for "oestrogens" to direct the chemical analysis. Thus, at each stage of fractionation, the aliquots generated are assessed for oestrogenic activity, and only those possessing activity are subjected to further fractionation. The aim is to purify the chemicals possessing oestrogenic activity to such an extent that they can be identified by procedures such as GC-MS. Such an approach enabled us (unpublished observations) to show that the majority (>80%) of the oestrogenic activity in the five different STW effluents analyzed was contributed by the natural oestrogens oestradiol, oestrone and, occasionally, the synthetic oestrogen ethinyl oestradiol. These oestrogens were present at concentrations in the tens of nanograms per litre range. Subsequent *in vivo* bioassays, in which trout and roach (a native cyprinid fish) were exposed to these steroids, at the concentrations found in effluents, showed that such concentrations were indeed strongly oestrogenic to fish (unpublished observations). Thus, it appears that "real" oestrogens, rather than xenoestrogens, provide much of the oestrogenic activity of many typical STW effluents in the U.K. It is likely, however, that in particular situations, industrial chemicals do contribute significantly to the oestrogenic activity of some STW effluents (see earlier discussion on the caged fish studies, and Harries et al , 1997).

Conclusions

Endocrine disruption is a new and exciting area of research, which embraces a very wide range of scientific disciplines (including biology, biochemistry and chemistry). It also involves a wide range of scientists, who approach the issue not only from an academic perspective, but also from regulatory and industrial perspectives. Presently, despite the very high profile of this field of research, I consider it impossible to judge how important an issue endocrine disruption is. There seems little doubt that in localized areas (such as some polluted lakes in North America, or some lowland rivers in the UK receiving high proportions of STW effluent), both man-made and natural chemicals are causing adverse effects on a range of animals. What is much less clear presently is how general these adverse effects are; for example, are a high proportion of fish in many UK rivers affected, or are some populations living immediately downstream of large STW discharges affected? It is also unclear how adverse the reported effects are, or if, indeed, they are adverse. To date, no study has translated the reported effects on individuals to the population level, to discover whether adverse population level effects occur. Only considerable further research will answer this most important of questions.

Acknowledgements

Nearly all of the research on U.K. effluents was conducted by colleagues and collaborators. I am therefore especially grateful to Drs G Brighty, C Desbrow, JE Harries, S Jobling, P Matthiessen, EJ Routledge, CR Tyler, DA Sheahan and M Waldock, who between them carried out all the laboratory studies and fieldwork on which this review is based. I also thank the Natural Environment Research Council (NERC), the Environment Agency, and the Department of the Environment for funding and continually supporting our studies.

References

Blackburn MA, Waldock MJ (1995) Concentrations of alkylphenols in rivers and estuaries in England and Wales. Water Res 29:1623-1629

Bortone SA, Davis WP (1994) Fish intersexuality as an indicator of environmental stress. Bioscience 44:165-172

Folmar LC, Denslow ND, Rao V, Chow M, Crain DA, Emblom J, Marcino J, Guillette LJ (1996) Vitellogenin induction and reduced serum testosterone concentrations in feral male carp (*Cyprinus carpio*) captured near a major metropolitan sewage plant. Environ Health Perspect 104:1096-1101

Fry DM (1995) Reproductive effects in birds exposed to pesticides and industrial chemicals. Environ Health Perspect 103(Suppl 7):165-171

Guillette LJ, Gross TS, Masson GR, Matter JM, Percival HF, Woodward AR (1994) Developmental abnormalities of the gonad and abnormal sex hormone concentrations in juvenile alligators from contaminated and control lakes in Forida. Environ Health Perspect 102:680-688

Harries JE, Sheahan DA, Jobling S, Matthiessen P, Neall P, Routledge E, Rycroft R, Sumpter JP, Tylor T (1996) A survey of estrogenic activity in United Kingdom inland waters. Environ Toxicol Chem 15:1993-2002

Harries JE, Sheahan DA, Jobling S, Matthiessen P, Neall P, Sumpter JP, Tylor T, Zaman N (1997) Estrogenic activity in five United Kingdom rivers detected by measurement of vitellogenesis in caged male trout. Environ Toxicol Chem 16.3:534-542

Kelce WR, Stone CR, Laws SC, Gray LE, Kemppainen JA, Wilson EM (1995) Persistent DDT metabolite p,p'-DDE is a potent androgen receptor antagonist.Nature 375:581-585

Naylor CG, Mieure JP, Adams WJ, Weeks JA, Castaldi FJ, Ogle LD, Romano RR (1992) Alkylphenol ethoxylates in the environment. JOACS 69:695-703

Purdom CE, Hardiman PA, Bye VJ, Eno NC, Tyler CR, Sumpter JP (1994) Estrogenic effects of effluents from sewage treatment works. Chemistry & Ecology 8:275-285

Routledge EJ. Sumpter JP (1997) Structural features of alkylphenolic chemicals associated with estrogenic activity. J Biol Chem 272.6:3280-3288

Sumpter JP, Jobling S (1995) Vitellogenesis as a biomarker for estrogenic contamination of the aquatic environment. Environ Health Perspect 103(Suppl 7):173-178

Tran DQ, Klotz DM, Ladlie BL, Ide CF, McLachlan JA, Arnold SF. (1996) Inhibition of progesterone receptor activity in yeast by synthetic chemicals. Biochem Biophys Res Comm 229:518-523

Tyler CR, Sumpter JP (1996) Oocyte growth and development in teleosts. Rev Fish Biol 6:287-318

Wheeler TF, Heim JR, LaTorre MR, Janes AB (1997) Mass spectral characterization of p-nonylphenol isomers using high-resolution capillary GC-MS. J Chromatographic Sci 35,l:19-30

Effects of Chemical-Induced DNA Damage on Male Germ Cells

J.A. Holme, C. Bjørge, M. Trbojevic, A.-K. Olsen, G. Brunborg, E.J. Søderlund, M. Bjørås[1], E. Seeberg[1], T. Scholz[2], E. Dybing and R. Wiger
Department of Environmental Medicine, National Institute of Public Health, P. O. Box 4404 Torshov, N-0403 Oslo, Norway,
[1] Department of Microbiology, The National Hospital, Oslo, Norway and
[2] Institute for Surgical Research and Surgical department B, The National Hospital, Oslo, Norway

Introduction

Several recent studies indicate declines in sperm production, as well as increases in the incidence of genitourinary abnormalities such as testicular cancer, cryptorchidism and hypospadias (Toppari et al., 1996). It is not known if these effects are due to exposure to chemical pollutants or if other ethiological factors are involved. Animal studies indicate that chemicals will induce such effects by various genetic, epigenetic or non-genetic mechanisms. Recently, much attention has been focused on embryonic/fetal exposure to oestrogen-mimicking chemicals (Toppari et al., 1996). However, the possibility that chemicals may cause reproductive toxicity by other mechanisms such as interactions with DNA, should not be ignored. DNA damage in germ cells may lead to the production of mutated spermatozoa, which in turn may result in spontaneous abortions, malformations and/or genetic defects in the offspring. Regarding the consequences of DNA alterations for carcinogenesis it is possible that genetic damage may occur germ cells, but the consequences are not expressed until certain genetic events occur in postnatal life. Transmission of genetic risk is best demonstrated by cancer-prone disorders such as hereditary retinoblastoma and the Li-Fraumeni syndrome. A number of experiments indicate that germ cells and proliferating cells may be particularly sensitive to DNA damaging agents compared to other cells (Masters et al., 1993; Holme et al., 1997). Furthermore, several lines of evidence have indicated that one of the best documented male reproductive toxicants, 1,2-dibromo-3-chloropropane (DBCP), causes testicular toxicity through DNA damage (Dybing et al., 1989). It is possible that testicular cells at certain maturational stages are more subject to DNA damage, have less efficient DNA repair, or have different thresholds for initiating apoptosis following DNA damage than other cell types.

To further explore possible roles of DNA damage in DBCP toxicity, and to characterise DNA damage and repair in the testes, we have used cells and seminiferous tubules isolated from human as well as animal testis. We here report on the sensitivity of testicular cells from human and rat, and cells from different stages of spermatogenesis, with respect to chemically-induced DNA damage and their DNA repair capacity. Using a transillumination technique on seminiferous tubules, we have isolated defined spermatogenic stages in the rat and studied the effect of chemicals on DNA synthesis, meiosis and spermiogenesis.

Materials and Methods

Testes were obtained from sexually mature male Wistar rats (MOL: WIST) or from human organ transplant donors, and isolation of cells were performed by enzymatic digestion of tissue, essentially as described by Bradley and Dysart (1985) with some modifications (Søderlund et al., 1988). Germ cells stained with Hoechst 33258 (Bjørge et al., 1995) or with anti-vimentin (Hittmaier et al., 1994) were characterised by flow cytometry and microscopic analysis. Segments of seminiferous tubules from defined stages of the epithelial cycle were isolated by transillumination-assisted microdissection (Kangasniemi et al., 1990). The formation of single strand DNA breaks and alkali-labile sites (ssDNA breaks) were measured by an automated alkaline filter elution system (Brunborg et al., 1988), or by single cell gel electrophoresis (SCGE) as described elsewhere (Bjørge et al., 1995). The activity of DNA repair enzymes were measured in protein extracts (Bjelland et al., 1995). DNA substrates with ^{3}H-labelled methylated bases (Riazuddin and Lindahl, 1978), ^{3}H-labelled formamidopyrimidine residues (FaPy; Boiteux et al., 1984) and synthetic oligonucleotides (Eide et al., 1996) were prepared as described. The enzymatic reactions were carried out according to Eide and coworkers (1996).

Results and Discussion

Composition of testicular cells from rats and humans. Compared to rats, the interindividual variation in the composition of testicular cells is greater in human (Bjørge et al., 1996a). Among the approximately 40 organ donor testes that have been analysed so far, a relatively high inter-individual variation in the cellular composition appears to be a common feature as determined by both flow cytometric and histological analyses. The variation in cell type composition has been observed in all age classes of the donors. However, despite the observed compositional differences among the donors, only minor differences in the induced ss DNA breaks in crude human testicular cell preparations were observed (Bjørge et al., 1996b).

Chemically-induced DNA damage in testicular cells from rats and humans. During repair of DNA damage, ssDNA breaks are often formed, thus most types of DNA damage may be detected by techniques such as alkaline filter elution and SCGE. The recorded level of ssDNA breaks depends on the dynamic balance between the rates of incision and polymerisation/ligation. By adding inhibitors of the latter process such as hydroxyurea and cytosine-1-β-D-arabinofuranoside (Ara-C) it is possible to increase the level of ssDNA breaks.

In a previously published study (Bjørge et al., 1996b), we exposed testicular cells from human organ donors or rats to various categories of chemicals, most of which were direct acting genotoxicants. In addition, a few indirect genotoxicants and non-genotoxic reproductive toxicants were included. Six of the chemicals did not induce significant levels of ssDNA breaks in either human or rat testicular cells: methoxychlor, benomyl, thiotepa, cisplatin, Cd^{2+} and acrylonitrile. Four of these induced significant levels of ssDNA breaks in testicular cells from both species: styrene oxide, 1,2-dibromoethane, thiram and chlordecone. Finally, five chemicals induced ssDNA breaks in only one of the two species. Four chemicals induced significant ssDNA breaks in rat testicular cells only: DBCP, dinitrobenzene, Cr^{6+} and aflatoxin B_1 ; the last two of these produced only a minor positive response. One chemical, acrylamide, induced a marginal increase in ssDNA breaks in human, but not in rat testicular cells. Although based on a limited number of donors, the data indicate a close correlation between the induction of DNA damage in human and rat testicular cells *in vitro*. For some chemicals, however, there appear to be differences in the susceptibility to chemically-induced ssDNA breaks in isolated testicular cells from the two species. The parallel use of human and rat testicular cells provides a valuable tool in the assessment of human testicular toxicity.

For DBCP the species difference was marked. DBCP induced significant levels of ssDNA breaks in rat testicular cells, which is in accordance with our earlier studies (Søderlund et al., 1988; Dybing et al., 1989; Holme et al., 1989). Conjugation of DBCP with glutathione and the subsequent formation of a reactive episulfonium ion is suggested to be a major activation pathway in these cells (Søderlund et al., 1988; Omichinski et al., 1987, 1988). It is interesting to note that the episulfonium ion forms N7-guanine adducts and appears to be a DNA cross-linking agent (Humphreys et al., 1991). Repair of such damage could create double strand (ds) DNA breaks which may be of major importance for DBCP-induced cell death. Several lines of evidence suggest that DNA damage is important for DBCP-induced testicular toxicity (Dybing et al., 1989). DNA damage was observed shortly after administration of DBCP and at doses lower than those causing testicular toxicity. Structure-activity studies with halogenated and methylated analogs have shown that testicular DNA damage *in vivo* correlated well with the toxic effects. Furthermore, there are good correlations between species differences in testicular necrosis on the one hand, and DNA damage on the other. The low level of ssDNA breaks observed in human testicular cells with DBCP was thus unexpected, since occupationally exposed humans appear to be rather sensitive to DBCP-induced testicular toxicity. Furthermore, based on the level of covalent binding, one would have expected significant levels of induced ssDNA breaks in humans (Bjørge et al., 1996a). The apparent lack of ssDNA breakage in human testicular cells following DBCP exposure could indicate that the spectrum of DBCP-induced DNA lesions or their repair may differ in rat and human testicular cells. Another possible explanation may be that specific cell types representing a minor fraction of the whole cell population could be important target cells for DBCP-induced toxicity. These possibilities should be further explored.

Yield and purity of enriched germ cell fractions following centrifugal elutriation. Cells were separated by centrifugal elutriation based on the method of Meistrich and co-workers (1981), with modifications (Bjørge et al., 1995). Three different fractions of enriched germ cells were obtained, namely spermatocytes, round spermatids and elongating/elongated spermatids. Data from flow cytometric and microscopic analysis of the various rat testicular cell fractions showed that tetraploid spermatocytes accounted for approximately 75% of the enriched spermatocyte fraction. Contamination consisted mostly of round spermatids with some elongating/elongated spermatids, and approximately 2% were somatic cells as judged by vimentin staining. The purity of the enriched round spermatid fraction was more than 80% with some contamination with elongating/elongated spermatids. The enriched fractions containing round or elongating/elongated spermatids were further purified using metrizamide gradient centrifugation, which resulted in a purity of 90-95%. Examination of Giemsa, Hoechst and vimentin-stained smears verified the composition data obtained by flow cytometry. Similar data with human testicular cells were obtained. These enriched germ cell fractions represent a valuable material for characterisation of chemical induced DNA damage and repair processes in various stages of spermatogenesis.

Induction of DNA damage in various populations of testicular cells. Enriched fractions of various germ cells can be exposed to DNA damaging agents and the relative amount of induced DNA damage can be measured by a number of techniques. In some experiments previously published (Bjørge et al., 1995), ssDNA breaks measured by alkaline filter elution were detected in testicular cells after 30 min incubation with low concentrations of DBCP ($\geq$ 10 μM). ssDNA breaks appeared to be present at higher levels in round spermatids and in low levels in elongating/elongated spermatids. An intermediate level of ssDNA breaks was detected in spermatocytes and Sertoli cells. The SCGE assay, which allows the identification of ssDNA breaks in individual cells, confirmed that round spermatids had higher levels of ssDNA breaks than the other cell types (Bjørge et al., 1995).

Differences in metabolic activation as judged by the covalent binding of ^{14}C-DBCP to macromolecules may partly explain these differences (Bjørge et al., 1995). The low level of DBCP-induced ssDNA breaks observed in elongating/elongated spermatids, however, may also be partly due to the replacement of chromosomal histones first by transition proteins and later by protamines in these cells (Joshi et al., 1990; Oko et al., 1996). The resulting increased compactness of the nuclear material is likely to reduce both the induced level of DNA damage as well as its repair. Other studies have reported that ionizing radiation caused fewer ssDNA breaks in elongating/elongated spermatids compared to earlier stages of spermatogenic cells (Joshi et al., 1990; Van Loon et al., 1991,1993).

DNA repair in testicular cells. To become protected against DNA damage, organisms have evolved an elaborate array of DNA repair mechanisms such as direct lesion removal, DNA base excision repair (BER), nucleotide excision repair (NER), mismatch and postreplication repair (Barnes et al., 1993). During recent years a large number of genes involved in these DNA repair systems have been cloned and the gene products purified. Studies of tissue-specific differences in the expression of various genes have started by hybridizing total RNA from various organs with specific cRNA or cDNA probes. To provide insight into the cellular sites of expression, *in situ* hybridization has been applied successfully. Purified proteins have been used to generate antibodies which can be employed to detect repair proteins either by Western blotting or by immunohistochemistry. The expression of a number of genes/enzymes that are, or may be, involved in DNA repair synthesis has been reported in the testis, such as O^6-alkylguanine-DNA-alkyltranferase (ATase; Wilson et al., 1994), apurinic/apyrimidinic (AP) endonuclease (Wilson et al., 1996), *XRCC1* (Walter et al., 1994), *MHR23A* and *B* (van der Spek et al., 1996), *hRAD50* (Dolgan et al., 1996), *MRAD51* (Yamoto et al., 1996), a human homologue of *RAD52* (Muris et al., 1994), and *HHR6A* and *B* (Roest et al., 1996). Studies so far have indicated that the expression of these enzymes, with the exception of ATase, appears to be elevated in the testis compared to other organs. Interestingly, in response to indium-114m, hepatic ATase was induced fivefold, whereas no induction was observed in the testes (Wilson et al., 1994). Furthermore, AP endonuclease, responsible for the repair of AP-sites in DNA, showed a specific sub-tissue expression in the testes (Wilson et al., 1996).

In the present study we have observed that protein extracts from rat spermatocytes (SC), round spermatids (RST) and elongating/elongated spermatids (EST) are all proficient in excising uracil to the same extent as extracts from human somatic cells like lymphocytes and hepatocytes. *E. coli* enzyme AlkA and the human homologue methyl purine DNA-glycosylase (MPG) are able to release a wide range of damaged bases such as methylated bases, inosine and 1,N^6-ethenoadenine (EthA). Preliminary experiments indicate that methylated bases are released by protein extracts from unfractionated testis. Inosine and EthA are released by the extracts from SC, RST and EST. The ability to remove oxidative damages such as formamidopyrimidines (FaPy) and 7,8-dihydro-8-oxoguanine (8oxoG) from DNA is also found to be present in extracts from SC, RST and EST. Furthermore, in the present study we found that extracts from the different enriched germ cell fractions were able to repair AP-sites in DNA to a similar degree as extracts from human lymphocytes and hepatocytes. Overall, protein extracts from enriched fractions of spermatogenic cells from the rat are able to release a number of altered or damaged bases from DNA. This is in accordance with the fact that most DNA repair genes have "housekeeping" functions, and much differential expression would not be expected. It is, however, striking that protein extracts from elongating/elongated spermatids are able to excise these DNA lesions despite the notably reduced level of transcription and translation in these cells. This could indicate the importance of correcting these types of DNA errors in order to produce genetically stable mature sperm. It will be of importance to further explore this possibility by studies at the cellular level.

Studies in which DNA repair is measured as increased incorporation of radio-labelled thymidine have indicated that elongated spermatids and spermatozoa do not repair ssDNA breaks induced by UV and chemicals (Bentley and Working, 1988). In order to further characterise the DNA repair capacity in testicular cells, we have exposed such cells from human and rat to various DNA damaging agents and followed the formation and removal of ssDNA breaks measured by alkaline elution as an indicator of repair. No marked differences in the repair capacity between the two species were observed. Whereas the apparent $t_{1/2}$ for removal of ssDNA breaks induced by X-rays (rat) was in the order of 1/4-1/2 hour and for those induced by 4-nitroquinoline N-oxide (4-NQO) and DBCP (rat) was 1-2 hours, the $t_{1/2}$ for ssDNA breaks induced by styrene oxide appeared to be longer (4-8 hours). To elucidate whether the various spermatogenic cell types have different repair capacities, we determined the relative amount of induced ssDNA breaks and its repair in spermatocytes, round spermatids and elongating/elongated spermatids following exposure to methylmethane sulphonate (MMS), 4-NQO and X-rays. The relative amount of MMS-, 4-NQO-, and X-ray-induced ssDNA breaks was highest in spermatocytes and lowest in the elongating spermatocytes. Preliminary data following 2 hours post-incubation indicate some removal of ssDNA breaks in all cell types, but the greatest reduction was seen in spermatocytes.

Interestingly, we have observed that testicular cells, either from rats (Brunborg et al., 1995) or humans, accumulate very few ssDNA breaks when incubated with ssDNA break repair inhibitors following low doses of UV-C (0.1-0.5 J/m^2). Using the cyclobutyl pyrimidine dimer specific enzyme T4endoV in a modified SCGE assay, we found that the rate of removal of this lesion during prolonged repair incubation (24 hours) seemed to be low in testicular cells.

The effects of DNA damage on DNA synthesis, meiosis, spermiogenesis, and their possible consequences. Specific stages of seminiferous tubule segments can be separated by transilluminated-assisted microdissection. The composition of specific stages can be analysed by using flow cytometry and fluorescent microscopy. The culturing of seminiferous tubule segments has proven to be valuable in studying the effects of chemicals on DNA synthesis, meiosis and apoptosis. In recent, unpublished experiments, we found that relatively high concentrations of DBCP (30 μM) did not cause any significant reduction in the premitotic DNA synthesis in type B spermatogonia or in premeiotic DNA synthesis in preleptotene spermatocytes. Thus, it appears that DBCP does not cause any marked arrest in G_1S and/or a slowing down of DNA synthesis as is often seen in somatic cells after DNA damage. Previously, ethyl methanesulphonate (EMS) was shown to reduce DNA synthesis in testicular cells, whereas X-rays were found to have only a slight inhibiting action (Lähdetie et al., 1983). Exposure to daily renewed 30 μM DBCP led to a small, but significant reduction in meiotic divisions occurring in late pachytene/diplotene spermatocytes following a 65 hours incubation. Furthermore, a similar exposure of spermatids (steps 11 and 12) during the phase of chromatin reorganisation resulted in an increased frequency of characteristic thin, spiral-shaped elongated spermatids. Thus, DBCP induced some effects on meiosis and chromatin packing. These effects are most probably due to its DNA damaging properties; an apparent rapid repair of DNA damage may partly explain why more marked effects on DNA synthesis, meiosis and spermiogenesis were not seen. Furthermore, it is possible that the DBCP-induced DNA damage is bypassed during DNA replication and meiosis, thus resulting in mutations, genetic instability and cell death.

In general, replicating cells appear to be more susceptible to death induced by DNA damage than non-replicating cells. The level of DNA damage in G_1 cells when entering the S-phase, mitosis and meiosis is considered to be of major importance for continued growth and survival of the cells. In HL-60 cells, we have found that low concentrations of DBCP cause an accumulation of cells first in the S-phase and later at the G_2M boundary. Higher concentrations caused selective death of cells in the S-phase. Flow cytometric and morphological analysis revealed some necrotic, but mostly apoptotic cells/bodies (Holme at al., 1997). Apoptosis also becomes more prevalent following exposure to DNA damage in the testes *in vitro* as well as *in vivo*. Recently, isolated seminiferous tubule segments were used to quantify stage-specific apoptosis following exposure to irradiation (Henriksén et al., 1996). Furthermore, there are studies suggesting that testicular cells may have a lower threshold for initiating apoptosis than other cell types. Recent findings indicate a hypersensitivity of human testicular tumours to chemotherapeutic drugs which is associated with functional p53 and a high Bax:Bcl-2 ratio and an increased susceptibility to DNA-damage-induced apoptosis (Chresta et al., 1996). Finally, it has been suggested that during normal spermatogenesis more than half of the germ cells undergo apoptosis, providing a mechanism through which superfluous and defect germ cells may be removed (Hsueh et al., 1996). Thus, to which extent DBCP and other DNA damaging chemicals change gene expression, increase the amount of apoptosis and ultimately lead to a decrease in sperm production should be further explored.

Conclusions

Enriched populations of various cell types and isolated seminiferous tubule segments from human and animal testis constitute a useful and important system for analysing the possible role of DNA damage and DNA repair in testicular toxicity. Information from such *in vitro* studies, combined with *in vivo* animal studies and information from epidemiological studies of exposed human populations, may provide a better basis for evaluating the role of chemically-induced DNA damage on male reproduction.

Acknowledgement

The present study is supported by European Commission, contract No.: ENV4-CT95-0204.

References

Barnes DE, Lindahl T, Sedgwick B (1993) DNA Repair. Curr Opinion Cell Biol 5:424-433

Bjelland S, Eide L, Time RW, Stote R, Eftedal I, Volden GM, Seeberg E (1995) Oxidation of thymine to 5-formyluracil in DNA: mechanisms of formation, structural implications, and base excision by human cell free extracts. Biochemistry 34:147598-14764

Bjørge C, Brunborg G, Wiger R, Holme HA, Scholz T, Dybing E, Søderlund EJ (1996b) A comparative study of chemically induced DNA damage in isolated human and rat testicular cells. Reprod Toxicol 10:509-519

Bjørge C, Wiger R, Holme JA, Brunborg G, Andersen R, Dybing E, Søderlund EJ (1995) *In vitro* toxicity of 1,2-dibromo-3-chloropropane (DBCP) in different testicular cell types from rats. Reprod Toxicol 9:461-473

Bjørge C, Wiger R, Holme JA, Brunborg G, Scholz T, Dybing E, Søderlund EJ (1996a) DNA strand breaks in testicular cells from humans and rats following *in vitro* exposure to 1,2-dibromo-3-chloropropane (DBCP). Reprod Toxicol 10:51-59

Boiteux S, Belleney J, Roques BP, Laval J (1984) Two rotameric forms of open ring 7-methylguanine are present in alkylated polynucleotides. Nucl Acid Res 12:5429-5439

Bradley MD, Dysart G (1985) DNA single-strand breaks, double-strand breaks and crosslinks in rat testicular germ cells: measurement of their formation and repair by alkaline and neutral filter elution. Cell Biol Toxicol 1:181-5

Brunborg G, Holme JA, Hongslo JK (1995) Inhibitory effects of paracetamol on DNA repair in mammalian cells. Mutat Res 342:157-170

Brunborg G, Holme JA, Søderlund EJ, Omichinski JG, Dybing E (1988) An automated alkaline elution system. DNA damage induced by 1,2-dibromo-3-chloropropane in vivo and *in vitro*. Anal Biochem 174:522-36

Chresta CM, Masters JRW, Hickman JA (1996) Hypersensitivity of human testicular tumors to etoposide-induced apoptosis is associated with functional p53 and a high Bax:Bcl-2 ratio. Cancer Res 56:1834-1841

Dolganov GM, Maser RS, Novikov A, Tosto L, Chong S, Bressan DA, Petrini JHJ (1996) Human Rad50 is physically associated with human Mre11: identification of a conserved multiprotein complex implicated in recombination DNA repair. Mol Cell Biol 16: 4832-4841

Dybing E, Omichincki JG, Søderlund EJ, Brunborg G, Låg M, Holme JA, Nelson SD (1989) Mutagenicity and organ damage of 1,2-dibromo-3-chloropropane (DBCP) and tris(2,3-dibromopropyl)phosphate (Tris-BP): role of metabolic activation. In Hodgson E, Bend JR, Philpot RM, (eds.) Reviews in biochemical toxicology. Vol. 10, Elsevier Science Publishing Co., New York; pp 139-187

Eide L, Bjørås M, Pirovano M, Alseth I, Berdal KG, Seeberg E (1996) Base excision of oxidative purine and pyrimidine DNA damage in *Saccharomyces cerevisiae* by a DNA glycosylase with sequence similarity to endonuclease III from *Escherichia coli*. Proc Natl Acad Sci 93:10735-10740

Henriks(n K, Kulmala J, Toppari J, Mehrotra K, Parvinen M (1996) Stage-specific apoptosis in the rat seminiferous epithelium: quantitation of irradiation effects. J Androl 17:394-402

Hittmair A, Rogatsch H, Mikuz G, Feichtinger H (1994) Quantification of spermatogenesis by dual-parameter flow cytometry. Fertil Steril 61:746-750

Holme JA, Søderlund EJ, Brunborg G, Omichinski JG, Bekkedal K, Trygg B, Nelson SD, Dybing E (1989) Different mechanisms are involved in DNA damage, bacterial mutagenicity and cytotoxicity induced by 1,2-dibromo-3-chloropropane in suspensions of rat liver cells. Carcinogenesis 10:49-54

Holme JA, Wiger R, Hongslo JK, Søderlund EJ, Brunborg G, Dybing E (1997) Cell death via interactions of agents with DNA. Adv Mol Cell Biol 20:145-182

Hsueh AJW, Eisenhauer K, Chun S-Y, Hsu S-Y, Billig H (1996) Gonadal cell apoptosis. Rec Prog Horm Res 51:433-456

Humphreys WG, Kim DH, Guengerich FP (1991) Isolation and characterization of N7-guanyl adducts derived from 1,2-dibromo-3-chloropropane. Chem Res Toxicol 4:445-453

Joshi DS, Yick J, Murray D, Meistrich ML (1990) Stage dependent variation in the radiosensitivity of DNA in developing germ cells. Radiat Res 121:274-281

Kangasniemi M, Kaipia A, Mali P, Toppari J, Huhtaniemi I, Parvinen M (1990) Modulation of basal and FSH-dependent cyclic AMP production in the rat seminiferous tubules staged by an improved transillumination technique. Anat Rec 227:62-76

Lähdetie J, Kaukopuro S, Parvinen M (1983) Genetoxic effects of ethyl methanesulfonate and X-rays at different stages of rat spermatogenesis, studied by inhibition of DNA synthesis and induction of DNA repair *in vitro*. Hereditas 99:269-278

Masters JRW, Osborne EJ, Walker MC, Parris CN (1993) Hypersensitivity of human testis-tumour cell lines to chemotherapeutic drugs. Int J Cancer 53:340-346

Meistrich ML, Longtin J, Brock WA, Grimes SR JR, Mace ML (1981) Purification of rat spermatogenic cells and preliminary biochemical analysis of these cells. Biol Reprod 25:1065-1077

Muris DFR, Bezzubova O, Buerstedde J-M. Vreeken K, Balajee AS, Osgood CJ, Troelstra C, Hoeijmakers JHJ, Ostermann K, Schmidt H, Natarajan AT, Eeken JCJ, Lohman PHM, Pastink A (1994) Cloning of human and mouse genes homologous to RAD52, a yeast gene involved in DNA repair and recombination. Mutat Res 315:295-305

Oko RJ, Jando V, Wagner CL, Kistler WS, Hermo LS (1996) Chromatin reorganization in rat spermatids during the disappearance of testis-specific histone, H1t, and the appearance of transition proteins TP1 and TP2. Biol Reprod 54:1141-1157

Omichinski JG, Søderlund EJ, Bausano JA, Dybing E, Nelson SD (1987) Synthesis and mutagenicity of selectively methylated analogs of 1,2-dibromo-3-chloropropane and tris(2,3-dibromopropyl)phosphate. Mutagenesis 2:287-292

Omichinski JG, Brunborg G, Holme JA, Søderlund EJ, Nelson SD, Dybing E (1988) The role of oxidative and conjugative pathways in the activation of 1,2 dibromo-3-chloropropane to DNA damaging products in testicular cells. Mol Pharmacol 34:74-79

Riazuddin S, Lindahl T (1978) Properties of 3-methyl-adenine-DNA-glycosylase from *Escherichia coli*. Biochemistry 17:2110-2118

Roest HP, Van Klaveren J, De Wit J, Van Gurp CG, Koken MHM, Vermey M, Van Roijen JH, Hoogerbrugge JW, Vreeburg JTM, Baarends WM, Bootsma D, Grootegoed JA, Hoeijmakers JHJ (1996) Inactivation of the HR6B ubiquitin-conjugating DNA repair enzyme in mice causes male sterility associated with chromatin modification. Cell 86: 799-810

Suter L, Koch E, Bechter R, Bobadilla M (1997) Three-parameter flow cytometric analysis of rat spermatogenesis. Cytometry 27:161-168

Søderlund EJ, Brunborg G, Omichinski JG, Holme JA, Dahl JE, Nelson SD, Dybing E (1988) Testicular necrosis and DNA damage caused by deuterated and methylated analogs of 1,2-dibromo-3-chloropropane in the rat. Toxicol Appl Pharmacol 94:437-447

Toppari J, Larsen JC, Christiansen P, Giwercman A, Grandjean P, Guillette Jr. LJ, Jegou B, Jensen TK, Jouannet P, Keiding N, Leffers H, McLachlan JA, Meyer O, Muller J, Rajpert De Meyts W, Scheike T, Sharpe R, Sumpter J, Skakkebæk NE (1996) Male reproduction health and environmental xenoestrogens. Environ Health Perspect 104(Suppl 4):741-803

Walter CA, Lu J, Bhakta M, Zhou Z-Q, Thompson LH, McCarrey JR (1994) Testis and somatic Xrcc-1 DNA repair gene expression. Somat Cell Mol Genet, 20: 451-461

Van Der Spek PJ, Visser CE, Hanaoka F, Smit B, Hagemeuer A, Bootsma D, Hoeijmakers JHJ (1996) Cloning, comparative mapping, and RNA expression of the mouse homologues of the *Saccharomyces cerevisiae* nucleotide excision repair gene RAD23. Genomics 31:20-27

Wilson RE, Hoyes KP, Morris ID, Sharma HL, Hendry JH, Margison GP (1994) *In Vivo* Induction of O^6-alkylguanine-DNA-alkyltransferase in response to indium-114m. Rad Res 138:26-33

Wilson TM, Rivkees SA, Deutsch WA, Kelley MR (1996) Differential expression of the apurinic/apyrimidinic endonuclease (APE/ref-1) multifunctional DNA base excision repair gene during fetal development and in adult rat brain and testis. Mutat Res 362:237-248

Yamamoto A, Taki T, Yagi H, Habu T, Yoshida K, Yoshimura Y, Yamamoto K, Matsuhiro A, Nishimune Y, Morita T (1996) Cell cycle-dependent expression of the mouse *Rad51* gene in proliferating cells. Mol Gen Genet 251:1-12

Biomarkers of Exposure

(Chair: H.-G. Neumann, Germany, and
E. Baatrup, Denmark)

Animal Locomotor Behaviour as a Health Biomarker of Chemical Stress

Erik Baatrup and Mark Bayley
Institute of Biological Sciences, Aarhus University, Building 135, DK-8000 Aarhus, Denmark

Introduction

The identification and development of biomarkers to assess the biological impact of chemical pollutants has become an important objective in ecotoxicology. Originally viewed as measurable responses at the suborganismal level, Depledge (1994) expanded the definition of an ecotoxicological biomarker to: *a biochemical, cellular, physiological or behavioural variation ... at the level of whole organisms (either individuals or populations) that provides evidence of exposure* (exposure biomarkers) *to, and/or effects* (health biomarker) *of, one or more chemical pollutants.* Employing this concept, biomarker responses to pollutants can be measured anywhere along the hierarchy of increasing biological complexity from the level of the molecule to the population (Fig. 1). The scientific value and applicability of selected biomarkers depend largely on how closely their responses are related to measured effects at both lower and higher levels of organization. Thus, *we must try to establish the mechanistic links by which pollutant effects at one level of organization give rise to higher level effects in order to develop meaningful biomarkers* (Depledge, 1994).

When a chemical is released into the environment, a chain of events through the levels of biological organization is initiated. The parent compound, or residues, will be absorbed by the animal from food or across outer surfaces. The initial effects will always be at the molecular level, and if the chemical is not neutralised by the organism, these molecular changes may have deleterious consequences for cell function and survival. If many, or critical, cells are affected, alterations in vital physiological processes will weaken the condition of the animal. This will almost certainly be expressed in an altered behaviour. For animals in the wild, even minor changes in behaviour may reduce their fitness through, e.g., their ability to obtain food, avoid predation and reproduce successfully.

In this context, quantitative measurements of animal behaviour are potentially suitable as biomarkers of chemical stress. As is evident from Fig. 1, animal behaviour has a central position in the flow of pollutant-induced effects through levels of increasing biological complexity, with links in both directions. On the one hand, pollutant-induced changes of behaviour are intrinsically linked to impaired biochemical, physiological or general metabolic processes within the animal. In the other direction, the ecological importance of behaviour in population maintenance is intuitively obvious.

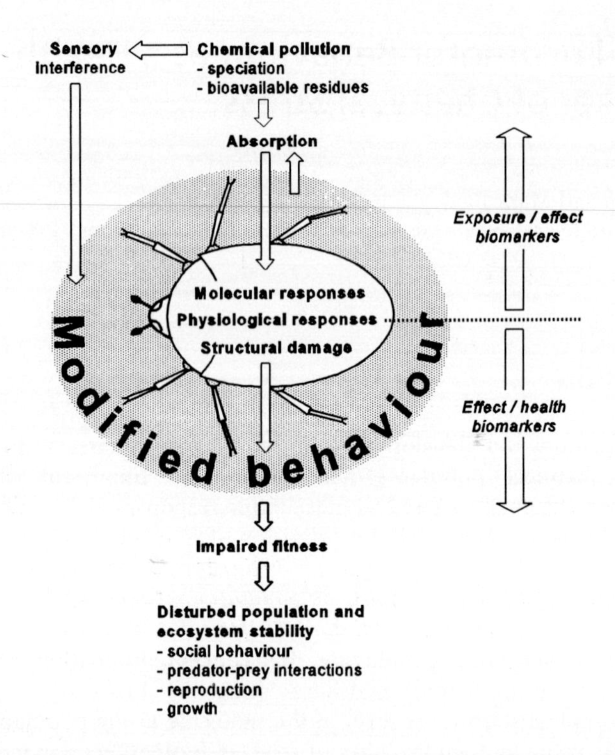

Fig. 1. Schematic representation of the flow of pollutant-induced effects.
When a chemical is released into the environment, a chain of events through the levels of biological organisation is initiated. The parent compound or residues will be absorbed by the animal from the food or across outer surfaces. The initial effects will always be at the molecular level, and if the chemical is not neutralised by the organism, these molecular changes may have deleterious consequences for cell function and survival. If many or critical cells are affected, alterations in vital physiological processes will weaken the condition of the animal. This will almost certainly be expressed in an altered behaviour. For animals in the wild, even minor changes in behaviour may reduce their fitness through for example their ability to obtain food, avoid predation and successfully reproduce.

One of the first proponents of behavioural toxicology was Warner et al.(1966) who noted that animal behaviour is the final integrated expression of the internal physiological processes, and is sensitive to changes in the steady state of an organism. Also Little (1990) concluded that *observations of behaviour provide a unique toxicological perspective, one that links the biochemical and ecological consequences of environmental contamination.* A further advantage of behavioural measurements is their non-invasive nature. Besides the ethical aspect, this means that the same individual can be measured repetitively (e.g. before and after a treatment) and can hereby be its own statistical control. Finally, behavioural measurements can be combined with any other test on the individual animal.

Despite these advantages, behavioural studies have played a modest role in ecotoxicological research, risk assessment and chemical authorisation procedures. Traditional behavioural studies are often time-consuming and require extensive biological insight, which makes them poorly suited to routine testing. In addition, behavioural studies are often qualitative rather than quantitative in nature and therefore difficult to compare and reproduce. Finally, it is often stated that even simple and well defined behavioural parameters have a high level of inherent variability.

In this communication we summarize recent results supporting the utilization of animal behaviour as a *health biomarker* in the assessment of the effects of environmental pollutants. The general hypothesis has been that animal behaviour represents the functional interface between the individual and its surrounding environment (cf. Fig. 1) and that it can be considered as the individual's integrated expression of its well-being. The majority of examples presented here are taken from our own research on the locomotor behaviour of invertebrate animals in the terrestrial environment.

Locomotor Behaviour is Quantifiable

The ability to move is a fundamental property of almost all animals. It is inherent in most behavioural patterns and is central to important life processes such as migration, territory maintenance, resource searching behaviour, predation and escape responses. Accordingly, locomotor behaviour contains a wealth of information about an animal's ecological, physiological and toxicological status.

Animal locomotion is especially well suited for quantitative analyses in that its constituent elements can be identified and expressed numerically. Existing methods for automated measurements of animal movements include treadmills, photocell arrays and locomotor compensators. Unfortunately, these methods have a number of inherent disadvantages such as restrictions on the size range of test organism, unnatural walking substrates and an often poor resolution in space or time. Digital video analysis, in contrast, is free of considerations with respect to animal size and has high temporal and spatial resolutions. Further, experiments can be carried out under a wide range of experimental conditions (temperature, lighting conditions, humidity etc.) with the animals moving on their natural substrate (soil, sand etc.) or in water in the case of aquatic species. The measurements can be carried out in the laboratory or under semi-field conditions.

Tracking Animal Locomotion in Digital Images

The computer-aided video tracking system used in the experiments described below allows simultaneous monitoring of up to six animals. The animals are placed in individual test arenas and tracked automatically for periods of minutes to days. The camera image of the six test arenas is displayed on a monitor for remote observation (Fig. 2). Each captured image (frame) is simultaneously digitised by a frame grabber in the computer. This process divides the image into 512 x 512=262144 pixels (small rectangles), where each pixel is assigned a value of between zero and 255 in each of the primary colours, red, green and blue (RGB colour definition). In the resultant digitised image, every pixel is thus defined by its position and its colour. Accordingly, an object such as an animal

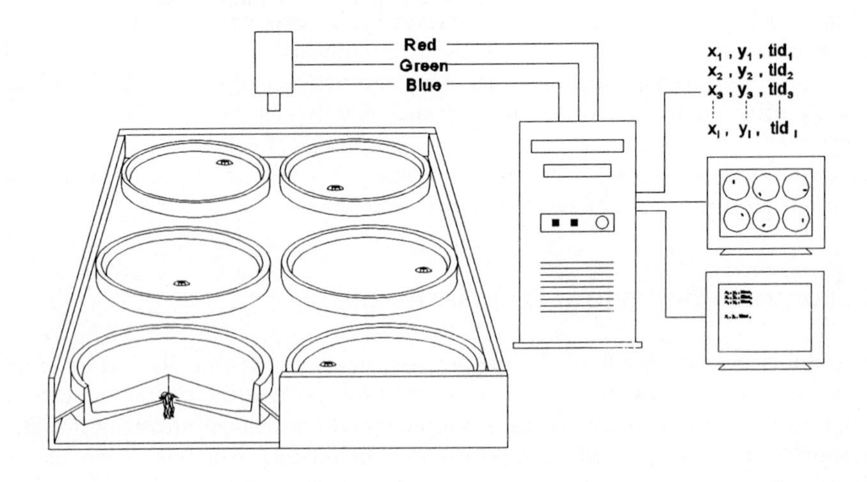

Fig. 2. Schematic representation of the computer-automated video-tracking system used in the studies.
The colour images of the test arenas are captured by the camera and digitised by a 4 Mb frame-grabber at 25 Hz. At user-defined intervals, the positions of the animals in the frame of pixels are determined by software-defined colour and size identification. The resulting time-series of X,Y positions form the basis of various graphic representations and the calculations, which dissect and analyse the components of the animals' locomotor behaviour.

can be characterised by its colour range (expressed by ranges in the three primary colours), its position (X-Y coordinate) and its size (number of pixels covered by the object). In the consecutive sequence of digitised images the computer searches each image for those areas that fulfill preset colour and size criteria of the animals, after which the centre of gravity of these areas is calculated. The time-series of Cartesian coordinates representing animal positions are written to file. Subsequently, the *X, Y, time* data is transformed into consecutive vectors from which a number of locomotor parameters can be calculated, including path length, average movement velocity, maximum velocity, movement continuity, number and duration of rest periods and turning behaviour.

Locomotor Behaviour is Sensitive to Chemical Stress

A biomarker may provide relevant new information, but unless it is sensitive relative to traditional endpoints such as mortality, its practical applicability will be limited (Huggett et al., 1992). In a review of the use of swimming behaviour in fish as an indicator of sublethal toxicity, Little and Finger (1990) found that alterations in swimming behaviour were consistently reported at concentrations as low as 0.7% of the LC_{50} value, and recommended that swimming activity be included in standard protocols for aquatic toxicity assessment. Similarly, video tracking studies at our laboratory showed that as little as 150 ng/g of the pyrethroid pesticide cypermethrin induced knock-down when applied topically to a wolf spider (Baatrup and Bayley, 1993). In subsequent studies of the effect of the organophosphate dimethoate on the locomotor behaviour of the woodlouse *Porcellio scaber* (Bayley, 1995) and the collembolan *Folsomia candida* (Sørensen et al, 1995), it was found that the sensitivity and type of locomotor response were both sex and species specific. In the woodlouse, low doses of dimethoate caused little change in the locomotor patterns of the females, but caused significant hyperactivity in males. These males remained hyperactive several weeks after the dimethoate treatment. The collembolan, on the other hand, reacted by decreasing their activity at concentrations as low as one tenth of those causing changes in reproductive output.

Locomotor Behaviour and Chemical Uptake

The uptake of chemical residues from soil and plant surfaces constitutes a major source of xenobiotic exposure to animals in the terrestrial environment. The more active an animal, the more chemical the animal will encounter and the more the animal can absorb. Hence, the locomotor behaviour of a terrestrial animal is not only a potential indicator of chemical stress, but also an important factor in determining the chemical dose acquired by the animal. This relationship has been highlighted in a number of modelling studies (Salt and Ford, 1984; Jepson et al., 1990; Jagers op Akkerhuis and Hamers, 1992). The relationship was experimentally demonstrated for woodlice walking on soil contaminated with ^{14}C-labelled dimethoate (Bayley and Baatrup, 1996). The locomotor behaviour of 60 woodlice was monitored using computer-automated video tracking during a 22 hrs exposure period on a soil surface treated with three sublethal doses of dimethoate. The amount of dimethoate absorbed by the animals was subsequently measured using liquid scintillation counting. Figure 3 shows the dimethoate uptake in the woodlice as a function of the distance they walked during the experiment. This experiment is the first direct experimental evidence of the link between locomotor activity and pollutant uptake. In future experimental and modelling studies such information may improve the understanding of inter-species differences in sensitivity to environmental contaminants and of differences in bioavailability from different substrate types.

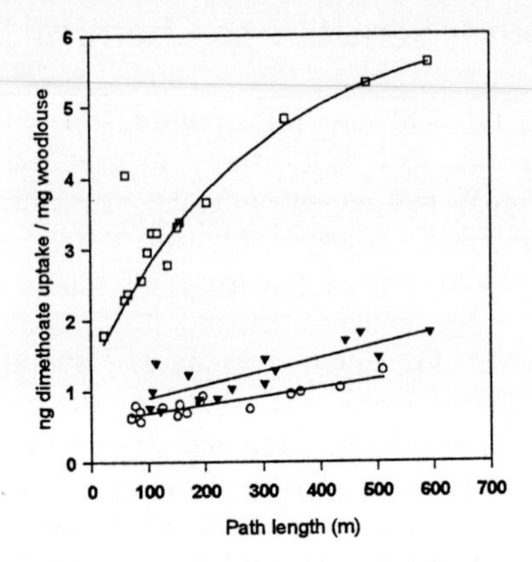

Fig. 3. When woodlice walked on soil surfaces contaminated with three concentrations of ^{14}C-labelled dimethoate, a strong correlation was found between locomotor activity and residue uptake.
□ 560 g/ha; ▼ 280 g/ha; ❍ 140 g/ha.

Locomotor Behaviour can Reveal Long-Term Effects

The response time for changes in locomotor behaviour has been found to vary from minutes in experiments where pesticide was placed directly on the animal's skin (Baatrup and Bayley, 1993), to hours in experiments where animals walked on pesticide-contaminated soil (Bayley, 1995; Sørensen et al., 1995). Thus, the response time of behaviour to absorbed xenobiotics can be rapid considering its late occurrence in the flow of events in Fig. 1. More surprisingly, we found that changes in behaviour can remain long after exposure to the pollutant has ceased and the toxic residues have been metabolized by the animal. In a laboratory experiment, the locomotor behaviour of male woodlice was measured before and after a 48 hrs exposure to dimethoate-contaminated soil (Bayley, 1995). Over the period of exposure, this acetylcholinesterase inhibitor induced gradually increasing activity in terms of time in movement, mean velocity and walked distance and a suppression of turning rate when compared with a control group. This excited state of locomotor activity was maintained in the exposed group even after 21 days of recovery on uncontaminated soil. Such behavioural changes can be interpreted as a *de facto* modification of the animal's fitness. The frequency distribution of velocities employed by the individual male woodlice (Fig. 4) emphasized that dimethoate maintained an excited (chemically stressed?) state in exposed animals.

Further evidence of long-term or permanent effects of chemicals on animal locomotor behaviour was obtained in a subsequent experiment demonstrating that a toxic action early in life can be conserved through metamorphosis to be finally expressed in an altered behaviour of the adult (Bayley et al., 1995). Larvae of the carabid beetle *Pterostichus cupreus* were raised on copper-contaminated food and soil, whilst a control group was raised under the same conditions, but without the addition of copper. At pupation all animals were transferred to clean soil and after emergence from the pupae as adult beetles both groups were fed uncontaminated food. The two groups had the same development time from egg to adult and the same adult emergence weight, but displayed differences in their locomotor behaviour. Thus, the group of beetles, which had been exposed to copper during larval development moved 30% slower than controls, covered 58% shorter distances during the period of measurement, had 43% more stops, and showed a 75% reduction in turn rate. Copper analysis of these adults revealed that copper levels were the same or only slightly elevated in comparison

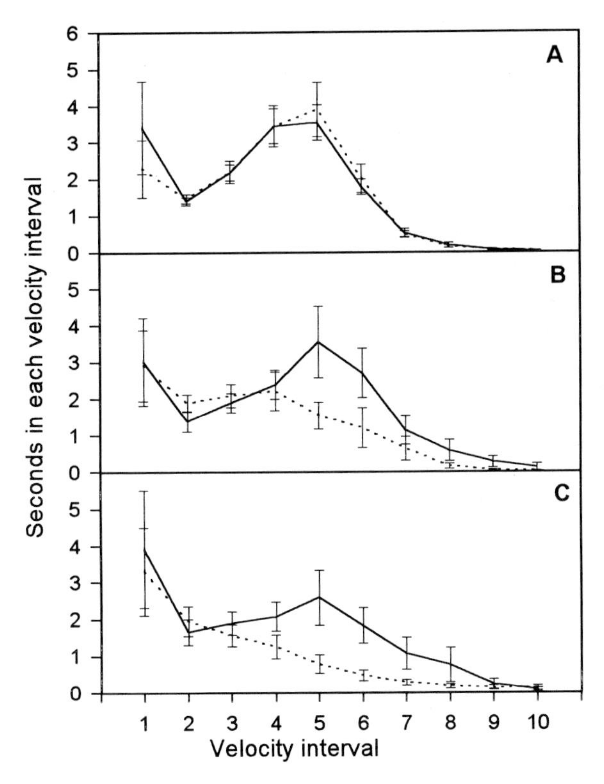

Fig. 4 Mean velocity distribution of 20 dimethoate-exposed (solid lines) and 20 control (dashed lines) woodlice during 8 hrs measurements.
Each velocity interval represents one-tenth of the individual animal's range of velocities from 1 mm/s to its maximum velocity. A shows the velocity distribution for the exposed and control animals during the first 8 hrs after introduction to the test arenas. In these new and presumably stressful surroundings without shelters, the woodlice of both groups had a relatively high level of activity, expressed by the peak in the middle of the velocity distribution curve. As the control animals became familiarised with their surroundings after 48 hrs (B) and further after 23 days (C), the peak gradually declined, the animals now preferentially moving at low velocities. The dimethoate treated animals, on the other hand, maintained an exited (stressed ?) state and continued to utilise the higher movement speeds despite the fact that these animals were allowed to recover on uncontaminated soil for 21 days prior to the last measurement.

with controls. These findings suggest that the altered locomotor behaviour in the adults is associated with copper- induced internal and irreversible damage (probably structural) during larval development and therefore expresses a prolonged or permanent effect. Such a considerable alteration in the spontaneous locomotor behaviour is likely to reduce the Darwinian fitness of these animals.

Linking Behaviour to Other Biomarkers

An important prerequisite for the applicability of locomotor behaviour as a biomarker of chemical exposure and effects must be that the measured responses can be related to altered biochemical or physiological processes within an animal, or are related to parameters of importance to population survival and stability. In other words, it is important to establish the mechanistic links between alterations in behaviour and the lower and higher effect levels illustrated in Fig. 1. Linkages between biochemical biomarkers and higher level effects such as behavioural biomarkers are also necessary, if the easily measured biochemical endpoints are to have toxicological and predictive value (Bayne et al., 1985; Hodson, 1990; Depledge 1994).

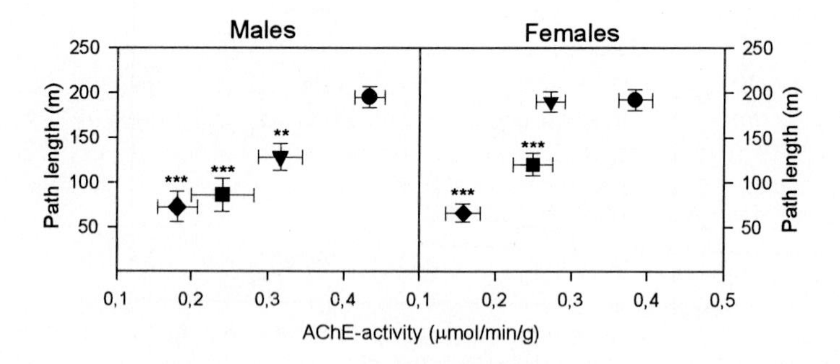

Fig. 5 Relationship between average whole-body AChE activity and walked distance (path length) of carabid beetles (*Pterostichus cupreus*) exposed topically to four dimethoate concentrations including controls. Error bars show standard error for behaviour and enzyme activity respectively. ● Vehicle only; ▼ 0.64 g dimethoate/g fw; ■ 1.35 g dimethoate/g fw; ◆ 3.32 g dimethoate/g fw. The significance levels between the exposed groups and the control group are indicated by * $p<0.05$; ** $p<0.01$; *** $p<0.001$

Acetylcholinesterase (AChE) inhibition resulting from organophosphate (eg. dimethoate) and carbamate poisoning is a well-established biochemical biomarker in ecotoxicology (Peakall, 1994). The enzyme acetylcholinesterase deactivates the chemical transmitter acetylcholine at the post-synaptic membrane, thus ensuring a controlled signal transmission between nerve cells and to the muscles. In a series of studies, the relationship between altered locomotor behaviour and AChE-inhibition was studied in carabid beetles (*Pterostichus cupreus*) exposed to sublethal doses of dimethoate (Jensen et al., 1997). Since animal locomotion depends on an intact nervous system, a functional relationship between the two biomarkers was hypothesised.

Groups of beetles were treated with four dimethoate concentrations, and after 24 hrs, their locomotor behaviour was measured for 4 hrs. Immediately hereafter, the animals were euthanatised by freezing and the AChE activity measured in each individual. A linear relationship between dimethoate dose and AChE-inhibition was observed in both sexes. Further, a strong correlation between AChE-inhibition and the locomotor behaviour of the beetles was displayed. Figure 5 shows the relationship between AChE-inhibition and the distance walked during the period of measurement for the four dimethoate doses. It is evident that the walked distance (path length) of male beetles increased almost linearly with increasing AChE activity (decreasing dimethoate dose). Similar relationships were found between AChE-inhibition and other locomotor parameters such as average movement velocity, time in activity and turning rates. This general pattern was repeated in the females. However, in contrast to the male beetles, female locomotion was unaffected at the lowest dimethoate dose despite the fact that AChE activity was inhibited by 29%. For yet unknown reasons, female *P. cupreus* apparently posses an ability to compensate behaviourally for a moderate reduction in AChE activity, a capability absent in males.

Two important conclusions can be drawn from this study. Firstly, with these new measuring techniques, the behavioural biomarker was found to be as rapid in response time and sensitivity to chemical stress as the biochemical biomarker AChE-inhibition. Secondly, the demonstrated relationship between AChE inhibition and behavioural changes provides new evidence for a mechanistic linkage between a molecular biomarker and a general biomarker of organism health, the latter with further strong links to population-level effects.

From Individual to Population

Alterations in animal behaviour are, as stated by Little (1990) intuitively linked to effects at the population level. The long term effects of chemical stressors on animal behaviour (Bayley 1995; Bayley et al., 1995) strongly suggest reductions in Darwinian fitness. However, there is still little scientific evidence that alterations in animal behaviour cause a direct impact on population dynamics. Effects of chemical stressors at the level of the population have been assessed directly through estimates of growth and reproduction and indirectly by measuring the energy reserves available for these processes (*scope for growth* estimates).

Whilst these methods are highly relevant from an ecological viewpoint, they are unfortunately expensive and time consuming. As a simple but determinative

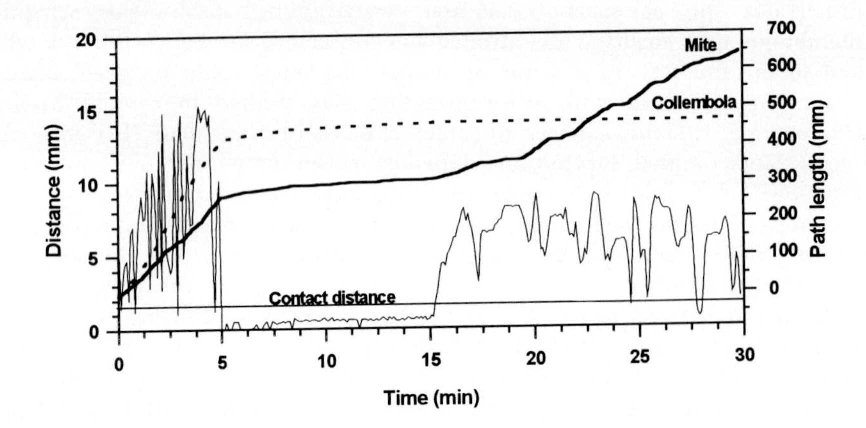

Fig. 6 Graphic representation of interactions between a mite (*Hypoaspis aculeifer*) and its collembolan prey (*Folsomia fimetaria*) during 30 min of measurement.

The jagged line shows the distance between the two animals for every 0.5 sec. When this line drops below the contact distance line, the mite and the collembola have been in physical contact. Further, the accumulated path lengths of the two animals are displayed. In the case shown here, there were four meetings before the mite captured and killed the collembola after 5 min. After about 15 min the mite left its prey, but returned twice during the remaining experimental period.

factor for population survival we chose to study the interactions between a predator and its prey. The ability of a predator to track and kill prey is vital to its own and therefore the population's well-being. Similarly, the ability of the prey to sense and avoid predators is central to its survival, or as expressed by Lima and Dill (1990), *few failures in the life of an organism are as unforgiving as the failure to avoid a predator*. The complex predator-prey interactions were quantified with new software routines, which can distinguish the two animals in the same arena, measure their individual locomotor activity and quantify a number of interactions, including number and duration of meetings, time of capture and the duration of capture and consumption. The predatory mite *Hypoaspis aculeifer* and one of its natural prey, the collembolan *Folsomia fimentaria*, were chosen as model organisms. Despite their small size, these animals are important in the ecology of the upper soil because of their high biomass. Before introducing xenobiotics to the system, the influence of natural variables cn the capture rate, such as age and size of the animals and the hunger level of the predator, were studied. In small test arenas, each of 245 mite-collembola pairs was monitored for 30 min. A simple graphic presentation of interactions between one such mite-collembola pair is shown in Fig. 6. Logistic regression was used to determine which of the measured factors were of importance for capture. Surprisingly, the results showed that the size of the prey animal and the average moving velocity of the predator were of paramount importance, whereas the age and hunger status of the mite and the activity level of the collembolan prey were less decisive.

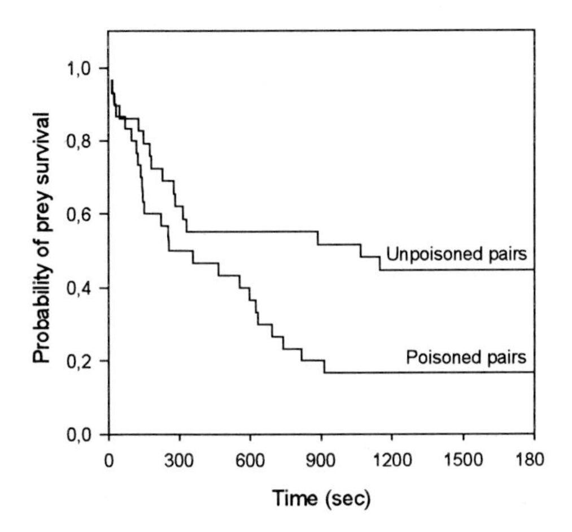

Fig. 7 **The temporal probability of prey survival in a predator-prey system can be described statistically with survival analysis.**
Groups of the predatory mite *Hypoaspis aculeifer* and its collembolan prey *Folsomia fimetaria* were incubated in clean and dimethoate-contaminated soil respectively. Subsequently, pairs of mites and collembola were video-tracked and the time of capture automatically monitored. In the group of unpoisoned pairs, the probability of a collembola surviving the 30 min experimental period was about 50%, whereas the corresponding probability in the weakly poisoned pairs was only 17%.

In subsequent experiments, groups of mites and collembola were exposed to low concentrations of dimethoate in soil for four days. The interactions of 30 mite collembola pairs were measured after this mild intoxication and compared with 30 control pairs from uncontaminated soil. Survival analysis (Kaplan-Meier) revealed that dimethoate reduced the probability of the collembola surviving the 30 min experimental period from 50% in controls to only 17% (Fig. 7). The data suggests that this change in prey killing can partly be eplained by a dimethoate-induced impairment of the collembolan escape response.

These experiments show that automated and quantitative measurements of animal locomotor behaviour can be linked to events of vital importance for the individual and the population and as such extend our understanding of the flow of effects towards increasing biological complexity. In addition, such experiments can provide valuable data for the mathematical modelling of predator-prey relationships.

Polluted Soil Alters the Behaviour of Woodlice

These previous studies were all carried out under controlled laboratory conditions with only one chemical influencing the locomotor behaviour. In their natural environment, animals are constantly affected by varying physical conditions such as humidity and temperature, and at polluted sites they are almost always exposed to complex chemical mixtures. It was our hypothesis that alterations in locomotor behaviour express an integrated response to the animals internal status, including chemical stress. We therefore wished to test whether animals from polluted locations differ in their locomotor behaviour when compared with individuals from clean environments. Results from such studies could contribute to the process of field validation of locomotor behaviour as a general biomarker of toxic stress and to assess the potential of using this behavioural biomarker as a screening tool in environmental management. The woodlouse *Oniscus asellus* was chosen for these field studies because it lives in, and eats polluted plant litter, is globally distributed and is present in both urban and natural habitats in large numbers (Hopkin et al., 1986). In addition, woodlouse locomotor behaviour is heavily involved in its water balance and therefore decisive for survival (Warburg, 1987). Three studies on field collected woodlice have so far been carried out, one in England (Sørensen et al., 1997) and two in Denmark (Bayley and Baatrup, 1997; Bayley and Baatrup, unpubl.).

A fire in 1991 at a plastics recycling factory in Thetford, England, resulted in 40 tons of molten plastic flowing into an adjacent beach wood. By 1995, this 2500 m^2 well defined area of pyrolyzed plastic was covered by a soil and leaf-litter layer, severely contaminated with heavy metals and a complex mixture of organic compounds. Woodlice were collected along a transect from the centre of the pollution to a point 100 m into the woods presumed to be relatively unpolluted. After measurement of their behaviour, the zinc, lead, cadmium and antimony content was measured. Also, the glycogen and protein content was determined in woodlice collected in parallel. The behavioural measurements showed that the animals collected from the centre of the pollution were less than half as active and had a more disrupted movement pattern than the animals from the reference area. Similarly, the polluted animals had reduced energy reserves in terms of glycogen level and were burdened with 2 to 13 times the concentrations of the measured metals than the reference animals. Whilst this study was our first indication that locomotor behaviour could reveal effects in populations of animals exposed to chemical mixtures in the field, it left many fundamental questions unanswered. In particular, what is the variation in locomotor behaviour *between* the reference sites with which the animals from contaminated areas are to be compared? Further, which elements in locomotor behaviour are the most efficient indicators of chemical stress?

These questions were addressed in a study in Denmark centered around a recently shut down iron foundry (Bayley and Baatrup, 1997). The question of variability in the behaviour of woodlice from distinct reference sites was approached by adopting an asymmetrical experimental design, where the behaviour of woodlice from the polluted site was compared with animals from *four* unpolluted reference sites. Those elements of the locomotor behaviour which are the best indicators of chemical stress were identified using linear discriminant analysis. This statistical technique combines the locomotor components into a linear equation which describes the function that best separates the groups. The locomotor elements most significant in group separation are given the greatest weighting in the equation. If such discriminant models have a sufficiently solid foundation, then they have a considerable predictive value and can be used to test animals from new localities. In the present study, the model was built using the locomotor parameters of the foundry group and three of the four control groups. The analysis showed that the control populations were remarkably similar in their locomotor behaviour despite their geographic separation of more than 300 km. The discriminant function showed that the foundry animals were significantly different from the controls in terms of their path length and movement velocities, whereas other parameters contributed little to group separation. The discriminant function correctly grouped 88 % of the control animals and 80% of the foundry animals. The fourth control population (not included in the model building) was used to validate the discriminant function and 89% of these animals were correctly identified as controls. This study contributes two important results to the development of locomotor behaviour as biomarker in environmental assessment. Firstly, the small variation between control populations of woodlice supports the use of this species as an indicator of chemically induced behavioural changes. Secondly, it is evident that woodlouse behaviour can be altered by complex chemical mixtures present in environmental pollution.

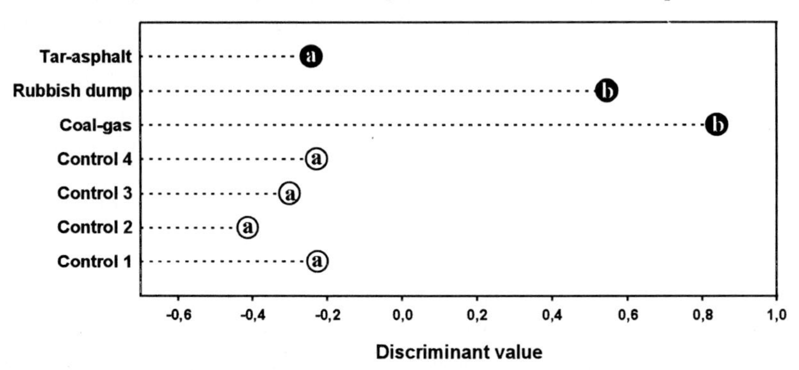

Fig. 8 The locomotor behaviour of the woodlouse *Oniscus asellus* is modulated by its chemical environment.

Employing discriminant analysis, all measured elements in the locomotor behaviour are weighed, making it possible to separate woodice living in polluted sites from individuals collected from unpolluted sites. In this study, the average discriminant value of 30 woodlice from each of four reference sites (Controls 1-4) were significantly different from the values of woodlice collected at a former industrial rubbish dump and a closed coal-gas works, but not different from animals collected from a renovated tar-asphalt works. Discriminant values marked by the same letter (a or b) are not significantly different (p>0.05) from each other (Tukey's honestly significant difference test).

In a subsequent study on the island of Fynen, the experimental design was expanded to include four clean reference sites separated by 50-100 km and three sites with different types of pollution. An industrial rubbish dump, closed since the sixties, was almost exclusively polluted by zinc, cadmium and lead, while the soil of a former coal gas works was heavily contaminated with cyanide and a range of organic pollutants. The third location was a partially renovated tar asphalt works still contaminated with high levels of PAHs in the soil. The results of this study confirmed that the locomotor behaviour of control populations was remarkably similar, and therefore independent of geographical location, at least on the scale available in Denmark. The locomotor behaviour of the woodlice collected from two of the polluted sites (the former rubbish dump and coal gas works) was again found to be significantly different from all controls, whereas the behaviour of the animals from the PAH-contaminated soil was not. It was not possible to discriminate the three very different pollution types from each other solely on the basis of the woodlouse locomotor behaviour. Possibly such a distinction can be achieved by including other biomarkers in the discriminant model.

Conclusions

As indicated by the studies selected for this condensed presentation, unbiased quantitative measurements of animal behaviour can provide scientifically sound and applicable biomarkers for the assessment and prediction of ecotoxicological consequences of chemical pollution. The computerized behaviour-monitoring system is an untiring observer with has a high spatial and temporal resolution and a superior data-handling capacity. This allows specific elements of composite behavioural patterns to be studied separately with high statistical strength. Hence, general aspects of an animal's behaviour, such as its locomotion, present an integration of the "well-being" of the animal, allowing the toxicologist to use the entire organism as the measuring instrument.

The measured alterations in the behaviour of terrestrial arthropods have been shown to be sensitive, rapid, and in some cases sustained responses to chemical impact. Also, the experiments demonstrated links between the integrated behavioural response of the individual and effects at other levels of biological organisation including the population level. Finally, the disrupted locomotor behavior of woodlice collected at polluted field sites suggests the applicability of standardized behavioural measurements as low cost and biologically meaningful test procedures in environmental management.

References

Baatrup E, Bayley M 1993. Effects of the pyrethroid insecticide Cypermethrin on the locomotor activity of the wolf spider *Pardosa amentata*. Quantitative analysis employing computer-automated video tracking. Ecotoxicol Environ Safety 26:138-152

Bayley M (1995). Prolonged effects of the insecticide Dimethoate on locomotor behaviour in the woodlouse *Porcellio scaber* Latr. (Isopoda). Ecotoxicology 4:79-90

Bayley M, Baatrup E, Heimbach U, Bjerregaard P (1995). Elevated copper levels during larval development cause altered locomotor behaviour in the adult carabid beetle *Pterostichus cupreus*. Ecotoxicol Environ Safety 32:166-170

Bayley M, Baatrup E (1996). Pesticide uptake and locomotor behaviour in the woodlouse: An experimental study employing video tracking and ^{14}C-labelling. Ecotoxicol 5:35-45

Bayley M, Baatrup E, Bjerregaard P (1997). Woodlouse locomotor behavior in the assessment of clean and contaminated field sites. Environ Toxicol Chem (in press).

Bayne B, Brown DA, Burns K, Dixon DR, Ivanovici A, Livingstone DR, Lowe DM, Moore MN, Stebbing ARD, Widdows J (1985). The Effects of Stress and Pollution on Marine Animals. Praeger, New York, USA

Depledge MH (1994). The rational basis for the use of biomarkers as ecotoxicological tools. In: Fossi MC, Leonzio C (eds) Nondestructive biomarkers in vertebrates. CRC Press, Boca Raton, pp 271-297

Hodson PV (1990). Indicators of ecosystem health at the species level and the example of selenium effects on fish. Environ Monit Assess 15:241-254

Hopkin SP, Hardisty GN, Martin MH (1986). The woodlouse *Porcellio scaber* as a biological indicator' of zinc, cadmium, lead and copper pollution. Environ Pollut (Ser B) 11:271-290

Huggett RJ, Kimrie RA, Mehrie PM, Bergman HL, Dickson KL, Fava JA, McCarthy JF, Parrish, R, Dorn, PB, McFarland, V, Lahvis, G (1992). Introduction. In: Huggett, R.J., Kimerle RA, Mehrie PM, Bergman H (eds) Biomarkers: Biochemical, Physiological and Histological Markers of Anthropogenic Stress. Lewis Publishers, Boca Raton, pp 1-3

Jagers op Akkerhuis GAJM, Hamers THM (1992). Substrate-dependant bioavailability of deltamethrin for the epigeal spider *Oedothorax apicatus* (Blackwall) (Araneae, Erigonidae). Pestic Sci 36:59-68

Jensen CS, Garsdal L, Baatrup E (1997). Acetylcholinesterase inhibition and altered locomotor behavior in the carabid beetle *Pterostichus cupreus*. A linkage between biomarkers at two levels of biological complexity. Environ Toxicol Chem (in press)

Jepson PC, Chaudry AG, Salt DW, Ford MG, Efe E, Chowdry ABMNU (1990). A reductionist approach towards short-term hazard analysis for terrestrial invertebrates exposed to pesticides. Funct Ecol 4:73-78

Lima SL, Dill LM (1990). Behavioural decisions made under the risk of predation: a review and prospectus. Can J Zool 68:619-640

Little EE (1990). Behavioral toxicology: Stimulating challenges for a growing discipline. Environ Toxicol Chem 9:1-2

Little EE, Finger SE (1990). Swimming behaviour as an indicator of sublethal toxicity in fish. Environ Toxicol Chem 9:13-19

Peakall DB (1994). Biomarkers. The way forward in environmental assessment. Toxicol Environ News 1:55-60

Salt DW, Ford MG (1984). The kinetics of insecticide action. Part III: the use of stochastic modelling to investigate the pick-up of insecticides from ULV-treated surfaces by larvae of *Spodoptera littoralis* Boisd. Pestic Sci 15:382-410

Sørensen FF, Bayley M, Baatrup E (1995). The effects of sublethal dimethoate exposure on the locomotor behaviour of the collembolan *Folsomia candida* (Isotomidae). Environ Toxicol Chem 14:1587-1590

Sørensen FF, Weeks JM, Baatrup E (1997). Altered locomotor behaviour in woodlouse (*Oniscus asellus* L.) collected at a polluted site. Environ Toxicol Chem 16(4):685-690

Warburg M R (1987). Isopods and their terrestrial environment. Adv Ecol Res 17:187-242
Warner RE, Peterson KK, Borgman L (1966). Behavioural pathology in fish: a quantitative study of sublethal pesticide toxication. J Appl Ecol 3 (suppl):223-247

Markers of Exposure to Aromatic Amines and Nitro-PAH

Hans-Günter Neumann, Iris Zwirner-Baier and Corinne van Dorp
Dept. of Toxicology, University of Würzburg, Versbacher Str. 9, 97078 Würzburg, Germany

Introduction

An essential step to protect human health from chemicals at the workplace or in the environment was to measure their concentration in the environment and to correlate them with adverse health effects. Tolerance values, like the maximum workplace concentrations (MAK-values), are established on that basis. A next step to improve exposure control was the introduction of human biomonitoring. Measuring the parent chemical or its stable metabolites in biological material, like blood or urine, allowed to assess the internal stress as compared to the external stress given by environmental monitoring. However, the biological activity of chemicals, of genotoxic carcinogens in particular, depends mostly on reactive intermediates generated in the course of metabolic activation. The biologically active dose cannot be assessed from stable metabolites usually measurable in biological metarial, but from the effects they produce. In contrast to „biomarkers of exposure" the expression „biomarkers of response" has been introduced. In this category two kinds of markers exist: biochemical effect markers and biological effect markers. The former markers are mainly protein and DNA adducts, the latter mutations, micronuclei, chromosomal aberrations or sister chromatid exchanges. Whereas the biochemical effects are not considered pathological *per se*, the biological effects can be seen one step further down to disease. However, as a dosimeter for the biologically active dose, adduct measurements are more specific and more sensitive than the biological endpoints. Moreover, they represent the strain of individuals best.

We have developed a method to analyse hydrolysable hemoglobin adducts formed by aromatic amines and nitroarenes, which can be applied widely for biochemical effect monitoring (Neumann 1984, Albrecht and Neumann 1985, Birner et al 1990). The results of several studies in human populations are reviewed and some conclusions are drawn.

The Hemoglobin Adducts

Aromatic amines are usually metabolically activated by N-oxidation to N-hydroxylamines. These are the common precursors for either the formation of DNA adducts or hemglobin (Hb)-adducts, i.e. esterification leads to an electrophile reacting with nucleophilic centers, among others of DNA, oxidation to a nitroso-derivative which adds to sulfhydryl groups (Eyer 1994). The latter reaction takes place in erythrocytes in the course of a cooxidation reaction leading to the nitroso-arene and methemoglobin. Since both oxidation products are reducible, a circle is established such that each hydroxylamine which enters the erythrocyte has several chances to react as the nitroso-derivative with the SH-group of cysteine in hemoglobin. The initial reaction product rearranges to give a sulfinamide, which is stable in vivo but can be hydrolysed readily in vitro. To analyse these hydrolysable hemoglobin adducts, 5 to 10 ml blood are collected, the erythrocytes are washed, the hemoglobin precipitated and extracted to remove non-covalently bound metabolites, the sulfinamides are hydrolysed, the resulting amines extracted, usually by solid phase extraction, derivatised and analysed by GC-MS.

Nitro-arenes enter the same pathway after reduction of a nitro-group (Suzuki et al 1989), which in many cases occurs in the gut by intestinal bacteria, but may also be achieved by liver enzymes. The nitro-group may be reduced all the way to the amine which can be reoxidised. The biological effects, however, depend on the bioavailability of the hydroxylamine and the nitroso-derivative, which is best represented by the adducts they form.

The method is also suitable to assess the biologically active dose of another group of chemicals, the azo-compounds (Birner et al 1990, Zwirner-Baier and Neumann 1994). Azo-groups are reductively cleaved by intestinal bacteria and the bioavailability of carcinogenic amines from azo-colourants has been demonstrated in this way.

Hemoglobin Adduct Analysis in Occupational Exposure Control

One of the first applications of the method was to establish biological tolerance values for occupational exposures. Originally, the tolerance value for aniline was set on the basis of methemoglobin formation (5% methemoglobin) or the concentration of free aniline in urine (1 mg/l). Both parameters are extremely time dependent and allow only to control most recent exposures, whereas aniline released from the hemoglobin adduct represents the integral uptake over the last 3-4 months, considering a lifetime of 120 days for human erythrocytes. A BAT-value (biological tolerance value) was established of 100 µg aniline released from hemoglobin of one liter whole blood. The same BAT-value applies to nitrobenzene, because the reduction product phenylhydroxylamine is considered the toxic metabolite as with aniline (DFG 1996). Although other BAT-values have not yet been formally announced, the exposure to many aromatic amines is controlled routinely in some industries by analysing hemoglobin adducts.

Exposure to Polycyclic Aromatic Nitro-Arenes (Nitro-PAH) in Pyrolysis Products

Coke oven workers: As part of an EU-coordinated project (ESCE Research Project, Contract n. 7280-01-014), we have tested the proposal to use hemoglobin adducts formed by polycyclic nitro-arenes as biochemical effect markers for exposures to pyrolysis products, just as DNA adducts from polycyclic aromatic hydrocarbons have been used for that purpose, hoping that the method would be more sensitive and more specific, as well as more easy to apply. Five different cleavage products were analysed obtained by hydrolysis of the respective mononitro-PAH-Hb-adducts in blood samples from coke oven workers and controls (Table 1, van Dorp 1997).

Table 1: Biomonitoring of coke oven workers (percentage of samples with detectable adduct levels).

Cleavage product	Exposed (N = 102)	Controls (N = 19)
1-Aminopyrene	95%	89%
2-Aminofluorene	91%	58%
9-Aminophenanthrene	89%	53%
3-Aminofluoranthene	56%	32%
6-Aminochrysene	51%	21%

Cleavage products from all the adducts studied were detectable in blood samples of most of the workers, but also in many of the controls. For a quantitative evaluation of the results, the workers were assigned roughly to three different workplaces, i.e. at the bottom of the coke oven, at the bench and on the topside. The adduct levels varied widely from below the detection limit (0.003 pmol/g Hb) to more than 2 pmol/g Hb within the groups and they overlapped between groups (Table 2).

The sum of all Hb-adducts of all workers was significantly different from controls ($P < 0.005$) and most of the values of those working in the bottom area were significantly different from controls ($P < 0.005$). It is therefore concluded that nitro-PAHs are present in this working environment as part of the pyrolysis products and contribute to the exposure to hazardous chemicals. The most remarkable result of this study was, however, that 40% of the workers exposed to the greatest extent in the bottom area had adduct levels as low as the lowest class found in controls, and only 30% had adduct levels clearly higher than the 95th percentile of the controls (Neumann et al 1995b). This means that external stress does not correlate with the individual strain, or, in other words, in a group of individuals exposed to a comparable extent, only some may be at increased risk.

Although this conforms with practical experience, biochemical effect monitoring may open the possibility to find out some of the reasons for individual susceptibility.

Table 2: Hemoglobin adduct levels of nitro-PAHs in coke oven workers and controls (95th percentile and median values in pmol adduct/g Hb).

Cleavage Product	Area of exposure							
	Bottom		Bench		Topside		Controls	
	95^{th}P	Med.	95th P	Med.	95th P	Med.	95th P	Med.
1-Aminopyrene	1.8	0.17	0.6	0.06	0.8	0.16	0.4	0.03
2-Aminofluorene	0.4	0.14	0.4	0.05	0.6	0.15	0.06	0.003
9-Aminophen.	0.3	0.03	0.1	0.02	0.1	0.03	0.09	0.006
3-Aminofluoran.	0.2	0.02	0.1	-	0.1	0.03	0.03	-
6-Aminochrysene	0.3	0.04	0.1	-	0.1	0.02	0.02	-

Diesel exhaust: In another coordinated EU-project (EG-BMH1-CT93-1309) the question was asked: is it possible to find biochemical effect markers for exposure to diesel exhaust and to assess human exposure? Nitro-PAHs are known constituents of diesel exhaust and 1-nitropyrene has been considered typical for that (IARC 1989, Scheepers et al 1995). Blood samples were collected from bus garage workers in Southampton. Individuals living in an urban area and in a putatively clean rural area served as controls. The same hemoglobin adducts were analysed (Table 3).

1-Aminopyrene and 2-aminofluorene were the most abundant cleavage products and were detectable in most of the blood samples. The adduct levels, however , did not differ significantly between groups (Table 4).

Table 3: Biomonitoring of nitro-PAHs: Hb-adducts from diesel exhaust (percentage of samples with detectable adduct levels).

Cleavage product	Exposed	Controls	
	Bus garage workers (N = 29)	Urban area (N = 20)	Rural area (N = 14)
1-Aminopyrene	86%	95%	100%
2-Aminofluorene	76%	70%	71%
3-Aminofluoranthene	35%	38%	7%
9-Aminophenanthrene	21%	17%	0%
6-Aminochrysene	7%	8%	0%

Table 4: Hb-adduct levels of nitro-PAHs in bus garage workers and controls (95. percentile and median values in pmol/g Hb).

Cleavage product	Bus garage workers (N = 29)		Urban controls (N = 20)		Rural controls (N = 14)	
	95th P	Median	95th P	Median	95th P	Median
1-Aminopyrene	0.29	0.13	0.58	0.16	0.58	0.10
2-Aminofluorene	0.11	0.04	0.17	0.06	0.12	0.03

Subsequently, we had a chance to obtain blood samples from miners engaged in clean-up work in a former potassium mine in East Germany, provided by Dr. Säverin (Berlin). Most of them were exposed to diesel exhaust from heavy buldozers. Moreover, we analysed blood samples from bus drivers as part from an ongoing study in Copenhagen (Autrup, pers. communication). Controls were not available in these cases. Again, 1-aminopyrene and 2-aminofluorene were most abundant, interindividual variability was great (10-15fold), but in both groups the average levels were higher than in the bus garage workers (Table 5).

Table 5: Hb-adduct levels in three different groups of workers occupationally exposed to diesel exhaust (median values in pmol/g Hb).

Cleavage product	Miners (N = 30)	Bus drivers (N = 40)	Bus garage workers (N = 29)
1-Aminopyrene	0.24	0.45	0.13
2-Aminofluorene	0.14	0.09	0.04
3-Aminofluoranthene	0.14	0.03	0.03

Consistently, the levels of different adducts do not correlate in the individuals, i.e. individuals high in 1-nitropyrene adducts are not necessarily also high in 1-nitrofluorene adducts. Although the composition of the external exposure mixture may differ and this may partly explain this finding, it will be interesting to find out to which extent it is caused by the individual's metabolism or pharmacokinetic properties in general.

Exposure to Wastes of TNT-Derived Explosives

Many areas of production and use in World Wars I and II of trinitrotoluene-(TNT) based explosives are still contaminated with waste material. The wastes contain a mixture of by-products from synthesis, purification and degradation. Among the most abundant components beside 2,4,6-TNT are the mono-reduction products 2-amino-4,6-DNT and 4-amino-2,6-dinitro-DNT, as well as 2,4- and 2,6-dinitrotoluene (DNT), and 1,3-dinitrobenzene. The content in soil may vary from the detection limit (10-50 µg/kg) to 1g/kg dry weight. All these monocyclic polynitro-arenes (nitro-MAH) are acutely toxic, mutagenic and either carcinogenic or suspected of being carcinogenic. 2,6-DNT is a comparatively strong liver carcinogen in rats (Leonhard et al 1987). An increased incidence of liver cancer among munition workers has recently been published (Stayner et al 1993),and 2,4,6-TNT-Hb-adducts have been found in munition workers (Sabbioni et al 1996) . Therefore the questions arose: is there a health hazard for people living in an area with contaminated soil, and what concentration of contaminants should be tolerated? To answer these questions it is necessary to find out whether these nitro-MAHs are suitable indicator substances and to determine how much of them is taken up under normal living conditions. We proposed to use hemoglobin adducts of the mono-reduction products of the above mentioned nitro-MAHs as a dosimeter for internal stress and individual strain (Zwirner-Baier et al 1994).

With polyfunctional nitro-arenes, the situation is more complicated, since more than one nitroso-derivative may be formed metabolically depending on which and how many nitro-groups are reduced and further metabolised. The cleavage products analysed are listed in Table 6, which shows the results of a first study with 34 potentially exposed individuals and 34 controls (van Dorp 1997, Neumann et al 1995a).

Table 6: Biomonitoring of nitro-MAHs from contaminated soil (percentage of samples with detectable adduct levels)

Cleavage product	Exposed (N = 34)	Controls (N = 34)
2-Amino-4,6-DNT	51%	56%
4-Amino-2,6-DNT	38%	32%
2-Amino-4-nitrotoluene	43%	44%
4-Amino-2-nitrotoluene	73%	82%
2-Amino-6-nitrotoluene	95%	100%
1-Amino-3-nitrobenzene	0	0

Except for 1-amino-3-nitrobenzene, all analysed amines were detected in at least some of the blood samples. Most surprisingly, the adducts are present to the same degree in controls, i.e. individuals living in the same city, but in an uncontaminated area. Even the two aminodinitrotoluenes were found in controls, although adduct levels were slightly, but not significantly, lower. Most remarkable are the high adduct levels of dinitrotoluenes in both groups (Table 7).

Table 7: Hb-adduct levels of nitro-MAHs of individuals living in a contamin-ated area and controls (95th percentile and median values in pmol/g Hb).

Cleavage product	Exposed (N = 34)		Controls (N = 34)	
	95th P	Median	95th P	Median
2-Amino-4,6-DNT	63	-	15	-
4-Amino-2,6-DNT	55	-	3	-
2-Amino-4-nitrotoluene	200	-	300	-
4-Amino-2-nitrotoluene	500	27	220	45
2-Amino-6-nitrotoluene	280	65	390	71

A number of conclusions can be drawn from this study with nitro-MAHs:

(1) The method is suitable to detect Hb-adducts of these nitro-MAHs in humans.

(2) The Hb-adducts of nitro-MAHs are not suitable to assess exposure to explosive waste because human populations are exposed to high levels of these chemicals, and the exposure from explosive wastes seems to add little to these backgrounds.

(3) A search for sources of nitro-MAHs is urgently needed.

(4) The same biologically active metabolites are formed in humans as in experimental animals.

Conclusions

The described studies confirm the advantages of using hemoglobin adducts for exposure control by biochemical effect monitoring. The method is specific. The adduct forming chemical is identified by retention time in the gas chromatogram and by mass in the mass spectrometric determination. The method is sensitive enough to detect background exposure levels. Several adducts can be determined simultaneously. The life time of 120 days for human erythrocytes allows to assess exposures of the last 3-4 months preceding blood sampling.

The advantage of biochemical effect monitoring over environmental monitoring lies in the possibility to assess the individual's strain. This fact links the improvement of exposure control with aspects of risk assessment (Neumann et al 1995b). The results of the present and other studies consistently reveal two important points:

(1) Even in situations in which considerable external exposures can be demonstrated, the internal stress and strain is much lower than one would have expected.

(2) Adduct levels of exposed and control individuals overlap strongly.

The latter aspect explains the difficulties to find, in epidemiological studies, significant differences on a group basis and to demonstrate a causal relationship between exposure and effect. Now, with the possibility to define reference values for internal stress and strain accounting for different sources and routes of uptake, it is easier to demonstrate if exposures in a specific situation add to background exposure, as shown by the example of the coke oven workers, and how big the excess is. But this example clearly shows also that only part of the workers at the „dirty" work place have adduct levels higher than their fellow citizens. Among the people living in the „dirty" area, nobody's strain parameter was increased above the reference value. Although biochemical effect markers do not allow to calculate any absolute risk, they provide an idea of the increase of relative risk, which is considered low in case of those coke oven workers with adduct levels above the reference value and minimal, if not absent, in case of living in the particular area contaminated with explosive waste.

References

Albrecht W, Neumann H-G (1985) Biomonitoring of aniline and nitrobenzene. Arch Toxicol 57:1-5

Birner G, Albrecht W, Neumann H-G (1990) Biomonitoring of aromatic amines III: hemoglobin binding of benzidine and some benzidine congeners. Arch Toxicol 64:97-102

DFG (1996) Deutsche Forschungsgemeinschaft List of MAK and BAT Values 1996, Commission for the Investigation of Health Hazards of Chemical Compounds in the Work area, Report No 32, VCH Verlagsgesellschaft, Weinheim

Eyer P (1994) Reactions of oxidatively activated arylamines with thiols: reaction mechanisms and biological implications. An overview. Environ Health Perspec 102 (Suppl 6): 123-132

IARC (1989) IARC Monographs on the Evaluation of Carcinogenic Risk to Humans: Diesel and Gasoline Engine Exhausts and Some Nitroarenes. Vol 46, International Agency for Research on Cancer, Lyon

Leonhard TB, Graichen ME, Popp JA (1987) Dinitrotoluene isomer-specific hepatocarcinogenesis in F344 rats. J Natl Cancer Inst 79:1313-1319

Neumann H-G (1984) Analysis of hemoglobin as a dose monitor for alkylating and arylating agents. Arch Toxicol 56:1-6

Neumann H-G, Albrecht O, van Dorp C, Zwirner-Baier I (1995) Macromolecular adducts caused by environmental chemicals. Clin Chem 41: 1835-1840

Neumann H-G, van Dorp C, Zwirner-Baier I (1995) The implications for risk assessment of measuring the relative contribution to exposure from occupation, environment and life-style: hemoglobin adducts from amino- and nitro arenes. Toxicol Letters 82/83:771-778

Sabbioni G, Wie J, Liu Y-Y (1996) Determination of hemoglobin adducts in workers exposed to 2,4,6-trinitrotoluene. J Chromatography B, 682:243-248

Scheepers PT, Martens MH, Velders DD, Fijneman P, van Kerkhoven M, Noordhoek J, Bos RP (1995) 1-Nitropyrene as a marker for the mutagenicity of diesel exhaust-derived particulate matter in workplace atmospheres. Environ Mol Mutagen 25:134-147

Stayner LT, Dannenberg AL, Bloom T, Thun M (1993) Excess hepatobiliary cancer mortality among munition workers exposed to dinitrotoluenes. J Occup Med. 35:291-296

Suzuki J, Meguro S-I, Morita O, Hirayama S, Suzuki S (1989) Comparison of in vivo binding of aromatic nitro compounds to rat hemoglobin. Biochem Pharmacol 38:3511-3519

van Dorp C (1997) Hämoglobinaddukte von Nitroaromaten als Dosimeter für die menschliche Beanspruchung durch Pyrolyseprodukte und Sprengstoffaltlasten, Thesis, Faculty of Chemistry and Pharmacy, University of Würzburg

Zwirner-Baier I, Neumann H-G (1994) Biomonitoring of aromatic amines IV: use of hemoglobin adducts to demonstrate the bioavailability of cleavage products from diarylide azo pigments in vivo. Arch Toxicol 68:8-14

Zwirner-Baier I, Kordowich F-J, Neumann H-G (1994) Hydrolyzable hemoglobin adducts of polyfunctional monocyclic N-substituted arenes: dosimeters of exposure and markers of metabolism. Environ Health Perspec 102 (Suppl 6):43-45

Hematological and Biochemical Parameters in Pollution-Exposed Mice

Miquel Borràs, Santiago Llacuna, Assumpta Górriz and Jacint Nadal
Department of Animal Biology - Vertebrates, Faculty of Biology, University of Barcelona, Diagonal 645, 08028 Barcelona, Spain

Introduction

The selection of a model in life sciences research, especially when dealing with environmental effects of pollution, poses the dilema of prioritizing one of the terms of the dialectic pair "realism vs. control of experimental parameters". This fact suggests the need for a multiple, diversificate approach.

Current research in this field has mainly focused on the exposure of laboratory rodents to artificially generated atmospheres under controlled conditions (Oguz and Baskurt, 1988; Baskurt, 1988; Baskurt et al., 1990; Dikmenoglu et al., 1991).

As part of a comprehensive study of the impact of a coal-fired power plant in Cercs (Catalonia, northeast Spain), we have assessed the effects of air pollutants (nitric and sulphur oxides and particulate material, considered as a complex mixture, submitted to chemical interactions and to the influence of climatic conditions) on animals captured in the wild in the contamined area compared with animals captured in a matched, clean one. These animals included rodents (Llacuna et al., 1993; Górriz et al., 1994; Górriz et al., 1996), passerine birds (Llacuna et al., 1995; Llacuna et al., 1996) and arthropoda (Salgado et al., 1996). This is a realistic approach, but it does not allow any control on sex, age, health status or genetic characteristics of the study subjects, nor on the real time of exposure or a possible selective pressure of predators.

In that context we considered it as an interesting alternative to undertake an intermediate approach, placing caged homogeneus, controlled laboratory rodents under field conditions into both the polluted and the control zones.

Materials and Methods

Study areas: The polluted zone was located 1 Km northwest of the power station (in the direction of predominant winds), at an altitude of 1000 m. The control area (Sant Jaume de Frontanyà) is 20 Km to the east, at the same altitude, and presents similar characteristics of climate, relief and vegetation.

Data of gaseous immision during the assay (emitted by the electric company, FECSA), are shown in Table 1. The corresponding measurements were made in the control area for a period of two weeks, with all the values being within the normal range.

Table 1. SO_2 and NO_x values ((g/m^3) measured from March 1995 to February 1996 at St.Corneli Station (Cercs). The large standard deviation observed was due to certain high pollution days.

Month	SO_2	NO	NO_2	NO_x
March	66±304	6±29	13±16	21±54
April	60±109	5±12	14±17	21±34
May	83±147	15±121	30±185	53±371
June	54±85	2±6	8±11	11±18
July	142±177	6±16	15±18	23±37
August	43±72	1±9	3±7	5±16
September	44±92	1±4	3±5	4±8
October	8±8	0±0	1±2	1±2
November	69±169	2±14	8±14	11±29
December	81±173	1±11	8±12	11±24
January	ND	ND	ND	ND
February	34±88	0±2	6±10	6±11

Animals: 40 multiparous female Ico:Swiss mice, 270 days old at the beginning of the asssay, were randomly allocated to both groups and maintained *in situ* for 6 months, in conventional makrolon cages, protected from rain. PANLAB A04 pellets and tap water (from the public supply of the city of Barcelona) were allowed *ad libitum.*

Weight development and water and food consumption: Body weigths and water and food intake (expressed as g or ml per animal and day) were determined every 10 days.

Blood and plasma samples: Blood was drawn (at 1 and 3 months after the starting of the study and at sacrifice) from the periorbital sinus by introducing a heparinized microhematocrit tube. Samples were placed in Ependorf tubes and kept cool until use for a maximum of 2 hours after extraction. Plasma samples were obtained by centrifuging blood, immediately after extraction, in a centrifuge Spotchem CF-9510 for three minutes at 4,000 rpm, and transported also in an ice container.

Blood and plasma parameters: The erythrocyte count (RBCC) was carried out in a COULTER counter. Hematocrit was determined by a standard microtechnique. Hemoglobin was measured by optical density at 540 nm, after tratment with Drabkin's reagent. Osmolality was determined with a freezing point-based microsmometer (Advanced Instruments 3MO). Total proteins, glutamate-oxalacetate transaminase (GOT), glutamate-pyruvate transaminase (GPT), blood urea nitrogen (BUN) and creatinine were analyzed using a SPOTCHEM system (Menarini).

Necropsy and Histopathology: At sacrifice animals underwent an anatomo-pathological study, the results of which are not reported in the present paper.

Statistical analysis: Comparison of weights and of hematological and biochemical values was made by means of unpaired Student's t test, after assessement of the homogeneity of the standard deviations (Barlett's test). For water and food intake the Student's t test for matched data was used, and the effectiveness of pairing assessed.

Results

Total body and organ weight values are shown in Table 2. No differences were seen in any case between animals from polluted and control zones, respectively.

Data of food and water intake are presented in Table 3. A statistically significant, although slight, increase was noticed in food consuption of pollution-exposed animals; on the contrary, their water intake was markedly decreased ($p=0.0003$).

Hematological data are summarized in Table 4. RBC counts, higher than normal due to the elevated altitude, do not show any differences related to pollution exposure. Hematocrit values show a slight increase in mice caged in the contamined zone, especially in the first month of the assay. Hemoglobin was not determined at the end of the study; at three months there is a slight decrease in control animals. No differences were observed in osmolality.

Table 2. Body and organ weight values (mean ± standard deviation)

Zone	St.Jaume Frontanyà	Cercs
Body weight	45.20 ± 6.6	45.29 ± 4.9
Liver weight	5.99 ± 0.49	5.75 ± 0.33
Spleen weight	0.345 ± 0.007	0.342 ± 0.009

Table 3. Food and water consumption values (mean ± standard deviation)

Zone	St.Jaume de Frontanyà	Cercs	P
Food intake	6.72 ± 0.64	7.10 ± 0.65	0.0474
Water intake	9.96 ± 1.49	8.27 ± 1.28	0.0003

Table 4. Hematological and serum osmolality values (mean ± standard deviation)

Zone		St.Jaume de Frontanyà	Cercs
Parameter	Period		
Hct(%)	1 month	52.61 ± 2.2	54.74 ± 3.2[a]
	2 months	49.95 ± 5.0	52.00 ± 1.7
	3 months	49.32 ± 2.1	49.90 ± 2.4
RBC (10^7/mm^3)	1 month	1.020 ± 0.5	1.000 ± 0.673
	2 months	1.015 ± 0.826	1.048 ± 0.368
	3 months	1.067 ± 0.742	1.029 ± 0.622
Hb(g/dl)	1 month	0.552 ± 0.3	0.555 ± 0.03
	2 months	0.510 ± 0.06	0.543 ± 0.02[a]
Osm	1 month	309.57 ± 5.13	307.63 ± 14.60
	2 months	298.86 ± 4.56	295.38 ± 10.81
	3 months	298.50 ± 7.12	294.40 ± 5.66

[a] $p < 0.05$

Plasma biochemical data are shown in Table 5. Total proteins and urea values are equivalent in both zones. A very small decrease in creatinine was observed in pollution-exposed animals at 1 month. There were high values of GOT and GPT among animals of the polluted zone at 1 month from the beginning of the assay, gradually decreasing at 3 months and reaching even lower-than-control levels at the end of the study.

Table 5. Plasma biochemical parameters (mean ± standard deviation)

Zone		St.Jaume de Frontanyà	Cercs
Parameter	Period		
GOT	1 month	81.86 ± 32.2	121.88 ± 68.6
	2 monts	83.28 ± 44.3	105.75 ± 40.8
	3 months	85.40 ± 20.3	51.27 ± 16.8[a]
GPT	1 month	30.86 ± 21.5	151.50 ± 84.2[a]
	2 months	30.86 ± 6.7	42.87 ± 15.1
	3 months	28.80 ± 12.3	18.18 ± 5.2[b]
BUN	1 month	22.04 ± 1.9	22.02 ± 5.7
	2 months	22.57 ± 1.8	25.00 ± 4.4
	3 months	21.50 ± 6.1	21.91 ± 1.5
Creatinine	1 month	14.51 ± 0.57	13.81 ± 0.47[b]
	2 months	14.81 ± 0.13	14.95 ± 0.18
	3 months	14.82 ± 0.20	14.70 ± 0.06
Protein	1 month	6.17 ± 0.42	5.85 ± 0.46
	2 months	5.99 ± 0.11	5.96 ± 0.24
	3 months	6.21 ± 0.43	6.01 ± 0.28

[a] $p < 0.005$
[b] $p < 0.05$

Discussion

The observed decrease in water intake (of about 2 ml per animal and day), consistent with data obtained in a previous study (Górriz et al., 1996), would require further research, considering that the physiological control of osmotic and volemic thirst is a complex mechanism involving many factors. Water administered to both groups was, however, of the same origin; thus, palatability differences could be ruled out.

Intravascular factors are probably not involved either, because osmolality was not altered (and taken together with normal concentration values of total proteins we may suggest that neither was the electrolyte fraction), and hematocrit even increased (however, a transient decrease in hematocrit, possibly due to dilution, had been previously observed, immediately after peaks of maximum pollution [Górriz et al., 1996]).

We may certainly not discard the possibility of undetected microclimatic factors; however, in our previous study (Górriz et al., 1996) the cages were located in quite different places within each zone. A possible explanation may be found in the fact that animals in the polluted zone presented a markedly reduced activity, as observed by video recording (Górriz et al., 1996). The role of porto-hepatic osmoreceptors in satiety-signalling is also to be considered in further experiences. Anyway, rodents are able to utilize very efficiently metabolic water, thus the slight increase in food consumption would be enough to maintain a normal ponderal evolution and general condition.

The slight increase in hematocrit at 1 month may be due to corpuscular volume; Kucera (1980, 1983, 1988) and Baskurt (1988) report also increses in hematocrit values after SO_2 and heavy metal exposure under field and laboratory conditions. At three months the difference in mean values is due toone single animal of the control zone which showed also very low RBCC and hemoglobin values, and which was culled before the end of the study due to a tumour in the uterus.

The slight decrease in hemoglobin values of control animals at 3 months is parallel to the RBCC and hematocrit evolution and does not seem to have any biological significance.

The renal profile was not significantly altered. However, it must be kept in mind that urea and creatinine are not very sensitive markers, and their levels often do not show any modification until kidney damage is severe.

The important increases detected in hepatic markers in the first part of the study are also consistent with previous observations in birds (Llacuna et al., 1996), and indicate some acute alteration of hepatic function that, however, leads eventually to a gradual adaptation.

Results of the histopathological study, that will be presented in detail elsewhere, include, for pollution-exposed animals, severe inflamatory processes in the respiratory tract, pigment deposition in Kupffer cells, interstitial nephritis and some degree of glomerular damage; focal necrosis surrounded by inflamatory reactions were seen in livers of animals from both zones. These preliminary results are compatible with the hematological and biochemical data presented here.

References

Baskurt OK, Levi E, Andac SO, Caglayan S (1990) Effects of sulfur dioxide inhalation on erythrocyte deformability. Clin Hemorheol 10(5):485-489

Baskurt OK (1988) Acute hematologic and hemorheologic effects of sulphur dioxide inhalation. Arch Environ Health 43(5):344-348

Dikmenoglu N, Baskurt OK, Levi E, Caglayan S, Guler S (1991) How does sulphur dioxide affect erythrocyte deformability. Clin Hemorheol 11(5):497-499

Górriz A, Llacuna S, Riera M, Nadal J (1996) Effects of air pollutin on hematologiacal and plasma parameters in *Apodemus sylvaticus* and *Mus musculus* Arch Environ Contam Toxicol 31:153-158

Górriz A, Llacuna S, Durfort M, Nadal J (1994) A study of the ciliar tracheal epithelium in passerine birds and small mammals subjected to air pollution: ultrastructural study. Arch Environ Contam Toxicol 27:137-142

Kucera E (1988) Effects of smelter emissions on the hemogram of the deer mouse (*Peromyscus maniculatus*) Environ Pollut 55:173-177

Kucera E (1980) Emissions distribution reflected in wild rodent tissues. Submission to the Clean Environment Comission Hearing in conection with Hudson Bay Mining and Smelting Co., Ltd. 15pp.

Kucera E (1983) Monitoring smelter emissions using small mammals as indicators. Manitoba Department of Environmental Workplace Safety and Health 83(14), 15pp.

Llacuna S, Górriz A, Riera M, Nadal J (1996) Arch Environ Contam Toxicol 31:148-152

Llacuna S, Górriz A, Sanpera C, Nadal J (1995) Metal accumulation in three species of passerine birds (*Emberiza cia, Parus major* and *Turdus merula*) Arch Environ Contam Toxicol 28:298-303

Llacuna S, Górriz A, Durfort M, Nadal J (1993) Effects of air pollution on passerine birds nd small mammals Arch Environ Contam Toxicol 24:59-66

Oguz K, Baskurt MD (1988) Acute hematologic and hemorheologic effects of sulfur dioxide inhalation Arch Environ Health 53(5):344-348

Salgado JM, Llacuna S, Górriz A, Borràs M, Nadal J (1996) Effects of a coal-fired power plant on arthropod biodiversity Toxicology Letters, suppl 1/88, 95

Current Molecular Approaches in Toxicology

(Chair: E. Nelson, Germany, and
E. Dybing, Norway)

Detection of Low Levels of DNA Damage Arising from Exposure of Humans to Chemical Carcinogens

Paul L Carmichael
Imperial College School of Medicine at St Mary's, Pharmacology and Toxicology, Norfolk Place, London, W2 1PG, U.K.

Introduction

The traditional approach to cancer epidemiology attempts to relate clinical outcome, i.e. cancer, with past exposure to a chemical carcinogen or putative chemical carcinogen. Using this approach the intermediate stages in the disease process are unknown. In contrast, new molecular approaches to cancer epidemiology attempt to bridge this gap by seeking to identify and quantify the successive biological stages that lead to tumour formation. For example, we can now measure and associate markers of exposure such as internal dose, effective dose and early biological effects with markers of disease such as altered cellular structure or function and use these measurements in a prognostic manner to assess potential cancer risk. Indeed, it is the early biological changes which are the most valuable factors for measurement in molecular epidemiology. In carcinogenesis the principal site for the majority of such early biological effects, is the nuclear DNA and there are a number of events or lesions that can be induced in DNA through the action of bioactivated chemical carcinogens (see figure 1).

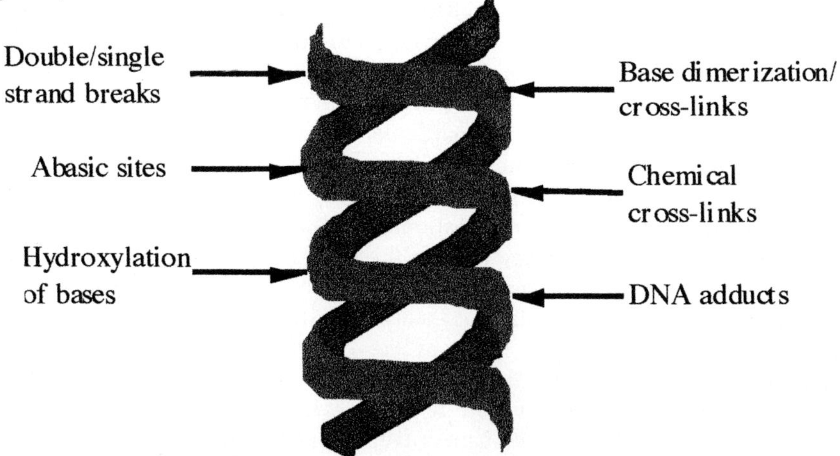

Figure 1. Carcinogen damage to DNA

Figure 1 shows some of the major forms of damage which can be induced in DNA by the action or interaction of chemical carcinogens; these include double or single strand breaks, the formation of abasic sites (both apurinic and apyrimidinic), the hydroxylation of bases (commonly formed through oxygen free-radical attack of DNA), base dimerization or base crosslinks, chemical crosslinks, and the formation of DNA adducts. A great many classes of environmental chemical carcinogens exert their effects through metabolism and covalent binding to form DNA adducts and there is considerable evidence that such damage is a key step in cancer induction. DNA adducts are therefore amongst the most useful biomarkers of exposure to low levels of carcinogens.

The formation of DNA adducts or indeed other carcinogen-DNA damage can have several consequences. The altered DNA within a cell may be repaired efficiently, resolving the problem, leading to a normal cell. If the carcinogen damage and alterations to the DNA are extensive, then a process of apoptosis and programmed cell death may be initiated. However, if a cell undergoes incorrect repair, or if the carcinogen-altered primary sequence remains unchanged, then following DNA replication and cell division, daughter cells may contain fixed mutations. Transcription and translation of that mutated DNA can result in the expression of incorrect or inappropriate proteins and if those mutations are at critical targets in the genome such as proto-oncogenes or tumour suppressor genes then the process of carcinogenesis may be initiated.

Methodologies

When measuring exposure of humans to carcinogens a number of possibilities present themselves. Firstly, one can determine the internal dose of a carcinogen by measuring levels of that carcinogen or its metabolites in body fluids. However, measurements of biological events which may have real consequences in the carcinogenic process are perhaps of greater use and may have greater prognostic value in assessing the potential cancer risk of an individual. One can determine biological events which may occur late in carcinogenesis, such as chromosome aberrations, gene mutations, deletions and rearrangements in normal or tumour tissues; however, measurement of early biological events in the carcinogenic process can be of particular value. Several methods enable us to detect and quantify DNA adducts in human tissues as biomarkers of carcinogen exposure. Exogenous chemical carcinogen exposure may arise from environmental exposure, occupational exposure, dietary components, carcinogen inhalation from smoking or the potential genotoxic hazard of some drugs. In addition, measurement of DNA adducts can also provide information on endogenous events, including oxygen free radical damage to DNA. However it is important to note that many of the carcinogen exposures that can be measured using DNA adduct determinations can also be confounding factors in molecular dosimetry studies and it is always important to document the smoking history, diet, medical history, drug intake, urban or rural dwelling, and potential occupational exposure for given individuals. The principal methods available for detection of DNA adducts in human tissues are immunoassays, fluorescence spectroscopy, HPLC-electrochemical detection, gas or liquid chromatography-linked mass spectrometry and DNA postlabelling utilising ^{35}S, ^{33}P or most commonly ^{32}P.

Immunoassays for the detection of DNA adducts include radioimmunoassay (RIA), enzyme-linked immunosorbent assay (ELISA) and immunohistochemistry which has the added advantage of allowing for the localisation of DNA adducts in tissues (Santella et al., 1990). However there are sensitivity limitations associated with immunoassays of DNA adducts, compared with other methods. In addition, immunoassays can require relatively large amounts of DNA, typically 50 µg or more, and such assays require adducts synthesised in sufficient quantity and purity to raise antibodies against. Furthermore, a common problem with the immunoassay of DNA adducts is the cross reactivity of antibodies. An antibody raised against one particular adduct can commonly recognise adducts derived from the DNA binding of other agents of the same class of compounds and this can often obscure the nature and level of the specific DNA adducts of interest (Poirier, 1991).

Fluorescence spectroscopy can be a highly sensitive method of DNA adduct detection and one can analyse intact DNA, DNA digests and hydrolysis products (Weston and Bowman, 1991). Unfortunately, not all DNA adducts are fluorescent and hence the technique, in general, is only suitable for the detection of DNA adducts arising from the covalent interaction of polycyclic aromatic hydrocarbons (PAH), aflatoxins, O^6- and N^7-methyldeoxyguanosine, and etheno adducts. In contrast, mass spectrometry has a much broader application of use and is performed on derivatised adducts in combination with liquid, capillary or gas chromatography (Shuker et al., 1993, Yamamoto et al., 1990). This method can provide a direct measurement of carcinogen-modified bases cleaved from DNA with potentially high levels of sensitivity. One particular advance in the use of this technique in the study of DNA adducts has been the development of accelerator mass spectrometry which can detect ultra-low levels of adducts (Turtletaub, 1993). However, the technique requires the use of radiolabelled carcinogens as it is a nuclear-physics technique which specifically counts nuclei of cosmogenic isotopes, rather than relying on decay. Consequently it is extremely sensitive and, for e.g. tracing ^{14}C, sensitivity is increased one million-fold relative to decay.

HPLC-electrochemical detection is a somewhat more limited technique available for the detection of specific DNA adducts such as those formed through oxygen free radical attack of DNA, in particular the detection of 8-hydroxydeoxyguanosine (Carmichael et al., 1995). In addition this technique may be used for detection of O^6-methylguanine (Park et al. 1989) and N^7-methylguanine in combination with immunoaffinity purification. This technique, in addition to the other physical methods described above, can be quite chemical-specific. However, their sensitivity largely depends on the adduct type and all of these physical techniques can require the use of quite large amounts of DNA, up to 1 mg. In addition, the methods require DNA-adduct standards for characterisation and quantitation.

^{32}P-Postlabelling

^{32}P-Postlabelling has a number of advantages over the methods previously described for the detection of DNA adducts. In particular, this technique is highly sensitive and frequencies as low as one adduct in 10^{10} normal nucleotides can be detected in human DNA. These analyses can be performed on relatively small quantities of DNA (typically 1 μg) and hence the technique is suitable for the analysis of human samples where a paucity of available tissue is often the norm. In addition, the technique has a wide range of applications and is appropriate for a very broad range of different classes of chemical carcinogen. Indeed, prior adduct characterization is not required and hence unknown adducts can be detected, although conversely the technique can provide little structural information in the absence of adduct standards.

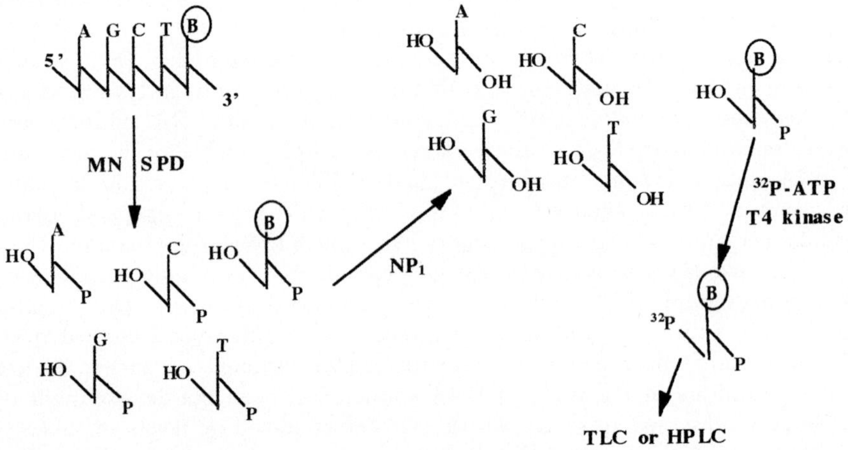

Figure 2. ^{32}P-Postlabelling (Nuclease P1 variation)

There are four main stages in the ^{32}P-postlabelling technique. These are, (i) the digestion of DNA, (ii) the enrichment or enhancement of DNA adducts, (iii) a kinase-catalysed labelling and (iv) the chromatography of those labelled DNA adducts. Figure 2 shows the principal steps in one of the most commonly used variations of the ^{32}P-postlabelling technique, that of the use of the enzyme nuclease P1 in the enrichment of DNA adducts. The figure demonstrates how two enzymes, micrococcal nuclease and spleen phosphodiesterase, are used to digest extracted DNA containing a base lesion or DNA adduct. This initial digestion yields nucleoside-3'-monophosphates which are further digested by treatment with nuclease P1. At this stage advantage is taken of the fact that nuclease P1 poorly recognises many classes of bulky-type DNA adducts and leaves them unchanged in the reaction mixture, but digests the vast majority of normal nucleotides to nucleotides i.e. removing the 3'-phosphates. Following the enrichment step, the DNA adducts with their intact 3'-monophosphates, remain substrates (unlike the normal nucleosides present in the reaction mixture) for the enzyme T4 polynucleotide kinase which, in the presence of ^{32}P-labelled ATP will 'postlabel' the DNA adducts by adding the radioactive phosphate at the 5'-position. The radiolabelled DNA adducts are then separated from the remaining reaction mixture by chromatography. This chromatography can take the form of either thin layer chromatography (TLC) or reverse-phase HPLC. In the case of TLC, a system of multi-directional polyethyleneimine-cellulose chromatography is most commonly used, coupled with autoradiography to reveal the presence of DNA adducts on the TLC plates as a radioactive spot which can be then quantified. Alternatively, the ^{32}P-postlabelled DNA adducts can be analysed by separation on reverse-phase HPLC using a flow-through radioactivity detector.

Human Biomonitoring of DNA Adducts

Sources of DNA for human biomonitoring studies can be from various sites. For example, ^{32}P-postlabelling studies have been performed utilising DNA from peripheral lymphocytes, buccal mucosa cells, cervical smears, exfoliated bladder cells recovered from urine, hair roots, spermatozoa, placenta, skin punch biopsies, cells recovered from lung washings, post-mortem tissues, surgical specimens, and organ or tissue biopsies of bone marrow, bladder, gut, uterus, prostate etc. White blood cells are clearly one of the most accessible sources of DNA for human biomonitoring studies and elevated levels of DNA adducts have been detected in lymphocytes arising from occupational exposure to polycyclic aromatic hydrocarbons in iron foundry workers, coke oven workers, aluminium plant workers, roofers, bus drivers and fire fighters (Phillips and Venitt 1995). In addition, the ^{32}P-postlabelling technique has been extensively used to characterize the elevated levels of DNA adducts which arise in human tissues due to smoking. Studies have measured significantly raised levels of cigarette smoke-derived DNA adducts in human lung, bronchus, bladder, cervix, oral mucosa and lymphocytes (Phillips and Venitt 1995, Phillips, 1988). Furthermore in conjunction with fluorescence and mass spectrometry techniques, ^{32}P-postlabelling has been used to characterize particular DNA adducts in human tissues including adducts arising from exposure to benzo[a]pyrene in human lung, placenta and white blood cells, the food mutagen PhIP in human colon and liver, plus adducts derived from 4-aminobiphenyl, styrene, nitrosamines and mitomycin C in various tissues.

There are, unfortunately, a number of pit-falls which have to be considered when studying DNA adducts, particularly with regard to environmental pollution and monitoring. As discussed above, this includes confounding factors such as smoking status, diet, occupational exposure and medical treatment, but other factors that need to be considered are the identity of the adduct and the nature of the tissue being studied. Some adducts can be detected at a level of 1 in

10^{10} bases using ^{32}P-postlabelling which is outside the range of chemical analysis; thus in certain cases, adducts can be observed, but their identity or origin cannot be determined. Typical measurements of oxidative DNA damage such as 8-hydroxydeoxyguanosine lie in the range of 1 adduct in 10^5 to 10^6 nucleotides and DNA damage arising from exposure to alkylating agents is typically at a level of 1 adduct in 10^6 to 10^7. Lower levels of adducts are more common for lipid peroxidation-derived etheno adducts which have been measured at levels of 1 adduct in 10^7 to 10^8, although typical levels of bulky PAH-type adducts arising from cigarette smoking, pollution or dietary exposure may be even lower at 1 adduct in 10^7 to 10^9. Hence methods for the detection of DNA adducts need to be highly sensitive as the levels of adducts encountered in human tissues are often very low. Most studies in animals have shown a linear correlation between tumour yield and adduct formation, but extrapolation of such data to humans is problematical. The detection of very low levels of DNA adducts in human tissues that cannot be characterized by existing physical methods presents a significant problem, and hence one of the greatest challenges in human biomonitoring is to identify the nature of the critical adducts involved in neoplasia, although considerable value can still be attached to assessments based on the total levels of DNA adducts. Thus, DNA adducts are one of the most useful indicators of exposure of humans to environmental carcinogens, but adduct formation alone cannot be used to undertake risk assessments.

References

Carmichael PL, Hewer A, Osborne MR, Strain AJ and Phillips DH (1995) Detection of bulky DNA lesions in the liver of patients with Wilson's disease and primary haemochromatosis. Mut. Res. 326: 235-243

Park JW, Cundy KC and Ames BN (1989) Detection of DNA adducts by high-performance liquid chromatgraphy with electrochemical detection. Carcinogenesis 10: 827-832

Phillips DH, Hewer A, Martin CN, Garner RC and King M (1988) Correlation of DNA adduct levels in human lung with cigarette smoking. Nature 336: 790-792

Phillips DH and Venitt S (1995) Environmental Mutagenesis, Bios Scientific Publications.

Poirier MC (1991) Development of Immunoassays for the detection of carcinogen-DNA adducts. In: Skipper PL and Groopman (eds) Molecular Dosimetry and Human Cancer, CRC Press, pp211-229

Santella RM, Yang XY, Hsieh LL and Young TL (1990) Immunologic methods for the detection of carcinogen adducts in humans. Prog. Clin. Biol. Res. 340C: 247-257

Shuker DEG, Prevost V, Friesen MD, Lin D, Ohshima H and Bartsch H (1993) Urinary markers for measuring exposure to endogenous and exogenous alkylating agents and precursors. Environ. Health Perspect. 99: 33-37

Turtletaub KW, Vogel JS, Frantz CE and Fultz E (1993) Studies on DNA adduction with heterocyclic amines by accelerator mass spectrometry: a new technique for tracing isotope-labelled DNA adduction. IARC Sci. Publ. 124: 293-301

Weston A and Bowman ED (1991) Fluorescence detection of benzo[a]pyrene-DNA adducts in human lung. Carcinogenesis 12: 1445-1449

Yamamoto J, Subramanian R, Wolfe AR and Meehan T (1990) The formation of covalent adducts between benzo[a]pyrenediol epoxide and RNA: structural analysis by mass spectrometry. Biochemistry 29: 3966-3972

Chemoprevention

(Chair: L.O. Dragsted, Denmark, and
H.E. Poulsen, Denmark)

Natural Antioxidants in Chemoprevention

Lars O. Dragsted
Institute of Toxicology, Danish Veterinary and Food Administration, Søborg, Denmark

It is well documented that diets rich in fruits and vegetables can reduce the risk of most common cancers, and that some food items from this class may be protective against heart disease. Several explanations have been offered, one of which relates to the natural presence of potent antioxidants in plant products. Destructive oxidation of lipids, proteins, DNA, and other important biomolecules, often involving radical chain reactions, affect vital cellular structures and their normal functions. Such processes are involved in the development of cancer as well as heart disease, and it seems logical to assume that antioxidants might be preventive. Large human trials with natural antioxidants have not provided a uniform support, however, for the hypothesis that antioxidation *per se* may prevent cancer or coronary heart disease (CHD)[1]. One reason is that other effects, unrelated to antioxidation, may compromise their preventive effects. Another reason may be that many potent antioxidants can also act as pro-oxidants under certain conditions. The interpretation of animal trials is likewise often compromised by the fact that most antioxidants have other physiological effects which might very well explain their protective action or lead to toxic side-effects. In addition, absorption, metabolism and distribution may profoundly influence their antioxidant actions, and not all cellular compartments are equally well protected. Furthermore, interactions between antioxidant systems are only partially understood.

Still, good evidence exists that antioxidants may reduce oxidative attacks on important structures *in vivo*. Furthermore, studies using biomarkers for oxidative damage have provided evidence in humans and animals that food items containing natural antioxidants can influence oxidative damage. Therefore, with increased knowledge about distribution and mechanistic actions of individual antioxidants, well-designed studies should be able to provide answers to some of the most important questions, i.e whether natural antioxidants can prevent disease by inhibiting oxidative damage to key biological structures, and whether natural antioxidants are indeed important factors in the dietary prevention of disease caused by fruits and vegetables.

---

[1]Abbreviations: CHD, coronary heart disease; LDL, low-density lipoprotein; MDA, malondialdehyde; 8-O-dG, 8-oxo-deoxyguanosine; ROS, reactive oxygen species; NCI, National Cancer Institute

Introduction

Natural primary antioxidants include vitamins E and C, carotenoids, polyphenols, coumarins, some di- and triterpenes, and some organosulphur compounds (Dragsted, 1993; Arouma, 1994; Halliwell, 1994). Selenium may additionally act as an antioxidant in animals by increasing glutathione peroxidase activity. Reactive oxygen species and other radicals are known to initiate oxidative chain reactions in biological molecules, and these processes are counteracted by numerous defence systems in the living organisms. The antioxidants are in the front line of defence by reacting with radicals to form less reactive, delocalized antioxidant radical species (Shahidi and Wanasundara 1992, Palozza and Krinsky 1994, Rice-Evans et al. 1996) which in turn may interact and terminate the process or pass their unpaired electrons on through a chain of antioxidants. Vitamin C has been shown to be at the end of such a chain in plasma. Vitamin C, like some other antioxidants, can accept two electrons and vitamin C is subsequently regenerated enzymatically. Oxidative deterioration of biomolecules is a natural part of the process of ageing (Ames et al. 1993) and is believed to be involved in the pathogenesis of cancer, atherosclerosis, and several other chronic diseases (Cerutti 1994, Halliwell 1994, Frei 1995, Giacosa and Filiberti 1996). The presence of substantial amounts of natural antioxidants in plant foods and the general observations of reduced cancer risk and reduced risk of coronary heart disease (CHD) in high fruit and vegetable consumers (reviewed in Steinmetz and Potter 1991, Ames et al. 1993, Machlin 1995, Potter and Steinmetz 1996) have led to the hypothesis that antioxidants are involved in the prevention of chronic disease. In the following, the evidence that natural antioxidants may be protective towards cancer or CHD is shortly reviewed, and the evidence from animal or human studies is discussed.

Oxidative Damage in Carcinogenesis

Carcinogenesis is a multistage process, involving at least four events or stages, namely initiation, promotion, conversion, and progression (Dragsted et al., 1993). According to this simplified model, the initiation and conversion steps involve genetic damage, whereas the promotion and progression stages are processes of clonal selection and cell dynamics.

Oxygen radicals and other reactive oxygen species (ROS) can interact with DNA to form oxidatively modified DNA bases. Hydroxyl radical in particular is deleterious to DNA, and is involved in the formation of at least 20 different oxidative products of all four bases in DNA. The best studied base modification, 8-oxo-deoxyguanosine (8-O-dG), is formed at a rate giving rise to a steady state of at least 5,000 modifications per cell in human lymphocytes (Collins et al. 1996). The presence of 8-O-dG has been shown to cause mismatch repair resulting in G→T transversions, and formation of mutated gene products (Moriya et al. 1991). Mutations in oncogenes or in tumour suppressor genes are believed to be the initiating event in tumourigenesis, and 8-O-dG derived mutational changes have been proposed to be present in Ki-*ras* and H-*ras* oncogenes and in the tumour suppressor gene, p53, found in human tumour tissues (Cerutti 1994). It seems therefore logical to assume that direct interaction of oxygen radicals with DNA may cause initiation of (or conversion in) tumourigenesis. Direct attempts to prove the initiating potential of 8-O-dG have often not been successful (Denda et al. 1995).

Several products of ROS attack on biomolecules, including malondialdehyde (MDA), are reactive towards DNA, and may be involved in tumour initiation. Oxidative attack on DNA has also been shown to cause sister chromatid exchanges, chromosome aberrations, and cell transformation *in vitro*, indicating that ROS may also be involved in the conversion step in carcinogenesis. Moreover, ROS has been implicated in the clonal selection processes involved in tumour promotion and progression, and induction of 8-O-dG in liver DNA by several different treatments is effective in tumour promotion in N-nitrosodiethylamine initiated rat liver tumourigenesis (Denda et al. 1995). The strong enhancement of promotion by local oxidant stress is supported by studies of dieldrin induced liver carcinogenesis (Klaunig et al. 1995). Consequently, antioxidants might theoretically be protective at all stages of carcinogenesis.

Oxidative Damage in Heart Disease

Atherosclerosis is characterised by progressive development of areas of lipid and cholesterol deposits in the smooth muscle wall of the arteries, from fatty streaks to atherosclerotic plaques to large atherosclerotic lesions. Such lesions can partially block the artery and increase the risk for blood clot formation with resulting risk for ischemia and, if overcome, reperfusion damage.

Oxidative damage to low-density lipoprotein (LDL) is believed by most researchers to play an important role in atherogenesis, although several important questions are still unresolved (Eaton, 1996). During atherosclerosis, increasing amounts of LDL are oxidised and deposited in the plaques in the form of immobilized macrophages (foam cells) containing the scavenged lipoprotein (Frei, 1995). The nature of the oxidants involved in LDL modification is currently a matter of controversy (Garner and Jessup 1996, Bucala 1996). Also the occurrence of oxidative damage during reperfusion of ischemic areas is controversial (Janero 1995).

According to a different theory of atherosclerosis formation, the smooth muscle cells of the intimal subendothelium are initiated to form a lesion, in much the same way as initiation occurs in carcinogenesis. This is supported by the observation of activated oncogenes in smooth muscle cells from aortic plaques, capable of transfecting 3T3 cells *in vitro*, and by observations that plaque muscle cell material is monoclonal in origin (Penn 1990). According to this latter hypothesis, a mutation leading to increased attraction of macrophages or a mutation causing increased oxidative damage to the LDL reaching the intima could be the initial step in atherogenesis.

In either case, a role for oxidative damage during the total process of atherosclerosis is suggested. Antioxidants might therefore be good candidates for chemoprevention of CHD.

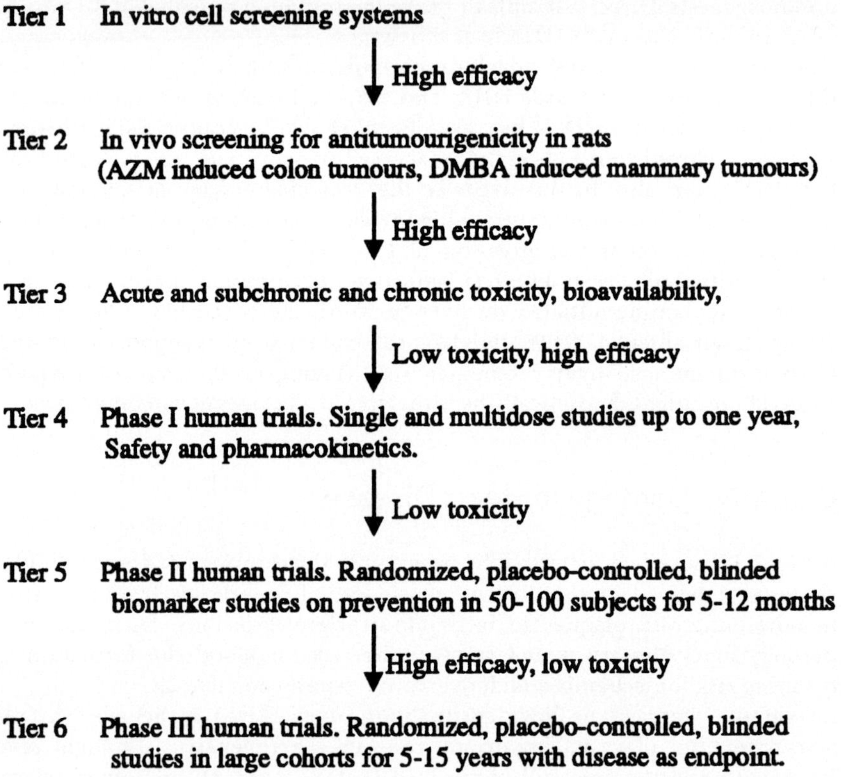

Tier 1 In vitro cell screening systems

↓ **High efficacy**

Tier 2 In vivo screening for antitumourigenicity in rats (AZM induced colon tumours, DMBA induced mammary tumours)

↓ **High efficacy**

Tier 3 Acute and subchronic and chronic toxicity, bioavailability,

↓ **Low toxicity, high efficacy**

Tier 4 Phase I human trials. Single and multidose studies up to one year, Safety and pharmacokinetics.

↓ **Low toxicity**

Tier 5 Phase II human trials. Randomized, placebo-controlled, blinded biomarker studies on prevention in 50-100 subjects for 5-12 months

↓ **High efficacy, low toxicity**

Tier 6 Phase III human trials. Randomized, placebo-controlled, blinded studies in large cohorts for 5-15 years with disease as endpoint.

Fig. 1 A simplified scheme of the process of selecting potential anticarcinogens in the NCI chemoprevention programme. Abbreviations: AZM, azoxymethane; DMBA, 7,12-dimethylbenz[*a*]anthracene.

Evaluation of Chemopreventive Agents

Confidence in the validity of the hypothesis that a natural antioxidant could act in the prevention of chronic disease can only be obtained by large-scale randomised placebo-controlled intervention studies. Such studies are exceedingly costly and careful selection of candidate compounds is therefore of major importance. A major program for identification and selection of candidate cancer chemopreventives has been launched some years ago at the NCI, but a similar program for natural CHD chemopreventives does not exist. The presence on the market of chemo-preventive medicines for individuals at increased risk for CHD may partly be responsible for this. Most major cancer chemoprevention studies with antioxidants initiated so far have included both cancer and CHD as endpoints. A detailed selection process of potential new chemopreventives is followed at the NCI, see figure 1 (Kelloff et al. 1994, Greenwald and Kelloff 1996).

This process generates systematically the same data which have been selected by IARC to be included into the evaluation process for cancer chemopreventives (Stewart et al. 1996). The compounds are initially selected by NCI based on reports of anticarcinogenic actions. Their ability to counteract various processes related to carcinogenesis *in vitro* is screened, followed by an evaluation of toxicity, bioavailability and antitumourigenicity in experimental systems involving organ cultures or whole animals. If, based on the totality of experimental evidence, a compound seems promising, phase I trials are initialised to evaluate the tolerated dose levels, bioavailability and severity of side effects in humans. If the compound is deemed acceptable, phase II trials are initialized, investigating its ability to counteract biomarkers of malignancies in high-risk populations. Promising agents may reach large phase III intervention trials. Intervention may be feasible only for high-risk populations or for people in general, but a discussion of this is beyond the scope of this paper. Only the vitamin antioxidants, vitamin C, E, and β-carotene, and the mineral, selenium, have so far been included in the NCI programme.

Evidence that Antioxidants have Chemopreventive Actions

A large number of natural antioxidants have been shown to protect against ROS *in vitro* in cell free systems or in cell cultures, where oxidative stress is induced by transition metals, peroxidases or toxicants capable of inducing ROS formation. Their antioxidant action may be determined as their ability to decrease fatty acid oxidation to thiobarbituric acid reactive substances, mainly MDA. Antioxidants have also been shown to counteract pro-oxidant induced damage to DNA, including 8-O-dG formation, mutagenesis and clastogenesis. Vitamin E and ellagic acid were found to protect against pentachlorophenol- or 2-nitropropane induced 8-O-dG formation in rat liver (Sai et al. 1995, Takagi et al. 1995) and Korkina et al. (1992) observed a dose-related inhibition by rutin or vitamin C of asbestos induced mutations in cultured human lymphocytes. Also ß-carotene and lycopene have been shown to counteract oxidant induced damage to chromatids *in vivo* or *in vitro* (Weitberg et al. 1985, Abraham et al. 1993, Kapitanov et al. 1994). Natural antioxidants also counteract the modification of LDL to forms which can be absorbed through the macrophage scavenger receptor or which are cytotoxic (Jialal et al. 1991, Negre Salvayre et al. 1995). Some further examples of their antioxidant actions against lipid, protein and DNA oxidation in rodents exposed to pro-oxidants are listed in table 1.

It must be noted, however, that several natural antioxidants can act as pro-oxidants under certain experimental conditions, notably vitamin C or flavonoids in the presence of transition metals (Stadler et al. 1995), β-carotene under high oxygen tension (Haila and Heinonen 1995, Palozza 1995), and possibly also

Table 1 Examples of antioxidant actions of natural antioxidants *in vivo*

Compound	Animal model	Endpoint	Result
Astaxanthin	CCl_4 dosed rats	lipid peroxidation	prevention[a]
α -tocopherol or catechin	PFO[b] dosed rats	lipid peroxidation	prevention[c]
α -tocopherol and Se	$CBrCl_3$ dosed, vitamin E deficient rats	oxidized haemoproteins	prevention[d]
quercetin, 1000 mg/kg	adult male rats	oxidized lysine in plasma albumin	30% decrease[e]
vitamin C, 100 mg/kg	irradiated rats	CAb in bone marrow	26% decrease[f]

[a] Nishigaki et al. (1994). [b] Abbreviations: CA, chromosome aberrations; PFO, peroxidized fish oil. [c] Byun et al. (1994). [d] Chen and Tappel (1995). [e] Dragsted et al. (unpublished). [f] El-Nahas et al. (1993).

vitamin E under certain experimental conditions (Myung et al. 1995). Some of these actions may also be evident *in vivo*, particularly at high doses. Furthermore, myricetin has been observed to modify LDL directly *in vitro* by a non-oxidative mechanism to forms readily absorbed by the macrophage scavenger receptor (Rankin et al. 1993), and both pro-oxidant and antioxidant effects of quercetin and myricetin on LDL oxidation *in vitro* have been reported (De Whalley et al. 1990, Myara et al. 1993, Miura et al. 1995).

Antioxidants in the Prevention of Tumourigenesis

There is a large number of studies to show that most natural antioxidants are capable of inhibiting experimentally induced carcinogenesis (Wattenberg 1992, Wattenberg 1996). Among natural antioxidants which have been found to counteract tumourigenesis are the flavonols quercetin, rutin and myricetin, and the flavanol, epigallocatechin gallate as well as green tea extracts which are very rich in flavanols. Also simple phenols like caffeic acid and chlorogenic acid and complex tannins have been shown to be antitumourigenic (Strube et al. 1993). Allylic polysulphides from *Allium* species are also powerful antitumourigens in experimental systems, and the same is the case for several non-vitamin A carotenoids. The high prevalence of natural antioxidants among the known antitumourigens is by no means proof that they act as anticarcinogens by antioxidation mechanisms.

Heterozygous and homozygous acatalasemic mice have an increased susceptibility to mammary cancers, indicating that ROS may be involved in their pathogenesis. The risk is restored to normal in these mice by vitamin E supplementation after adulthood (Ishii et al. 1996) in accordance with effects of ROS on later stages in mammary tumour development. Some antioxidants, notably carotenoids and flavonoids are active both as anti-promotors and as anti-initiators. The effects of β-carotene in animal studies are not uniformly preventive. In a search for the anticarcinogenic effects of carotenoids, thirtyfour reports on prevention of benign or malignant tumours of the oral cavity, skin,

Table 2 Actions of β-carotene in experimental cancer chemoprevention

Organ site	Number of experimental studies[a] showing:		
	positive effect	no effect	negative effect
oral cavity[b]	9	0	0
skin[c]	7	3	1
colon[d]	2	2	0
lung[e]	1	5	(2)
liver[f]	5	2	0

[a] Toxline (1968-1997), Medline (1986-1996), CAS (1968-1996) and BIOSIS (1990-1996) were searched. The effect evaluations are those of the authors.

[b] (Schwarz et al. 1986, 1989, 1990, 1991, Suda et al. 1986, Shklar and Schwartz 1988, Gijare et al. 1990, Shklar et al. 1993, Tanaka et al. 1994).

[c] (Mathews-Roth 1982, Santamaria et al. 1983, Jones et al. 1989, Lambert et al. 1990, 1994, Steinel and Baker 1990, Azuine et al. 1991, Murakoshi et al. 1992, Chen et al. 1993, Katsumura et al. 1996).

[d] (Colacchio and Memoli 1986, Temple and Basu 1987, Jones et al. 1989, Shivapurkar et al. 1995).

[e] (Beems et al. 1987, Castonguay et al. 1991, Moon et al. 1992, Murakoshi et al. 1992, Wolterbeek et al. 1995, Yun et al. 1995, Tsuda et al. 1996) none of the negative effects were statistically significant.

[f] (Jones et al. 1989, Nyandieka et al. 1989, 1990, Murakoshi et al. 1992, Sergeeva et al. 1992, Sarkar et al. 1995, Tsuda et al. 1996).

colon, lung and liver were identified (see table 2). The chemopreventive efficacy of β-carotene seems to be organ-specific with good chemopreventive effects against epithelial cancers of the oral cavity, skin or liver, variable effects in colon and no effect or even an enhancing effect of experimentally induced cancers of the bronchus or lung. Such effects can be due to differences in distribution to the various organs or to differences in the influence of oxidative stress in the various animal models. Beta-carotene is almost quantitatively transformed to retinol during absorption in rodents, and plasma levels of β-carotene have generally been low in these studies compared with ordinary human plasma levels, despite massive doses (van Vliet 1995). In the oral cavity and skin tumour studies, local application of β-carotene may have increased bioavailability in the target organs, increasing its potential to act there as an antioxidant.

Antioxidants in the Prevention of Heart Disease

Despite good experimental evidence that antioxidants could counteract several steps in the development and course of heart disease by quenching radicals, there is no clear proof of such an effect on atherosclerosis or on ischemia-reperfusion damage (Witztum 1995, Janero 1995, Eaton 1996). Alpha-tocopherol has been shown to counteract oxidative stress specifically in the heart, and cholesterol decreasing effects of antioxidants are also well known (Igarashi et al. 1995, Edes et al., 1986). There are, however, only a few published experimental studies, where natural antioxidants have been shown to prevent these individual processes in CHD, and again, quite different mechanisms might be involved. Beta-carotene counteracts atherosclerosis in cholesterol-fed rabbits in one study but not in another, whereas vitamin E was inactive in both (Keaney et al. 1993, Shaish et al. 1995). Curiously, only vitamin E inhibited oxidation of LDL *ex vivo* in both studies. Vitamin E has also been shown to be consumed in the process of reperfusion damage after ischemia (Yoshikawa et al. 1991) indicating that oxidative damage to ischemic tissue can be counteracted specifically by this antioxidant. Examples of natural antioxidants in the prevention of atherosclerosis in sensitive animals or in cholesterol-fed rodents are shown in table 3.

Human Intervention Trials with Natural Antioxidants

The results of most case-control and cohort studies are compatible with a hypothesis of a preventive effect of natural antioxidants, particularly vitamin C, E, and β-carotene (Peto 1981, National Research Council 1982, Omenn 1996). Generally the protective effects of fruits and vegetables were better than the derived effects of the antioxidants in these studies, but serum levels of antioxidants have also correlated negatively with risk in controlled studies (Steinmetz and Potter 1991). In at least one large prospective study, the protective effect of vitamin E was largest in individuals taking supplements (Machlin 1995). Several studies in phase I or II of natural antioxidants have been conducted during the later years (see Malone (1991), Hakama et al. (1996), Stewart et al. (1996) and Burri (1997) for recent reviews). Only a few major prospective intervention trials with antioxidant vitamins have been initialized, and results from five of them have been published (see table 4).

Table 3 Examples of natural antioxidants in experimental systems related to CHD chemoprevention

Compound	Animal model	Endpoint	Result
α -tocopherol	alcohol-induced rat heart damage	conjugated dienes	prevention[a]
quercetin or isorhamnetin	cholesterol fed rats	serum cholesterol	significant decrease[b]
β-carotene	cholesterol fed rabbits	atherosclerosis	decrease[c]
α -tocopherol or β-carotene	cholesterol fed rabbits	atherosclerosis	no effect[d]
morin or luteolin-7-glucoside	ligated rabbit heart	ischemia reperfusion damage	decrease[e]

[a] Edes et al. (1986). [b] Igarashi et al. (1995). [c] Shaish et al. (1995). [d] Keaney et al. (1993), Shaish et al. (1995). [e] Rump et al. (1994), Wu et al. (1995).

Unfortunately, the results are not generally in agreement with a chemopreventive effect of antioxidant vitamins on cancer or heart disease. Only the Linxian study on β-carotene, Se, and vitamin E as well as the relatively small CHAOS study conferred prevention to the study subjects. On the contrary, there is now good human evidence for an adverse effect of β-carotene on lung cancer in high-risk groups. The proposed explanations for this are many (Huttunen 1996), but it seems that rather high serum levels of supplied β-carotene are necessary for an adverse effect on lung cancer in heavy smokers (Mayne et al. 1996). The lack of an effect on lung cancer in the Physicians' Health Study (PHS) may be due to the low number of heavy smokers in this study as well as the high social class of the participants. The preventive effect of organic selenium is in sharp contrast with most other studies, and may give part of the explanation for the preventive effect observed in the Linxian study.

Table 4 Examples of blinded, randomized, placebo controlled human intervention trials with natural antioxidants.

Compound	Study group	Endpoint	Result
β-carotene, vit. E and organic Se (Linxian)	29,584 Chinese with low-antioxidant diet	Total cancer or stomach cancer	Decrease[a]
Vitamin C and molybdenum (Linxian)	29,584 Chinese with low-antioxidant diet	Total cancer or stomach cancer	No effect[a]
β-carotene (ATBC)	29,133 Finnish smokers	Lung cancer	Increase[b,c]
β-carotene and/or vitamin E (ATBC)	29,133 Finnish smokers	CHD deaths	No effect[c]
β-carotene (PHS)	22,071 US male physicians	Neoplasms and CHD	No effect[d]
β-carotene and retinol (CARET)	18,314 smokers and asbestos workers	1.Lung cancer and 2. CHD	1. Increase[e] 2. No effect
Vitamin E (CHAOS)	2002 coronary atherosclerosis patients	Myocardial infarction	Reduction[f]
Organoselenium	974 skin cancer patients	1. Skin cancers 2. Prostate cancer	1. No effect[g] 2. Prevention

[a] Blot et al. (1993). [b] Albanes et al. (1996). [c] The ATBC Cancer prevention study group (1994). [d] Hennekens et al. (1996). [e] Omenn et al. (1996). [f] Stephens et al. (1996). [g] Clarke et al. (1997).

Discussion

Although many different antioxidants are present in foods of plant origin, human studies have so far focused on the antioxidant vitamins and selenium. Of these, only selenium in the form of yeast selenium complexes has been uniformly protective in phase III trials, but the underlying mechanism was not identified in these studies. The mechanistic information is generally highest in studies conducted *in vitro* and decreases the more complex the system is, from

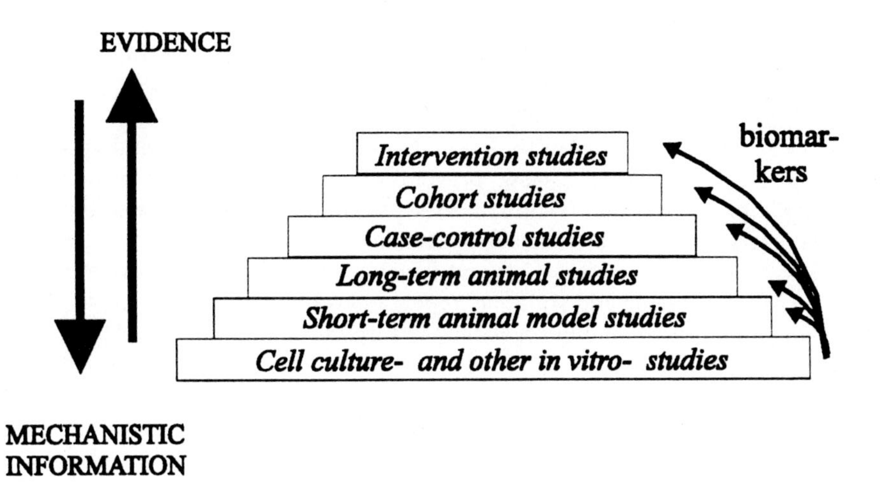

Fig. 2 The pyramid of evidence in evaluation of chemopreventives. The more complex test systems providing the stronger evidens also tend to give less information about the mechanisms involved. Biomarkers can partially help to overcome this problem.

cell-free systems to cell cultures to organ cultures to animal studies and finally to human studies. At the same time, the evidence for a true effect on humans increases (see fig. 2). The small numbers of studies on the top of the pyramid outweighs all the others in terms of evidence, but opens all kinds of speculations as to the mechanisms involved. The mechanistic information may be carried on from the less to the more complex systems by using appropriate biomarkers, and inclusion of biomarkers could have provided evidence for presence or absence of antioxidant actions on suspected target molecules of the putative chemopreventive regimens investigated. The lack of preventive effects of β-carotene on heart disease are perhaps the most unexpected, since evidence at all lower levels, including most *in vitro* studies, animal studies, and epidemiological studies were supporting the expectations of prevention. The most controversial issue in the underlying evidence is the question of whether β-carotene is actually an antioxidant *in vivo* (Rice-Evans et al. 1997).

The expectations of an anticarcinogenic effect in the lung, was solely relying on interpretations of epidemiological studies, and none of the animal studies carried out in the meantime support such an effect (*vide supra*). Neither are preventive effects on colon cancer in animals uniformly in agreement with human observations in smaller intervention trials (MacLennan et al. 1995, Faivre et al. 1996), whereas preventive effects of β-carotene on oral cavity and skin cancers are to be expected based on animal evidence. The preventive effects of β-carotene on regression of mouth plaques in human phase II trials are in agreement with this (Garewal et al. 1990, Garewal 1994). Admittedly, the uptake and disposition of β-carotene differs much between animals and humans, so the apparently coinciding organotropy may rely on quite different underlying mechanisms.

The evidence from *in vitro* studies and animal studies are in good agreement with a possible role in disease prevention of several natural antioxidants. Epidemiological evidence is difficult to use here since most natural antioxidants might correlate with intakes of fruits and vegetables. Some studies point specifically to effects of flavonols and flavones on heart disease prevention but not on cancer (Hertog and Hollman, 1996), and other antioxidants might also be unevenly distributed among vegetables and fruits so as to allow inferences about the possible presence or absence of preventive effects at dietary relevant levels (Dragsted et al. 1997).

The evidence from animal studies seem to support the interpretation that vitamin E, β-carotene and some polyphenols are able to act as antioxidants at certain sites. But evidence for an effect on LDL in the aortic intima or on bronchial DNA is lacking. This is not trivial since there is much evidence for the presence of discrete microenvironments with vastly different antioxidative defence potential (Smith et al. 1996, Daneshvar unpubl.). One compartment may be adequately supported by the existing protective systems, while another compartment is not. The plants have developed an array of antioxidants to suit specific needs in plant tissues, and we may not have any benefit at all from an antioxidant designed to protect leaves against singlet oxygen. Since the distribution and metabolism differs widely between the chemically diverse compounds having antioxidant activities, the possibility that some antioxidants could have chemopreventive effects by protecting sensitive targets against ROS or other radicals in the body is still open.

Acknowlegdements

The author wishes to thank Vibeke Breinholt for valuable comments during the preparation of this manuscript. The work was supported in part by a Danish Food Technology Grant (FØTEK2) to the author.

References

Abraham SK, Sarma L, Kesavan PC (1993) Protective effects of chlorogenic acid, curcumin and beta-carotene against gamma-radiation-induced *in vivo* chromosomal damage. Mutat Res 303:109-112

Albanes D, Heinonen OP, Taylor PR, Virtamo J, Edwards BK, Rautalahti M, Hartman AM, Palmgren J, Freedman LS, Haapakoski J, Barrett MJ, Pietinen P, Malila N, Tala E, Liippo K, Salomaa E-R, Tangrea JA, Teppo L, Askin FB, Taskinen E, Erozan Y, Greenwald P, Huttunen JK (1996) alpha-Tocopherol and beta-carotene supplements and lung cancer incidence in the alpha-tocopherol, beta-carotene cancer prevention study: Effects of base-line characteristics and study compliance. J Natl Cancer Inst 88:1560-1570

Ames BN, Shigenaga MK, Hagen TM (1993) Oxidants, antioxidants, and the degenerative diseases of aging. Proc Natl Acad Sci USA 90:7915-7922

Aruoma OI (1994) Nutrition and Health Aspects of Free Radicals and Antioxidants. Fd Chem Toxic 32:671-683

Azuine MA, Amonkar AJ, Bhide SV (1991) Chemopreventive efficacy of betel leaf extract and its constituents on 7,12-dimethylbenz(a)anthracene induced carcinogenesis and their effect on drug detoxification system in mouse skin. Indian Journal of Experimental Biology 29:346-351

Beems RB (1987) The effect of beta-carotene on BP-induced respiratory tract tumors in hamsters. Nutr Cancer 10:197-204

Blot WJ, Li JY, Taylor PR, Guo W, Dawsey S, Wang GQ, Yang CS, Zheng SF, Gail M (1993) Nutrition intervention trials in Linxian, China: Supplementation with specific vitamin/mineral combinations, cancer incidence, and disease-specific mortality in the general population. J. Natl Cancer Inst 85:1483-1492

Burri BJ (1997) Beta-Carotene and Human Health: A Review of Current Research. Nutrition Research 17:547-580

Byun DS, Kwon MN, Hong JH, Jeong DY (1994) Effects of flavonoids and alpha-tocopherol on the oxidation of n-3 polyunsaturated fatty acids: 2. Antioxidizing effect of catechin and alpha-tocopherol in rats with chemically induced lipid peroxidation. Bulletin of the Korean Fisheries Society 27:166-172

Castonguay A, Pepin P, Stoner GD (1991) Lung tumorigenicity of NNK given orally to A/J mice: its application to chemopreventive efficacy studies. Exp Lung Res 17:485-499

Cerutti PA (1994) Oxy-radicals and cancer. The Lancet 344:862-863

Chen H, Tappel AL (1995) Vitamin E, selenium, trolox C, ascorbic acid palmitate, acetylcysteine, coenzyme Q, beta-carotene, canthaxanthin, and (+)-catechin protect against oxidative damage to kidney, heart, lung and spleen. Free Radic Res 22:177-186

Chen LC, Sly L, Jones CS, Tarone R, De Long M (1993) Differential effects of dietary beta-carotene on papilloma and carcinoma formation induced by an initiation-promotion protocol in SENCAR mouse skin. Carcinogenesis 14:713-717

Clarke L, Krongrad A, Dalkin B, Witherington R, Herlong H, Carpenter D, Borosso C, Janosko E, Falk S, Rounder J, Turnbull B, Slate E, Combs G (1997) Decreased incidence of prostate cancer with selenium supplementation: 1983-1996 results of a double blind cancer prevention trial. Micronutrients and human Cancer Prevention, Aarhus, Denmark (Abstract)

Colacchio TA, Memoli VA (1986) Chemoprevention of colorectal neoplasms. Ascorbic acid and beta-carotene. Arch Surg 121:1421-1424

Collins AR, Dusinská M, Gedik CM, Stetina R (1996) Oxidative Damage to DNA: Do We Have a Reliable Biomarker? Environmental Health Perspectives 104:465-469

De Whalley CV, Rankin SM, Hoult JR, Jessup W, Leake DS (1990) Flavonoids inhibit the oxidative modification of low density lipoproteins by macrophages. Biochem Pharmacol 11:1743-1750

Denda A, Endoh T, Nakae D, Konishi Y (1995) Effects of oxidative stress induced by redox-enzyme modulation on rat hepatocarcinogenesis. Toxicology Letters 82/83:413-417

Dragsted LO, Strube M, Larsen JC (1993) Cancer-protective factors in fruits and vegetables: biochemical and biological background. Pharmacol Toxicol 72:116-135

Dragsted LO, Strube M, Leth T (1997) Dietary levels of plant phenols and other non-nutritive components: Could they prevent cancer? Eur J Cancer Prev (submitted)

Eaton JW (1996) Low-density lipoprotein oxidation and atherogenesis: we got the bandwagon, we got the band, but where's the music? Redox Report 2:81-82

Edes I, Toszegi A, Csanady M, Bozoky B (1986) Myocardial lipid peroxidation in rats after chronic alcohol ingestion and the effects of different antioxidants. Cardiovasc Res 20:542-548

El-Nahas SM, Mattar FE, Mohamed AA (1993) Radioprotective effect of vitamins C and E. Mutat Res 301:143-147

Faivre J, Hofstad B, Bonelli L, Rooney P, Couillault C (1996) European intervention trials of colorectal cancer prevention. IARC Scient Publ 136:45-51

Frei B (1995) Cardiovascular disease and nutrient antioxidants: role of low-density lipoprotein oxidation. Crit Rev Food Sci Nutr 35:83-98

Garewal H (1994) Chemoprevention of oral cancer: beta-carotene and vitamin E in leukoplakia. Eur J Cancer Prev 3:101-7

Garewal HS, Meyskens FLJ, Killen D, Reeves D, Kiersch TA, Elletson H, Strosberg A, King D, Steinbronn K (1990) Response of oral leukoplakia to beta-carotene. J Clin Oncol 8:1715-1720

Garner B, Jessup W (1996) Cell-mediated oxidation of low-density lipoprotein: the elusive mechanism(s). Redox Report 2:97-104

Giacosa A, Filiberti R (1996) Free radicals, oxidative damage and degenerative diseases. Eur J Cancer Prev 5:307-312

Gijare PS, Rao KV, Bhide SV (1990) Modulatory effects of snuff, retinoic acid, and beta-carotene on DMBA-induced hamster cheek pouch carcinogenesis in relation to keratin expression. Nutr Cancer 14:253-259

Greenwald P, Kelloff GJ (1996) The role of chemoprevention in cancer control. IARC Scientific Publications 139:13-22

Haila K, Heinonen M (1994) Action of beta-Carotene on Purified Rapeseed Oil During Light Storage. Lebensmittel-Wissenschaft & Technologie 27:573-577

Hakama M, Beral V, Buiatti E, Faivre J, Parkin DM (1996) Chemoprevention in cancer control. IARC Scientific Publ 136, International Agency for Research on Cancer, Lyon pp. 1-140

Halliwell B (1994) Free radicals, antioxidants, and human disease: curiosity, cause, or consequence? The Lancet 344:721-724

Hertog MGL, Hollman PCH (1996) Potential health effects of the dietary flavonol quercetin. European Journal of Clinical Nutrition 50:63-71

Huttunen JK (1996) Why did antioxidants not protect against lung cancer in the Alpha-Tocopherol, Beta-Carotene Cancer Prevention Study? IARC Scient Publ 136:63-66

Igarashi K, Ohmuma M (1995) Effects of isorhamnetin, rhamnetin, and quercetin on the concentrations of cholesterol and lipoperoxide in the serum and liver and on the blood and liver antioxidative enzyme activities of rats. Biosci Biotechnol Biochem 59:595-601

Ishii K, Zhen L-X, Wang D-h, Funamori Y, Ogawa K, Taketa K (1996) Prevention of Mammary Tumorigenesis in Acetalasemic Mice by Vitamin E Supplementation. Jpn J Cancer Res 87:680-684

Janero DR (1995) Ischemic Heart Disease and Antioxidants: Mechanistic Aspects of Oxidative Injury and its Prevention. Crit Rev Food Sci Nutr 35:65-81

Jialal I, Norkus EP, Cristol L, Grundy SM (1991) Beta-carotene inhibits the oxidative modification of low-density lipoprotein. Biochimica et Biophysica Acta 1086:134-138

Jones RC, Sugie S, Braley J, Weisburger JH (1989) Dietary beta-carotene in rat models of gastrointestinal cancer. J Nutr 119:508-514

Kapitanov AB, Pimenov AM, Obukhova LK, Izmailov DM (1994) [Radiation-protective effectiveness of lycopene]. Radiats Biol Radioecol 34:439-445

Katsumura N, Okuno M, Onogi N, Moriwaki H, Muto Y, Kojima S (1996) Suppression of mouse skin papilloma by canthaxanthin and β-carotene *in vivo*: possibility of the regression of tumorigenesis by carotenoids without conversion to retinoic acid. Nutr Cancer 26:203-208

Keaney JF, Jr., Gaziano JM, Xu A, Frei B, Curran CJ, Shwaery GT, Loscalzo J, Vita JA (1993) Dietary antioxidants preserve endothelium-dependent vessel relaxation in cholesterol-fed rabbits. Proc Natl Acad Sci USA 90:11880-11884

Kelloff GJ, Boone CW, Crowell JA, Steele VE, Lubet R, Sigman CC (1994) Chemopreventive drug development: perspectives and progress. Cancer Epidemiol Biomarkers Prev 3:85-98

Klaunig JE, Xu Y, Bachowski S, Ketcham CA, Isenberg JS, Kolaja KL, Baker TK, Walborg EF Jr., Stevenson DE (1995) Oxidative stress in nongenotoxic carcinogenesis. Toxicology Letters 82/83:683-691

Korkina LG, Durnev AD, Suslova TB, Cheremisina ZP, Daugel Dauge NO, Afanas'ev IB (1992) Oxygen radical-mediated mutagenic effect of asbestos on human lymphocytes: suppression by oxygen radical scavengers. Mutat Res 265:245-253

Lambert LA, Koch WH, Wamer WG, Kornhauser A (1990) Antitumor activity in skin of Skh and Sencar mice by two dietary beta-carotene formulations. Nutr Cancer 13:213-221

Lambert LA, Wamer WG, Wei RR, Lavu S, Chirtel SJ, Kornhauser A (1994) The protective but nonsynergistic effect of dietary beta-carotene and vitamin E on skin tumorigenesis in Skh mice. Nutr Cancer 21:1-12

Machlin LJ (1995) Critical assessment of the epidemiological data concerning the impact of antioxidant nutrients on cancer and cardiovascular disease. Crit Rev Food Sci Nutr 35:41-50

MacLennan R, Macrae F, Bain C, Battistutta D, Chapuis P, Gratten H, Lambert J, Newland RC, Ngu M, Russell A, Ward M, Wahlqvist ML (1995) Randomized Trial of Intake of Fat, Fiber, and Beta Carotene to Prevent Colorectal Adenomas. J Natl Cancer Inst 87:1760-1766

Malone WF (1991) Studies evaluating antioxidants and β-carotene as chemopreventives. Am J Clin Nutr 53:305S-313S

Mathews-Roth MM (1982) Antitumor activity of beta-carotene, canthaxanthin and phytoene. Oncology 39:33-37

Mayne ST, Handelman J, Beecher G (1996) β-Carotene and Lung Cancer Promotion in Heavy Smokers - a Plausible Relationship. J Natl Cancer Inst 88:1513-1515

Miura S, Watanabe J, Sano M, Tomita T, Osawa T, Hara Y, Tomita I (1995) Effects of Various Natural Antioxidants on the Cu^{2+}-Mediated Oxidative Modification of Low Density Lipoprotein. Biological & Pharmaceutical Bulletin 18:1-4

Moon RC, Rao KV, Detrisac CJ, Kelloff GJ (1992) Animal models for chemoprevention of respiratory cancer. Monogr Natl Cancer Inst 13:45-49

Moriya M, Du C., Bodepudi V, Johnson F, Takeshita M, Grollman A (1991) Site-specific mutagenesis using a gapped duplex vector: a study of translesion synthesis past 8-oxodeoxyguanosine i E coli. Mutat Res 254:281-288

Murakoshi M, Nishino H, Satomi Y, Takayasu J, Hasegawa T, Tokuda H, Iwashima A, Okuzumi J, Okabe H, Kitano H, et al (1992) Potent Preventive Action of alpha-Carotene against Carcinogenesis: Spontaneous Liver Carcinogenesis and Promoting Stage of Lung and Skin Carcinogenesis in Mice Are Supressed More Effectively by alpha-Carotene Than by beta-Carotene. Cancer Research 52:6583-6587

Myara I, Pico I, Vedie B, Moatti N (1993) A method to screen for the antioxidant effect of compounds on low-density lipoprotein (LDL): Illustration with flavonoids. J Pharmacol Toxicol Methods 30:69-73

Myung K, Sook HR, Hong SC (1995) Effect of tocopherols and beta-carotene on the oxidation of linoleic acid mixture in the solid model system. Journal of the Korean Society of Food and Nutrition 24:67-73

National Research Council (1982) Diet, Nutrition and Cancer. National Academy Press, Washington,DC

Negre Salvayre A, Mabile L, Delchambre J, Salvayre R (1995) Alpha-Tocopherol, ascorbic acid, and rutin inhibit synergistically the copper-promoted LDL oxidation and the cytotoxicity of oxidized LDL to cultured endothelial cells. Biological Trace Element Research 47:81-91

Nishigaki I, Dmitrovskii AA, Miki W, Yagi K (1994) Suppressive effect of astaxanthin on lipid peroxidation induced in rats. J Clin Biochem Nutr 16:161-166

Nyandieka HS, Wakhisi J, Kilonzo M (1989) Inhibition of AFB1-induced liver cancer and induction of increased microsomal enzyme activity by dietary constituents. East Afr Med J 66:796-803

Omenn GS (1996) Micronutrients (vitamins and minerals) as cancer-preventive agents.IARC Scientific Publications 139:33-45

Omenn GS, Goodman G, Thornquist M, Barnhart S, Balmes J, Cherniack M, Cullen M, Glass A, Keogh J, Liu D, Mayskens F, Jr., Perloff M, Valanis B, Williams J, Jr. (1996) Chemoprevention of lung cancer: the β-Carotene and Retinol Efficacy Trial (CARET) in high-risk smokers and asbestos-exposed workers. IARC Scientific Publications 136:67-85

Palozza P, Calviello G, Bartoli GM (1995) Prooxidant activity of beta-carotene under 100 percent oxygen pressure in rat liver microsomes. Free Radical Biology & Medicine 19:887-892

Palozza P, Krinsky NI (1994) Antioxidant Properties of Carotenoids. In: Livrea MA, Vidali G (eds) Retinoids: From Basic Science to Clinical Applications. Birkhäuser Verlag, pp 35-41

Penn A (1990) Mutational events in the etiology of arteriosclerotic plaques. Mutat Res 239:149-162

Peto R, Doll R, Buckley JD, Sporn MB (1981) Can dietary beta-carotene materially reduce human cancer rates? Nature (London) 290:201-208

Potter JD, Steinmetz K (1996) Vegetables, fruit and phytoestrogens as preventive agents. IARC Scientific Publications 139:61-90

Rankin SM, de Whalley CV, Hoult JRS, Jessup W, Wilkins GM, Collards J, Leake DS (1993) The modifications of low density lipoprotein by the flavonoids myricetin and gossypetin. Biochemical Pharmacology 45:67-75

Rice-Evans CA, Sampson J, Bramley PM, Holloway DE (1997) Why Do We Expect Carotenoids to be Antioxidants *in vivo*? Free Rad Res 26:381-398

Rump AF, Schussler M, Acar D, Cordes A, Theisohn M, Rosen R, Klaus W, Fricke U (1994) Functional and antiischemic effects of luteolin-7-glucoside in isolated rabbit hearts. Gen Pharmacol 25:1137-1142

Sai KK, Umemura T, Takagi A, Hasegawa R, Tanimura A, Kurokawa Y (1995) Pentachlorophenol-induced oxidative DNA damage in mouse liver and protective effect of antioxidants. Food and Chemical Toxicology 33:877-882

Santamaria L, Bianchi A, Arnaboldi A, Andreoni L, Bermond P (1983) Dietary carotenoids block photocarcinogenic enhancement by benzo (a)pyrene and inhibit its carcinogenesis in the dark. Experientia 39:1043-1045

Sarkar A, Mukherjee B, Chatterjee M (1995) Inhibition of 3'-methyl-4-ethylaminoazobenzene-induced hepatocarcinogenesis in rat by dietary beta-carotene: changes in hepatic anti-oxidant defense enzyme levels. Int J Cancer 61:799-805

Schwartz JL, Suda D, Light G (1986) Beta-carotene is associated with the regression of hamster buccal pouch carcinoma and the induction of tumor necrosis factor in macrophages. Biochem Biophys Res Commun 136:1130-1135

Schwartz JL, Sloane D, Shklar G (1989) Prevention and inhibition of oral cancer in the hamster buccal pouch model associated with carotenoid immune enhancement. Tumour Biol 10:297-309

Schwartz JL, Shklar G, Flynn E, Trickler D (1990) The administration of beta-carotene to prevent and regress oral carcinoma in the hamster cheek pouch and the associated enhancement of the immune response. Adv Exp Med Biol 262:77-93

Schwartz JL, Shklar G, Flynn E, et al (1991) The administration of beta-carotene to prevent and regress oral carcinoma in the hamster cheek pouch and the associated enhancement of the immune response. In: Bendich A, Philips M, Tengerdy RP (eds): Antioxidant Nutrients and Immune Functions. Plenum Press, New York, pp 77-93

Sergeeva TI, Vakulova LA, Zhidkova TA, Sergeev AV (1992) Inhibiting effect of domestic, synthetic beta-carotene and ascorbic acid on endogenous carcinogenesis in mice. Vopr Med Khim 38:12-14

Shahidi F, Wanasundara PKJPD (1992) Phenolic Antioxidants. Critical Reviews in Food Science and Nutrition 32:67-103

Shaish A, Daugherty A, O'Sullivan F, Schonfeld G, Heinecke JW (1995) Beta-carotene inhibits atherosclerosis in hypercholesterolemic rabbits. Journal of Clinical Investigation 96:2075-2082

Shivapurkar N, Tang Z, Frost A, Alabaster O (1995) Inhibition of progression of aberrant crypt foci and colon tumor development by vitaminE and beta-carotene in rats on a high-risk diet. Cancer Lett 91:125-132

Shklar G, Schwartz J, Trickler D, Cheverie SR (1993) The effectiveness of a mixture of beta-carotene, alpha-tocopherol, glutathione, and ascorbic acid for cancer prevention. Nutr Cancer 20:145-151

Shklar G, Schwartz J (1988) Tumor necrosis factor in experimental cancer regression with alpha-tocopherol, beta-carotene, canthaxanthin and algae extract. Eur J Cancer Clin Oncol 24:839-850

Smith CV, Jones DP, Guenthner TM, Lash LH, Lauterburg BH (1996) Compartmentation of Glutathione: Implications for the Study of Toxicity and Disease. Toxicology and Applied Pharmacology 140:1-12

Steinel HH, Baker RSU (1990) Effects of beta-carotene on chemically-induced skin tumors in HRA/Skh hairless mice. Cancer Lett (Shannon 51:163-168

Steinmetz KA, Potter JD (1991) Vegetables, fruit, and cancer. II. Mechanisms. Cancer Causes Control 2:427-442

Stephens NG, Parsons A, Schofield PM, Kelly F, Cheeseman K, Mitchinson MJ, Brown MJ (1996) Randomised controlled trial of vitamin E in patients with coronary disease: Cambridge Heart Antioxidant Study (CHAOS). The Lancet 347:781-786

Stewart BW, McGregor D, Kleihues P (1996) Principles of chemoprevention. IARC Sci Publ 139, International Agency for Research on Cancer, Lyon pp. 1-332

Strube M, Dragsted LO, Larsen JC (1993) Naturally occurring antitumourigens I. Plant phenols. Nordic Council of Ministers Working Reports 605 pp. 1-222.

Suda D, Schwartz J, Shklar G (1986) Inhibition of experimental oral carcinogenesis by topical beta-carotene. Carcinogenesis 7:711-715

Takagi A, Sai K, Umemura T, Hasegawa R, Kurokawa Y (1995) Inhibitory effects of vitamin E and ellagic acid on 8-hydroxy-deoxyguanosine formation in liver nuclear DNA of rats treated with 2-nitropropane. Cancer Lett 91:139-144

Tanaka T, Makita H, Ohnishi M, Hirose Y, Wang A, Mori H, Satoh K, Hara A, Ogawa H (1994) Chemoprevention of 4-nitroquinoline 1-oxide-induced oral carcinogenesis by dietary curcumin and hesperidin: comparison with the protective effect of beta-carotene. Cancer Res 54:4653-4659

Temple NJ, Basu TK (1987) Protective effect of beta-carotene against colon tumors in mice. J Natl Cancer Inst 78:1211-1214

The ATBC Cancer Prevention Study Group (1994) The Alpha-Tocopherol, Beta-Carotene Lung Cancer Prevention Study: Design, methods, participant characteristics, and compliance. Annals of Epidemiology 4:1-10

Tsuda H, Iwahori Y, Asamoto M, Baba-Toriyama H, Hori T, Kim DJ, Uehara N, Iigo M, Takasuka N, Murakoshi M, Nishino H, Kakizoe T, Araki E, Yazawa K (1996) Demonstration of organotropic effects of chemopreventive agents in multiorgan carcinogenesis models. IARC Sci Publ 139:143-150

van Vliet (1995) Intestinal absorption and cleavage of β-carotene in rat, hamster and human models. Ph.D. Thesis, University of Amsterdam, Holland

Wattenberg LW (1992) Inhibition of carcinogenesis by minor dietary constituents. Cancer Res Suppl 52:2085s-2091s

Wattenberg LW (1996) Inhibition of tumorigenesis in animals. IARC Sci Publ 139:151-158

Weitberg AB, Weitzman SA, Clark EP, Stossel TP (1985) Effects of antioxidants on oxidant-induced sister chromatid exchange formation. J Clin Invest 75:1835-1841

Witztum JL (1994) The oxidation hypothesis of atherosclerosis. The Lancet 344:793-795

Wolterbeek APM, Roggeband R, Baan RA, Feron VJ, Rutten AAJJL (1995) Relation between benzo(a)pyrene-DNA adducts, cell proliferation and p53 expression in tracheal epithelium of hamsters fed a high beta-carotene diet. Carcinogenesis 16:1617-1622

Wu TW, Fung KP, Zeng LH, Wu J, Hempel A, Grey AA, Camerman N (1995) Molecular properties and myocardial salvage effects of morin hydrate. Biochemical Pharmacology 49:537-543

Yoshikawa T, Yasuda M, Ueda S, Naito Y, Tanigawa T, Oyamada H, Kondo M (1991) Vitamin E in gastric mucosal injury induced by ischemia-reperfusion. Am J Clin Nutr 53:210S-214S

Yun TK, Kim SH, Lee YS (1995) Trial of a new medium-term model using benzo(a)pyrene induced lung tumor in newborn mice. Anticancer Res 15:839-845

Development of In Vitro Models for Cellular and Molecular Studies in Toxicology and Chemoprevention

K. Macé, E.A Offord, C.C Harris and A.M.A Pfeifer
Nestlé Research Center, P.O. box 44, CH-1000 Lausanne 26, Switzerland

Abstract

Many natural dietary phytochemicals found compounds found in fruits, vegetables, spices and tea have been shown in recent years to be protective against cancer in various animal models. In the light of the potential impact of these compounds on human health it is important to elucidate the mechanisms involved. We therefore developed and characterized relevant in vitro models using immortalized human epithelial cell lines derived from target tissues in carcinogenesis, such as lung, liver and colon. Assays were established, allowing the evaluation of the cytotoxic and genotoxic effects of various procarcinogens, including nitrosamines, mycotoxins and heterocyclic amines on these metabolically-competent human epithelial cell lines. These cellular models appeared to be a useful tool to study the capacity of certain food components to block the initiation stage of carcinogenesis. The ability of carnosol and carnosic acid from rosemary as well as the synthetic dithiolethione, oltipraz, to block the formation of DNA adducts, and their effects on the expression of phase I and phase II enzymes was investigated. We have observed that both rosemary extracts and oltipraz inhibited benzo(a)pyrene- or aflatoxin B_1-induced DNA adduct formation by strongly inhibiting CYP450 activities and inducing the expression of glutathione S-transferase. These results in human cell models give some insight into the different mechanisms involved in the chemopreventive action of both natural and synthetic compounds in relation to phase I and phase II enzymes.

Introduction

The development of in vitro models for toxicity and carcinogenicity testing has increased rapidly in the last decade. Although the majority of human tumours arise from epithelial cells and many carcinogens exhibit organ specificity, most of the in vitro tests utilize non-epithelial cell types (e.g. animal or human fibroblasts) which have fundamentally different metabolism, differentiation pattern and sensitivity to carcinogens. This can complicate extrapolation to man for risk assessment and chemoprevention of epithelial diseases such as cancer. Therefore, progress in developing methods for culturing functionally active human epithelial cells or tissues is of particular value in establishing more relevant in vitro test systems. The important contributions of human explant cultures from various tissues to validate extrapolations from laboratory animals to man in carcinogenesis research has been reviewed (DiPaolo et al. 1986; Harris et al. 1982). Short-term explant cultures and isolated epithelial cells derived from many normal human tissues, e.g. bronchus, colon, oesophagus, bladder, buccal mucosa, epidermis, kidney and fetal liver can be maintained or grown under defined serum-free conditions (Brash et al. 1987; Grafstrom 1990). However, the limited access to human tissue and the short life span of primary cells make it difficult to use such an approach for standardized toxicity and mutagenicity studies. Immortalization offers a feasible approach to develop functional human cell lines. In contrast to rodent cells, normal human cells are relatively resistant in vitro to transformation by known carcinogenic agents, probably due to their higher genetic stability (Newbold et al. 1982). SV40 T-antigen, adenovirus E1a, E1b and papilloma virus 16 and 18 have been most commonly employed in human epithelial cell immortalization and transformation (Brash et al. 1987; Grafstrom 1990; Linder and Marshall 1990). Cells immortalized by introduction of oncogenes often show reduced differentiation and loss of tissue-specific functions, including enzymes responsible for the activation of mutagens and carcinogens. However, genetic manipulation can efficiently restore metabolic deficiencies.

Here, we will describe the generation of our metabolically-competent human bronchial and liver epithelial cell lines and their applications in toxicology, carcinogenicity and chemoprevention.

Development of Metabolically-Competent Human Cell Lines.

Bronchial Epithelial Cells: The SV40 T-antigen immortalized human bronchial epithelial cell line (BEAS-2B) (Reddel et al. 1988) retains many characteristics of normal cell growth and differentiation, such as growth inhibitory response to $TGF\beta_1$, and proved to be a useful model to study activated proto-oncogenes and inactivated tumour suppressor genes in human carcinogenesis (Pfeifer et al. 1989; Gerwin et al. 1992). However, although the BEAS-2B cells retain the expression of most of the phase II enzymes and oxidant defence enzymes, such as glutathione S-transferase π (GSTπ), epoxide hydrolase (EH), catalase (CAT), superoxide dismutase (SOD) and glutathione peroxidase (GPX), they show limited expression of phase I enzymes (Macé et al. 1994). Among the different cytochromes P450 (CYP450s), namely CYP1A1, CYP2A6, CYP2B6, CYP2C8, CYP2C18, CYP2E1 and CYP3A5, detected in normal human bronchial tissues (Macé et al. 1997b), only CYP1A1 mRNA was expressed in the BEAS-2B cells. Moreover, the BEAS-2B cells exhibit polycyclic aromatic hydrocarbon-induced CYP1A1 activity (Offord et al. 1995). In order to restore the metabolic capacity of the BEAS-2B cells, human CYP450 cDNAs were cloned into the pCMVneo vector (Macé et al. 1997d) and introduced into the cells by lipofection. A panel of different cell lines, expressing either CYP1A2, CYP2A6, CYP2B6, CYP2D6, CYP2E1, CYP3A4 or CYP3A5 were obtained (Macé et al. 1994; Macé et al. 1997e). Western blot analysis and measure of specific catalytic activities proved that these different cell lines exhibit high and stable expression of individual CYP450 (Macé et al. 1997e).

Liver Epithelial Cells: Successful immortalization of human epithelial liver cells by infection with a SV40-T antigen recombinant retrovirus was described by our group (Pfeifer et al. 1993; Pfeifer et al. 1995; Macé et al. 1996). At least 7 different non tumorigenic cell lines (THLE cells), as assessed by subcutaneous injection in athymic nude mice, were established. The metabolic potential of the THLE cells was demonstrated by the detection of DNA-adduct formation of two different chemical classes of carcinogens, benzo(a)pyrene (B(a)P) and aflatoxin B_1 (AFB_1) (Pfeifer et al. 1993). The expression of phase II and oxidant defence enzymes in the different THLE cell lines as well as in the hepatocellular carcinoma HepG2 cells was determined by Northern blot analysis. As shown in Table 1, all the cell lines expressed GSTμ whereas only the T-9A3 and T-6 cells expressed GSTπ, normally not expressed in human hepatocytes. On the other hand the THLE cells did not show detectable levels of GSTα mRNA. The expression of the quinone reductase (QR), epoxide hydrolase (EH) as well as SOD, GPX and CAT, was detected in all the cell lines.

Table 1 Expression of phase II and oxidant defence enzyme mRNA in different THLE cell lines and in HepG2 cells

	T-5B cl5	T-5B cl8	T-9A3	T-9A5	T-10b5	T-6	HepG2
GSTμ	++	++	+	+	+	++++	+++
GSTπ	-	-	+++	-	-	+++	-
QR	++	++	++	++	++	++	++
EH	++	++	++	++	+	+	++
SOD	++	++++	++	++	+	+	+++
GPX	+	+	+	++	+	++	++
CAT	++	++	++	++	++	+	++

Fifteen micrograms of total RNAs were subjected to electrophoresis and transferred on nitrocellulose. Hybridization was performed with α-^{32}P-labelled specific probes as previously described (Macé et al. 1994; Offord et al. 1995).GST, glutathione S-transferase; QR, quinone reductase; EH, epoxide hydrolase; SOD, superoxide dismutase; GPX, glutathione peroxidase; CAT, catalase.

RT-PCR analysis using specific primers (Macé et al. 1997b) showed that all the cell lines expressed CYP1A2, CYP2C, CYP2E1 and CYP3A5 mRNA but have lost the expression of CYP2A6 (except the T-9A5 cells), CYP2B6, CYP2D6 (except the

HepG2 cells) and CYP3A4 (Table 2). Although some phase I enzymes were retained in the immortalized cell lines, analysis at the protein levels indicated reduced CYP450 expression as demonstrated by the 10-times smaller CYP1A2 protein expression in the T-5B cl5 cells in comparison with normal human hepatocytes (Macé et al. 1996). Restoration of the CYP450 activities in the T-5B cl5 cells was accomplished by genetic manipulation as described above for the BEAS-2B cells. CYP1A2-, CYP2A6-, CYP2B6-, CYP2E1- and CYP3A4-expressing cell lines were established exhibiting liver-like CYP450 expression, as determined by the measurement of catalytic activities and the comparison with normal human hepatocytes (Macé et al. 1997a; Barcelo et al. 1997).

Table 2 Expression of CYP450 mRNA in different in different THLE cell lines and in HepG2 cells

	T-5B cl5	T-5B cl8	T-9A3	T-9A5	T-10b5	T-6	HepG2
CYP1A2	+	+	+	+	+	+	+
CYP2A6	-	-	-	+	-	-	-
CYP2B6	-	-	-	-	-	-	-
CYP2C	+	+	++	++	+	+	+
CYP2D6	-	-	-	-	-	-	+
CYP2E1	+	+	+	+	+	+	++
CYP3A4	-	-	-	-	-	-	-
CYP3A5	+++	+	+++	+++	+++	+	+++

Synthesis of single-stranded cDNA and RT-PCR analysis were performed as previously described (Macé et al. 1997b).

Toxicological Applications

Carcinogen activation can be studied easily in the CYP450-expressing BEAS-2B cells (B-CMVCYP450) by direct measurement of relevant toxicological end-points such as cytotoxicity, DNA adduct formation and mutagenesis. Colony-forming efficiency assays indicated that the B-CMV2E1 cells were able to specifically activate N-nitrosodimethylamine (NDMA) to cytotoxic metabolites whereas the B-CMV1A2, B-CMV2A6 and B-CMV3A4 cells were sensitive to the cytotoxic effects induced by aflatoxin B_1 (AFB_1) exposure (Macé et al. 1997e). The AFB_1-induced cytotoxicity was accompanied by the formation of the AFB_1-N^7-guanine adduct in the B-CMV1A2 cells and a 7-fold increase in the numbers of mutants at the *hprt* locus in comparison with the control cells (Macé et al. 1997e).

A rapid 96-well screening cytotoxicity test was developed for the toxicological applications of the CYP450-expressing THLE cells (T5-CYP450). The cells were seeded at densities between 10 to 25 x 10^3 cells/well, depending on the plating efficiency of the different T5-CYP450 cell lines. After 1 day, the cells were treated with different concentrations of cytotoxic agents for 20 hours, washed and incubated 2 hours with MTT to determine the cell viability, as previously described (Alley et al. 1988). In this way, the cytotoxic effects of several mycotoxins, such as cyclopiazonic acid (CA), deoxylivalenol (DON), fumonisin B_1 (FB_1), patulin, sterigmatocystin (ST) and T2-toxin were evaluated in the T5-1A2, T5-2A6, T5-2B6, T5-2E1 and T5-3A4 cells. As shown in figure 1, the different T5-CYP450 cell lines displayed large differerences in the sensitivity levels to the cytotoxic effects induced by ST exposure. The T5-3A4 and T5-2B6 cells were the most sensitive with a CD_{50} (dose inducing 50% of cytotoxicity) of 0.05 and 0.08 μM ST, respectively. CD_{50} of 0.25 and 0.7 μM ST were observed for the T5-1A2 and T5-2A6 cells, whereas the cytotoxicity of the T5-2E1 cells did not differ from that of the control T5-neo cells which do not express CYP450 cDNA (Table 3). These results indicated that several CYP450s, including CYP3A4, CYP2B6, CYP1A2 and CYP2A6, are able to activate ST to cytotoxic metabolites.

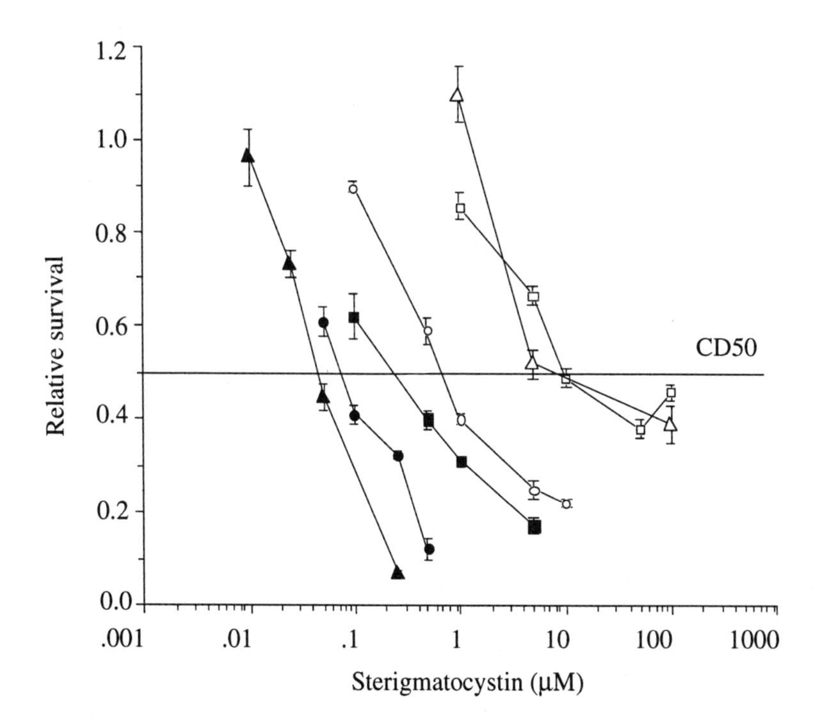

Fig. 1 Sterigmatocystin cytotoxicity in different T5-CYP450 cell lines
The cells were exposed to the indicated concentrations of sterigmatocystin for 20 hours. Cell viability was determined by the MTT assay and expressed as survival relative to the corresponding untreated cells. Open squares, T5-neo (control cells); closed squares, T5-1A2; open circles, T5-2A6; open triangles, T5-2E1; closed triangles, T5-3A4.

As expected, the CYP450 enzymes do not seem to be involved in the activation of the other mycotoxins tested such as CPA, DON, FB_1, Patulin and T2-toxin, as no significant difference in the CD_{50} values was observed between the different T5-CYP450 cell lines and the control T5-neo cells (Table 3).

Table 3 Cytotoxicity effects of mycotoxins in CYP450-expressing human liver epithelial cell lines

	CPA	DON	FB_1	Patulin	ST	T2-toxin
T5-neo	30	2.0	300	2.0	10	0.02
T5-1A2	30	2.0	150	2.0	0.25	0.02
T5-2A6	30	2.0	300	2.0	0.70	0.02
T5-2B6	30	2.0	300	2.0	0.08	0.02
T5-2E1	30	2.0	300	2.0	10	0.02
T5-3A4	30	2.0	300	2.0	0.05	0.02

Expressed as the dose (μM) inducing 50% of cytotoxicity.
CPA, cyclopiazonic acid; DON, deoxynivalenol; FB_1, fumonisin B_1; ST, sterigmatocystin

Mutagenicity assays, using RFLP-PCR technology, were performed on the T5-CYP450 cell lines to demonstrate that both CYP3A4 and CYP1A2 contribute to the formation of AFB_1-induced p53 tumor suppressor gene mutations (Macé et al. 1997a).

Chemoprevention Applications

Carcinogenesis is a multistep process involving tumour initiation, promotion and progression. Many naturally occurring phytochemicals have been shown to protect against these different stages of carcinogenesis in animal models (Huang et al. 1994). The synthetic dithiolthione, oltipraz, is the best characterised chemopreventive agent and is now being tested in human clinical trials (Kensler and Helzlouer 1995). The bronchial and liver cell lines described in this paper are particularly suited to study the initiation stage of carcinogenesis due to their metabolic competence in phase I and phase II enzymes. The procarcinogens benzo(a)pyrene (B(a)P) and aflatoxin B_1 (AFB_1) require metabolic activation by CYP450 enzymes to form their genotoxic metabolites and are detoxified by the phase II enzyme glutathione S-transferase (GST) (Pelkonen and Nebert 1982; Eaton and Gallagher 1994).

The common spice, rosemary, is a rich source of polyphenolic antioxidants, wich are effective both in food stabilization and in biological systems, the most potent components being carnosol and carnosic acid (Aruoma et al. 1992; Offord et al. 1997a). Rosemary extract or its active constituents show antimutagenic activity in bacteria and anticarcinogenic properties in animal models (Minnuni et al. 1992; Ho et al. 1994). We used the CYP450-expressing bronchial and liver cells to study the effect of rosemary components on the formation of DNA adduct formation by measuring the incorporation of tritiated metabolites of B(a)P or AFB_1 into cellular DNA (Offord et al. 1995; Offord et al. 1997b). Rosemary extract (RE-S, 5 mg/ml) or an equivalent concentration of carnosol or carnosic acid (1 mg/ml or 3 mM) strongly inhibited B(a)P-DNA adduct formation (80%) in the bronchial BEAS-2B cells (Offord et al. 1995). Significant inhibition of AFB_1-DNA adduct formation was also observed in the liver T5.1A2 and T5.3A4 cells (Offord et al. 1997b). The rosemary components strongly inhibited CYP1A1 activity, the principal enzyme involved in B(a)P activation as well as CYP1A2 and CYP3A4, the major enzymes involved in AFB_1 activation, thus blocking the formation of genotoxic metabolites. Oltipraz was also found to be a potent inhibitor of CYP450 activities in this cellular system. Furthermore, both agents were strong inducers of the detoxifying enzymes, GST and quinone reductase. Thus the rosemary polyphenols act in a similar way to the well-known chemopreventive agent oltipraz in blocking initiation of carcinogenesis by a dual mechanism (i) prevention of metabolic activation of procarcinogens by inhibition of CYP450 enzymes (ii) induction of detoxifying enzymes.

Cruciferous vegetables contain phytochemicals such as isothiocyanates which have anticancer properties (Hecht 1996). The broccoli component, sulforaphane (1-isothiocyanate-4-methylsulfinylbutane) is a potent inducer of phase II enzymes and blocks carcinogen-induced mammary tumorigenesis in rodents (Zhang et al. 1994). Interestingly, Barcelo et al have used the T5-1A2 and T5-2E1 cells to show inhibition of DNA strand breaks induced by N-nitrosodimethylamine (NDMA) and 2-amino-3-methylimidazo[4,5-f]quinoline (IQ), respectively, by sulforaphane (Barcelo et al. 1997).

Conclusions and Perspectives

Due to interspecies differences in the expression of xenobiotic-metabolizing enzymes, the development of sensitive and specific human in vitro models is an essential step for relevant risk extrapolation from animal studies to humans. Our cellular systems, and more specifically the CYP450-expressing THLE cells and BEAS-2B cells, have several practical and functional properties that give them a strong potential for toxicological and genotoxicity studies. The cells are non-tumorigenic and are therefore suitable for transformation assays. They are of epithelial origin and are derived from target organs for human carcinogens. In addition these cell lines grow well under serum-free conditions, eliminating potential interactions of test compounds with unknown factors in serum. Finally, these cellular models represent a useful tool for the rapid screening of chemopreventive agents against cytotoxic, genotoxic and mutagenic compounds.

Recently, we developed a SV40-T antigen immortalized human adult colonic cell line (HCEC) showing expression of the xenobiotic-metabolizing enzymes comparable to normal colonic tissues (Blum et al. 1997). This non tumorigenic cell model is a promising tool for studying colon specific mutagens and carcinogens. Among these compounds, the heterocyclic amine PhIP (2-amino-1-methyl-6-phenylimidazo(4,5-b)pyridine) is suspected to be implicated in the etiology of human colorectal cancers (Schiffman and Felton 1990). Whereas PhIP-DNA adducts were preferentially found in the colon, activation of PhIP through N-hydroxylation occurs mainly in the liver (Kaderlik et al. 1994). In order to mimic this process, a co-culture model using the T5-1A2 cells as "pre-activating" system and the HCEC cell line as target cells has been developed. In the presence of the T5-1A2 cells high levels of dG-C8-PhIP adducts were detected in the colonic cell line (Macé et al. 1997c). In the future, this in vitro co-culture model will be used to study potential colon-specific chemopreventive agents such as butyrate and probiotics.

Acknowledgments

This work was supported in part by funds provided by Nestlé and by the Swiss Federal Office of Education and Science, in relation to the EU project AIR2-CT93-0860.

References

Alley MC, Scudiero DA, Monks A, Hursey ML, Czerwinski MJ, Fine DL, Abbott BJ, Mayo JG, Shoemaker RH, Boyd MR (1988) Feasibility of drug screening with panels of human tumor cell lines using a microculture tetrazolium assay. Cancer Res. 48:589-601

Aruoma OI, Halliwell B, Aeschbach R, Lõliger J (1992) Antioxidant and pro-oxidant properties of active rosemary constituents. Xenobiotica 22:257-268

Barcelo S, Macé K, Pfeifer AMA, Chipman JK (1997) Production of DNA strand breaks by N-nitrosodimethylamine (NDMA) and 2-amino-3-methylimidazo[4,5-f]quinoline (IQ) in THLE cells expressing human CYP isoenzymes and inhibition by sulforaphane. Mut Res In press.

Blum S, Offord E, Macé K, Servin A, Tromvoukis Y, Zbinden I, Pfeifer AMA (1997) Expression of intestine specific functions and metabolic competence in a SV40-immortalized human colon cell line. Gut, Submitted

Brash DE, Mark GE, Farrell MP, Harris CC (1987) Overview of human cells in genetic research: altered phenotypes in human cells caused by transferred genes. Somat. Cell Mol Genet 13:429-440

DiPaolo JA, Burkhart A, Doniger J, Pirisi L, Popescu NC, Yasumoto S (1986) In vitro models for studying the molecular biology of carcinogenesis. Toxicol Pathol 36:221-231

Eaton DL, Gallagher EP (1994) Mechanism of aflatoxin carcinogenesis. Annu Rev Pharmacol Toxicol 34:135-172

Gerwin BL, Spillare E, Forrester K, Lehman TA, Kispert J, Welsh JA, Pfeifer A, Lechner JF, Baker SJ, Vogelstein B, Harris CC (1992) Mutant p53 can induce tumorigenic conversion of human bronchial epithelial cells and reduce their responsiveness to a negative growth factor, transforming growth factor β_1. Proc Natl Acad Sci USA 89:2759-2763

Grafstrom RC (1990) Carcinogenesis studies in human epithelial tissues and cells in vitro: emphasis on serum-free culture conditions and transformation studies. Acta Physiol Scand Supp 592:93-133

Harris CC, Trump BF, Grafstrom R, Autrup H (1982) Differences in metabolism of chemical carcinogens in cultured human epithelial tissues and cells. J Cell Biochem 18:285-294

Hecht SS (1996) Chemoprevention of lung cancer by isothiocyanates. In: Back N, Cohen IR, Kritchevsky D, Lajtha A, Paoletti R (eds) Dietary phytochemicals in cancer prevention and treatment, vol 401. Plenum press, New York, pp 1-11

Ho C-T, Ferraro T, Chen Q, Rosen RT, Huang M-T (1994) Phytochemicals in teas and rosemary and their cancer-preventive properties. In: Ho C-T, Osawa T, Huang M-T, Rosen RT (eds) Food phytochemicals for cancer prevention II. American Chemical Society, Washington,D.C, pp 2-19

Huang M-T, Ferraro T, Ho C-T (1994) Cancer chemoprevention by phytochemicals in fruits and vegetables. In: Huang M-T, Osawa T, Ho C-T, Rosen RT (eds) Food chemicals for cancer prevention I. American Chemical Society, Washington, DC, pp 2-16

Kaderlik KR, Minchin RF, Mudler GJ, Ilett KF, Daugaard-Jenson M, Teitel CH, Kadlubar FK (1994) Metabolic activation pathway for the formation of DNA adducts of the carcinogen 2-amino-1-methyl-6-phenylimidazo[4,5-b]pyridine (PhIP) in rat extrahepatic tissues. Carcinogenesis 15:1703-1709

Kensler TW, Helzlouer KJ (1995) Oltipraz: Clinical opportunities for cancer chemoprevention. J Cell Biochem 22s:101-107

Linder S, Marshall H (1990) Immortalization of primary cells by DNA tumour viruses. Exp Cell Res 191:1-7

Macé K, Gonzalez FJ, McConnell IR, Garner RC, Avanti O, Harris CC, Pfeifer AMA (1994) Activation of promutagens in a human bronchial epithelial cell line stably expressing human cytochrome P450 1A2. Mol Carcinog 11:65-73

Macé K, Harris CC, Lipsky MM, Pfeifer AMA (1996) Human hepatocytes. In: Freshney RI, Freshney MG (eds) Culture of immortalized cells. Wiley-Liss, New York, pp 161-181

Macé K, Aguilar F, Wang J-S, Vautravers P, Gomez-Lechon M, Gonzalez FJ, Groopman J, Harris CC, Pfeifer AMA (1997a) Aflatoxin B_1-induced DNA adduct formation and p53 mutations in CYP450-expressing human liver cell lines. Carcinogenesis 18:1291-1297

Macé K, Bowman ED, Vautravers P, Shields PG, Harris CC, Pfeifer AMA (1997b) Characterization of phase I and phase II enzyme expression in human bronchial mucosa and peripheral lung tissues. Eur J Cancer Submitted

Macé K, Mauthe RJ, Blum S, Turteltaub K, Pfeifer AMA (1997c) 2-amino-1-methyl-6-phenylimidazo (4,5-b)pyridine (PhIP)-induced DNA adduct formation in an SV40-T antigen immortalized colonic cell line co-cultivated with a metabolically competent liver cell line. Eighty-eighth annual meeting of American Association for Cancer Research., vol 38, San Diego, USA, pp 449

Macé K, Offord EA, Pfeifer AMA (1997d) Drug metabolism and carcinogen activation studies with human genetically engineered cells. In: Castell JV, Gomez-Lechon MJ (eds) In vitro Methods in Pharmaceutical Research. Academic Press Ltd, London, pp 433-456

Macé K, Vautravers P, Granato D, Gonzalez FJ, Harris CC, Pfeifer AMA (1997e) Development of CYP450-expressing human bronchial epithelial cell lines for in vitro pharmacotoxicological applications. In Vitro Toxicol 10:85-92

Minnuni M, Wolleb U, Mueller O, Pfeifer AMA, Aeschbacher HU (1992) Natural oxidants as inhibitors of oxygen species induced mutagenicity. Mut Res 269:193-200

Newbold RF, Overell RW, Connell JR (1982) Induction of immortality is an early event in malignant transformation of mammalian cells by carcinogens. Nature 299:633-635

Offord EA, Macé K, Ruffieux C, Malnoë A, Pfeifer AMA (1995) Rosemary components inhibit benzo(a)pyrene-induced genotoxicity in human bronchial cells. Carcinogenesis 16:2057-2062

Offord EA, Guillot F, Aeschbach J, Löliger J, Pfeifer AMA (1997a) Antioxidant and biochemical properties of rosemary components: Implications for food and health. In: Shahidi F (ed) Natural antioxidants: Chemistry, health effects and applications. AOCS Press, Champaign (IL)

Offord EA, Macé K, Avanti O, Pfeifer AMA (1997b) Mechanisms involved in the chemoprotective effects of rosemary extract studied in human liver and bronchial cells. Cancer Lett. 114:275-281

Pelkonen O, Nebert DW (1982) Metabolism of polycyclic aromatic hydrocarbons: Etiologic role in carcinogenesis. Pharmacol Rev 34:189-222

Pfeifer A, Mark GE, Malan-Shibley L, Graziano SL, Amstad P, Harris CC (1989) Cooperation of *c-ras-1* and *c-myc* protooncogenes in the neoplastic transformation of SV40 T-antigen immortalized human bronchial epithelial cells. Proc Natl Acad Sci USA 86:10075-10079

Pfeifer AMA, Cole KE, Smoot DT, Weston A, Groopman JD, Shields PG, Vignaud J-M, Juillerat M, Lipsky MM, Trump BF, Lechner JF, Harris CC (1993) Simian virus 40 large tumor antigen-immortalized normal human liver epithelial cells express hepatocyte characteristics and metabolize chemical carcinogens. Proc Natl Acad Sci USA 90:5123-5127

Pfeifer AMA, Macé K, Tromvoukis Y, Lipsky MM (1995) Highly efficient establishment of immortalized cells from adult human liver. Meth Cell Science 17:83-89

Reddel RR, Ke Y, Gerwin BI, McMenamin MG, Lechner JF, Su RT, Brash DE, Park JB, Rhim JS, Harris CC (1988) Transformation of human bronchial epithelial cells by infection with SV40 or adenovirus-12 SV40 hybrid virus, or transfection via strontium phosphate coprecipitation with a plasmid containing SV40 early region genes. Cancer Res 48:1904-1909

Schiffman MH, Felton JS (1990) Fried foods and the risk of colon cancer. A. J. Epidemiol 131:376-378

Zhang Y, Kensler TW, Cho CG, Posner GH, Talalay P (1994) Anticarcinogenic activities of sulforaphane and structurally related synthetic norbornyl isothiocyanates. Proc Natl Acad Sci U S A 91:3147-50

Bioavailability and Health Effects of Dietary Flavonols in Man

Peter C.H. Hollman[1] and Martijn B. Katan[2]
[1] DLO State Institute for Quality Control of Agricultural Products (RIKILT-DLO). Bornsesteeg 45, NL-6708 PD Wageningen, The Netherlands.
[2] Agricultural University, Department of Human Nutrition. Bomenweg 2, NL-6703 HD Wageningen, The Netherlands

Abstract

Flavonoids are polyphenolic compounds that occur ubiquitously in foods of plant origin. Over 4000 different flavonoids have been described, and they are categorized into flavonols, flavones, catechins, flavanones, anthocyanidins, and isoflavonoids. Flavonoids have a variety of biological effects in numerous mammalian cell systems, as well as *in vivo*. Recently much attention has been paid to their antioxidant properties and to their inhibitory role in various stages of tumour development in animal studies.

Quercetin, the major representative of the flavonol subclass, is a strong antioxidant, and prevents oxidation of low density lipoproteins *in vitro*. Oxidized low density lipoproteins are atherogenic, and are considered to be a crucial intermediate in the formation of atherosclerotic plaques. This agrees with observations in epidemiological studies that the intake of flavonols and flavones was inversely associated with subsequent coronary heart disease. However, no effects of flavonols on cancer were found in these studies.

The extent of absorption of flavonoids is an important unsolved problem in judging their many alleged health effects. Flavonoids present in foods were considered non-absorbable because they are bound to sugars as β-glycosides. Only free flavonoids without a sugar molecule, the so-called aglycones were thought to be able to pass through the gut wall. Hydrolysis only occurs in the colon by microorganisms, which at the same time degrade flavonoids. We performed a study to quantify absorption of various dietary forms of quercetin. To our surprise, the quercetin glycosides from onions were absorbed far better than the pure aglycone. Subsequent pharmacokinetic studies with dietary quercetin glycosides showed marked differences in absorption rate and bioavailability. Absorbed quercetin was eliminated only slowly from the blood.

The metabolism of flavonoids has been studied frequently in various animals, but very few data in humans are available. Two major sites of flavonoid metabolism are the liver and the colonic flora. There is evidence for O-methylation, sulfation and glucuronidation of hydroxyl groups in the liver. Bacterial ring fission of flavonoids occurs in the colon. The subsequent degradation products, phenolic acids, can be absorbed and are found in urine of animals. Quantitative data on metabolism are scarce.

Introduction

A large number of epidemiological studies show a protective effect of vegetables and fruits against cancer (Steinmetz and Potter, 1991; Block et al. 1992). Although not studied as extensively as for cancer, epidemiological studies also suggest a strong protective effect of vegetables and fruits for stroke, and a weaker protective effect for coronary heart disease (Ness and Powles, 1997). Various hypotheses have been suggested to explain these beneficial effects of an increased consumption of vegetables and fruits. An attractive hypothesis is that vegetables and fruits contain compounds that have a protective effect independent of that of known nutrients and micronutrients. This is supported by *in vitro* and *in vivo* studies which show that naturally occurring plant compounds may inhibit various stages in the cancer process (Wattenberg, 1992). In these studies flavonoids have also been studied extensively. Reduced risk of cardiovascular disease is possibly associated with high intakes of dietary antioxidants, of which vitamins have been most frequently studied (Jha et al. 1995).

Flavonoids are polyphenolic compounds that occur ubiquitously in foods of plant origin. Variations in the heterocyclic ring C give rise to flavonols, flavones, catechins, flavanones, anthocyanidins, and isoflavonoids (Fig. 1). In addition, the basic structure of flavonoids allows a multitude of substitution patterns in the benzene rings A and B within each class of flavonoids: phenolic hydroxyls, O-sugars, methoxy groups, sulfates and glucuronides. Over 4000 different naturally occurring flavonoids have been described(Middleton and Kandaswami, 1994). Flavonoids are common substances in the daily diet (Table 1).

Due to their antioxidant properties *in vitro* and to their inhibitory role in various stages of tumour development in animal studies, flavonoids may contribute to the protective effects of vegetables and fruits and dietary antioxidants.

Biological Effects of Flavonoids

A multitude of *in vitro* studies has shown that flavonoids can inhibit, and sometimes induce, a large variety of mammalian enzyme systems. The effects of mainly flavones and flavonols on 24 different enzymes or enzyme systems were described in a review (Middleton and Kandaswami, 1994). Some of these enzymes are involved in important pathways that regulate cell division and proliferation, platelet aggregation, detoxification, and inflammatory and immune response. Thus it is not surprising that effects of flavonoids have been found in cell systems and animals, on different stages in the cancer process, on the immune system, and on haemostasis (Middleton and Kandaswami, 1994). Worries about the mutagenicity of flavonoids in bacterial systems (Sugimura et al. 1977) triggered much research on the flavonol quercetin. However, mutagenicity of flavonoids *in vivo* in mammals was never found (Aeschbacher et al. 1982; MacGregor et al. 1983). Animal studies of the carcinogenicity of quercetin were also negative except in one case (Middleton and Kandaswami, 1994). However, the anticarcinogenic and antiproliferative effects of quercetinand other flavonoids are becoming increasingly evident (Huang and Ferraro, 1992).

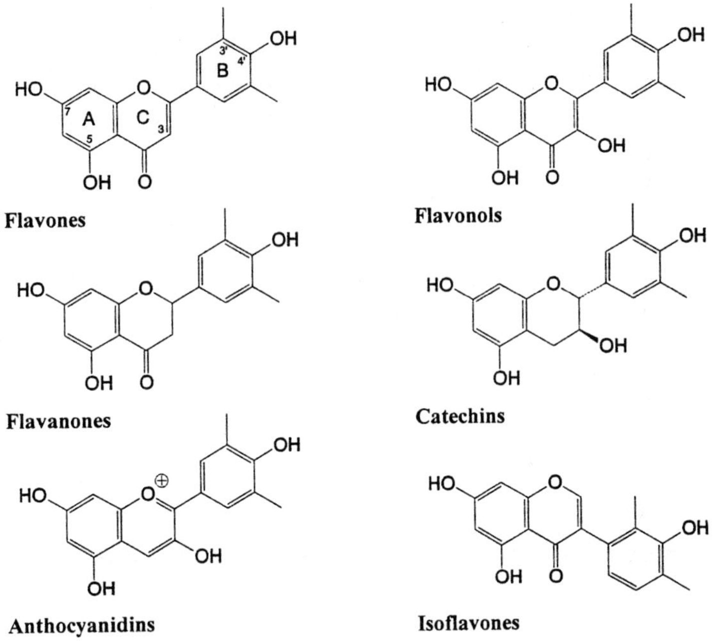

Flavones

Flavonols

Flavanones

Catechins

Anthocyanidins

Isoflavones

Fig. 1. Subclasses of flavonoids. Classification is based on variations in the heterocyclic ring C.

Recently, it has been hypothesised that their antioxidant properties (Kandaswami and Middleton, 1994) may protect tissues against oxygen free radicals and lipid peroxidation. Oxygen free radicals and lipid peroxidation might be involved in several pathological conditions such as atherosclerosis, cancer, and chronic inflammation (Halliwell, 1994). Quercetin is a powerful antioxidant. The Trolox equivalent antioxidant capacity (TEAC) of quercetin is 4 fold higher than that of the antioxidant (pro)vitamins (Rice-Evans and Miller, 1996). Quercetin prevents oxidation of low density lipoproteins (LDL) *in vitro* (de Whalley et al. 1990). Oxidized LDL has been found in atherosclerotic lesions of humans (Shaikh et al. 1988), and increased plasma concentrations of autoantibodies against oxidized LDL occur in patients with atherosclerosis (Salonen et al. 1992; Bergmark et al. 1995). Quercetin may therefore contribute to the prevention of atherosclerosis (Steinberg et al. 1989), cancer and chronic inflammation (Halliwell, 1994).

Table 1. Occurrence of flavonoids in common foods (Kühnau, 1976; Hertog et al. 1992; Hertog et al. 1993b)

Flavonoid Subclass	Major Food Sources
Flavonols	- onions, kale, broccoli - apples, cherries, berries - tea, red wine
Flavones	- parsley, thyme
Flavanones	- citrus
Catechins	- apples - tea
Anthocyanidins	- cherries, grapes
Isoflavones	- soya beans, legumes

Flavonols in Cancer and Cardiovascular Disease

Reliable data on flavonoid contents of common vegetables and fruits are needed to be able to study the potential role of dietary flavonoids in cancer and coronary heart disease prevention Such a database did not exist for the Netherlands, and data produced in other countries were fragmentary. In addition, the quality of these data was questionable, because they were obtained with methods now considered obsolete. We undertook to produce a database on flavonoid contents of vegetables and fruits commonly consumed in the Netherlands. We focused on the subgroups of flavonols and flavones, because these flavonoids, including the flavonol quercetin (3,5,7,3',4'-pentahydroxyflavone) occur ubiquitously in plant foods and were the ones most frequently studied in model systems.

We determined the flavonols quercetin, kaempferol, myricetin, and the flavones luteolin and apigenin in 28 vegetables, 9 fruits, and 10 beverages commonly consumed in the Netherlands (Hertog et al. 1992; Hertog et al. 1993b). Quercetin was by far the most important flavonol, followed by kaempferol (3,5,7,4'-tetrahydroxyflavone). Flavones were only found in a few products. These data have been used in a number of prospective cohort studies and in one prospective cross-cultural study on the relation between flavonol and flavone intake and cancer and cardiovascular disease.

Cancer. The intake of flavonols and flavones was calculated in a population of elderly men in the Dutch town of Zutphen, the Zutphen Elderly Study. In 1985 their food consumption was assessed using a dietary history method. A total number of 805 men aged 65-84 years, entered the study. The intake of flavonols and flavones was on average 26 mg/day. Major sources of flavonols and flavones were tea (61%), onions (13%) and apples (10%). After 5 years, in 1990, their health records were collected, and morbidity and mortality data were studied. Differences in baseline characteristics of these men between tertiles of flavonol and flavone intake were evaluated, and relative risks were calculated. No associations were found between flavonol and flavone intake and total cancer mortality. Also specific forms of cancer, such as lung cancer were not associated with flavonols and flavones (Hertog et al. 1994).

In a large cohort study, The Netherlands Cohort study, consisting of 120 850 men and women aged 55-69 years, also no association was found with flavonol and flavone intake and stomach cancer, colon cancer and lung cancer during 4.3 years of follow-up (Goldbohm et al. 1995).

The Zutphen Study cohort is one of the cohorts of the Seven Countries Study, a cross-cultural study of diet, lifestyle and disease. In 1987 the foods that represented the baseline diet as per 1960 of each cohort were bought locally. The foods were combined into food composites that represented the average daily food intake of each cohort. In these food composites flavonols and flavones were determined. The intake of flavonols and flavones ranged from 3 mg/day in a Finnish cohort to 70 mg/day in a Japanese cohort. The major dietary sources of flavonols and flavones varied substantially between cohorts. In the Japanese and Dutch cohorts the major source was tea, while red wine was the major source in Italy. Onions and apples were the predominant sources in the United States, Finland, Greece and former Yugoslavia. Again, no association with cancer mortality was found (Hertog et al. 1995). Thus, no association with cancer mortality was found in these three studies (Table 2).

Cardiovascular Disease. As for cancer, the only studies relating the intake of dietary flavonoids to risk of cardiovascular disease have been observational in nature. We determined the average dietary flavonol and flavone intake as it was around 1960 in 16 cohorts participating in the Seven Countries Study. The average flavonol and flavone intake was inversely correlated to mortality rates of coronary heart disease after 25 years of follow-up (Hertog et al. 1995). The intake of flavonols and flavones, together with smoking and the intake of saturated fat, explained about 90% of the variance in coronary heart disease mortality rates across the 16 cohorts.

Five prospective within-population cohort studies have been carried out. Coronary heart disease mortality was strongly inversely associated with flavonol and flavone intake in the Zutphen Elderly Study (Hertog et al. 1993a) with a reduction in mortality risk of more than 50% being recorded in the highest tertile of flavonol intake. Average flavonol intake in the highest tertile was 42 mg/day, and in the lowest 12 mg/day. Recently, the ten year follow-up of the Zutphen Elderly Study was completed with results strengthening the findings of the five year follow-up (Hertog et al. 1997a). Unlike the findings of the five year follow-up, a clear dose-response relationship between flavonol intake and coronary heart disease mortality was now recorded.

Table 2. Summary of epidemiological prospective studies on flavonol and flavone intake and cancer risk

Population	Age (y)	Follow-up (y)	Relative Risk[a] (95% Confidence Interval)
Cohort studies			
805 men; Zutphen (The Netherlands) (Hertog et al. 1994)	65-84	5	1.2 (0.7 - 2.2)
120 852 men + women; Netherlands Cohort Study (Goldbohm et al. 1995)	55-69	4	1
Cross-cultural study			
12 763 men; Seven Countries Study (Hertog et al. 1995)	40-59	25	r = 0.39 (P = 0.14)

[a] Relative risk of highest versus lowest flavonol intake group, adjusted for age, diet and other risk factors for cancer.

The association between flavonol and flavone intake and risk of stroke was studied in a cohort of 550 middle-aged men (Keli et al. 1996). These men were followed for 15 years, and the men in the highest quartile of flavonol and flavone intake (>30 mg/ day) showed a considerably reduced risk of the disease of about 60%.

Mortality from coronary heart disease was weakly inversely associated with flavonol and flavone intake in a cohort of 5130 Finnish men and women aged 30 - 69 years followed over a 20 years period (Knekt et al. 1996). The relative risks of mortality from coronary heart disease between the highest (>5 mg/day) and lowest quartiles (<2.5 mg/day) of flavonol and flavone intake were 0.73 for women and 0.67 for men.

Recently, in male US health professionals a modest, but non-significant, inverse association between flavonol and flavone intake and coronary mortality was found only in men with previous history of coronary heart disease (Rimm et al. 1996). Median flavonol intake in the highest quintile was 40 mg/day and 7 mg/day in the lowest.

In contrast to the above studies, increased mortality of ischaemic heart disease was found in Welsh men (Hertog et al. 1997b) in all quartiles of high flavonol intake compared to the lowest quartile. Mean flavonol intake in the highest quartile was 43 mg/day, and 14 mg/day in the lowest quartile.

To summarize (Table 3), a protective role for flavonols in cardiovascular disease was found in 3 out of 5 prospective cohort studies, in addition to one cross-cultural study. One prospective cohort study showed no association, and one a weakly positive association between flavonol intake and coronary heart disease. So far, the epidemiological evidence points to a protective effect of antioxidant flavonols in cardiovascular disease but it is not conclusive.

Table 3. Summary of epidemiological prospective studies on flavonol and flavone intake and coronary heart disease (CHD) and stroke risk

Population	Age (y)	Follow-up (y)	Relative Risk[a] (95% Confidence Interval)
Cohort studies			
CHD, 805 men; Zutphen (The Netherlands) (Hertog et al. 1993a)	65 - 84	5	0.32 (0.15 - 0.71)
CHD, 5133 men + women; Finland (Knekt et al. 1996) :	30 - 69	20	0.73 (0.41 - 1.32) 0.67 (0.44 - 1.00)
CHD, 34 789 men Health Professionals (U.S.A.) (Rimm et al. 1996)	40 - 75	6	1.08 (0.81 - 1.43)
CHD, 1900 men Caerphilly (U.K.) (Hertog et al. 1997b)	49 - 59	14	1.6 (0.9 - 2.9)
Stroke, 552 men; Zutphen (The Netherlands) (Keli et al. 1996)	50 - 69	15	0.27 (0.11 - 0.70)
Cross-cultural study			
CHD, 12 763 men Seven Countries Study (Hertog et al. 1995)	40 - 59	25	r = -0.50 (*P* = 0.01)

[a] Relative risk of highest versus lowest flavonol intake group, adjusted for age, diet and other risk factors for coronary heart disease

Absorption and Metabolism of Flavonoids

These epidemiological data support a role of flavonols as antioxidants in coronary heart disease prevention. However, absorption from the diet is a prerequisite for a causal relation between flavonols and coronary heart disease. In addition, metabolism of flavonols after absorption should not substantially inhibit their antioxidant capacity. The absorption and subsequent distribution, metabolism and excretion of flavonoids in humans is little studied. Absorption of flavonoids from the diet was long considered to be negligible, as they are present in foods bound to sugars as β-glycosides (with the exception of catechins). Only free flavonoids without a sugar molecule, the so-called aglycones, were considered to be able to pass the gut wall, and no enzymes that can split these predominantly β-glycosidic bonds are secreted into the gut or present in the intestinal wall (Kühnau, 1976; Griffiths, 1982). Hydrolysis only occurs in the colon by microorganisms, which at the same time degrade dietary flavonoids extensively (Kühnau, 1976). Thus, only a marginal absorption of dietary flavonoids is to be expected. However, research on the mechanisms for aglycone transfer across the gut wall is lacking. Although only flavonoid aglycones were considered to be able to pass the gut wall, the orally administered aglycone of quercetin was poorly absorbed ($< 1\%$) in a human trial (Gugler et al. 1975). In contrast, absorption of about 20% was demonstrated in rats after oral administration of quercetin aglycone (Ueno et al. 1983). We recently confirmed these results in a human study with ileostomy subjects: absorption of orally administered quercetin aglycone was 24% (Hollman et al. 1995). The absorption of quercetin glycosides from onions in this study was 52%, and 17% for pure quercetin rutinoside, a common glycoside in foods (Hollman et al. 1995). These data show that humans absorb appreciable amounts of quercetin and that absorption of glycosides in the small intestine is possible.

After absorption of flavonoids, the subsequent metabolism of flavonoids is rather well known from animal studies (Griffiths, 1982; Hackett, 1986), but practically no data in humans are available. Hydroxyl groups are conjugated with glucuronic acid or sulfate in the liver. In addition O-methylation may occur. Excretion in bile of glucuronides and sulfates seems to be important. Bacteria in the colon hydrolyse conjugates which is supposed to enable absorption of the liberated aglycones. However, these microorganisms also substantially degrade the flavonoid moiety by cleavage of the heterocyclic ring. Three main types of ring scission for catechins, flavonols, and flavones and flavanones, each leading to various phenolic acids or their lactones, have been postulated. These primarily produced phenolic acids are prone to secondary reactions such as β-oxidation, reduction, demethylation, dehydroxylation, and decarboxylation. The phenolic acids are absorbed and excreted in the urine. The significance of these results for humans is not clear.

Bioavailability of Quercetin Glycosides from Foods

We studied the time course of the plasma quercetin concentration in healthy subjects after supplementation of major food sources of quercetin: onions, apples and pure quercetin rutinoside. Quercetin rutinoside is a major glycoside of tea. The subjects ingested a single dose of about 4 times the average intake of flavonols and flavones in the Netherlands (Hollman, 1997). Peak plasma levels of 225 ng/ml (= 0.8 μM) were reached after 0.7 h for the onions supplement, 90 ng/ml after 2.5 h for the apples, and 90 ng/ml after 9 h for the rutinoside. Disposition of quercetin in plasma was biphasic for onions and apples, with an elimination half-life of about 25 h. This implies that repeated dietary intake of quercetin throughout the day would lead to a build-up of the concentration in plasma. The bioavailability of quercetin from apples and the rutinoside was about one third of that from onions. Thus, dietary quercetin glycosides are absorbed in man. Absorption kinetics and bioavailibility is probably governed by the type of glycoside. The dietary antioxidant quercetin could significantly increase the antioxidant capacity of blood plasma.

Conclusions

Flavonoids are bioactive polyphenols that occur ubiquitously in plant foods. Animal studies and *in vitro* studies suggest that dietary flavonols could inhibit cancer in humans. However, so far no association with cancer mortality was found in three epidemiological studies. In contrast, intake of flavonols and flavones was inversely associated with cardiovascular disease in three prospective cohort studies and in a prospective cross-cultural study. However, in one large prospective cohort study no association with coronary heart disease was found. Antioxidant effects of flavonoids possibly could explain these results. Dietary quercetin, the major flavonol in foods, is absorbed in humans and is only slowly eliminated throughout the day. Quercetin could thus contribute significantly to the antioxidant defences present in blood plasma. The metabolism of flavonoids in humans is little studied and pharmacokinetic data are scarce, probably because selective and sensitive analytical methods for these compounds in body fluids are lacking.

The role of dietary flavonols and flavones in cardiovascular disease prevention is promising. Epidemiological research in other countries and cultures, studies on biological mechanisms and bioavailability, and intervention studies are needed to fully evaluate their role in human health.

Acknowledgements

We thank Prof. D. Kromhout, who took the initiative for the epidemiological studies. We are grateful to Dr. M.G.L. Hertog, Dr. E. Feskens, Prof. J.G.A.J. Hautvast, and J.H.M. de Vries for valuable discussions, and to John M.P. van

Trijp, M.N.C.P. Buysman, D.P. Venema, and B.v.d. Putte for technical assistance.
This work was supported by grants from the Foundation for Nutrition and Health Research and the Netherlands Heart Foundation (94.128).

References

Aeschbacher HU, Meier H, Ruch E (1982) Nonmutagenicity *in vivo* of the food flavonol quercetin. Nutr Cancer 4:90-98

Bergmark C, Wu R, de Faire U, Lefvert AK, Swedenborg J (1995) Patients with early-onset peripheral vascular disease have increased levels of autoantibodies against oxidized LDL. Arterioscler Thromb Vasc Biol 15:441-445

Block G, Patterson B, Subar A (1992) Fruit, vegetables, and cancer prevention: a review of epidemiological evidence. Nutr Cancer 18:1-29

Goldbohm RA, van den Brandt PA, Hertog MGL, Brants HAM, van Poppel G (1995) Flavonoid intake and risk of cancer: a prospective cohort study. Am J Epidemiol 141(Suppl):s61

Griffiths LA (1982) Mammalian metabolism of flavonoids. In: The Flavonoids: Advances in Research (Harborne J, Mabry T eds), London: Chapman and Hall, pp. 681-718

Gugler R, Leschik M, Dengler HJ (1975) Disposition of quercetin in man after single oral and intravenous doses. Eur J Clin Pharmacol 9:229-234

Hackett AM (1986) The metabolism of flavonoid compounds in mammals. In: Plant flavonoids in biology and medicine. Biochemical, pharmacological, structure-activity relationships (Cody V, Middleton E, Harborne J eds), New York: Alan R. Liss, Inc., pp. 177-194

Halliwell B (1994) Free radicals, antioxidants, and human disease: curiosity, cause, or consequence? Lancet 344:721-724

Hertog MGL, Hollman PCH, Katan MB (1992) Content of potentially anticarcinogenic flavonoids of 28 vegetables and 9 fruits commonly consumed in the Netherlands. J Agric Food Chem 40:2379-2383

Hertog MGL, Feskens EJM, Hollman PCH, Katan MB, Kromhout D (1993a) Dietary antioxidant flavonoids and risk of coronary heart disease: the Zutphen Elderly Study. Lancet 342:1007-1011

Hertog MGL, Hollman PCH, van de Putte B (1993b) Content of potentially anticarcinogenic flavonoids of tea infusions, wines, and fruit juices. J Agric Food Chem 41:1242-1246

Hertog MGL, Feskens EJM, Hollman PCH, Katan MB, Kromhout D (1994) Dietary flavonoids and cancer risk in the Zutphen Elderly Study. Nutr Cancer 22:175-184

Hertog MGL, Kromhout D, Aravanis C, Blackburn H, Buzina R, Fidanza F, Giampaoli S, Jansen A, Menotti A, Nedeljkovic S, Pekkarinen M, Simic BS, Toshima H, Feskens EJM, Hollman PCH, Katan MB (1995) Flavonoid intake and long-term risk of coronary heart disease and cancer in the Seven Countries Study. Arch Intern Med 155:381-386

Hertog MGL, Feskens EJM, Kromhout D (1997a) Antioxidant flavonols and coronary heart disease risk: ten year follow-up of the Zutphen Elderly Study. Lancet 349: 699

Hertog MGL, Sweetnam PM, Fehily AM, Elwood PC, Kromhout D (1997b) Antioxidant flavonols and ischaemic heart disease in a Welsh population of men. The Caerphilly Study. Am J Clin Nutr 65:1489-1494

Hollman PCH, de Vries JHM, van Leeuwen SD, Mengelers MJB, Katan MB (1995) Absorption of dietary quercetin glycosides and quercetin in healthy ileostomy volunteers. Am J Clin Nutr 62:1276-1282

Hollman PCH (1997) Determinants of the absorption of the dietary flavonoid quercetin in man. Wageningen: Thesis, Agricultural University

Huang M-T, Ferraro T (1992) Phenolic compounds in food and cancer prevention. In: Phenolic compounds in food and their effects on health II. Antioxidants & cancer prevention (Huang M-T, Ho C, Lee CY eds), Washington DC: American Chemical Society, pp. 8-34

Jha P, Flather M, Lonn E, Farkouh M, Yusuf S (1995) The antioxidant vitamins and cardiovascular disease. A critical review of epidemiologic and clinical trial data. Ann Intern Med 123:860-872

Kandaswami C, Middleton E (1994) Free radical scavenging and antioxidant activity of plant flavonoids. Adv Exp Med Biol 366:351-376

Keli SO, Hertog MGL, Feskens EJM, Kromhout D (1996) Flavonoids, antioxidant vitamins and risk of stroke. The Zutphen study. Arch Intern Med 156:637-642

Knekt P, Järvinen R, Reunanen A, Maatela J (1996) Flavonoid intake and coronary mortality in Finland: a cohort study. Br Med J 312:478-481

Kühnau J (1976) The flavonoids. A class of semi-essential food components: their role in human nutrition. World Rev Nutr Diet 24:117-191

MacGregor JT, Wehr CM, Manners GD, Jurd L, Minkler JL, Carrano AV (1983) *In vivo* exposure to plant flavonols. Influence on frequencies of micronuclei in mouse erythrocytes and sister-chromatid exchange in rabbit lymphocytes. Mutation Res 124:255-270

Middleton E, Kandaswami C (1994) The impact of plant flavonoids on mammalian biology: implications for immunity, inflammation and cancer. In: The Flavonoids: advances in research since 1986 (Harborne JB ed), London: Chapman & Hall, pp. 619-652

Ness AR, Powles JW (1997) Fruit and vegetables, and cardiovascular disease: a review Int J Epidemiol 26:1-13

Rice-Evans CA, Miller NJ (1996) Antioxidant activities of flavonoids as bioactive components of food. Biochem Soc Trans 24:790-795

Rimm EB, Katan MB, Ascherio A, Stampfer MJ, Willett WC (1996) Relation between intake of flavonoids and risk for coronary heart disease in male health professionals. Ann Intern Med 125:384-389

Salonen JT, Ylä-Herttuala S, Yamamoto R, Butler S, Korpela H, Salonen R, Nyyssönen K, Palinski W, Witztum JL (1992) Autoantibody against oxidised LDL and progression of carotid atherosclerosis. Lancet 339:883-887

Shaikh M, Martini S, Quiney JR, Baskerville P, La Ville AE, Browse NL, Duffield R, Turner PR, Lewis B (1988) Modified plasma-derived lipoproteins in human atherosclerotic plaques. Atherosclerosis 69:165-172

Steinberg D, Parthasarathy S, Carew TE, Khoo JC, Witztum JL (1989) Beyond cholesterol: modifications of low density lipoprotein that increase its atherogenicity. N Engl J Med 320:915-924

Steinmetz KA, Potter JD (1991) Vegetables, fruit and cancer. I. Epidemiology. Cancer Cause Control 2:325-357

Sugimura T, Nagao M, Matsushima T, Yahagi T, Seino Y, Shirai A, Sawamura M, Natori S, Yoshibira K, Fukuoka M, Kuroyanagi M (1977) Mutagenicity of flavone derivatives. Proc Jpn Acad 53:194-197

Ueno I, Nakano N, Hirono I (1983) Metabolic fate of [^{14}C]quercetin in the ACI rat. Jpn J Exp Med 53:41-50

Wattenberg LW (1992) Inhibition of carcinogenesis by minor dietary constituents. Cancer Res 52(Suppl):2085S-2091S

de Whalley C, Rankin SM, Hoult JRS, Jessup W, Leake DS (1990) Flavonoids inhibit the oxidative modification of low density lipoproteins by macrophages. Biochem Pharmacol 39:1743-1750

Epidemiological Studies on Antioxidants, Lipid Peroxidation and Atherosclerosis

Jukka T. Salonen,
Research Institute of Public Health, University of Kuopio, Box 1627, 70211 Kuopio, Finland

The hypothesis that oxidative modification increases the atherogenicity of low density lipoprotein (LDL) is actively studied. In two separate studies we observed an association between a high ratio of antibodies to oxidized LDL/native LDL and accelerated progression of carotid atherosclerosis. This has been confirmed in three cross-sectional studies. We have also studied the association of lipid oxidation products with the progression of early carotid atherosclerosis. The strongest predictor was serum 7β-OH-cholesterol, a stable oxidation product of cholesterol.

Redox-active forms of transition metals (e.g. iron and mercury) catalyse the formation of free radicals. In our prospective study, the KIHD, both elevated serum ferritin ($\geq$ 200 μg/l, marker of body iron stores) and the accumulation of methylmercury (hair mercury $\geq$ 2.0 μg/g) were associated with an increased, and donating blood with a reduced risk of myocardial infarction (MI). Prospective epidemiologic studies suggest that high intake of vitamin E is associated with reduced CHD risk. In our prospective studies, deficiencies in selenium and vitamin C were associated with increased risk of MI.

No conclusive clinical trials concerning the preventive effect of antioxidants on atherosclerosis have so far been reported. In three large trials, β-carotene had no effect on cardiovascular deaths. In the CHAOS study, large doses (400 or 800 IU/d) of vitamin E reduced the incidence of non-fatal MI but not cardiovascular mortality in CHD patients. In the Linxian study, a vitamin/mineral supplement reduced stroke mortality by 68% in men and 7% (ns) in women. In a recent selenium supplementation trial, total mortality was 21% smaller (p=0.07) in the selenium than in the placebo group.

Our "Antioxidant Supplementation in Atherosclerosis Prevention" ("ASAP") study is a 2x2 factorial double-masked placebo-controlled randomized 3-year trial, testing the effect of 200 mg/d of d-α-tocopherol acetate and 500 mg/d of slow-release vitamin C on the progression of carotid atherosclerosis, blood

---

Abbreviations: LDL, Low Density Lipoproteins; LDLC, LDL Cholesterol; CHD, Coronary Heart Disease; (A)MI, (Acute) Myocardial Infarct; OxLDL, Oxidised LDL; TBARS, ThioBarbituric Acid Reactive Substance; VLDL, Very Low Density Lipoproteins; COPS, Cholesterol Oxidation Products; KIHD, Kuopio Ischaemic Disease Risk Factor Study; GSHPx, Glutathioneperoxidase; MDA, Malondialdehyde; DPPD, N,N'-Diphenyl-p-phenylenediamine; CVD, CardioVascular Disease

pressure and lipid peroxidation in 500 men and women aged 45-69 years. This trial and other on-going preventive trials will eventually provide further necessary information about the role of antioxidants in atherosclerosis and in the prevention of atherosclerotic cardiovascular diseases.

Introduction

Atherosclerosis is a degenerative process in the arteries which is characterized in its early phase by endothelial cell injury, the accumulation of lipid and proliferation of smooth muscle cells and consequent thickening of the intimal and medial layers of the arterial wall. These and other pathophysiologic phenomena in atherogenesis are assumed to be influenced by free radicals and oxidised lipids (Steinberg et al. 1989; Salonen et al. 1992a; Witztum 1994).

Low density lipoprotein (LDL) particles contain polyunsaturated fatty acids and cholesterol that can be oxidised by oxidative free radicals. Oxidative and other chemical modifications of LDL make it immunogenic and are thought to increase its atherogenicity (Steinberg et al. 1989, Salonen et al. 1992a, Witztum 1994). The oxidative modification of LDL in the human arterial wall is influenced by the availability of antioxidative mechanisms as well as the presence of transition metals such as iron and copper, which act as catalysts of oxidation. Among the key physiological antioxidants inhibiting the oxidative modification of LDL are vitamin E, ascorbic acid and free radical scavenging enzyme systems including selenium-dependent glutathione peroxidase (Gey 1986, Halliwell and Gutteridge 1989, Steinberg et al. 1992, Esterbauer et al. 1992, Ames et al. 1993, Halliwell 1994).

There is some epidemiological data which is consistent with this. We and others have found that the presence of oxidized LDL, as expressed as autoantibodies against oxidized LDL, associates with an accelerated progression of atherosclerosis (Salonen et al. 1992a, Salonen et al. 1992b, Maggi et al. 1993, Salonen 1995).

Lipid Peroxidation and Atherosclerotic Progression

We have investigated risk factors for the progression of common carotid atherosclerosis for over a decade (Salonen and Salonen 1993). We use high-resolution B-mode ultrasonography to assess atherosclerosis. This methodology is superior to angiography, which provides information only on the obstruction of the arterial lumen but not about the layers of the arterial wall. Early atherosclerosis is detectable as thickening of the intimal and medial layers of the vessel wall. Obstruction of the lumen and changes in blood flow appear at a much more progressed stage and are only detectable later. In addition, arterial ultrasonography is safe and inexpensive (Salonen and Salonen 1993).

Oxidative modification of LDL renders it immunogenic, and autoantibodies to epitopes of oxidized LDL (Ox-LDL) have been shown in humans. To determine whether these antibodies are related to atherosclerotic progression we compared the ratio of serum titre of autoantibodies to malondialdehyde (MDA)-modified LDL to that for native LDL in baseline sera of 30 eastern Finnish men with accelerated 2-year progression of carotid atherosclerosis and 30 age-matched controls without progression (Salonen et al. 1992a). IgG antibody titre was determined by solid phase RIA. To ensure the specificity of the measure, the ratio of antibody titres binding to MDA-LDL/binding to native LDL was used. Cases had significantly higher titre to MDA-LDL (2.67 vs. 2.06, $p=0.003$). Cases also had greater proportion of smokers (37 vs. 3%), higher LDLC (4.2 vs. 3.6 mmol/l), and higher serum copper concentration (1.14 vs. 1.04 mg/l). Even after adjusting for these variables and baseline atherosclerosis severity in a multivariate logistic model, the difference in antibody titre remained significant ($p=0.031$) (Table 1). Thus, the titre of auto-antibodies to MDA-LDL was an independent predictor of the progression of carotid atherosclerosis in a prospective study in Finnish men. This was the first epidemiologic study to show a role of oxidatively modified LDL in atherogenesis.

In a second study, conducted in collaboration with Professor Esterbauer, we investigated the relationship between the presence of autoantibodies against Cu-oxidized LDL and the progression of carotid and femoral atherosclerosis in 212 eastern Finnish men with serum LDLC $\geq$ 4.0 mmol/l (Salonen et al. 1992b). In multivariate models adjusting for age, smoking, serum LDLC and triglyceride concentrations, triglyceride and protein content of LDL, systolic blood pressure, and the 12-month change in serum LDLC, the 12-month increase of IMT (mean of all 6 arterial segments measured) was significantly greater in men with elevated (mean 0.29 mm, SD 0.15 mm) as compared to men with borderline (0.17 mm, SD 0.18 mm) and low titre (0.17 mm, SD 0.13 mm) ($p=0.002$ for heterogeneity). These data suggested that Ox-LDL-antibodies associate with accelerated atherosclerotic progression also in hypercholesterolemic men and provide further evidence for the role of in vivo oxidation of LDL in atherosclerosis.

Maggi and coworkers confirmed our findings in a cross-sectional case-control study (Maggi et al. 1993). They reported a higher titre ratio of antibodies to oxidized/native LDL in 94 patients with CHD than in 42 healthy controls.

We observed in a 2-year follow-up study of 126 men that both low serum selenium level and low plasma vitamin C concentration were associated with accelerated progression of common carotid atherosclerosis (Salonen et al. 1992c).

We also studied the association of cholesterol oxidation products (COPS) with the progression of carotid atherosclerosis (Salonen et al. 1997). Twenty subjects with a fast progression and 20 with no progression of carotid atherosclerosis in three years were selected from over 400 participants of Kuopio Atherosclerosis Prevention Study. Progression of carotid atherosclerosis was assessed by high-resolution B-mode ultrasonography. Serum 7β-hydroxycholesterol, a major oxidation product of cholesterol in membranes and lipoproteins, and seven other cholesterol oxidation products were measured by isotope dilution mass spectrometry, lipid hydroperoxides in low density lipoprotein (LDL) fluorometrically as thiobarbituric acid reactive substances and oxidation susceptibility of LDL and very low density lipoprotein (VLDL) kinetically.

High concentrations of serum 7β-hydroxycholesterol (standardized coefficient 0.47, $p=0.0005$), cigarette smoking (0.35, $p=0.0167$), LDL TBARS (0.23, $p=0.0862$) and an increased oxidation susceptibility of VLDL+LDL (0.22, $p=0.1114$) were the strongest predictors of 3-year increase in carotid wall thickness of over 30 variables tested in step-up least squares regression models. The model of 10 variables explained 60 percent of the atherosclerotic progression. In a multivariate logistic model, the risk of experiencing a fast progression increased by 80% ($p=0.013$) per unit (μg/l) of 7β-hydroxycholesterol. Antibodies are an indirect measurement of lipid peroxidation and they also have the problem of cross-reactivity between various antigens and epitopes. These problems do not concern lipid oxidation products and the oxidation resistance of lipoproteins. These findings provide the first direct evidence of an association between lipid oxidation and atherogenesis in humans.

Pro-Oxidants and the Risk of Coronary Heart Disease

In prospective population studies or follow-up studies, risk factors or protective factors are measured in a population sample and the incidence or mortality in this cohort is followed after the baseline measurements. The advantage of this design is that the disease does not usually affect the risk factor measurements. Systematic measurement errors in the risk factor measurements can cause biased associations, but this is rare. The most typical bias in a prospective study is a negative one: a truly existing association is not observed (Salonen 1987). In studies concerning pro- and antioxidants and chronic diseases, the main reasons for this are the random error in nutrient and other risk factor measurements due to intraindividual variation over time, and the limited range of nutrient intake within a population. There are ways to make a statistical correction due to the "regression dilution bias", but these are seldom applied (Salonen 1992).

Iron

Iron is a transition metal which can easily become oxidized and thus act as an oxidant. Halliwell and Gutteridge (1989) have proposed that the general effect of catalytic iron is to convert poorly reactive free radicals such as H_2O_2 into highly reactive ones, such as the hydroxyl radical. To be able to promote free radical production in the human body, iron needs to be liberated from proteins such as haemoglobin, transferrin or ferritin. It is thought that oxidative stress itself can provide the iron necessary for formation of reactive oxygen species (Halliwell and Gutteridge 1990). For example, superoxide radical can mobilize iron from ferritin.

Oxidative stress due to oxygen free radicals promotes the oxidation of lipids. Redox-active forms of iron catalyse free radical production and can thus promote the oxidation of lipids. Transition metal ions have been found to be required for the peroxidation of LDL by neutrophils, monocyte/macrophages and smooth muscle cells (see Salonen 1993 and Salonen et al. 1992d for review).

There is little clinical data concerning the effect of iron on lipid peroxidation. We observed in 60 eastern Finnish men a positive association between blood haemoglobin concentration and titre of autoantibodies against MDA-modified LDL ($r=0.27$, $p<0.05$), suggestive of a role of heme iron or the haemoglobin itself in lipid peroxidation in vivo in men (Salonen et al. 1992a).

We investigated the association of *serum ferritin concentration and dietary iron* intake with the risk of acute myocardial infarction in 1,931 randomly selected men aged 42, 48, 54 or 60 years who had no symptomatic coronary heart disease at entry, examined in the Kuopio Ischaemic Heart Disease Risk Factor Study (KIHD) in 1984-1989 (Salonen et al. 1992d). In a multivariate model adjusting for the major coronary risk factors, serum ferritin $\geq$ 200 µg/l was associated with a 2.2-fold risk of acute myocardial infarction (95% confidence interval 1.2-4.0, $p<0.01$). This association was stronger in men with serum LDLC-concentrations of $\geq$ 5.0 mmol/l (193 mg/dl) than in others. In another model, a high dietary intake of iron associated with an increased risk.

The original finding was repeated later in a more thorough statistical analysis that was based on two years longer follow-up time (Salonen et al. 1994a). The result was virtually identical. Other nutrients and inflammation as possible confounders of the observed association were ruled out. Our data suggest that high stored iron level is a risk factor for coronary heart disease.

Our findings concerning the role of dietary iron intake was recently confirmed in the US Health Professionals Study in 45,720 men aged 40-75 with no history of cardiovascular disease (Ascherio et al. 1994). In this large prospective study, dietary intake of heme iron (but not that of total iron) had a consistent and statistically significant association with an increased risk of myocardial infarction.

In a prospective population study from Canada, Morrison and coworkers reported an association between serum iron and the risk of fatal AMI in 9,920 men and women (Morrison et al. 1994). Also in this Canadian cohort, high body iron and high serum cholesterol had a synergistic relationship with CHD. At least one prospective study has been reported, in which no association was observed between serum iron saturation and coronary mortality (Sempos et al. 1994), most likely due to the large analytical variability in the assessment of iron saturation (Salonen 1993, Salonen et al. 1994a).

Hypothetically, the possible coronary disease promoting effect of high iron stores could be through increased susceptibility to oxidation of lipids by elevated iron stores and consequent chronic exposure of lipids to redox-active iron (Salonen 1993). This theory receives some empirical support from our finding in a small clinical randomised cross-over trial, in which three blood donations in 4-5 months reduced the susceptibility of VLDL+LDL to oxidation (Salonen et al. 1995a). There was also an increase in serum HDL cholesterol during the blood letting periods (Nyyssönen et al. 1994a).

There are less studies on the role of body iron stores directly in atherosclerosis in humans. An Austrian group reported an association between serum ferritin concentration and the severity of carotid atherosclerosis in a cross-sectional study in 847 men and women aged 40-79 (Kiechl et al. 1994). They also observed a synergism between high ferritin and high serum cholesterol.

We also investigated the association of donating blood with the risk of acute myocardial infarction in a random population sample of middle-aged men using a prospective study design (Tuomainen et al., 1997b). During the follow-up of 5 1/2 years, one of the 153 donors (0.7%) compared to 226 of the 2529 non-donors (9.8%), experienced an acute myocardial infarction ($p<0.001$ for difference). In a multivariate Cox model adjusting for the main coronary risk factors, blood donors had a 86 % reduced risk (relative risk 0.14, 95 % CI 0.02 to 0.97, $p = 0.047$) of acute myocardial infarction compared to non-donors. An additional adjustment for a large number of measurements of medical history, health status, health practices and psychosocial characteristics attenuated this association only marginally.

This is the first study to report a reduced risk of coronary events in male blood donors. The mechanism through which donating blood could reduce the risk of coronary events could be the depletion of body iron stores. Iron depletion could result in a decrease in the amount of injury-promoting iron in the myocardium, in altered activity of iron-dependent enzymes, in increased plasma antioxidative capacity and in a decrease in lipid peroxidation, both in the circulation and in the vessel wall (Sullivan 1989, Salonen et al. 1992d, Salonen 1993, 1996). Harmfulness of high iron, or benefits of low iron, for the risk of coronary events and/or for atherosclerotic progression has experimental, clinical and epidemiological support (Salonen 1993, 1996). The lack of consistency in epidemiologic studies is probably explained by large measurement variability in estimates of iron stores and iron intake and by the diversity of study outcomes (Salonen 1996).

The loss of heme iron with each blood donation might well be the explanation behind the observed risk reduction. A contributing mechanism could be an improvement in insulin sensitivity by blood donation, as suggested by our cross-sectional data (Tuomainen et al. 1997a). However, voluntary blood donors appeared to be generally more health conscious and more healthy than those who do not donate blood, and this may cause self-selection bias. In our study the association between donating blood and reduced risk for myocardial infarction was weakened but remained strong and statistically significant after adjustment for the main coronary risk factors.

Our finding needs to be confirmed in other prospective population studies. Further, trials of the impact of iron depletion on atherosclerotic progression or coronary events are eventually necessary to verify or refute the theory concerning the role of excess iron and iron depletion in atherogenesis and in ischaemic heart disease.

Mercury

The human body (with an average weight of 70 kg) contains on the average 13 mg of mercury (Linder 1991). No metabolic functions in the human body are known for which mercury is required, so it can be considered an environmental poison.

At high concentrations mercury is known to cause liver and kidney damage as well as neurological symptoms (IPCS 1990). The long-term toxicological effects of low exposure to mercury in humans have not been studied much.

We studied the relationship of the dietary intake of mercury, as well as hair content and urinary excretion of mercury, with the risk of acute myocardial infarction (AMI) and death from CHD, cardiovascular disease (CVD) and any cause in 1,833 men aged 42-60 years, free of clinical CHD, stroke, claudication and cancer (Salonen et al. 1995b). Of these, 73 experienced an AMI in 2-7 years. Of the 78 deceased men, 18 died of CHD and 24 of CVD.

In multivariate models with the major cardiovascular risk factors as covariates, dietary intake of mercury was associated with significantly increased risk of AMI and CHD, CVD and any death. Men in the highest tertile ($\geq$ 2.0 μg/g) of hair mercury content had a 2.0-fold (95% CI 1.2-3.1, p=0.005) age- and CHD-adjusted risk of AMI and 2.9-fold (95% CI 1.2-6.6, p=0.014) adjusted risk of cardiovascular death compared with those with a lower hair mercury content. In a nested case-control subsample, the 24-hour urinary mercury excretion had a significant (p=0.042) independent association with the risk of AMI. Both the hair and urinary mercury associated significantly with titres of immune complexes containing oxidized LDL (Salonen et al. 1995b).

These data suggest that a high dietary intake of mercury and the consequent accumulation of mercury in the body is associated with an excess risk of AMI as well as death from CHD, CVD and any cause in eastern Finnish men and this increased risk may be due to the promotion of lipid peroxidation by mercury.

The notion that mercury promotes free radical generation was first presented by Ganther (1980), based on the observation that vitamin E and the antioxidant DPPD protected against methylmercury poisoning in rats (Welsh 1979). In in vitro studies, mercury (II) ions, in micromolar concentrations, have increased the production of superoxide anions in human neutrophils (Jansson and Harms-Ringdahl 1993) and the mitochondrial H_2O_2 production (Miller et al. 1991).

A study in vivo revealed a significant concentration-related depolarization of the inner mitochondrial membrane, increased H_2O_2 formation, glutathione depletion and formation of thiobarbituric acid reactive substances following the addition of mercury (II) to mitochondria isolated from kidneys of untreated rats (Lund et al. 1993).

Mercury has a very high affinity to sulfhydryl groups (Halliwell and Gutteridge 1989), which in plasma proteins have been estimated to account for as much as 10-50 % of the antioxidative capacity of plasma (Wayner et al. 1987). By binding to sulfhydryl groups mercury inactivates antioxidative thiolic compounds such as glutathione (Naganuma et al. 1980). Mercury poisoning, which is associated with increased lipid peroxidation in the liver and in the kidneys, also results in inactivation of superoxide dismutase and catalase (Naganuma et al. 1980), two important enzymes that scavenge H_2O_2. Thiol antidotes, such as DMPS and D-penicillamine, chelate mercury, and protect against mercury-induced lipid peroxidation (Benov et al. 1990).

Mercury forms an insoluble complex with selenium, mercury selenide (Cuvin-Aralar and Furness 1991), thus binding selenium in an inactive form that cannot serve as a cofactor for glutathione peroxidase, an important scavenger of H_2O_2 and lipid peroxides. Ganther and coworkers have demonstrated that selenium protects against methylmercury toxicity (Ganther 1980). Selenium has been observed to protect against the peroxidative liver injury caused by mercury (Cuvin-Aralar and Furness 1991).

All these pathways reduce the antioxidative capacity in both plasma and intracellularly and promote free radical stress and lipid peroxidation in cell membranes and lipoproteins. The theory that mercury would elevate the risk of AMI through the promotion of lipid peroxidation receives empirical support from our finding that both a high hair mercury content and high urinary mercury excretion associated with elevated titres of immune complexes containing oxidized LDL in a subsample of our study subjects (Witztum 1994). Also, as we reported already earlier, serum selenium concentration associated inversely with autoantibodies against oxidized LDL (Salonen et al. 1992a).

Prospective Population Studies on Antioxidants and the Risk of Coronary Heart Disease

There are several recent reviews on epidemiologic studies concerning antioxidants and cardiovascular disease (Salonen 1995, Price and Fowkes 1997). The purpose of this presentation is to update those reviews and to express my personal interpretations.

Selenium
We reported 13 years ago from a large prospective population study a relationship between *selenium* deficiency and an excess risk of MI as well as death from CHD in Eastern Finland (Salonen et al. 1982). The finding was subsequently confirmed in another prospective population study (Suadicani et al. 1992) and in a case-control study (Kok et al. 1989). We have also observed an association between low serum selenium levels and accelerated progression of carotid atherosclerosis in eastern Finnish men (Salonen et al. 1992c). Also negative studies have been published, but these all have been poorly conducted or included only subjects with selenium levels above the GSHPx saturation threshold (see Salonen 1987 for detailed review).

Vitamin E
In two recently reported large American cohort studies, the self-reported use of *vitamin E* supplements associated with a reduced risk of coronary events (Stampfer et al. 1993, Rimm et al. 1993). In the Nurses' Health Study in 87,245 women aged 34-59, free of previous CHD, vitamin E supplement users had a 36% lower risk of CHD (Rimm et al. 1993). In the Health Professionals Follow-up Study, based on 45,720 men aged 40-75 with no history of CVD, vitamin E supplement-using men had 24% lower risk of CHD event than non-users (Rimm et al. 1993). The greatest decrement in CHD risk was for users of at least 100 I.U.

of vitamin E daily for at least two years. Even though statistical adjustment was made for several coronary risk factors, it is impossible to exclude the possibility that vitamin E supplement users were more health conscious than non-users and had also other, possibly unmeasured characteristics or behaviors that put them at a lower coronary risk.

In the Finnish "Mobile Clinic" cohort study, CHD mortality was 32% lower in men and 65% lower in women in the highest third of baseline dietary vitamin E intake, compared to the lowest third (Knekt et al. 1994). In women but not in men, both high vitamin C and carotene intakes enhanced this association.

There are also within-population cohort studies, in which plasma or serum α-tocopherol concentration has not associated with the risk of CHD (Salonen et al. 1985, Hense et al. 1993, Street et al. 1994). The lack of an observed relationship could have been, besides imprecision of the measurements, also due to the limited range of plasma vitamin E levels within populations and because of the strong correlation of plasma α-tocopherol with plasma LDLC concentration. It may be impossible to separate the effects of a-tocopherol and LDL cholesterol by any statistical means. As LDLC is almost uniformly associated with an increased risk of CHD, any other parameters that co-vary with it will have biased observed associations with CHD risk.

We observed in a 4-year follow-up study of a random population sample of about 1000 men an association between a low lipid-standardised plasma vitamin E and an increased risk of non-insulin dependent diabetes mellitus (Salonen et al. 1995c). An increase in insulin resistance could be one mechanism through which vitamin E deficiency could elevate the risk of CHD. Our finding needs, however, to be confirmed in randomised trials.

Vitamin C

In another recent sizeable cohort study, the National Health and Nutrition Examination Survey (NHANES I) epidemiologic Follow-up Study, the men and women with the highest *vitamin C* intakes (>50 mg/d) had 45% and 25% lower CHD mortality than subjects with the lowest vitamin C intake (<50 mg/d) (Enström et al. 1992). In the 12-year follow-up of the Prospective Basel Study, a low level of both plasma vitamin C and carotene associated with 2-fold risk of CHD ($p=0.022$) (Eichholzer et al. 1992, Gey 1993).

In the KIHD (see Salonen et al., 1995b, for study design and measurements), low plasma concentration of ascorbate, measured from fresh samples (Parviainen et al. 1986), were associated with increased risk of AMI in 1605 men, free of CHD at entry (Nyyssönen et al. 1997a). Vitamin C deficient men (plasma ascorbate < 2.0 mg/l or 11 μmol/l) had a 2.7-fold ($p<0.01$) risk of AMI in a multivariate survival model adjusting for age, examination year and season, cigarette-years, serum apolipoprotein B, triglyceride, HDL cholesterol and ferritin concentrations, systolic blood pressure, diabetes, blood hemoglobin, blood leucocyte count, alcohol intake and maximal oxygen uptake.

In the large US Health Professionals' and Nurses studies, however, the self-reported use of vitamin C supplements had no significant association with the risk of CHD (Rimm et al. 1993, Knekt et al. 1994).

Table 1. Randomized controlled trials of antioxidantive vitamins in cardiovascular diseaeses

Study	Participants	n	Duration	Vitamins (dose)	Outcome
Physicians Health Study 1990	Men with stable angina or coronary surgery	333	6 years	Beta-carotene (50 mg/altday)	44% reduction in major coronary events
ATBC Study 1994	Male smokers 50-69 years	29 000	5-8 years	Vitamin E (50 mg/day) Beta-carotene (20 mg/day)	No effect on coronary mortality, small reduction of angina incidence
CARET Study 1994	Men and women, heavy smokers and asbestos workers	18 000	2-3 years	Vitamin A (25 000 IU/day) Beta-carotene (30 mg/day)	Terminated 1996, 28% increase in lung cancer, 17% increase in deaths
CHAOS Study 1996	Men and women with coronary atheroma	2002	3-981 days (median 510)	Vitamin E (400 mg or 800 mg/day)	75% reduction in non-fatal MI, non-significant increase in cardiovascular death
Physicians Health Study	Healthy males 40-84 years	22 000	12 years	Beta-carotene (50 mg/altday)	No effect on either cancer or cardiovascular mortality
SU.VI.M.AX Trial	Healthy men and women	15 000	Ongoing	Small doses of vitamin E, Beta-carotene, Vitamin C and minerals	Cardiovascular endpoints
Womans Health Study	Healthy females 40-84 years	40 000	Ongoing (Begun 1992)	Vitamin E (600 mg/altday) Beta-carotene (50 mg/altday)	Cancer and cardiovascular endpoints (Beta-carotene t erminated 1996)
ASAP Trial	Healthy men and women 45-69 years, 50% smokers	520	Ongoing (Begun 1993)	Vitamin E (200 mg/day) and/or Vitamin C (500 mg/day)	Atherosclerotic progression, blood pressure, cataract pro gression
Heart Protection Study	Men and women at increased ris k of future MI	18 000	Ongoing (Begun 1994)	Vitamin E (600 mg/day) Beta-carotene (20 mg/day) Vitamin C (250 mg/day)	Incidence of coronary and all-cause mortality
WACVS	Women with coronary artery disease or three risk factors	8000	Ongoing (Begun 1995)	Vitamin E (400 mg/day) Beta-carotene (20 mg/day) Vitamin C (1 g/day)	Cardiovascular endpoints
HOPE Study	Men and women at increased risk of future MI	9000	Ongoing (Begun 1995)	Vitamin E or ACE inhibitor	Cardiovascular endpoints
GISSI	Men and women with MI	11 000	Ongoing (Begun 1996)	Vitamin E or fish oil	Cardiovascular endpoints

Carotene

In the Basle Prospective Study there was increased CHD and stroke mortality in men in the lowest quarter of both plasma vitamin C and lipid-standardized carotene (Gey 1993). Plasma vitamin E had no significant relation to CHD.

In the Lipid Research Clinics Coronary Primary Prevention Trial and Follow-up Study among 1899 hypercholesterolemic men, those in the highest quartile of serum carotenoid levels had a 36% reduced risk of CHD for all men and a 72% reduced risk of CHD among never-smokers (Morris et al. 1994).

In a large nested case-control study in the Washington County, low serum β-carotene concentration was associated with increased risk of myocardial infarction, and this relation was stronger in smokers than in nonsmokers (Street et al. 1994).

Flavonoids

In the Dutch cohort of the "Seven Countries Study" in 805 elderly men aged 65-84, the estimated dietary intake of *flavonoids* principally from tea, onions and apples was associated with a reduced mortality from CHD, even after controlling for the intakes of other antioxidants (Hertog et al. 1993).

Case-Control and Cross-Population Studies Concerning Antioxidants and Coronary Heart Disease

Case-control and cross-sectional studies are in general quicker and less expensive than prospective studies, but they provide only weak evidence of causal relationships. As the risk factor measurements are being carried out after the subjects already have acquired a disease, the disease process or consequent health habit changes may possibly have influenced the risk factor measurements.

In a cross-sectional study in 1132 randomly selected eastern Finnish men aged 54 years, those subjects with either symptomatic or symptomless CHD, as assessed by maximal exercise test, had reduced serum selenium levels and reduced blood glutathione peroxidase activity (Salonen et al. 1988).

In a population-based case-control study in Scotland, low plasma concentrations of both vitamin E and C were associated with increased risk of angina pectoris (Riemersma et al. 1991).

In the multicentre "Euramic" case-control study, adipose tissue β-carotene but not α-tocopherol was associated with reduced CHD (Kardinaal et al. 1993).

Of the risk factors for atherosclerotic cardiovascular diseases, low antioxidant levels have been associated with elevated blood pressure (Salonen 1991) and enhanced platelet aggregability (Salonen 1989).

A number of cross-population or other cross-area studies have been reported, in which associations between population means of various antioxidants with CHD incidence or mortality rates have been analyzed (Gey 1993, Gey et al. 1993, Bellizzi et al. 1994, Hensrud et al. 1994). These studies are inexpensive and have high publicity value. The problem with these ecological studies is, that populations always differ in an indefinite number of characteristics from each other, and all relevant aspects can never be measured in a single study. Thus, the control of confounding is virtually impossible, and consequently one cannot tell, whether the difference in e.g. mean vitamin E level in the population was the factor that caused the difference in mortality rates. This source of bias is called the "ecological fallacy" (Salonen 1985). Due to this typical bias in ecological studies, the cross-population comparisons provide only very weak evidence of causal associations. They merely produce hypotheses to be tested in further studies with stronger designs.

Even though the results of cross-population studies are greatly influenced by the choice of populations, a consistent finding in these studies appears to be the inverse association between population vitamin E intake (Bellizzi et al. 1994) and plasma levels (Gey 1993, Gey et al. 1993) and CHD mortality.

Preventive Trials

We observed in a pair-matched, randomised, placebo-controlled clinical trial in 80 male subjects a serum TBARS reducing and platelet activity inhibiting effect of a combination of α-tocopherol, β-carotene, ascorbic acid and selenium (Salonen et al. 1991). The lipid peroxidation reducing effect of antioxidant combinations has subsequently been shown in a number of other smaller trials using a variety of methods to assess lipid peroxidation (Jialal and Grundy 1993, Abbey et al. 1993, Nyyssönen et al. 1994b. We have also observed a blood pressure lowering effect of an antioxidant combination in a small double-masked, placebo-controlled randomised trial in men (Salonen et al. 1994b).

Even though randomised placebo-controlled trials are the only studies that can conclusively establish a cause-effect relationship, randomised double-masked trials testing the efficacy of various interventions in the prevention of coronary heart disease events have several problems. Large sample sizes and long follow-ups are necessary, and this makes these trials very expensive and hard to administrate. Due to the length of the follow-up, drop-out are always a problem. Most importantly, the requirements of statistical power enable the testing of only one or two preventive agents in a single trial. When the effects of antioxidants in the prevention of CHD are tested, this makes these trials unreal. Isolated effects of single antioxidants are tested, even though no antioxidant works in the human body in isolation. If a combination of antioxidants is studied and found to have effects, then it can not be deduced which was the compound that was effective. Also, exposures for decades to antioxidants may be necessary to produce a detectable effect on coronary events or deaths. For these reasons a lack of observed effect in a single trial does not disprove the hypothesis concerning the role of antioxidants in cardiovascular diseases.

No conclusive clinical trials concerning the preventive effect of antioxidants in CHD have so far been reported. In the Chinese Linxian trial, there was reduced total mortality in a group that received a combination of α-tocopherol, β-carotene and selenium, as compared with three parallel randomised groups receiving other vitamins and minerals (Blot et al. 1993).

In the Finnish α-tocopherol - β-carotene lung cancer prevention trial, neither antioxidant had any significant effect on cardiovascular mortality. In this trial, the supplementation of smoking men with either 50 mg of dl-α-tocopheryl acetate or 20 mg of β-carotene daily for 5-8 years did not reduce either CHD or cerebrovascular mortality significantly, although there was a nonsignificant trend towards reduced CHD (by 5%, 658 vs. 704 deaths) and ischemic stroke (by 16%) mortality in men who received α-tocopherol acetate (The Alpha-Tocopherol, Beta Carotene Cancer Prevention Group 1994). The α-tocopherol dose used in the Finnish trial may have been too small to represent a meaningful test of the effect of vitamin E in the prevention of CHD, especially, as the Finnish male population has very low average plasma vitamin C levels (Parviainen and Salonen 1990). Also, all participants in the study were smokers, who have even further reduced plasma vitamin C levels. There might thus not have been enough ascorbate in their cells to regenerate tocopheroxyl radicals back to tocopherol.

The Probucol Quantitative Regression Swedish Trial (PQRST) tested, whether treatment of hypercholesterolemic men and women below the age of 71 years with probucol affected femoral atherosclerotic progression (Walldius et al. 1994). Out of those who responded to probucol treatment, 303 were randomised for three years to either probucol or placebo. All subjects received cholesterol-lowering diet and cholestyramine. After three years, the probucol-treated subjects had 12% lower LDL cholesterol and 24% lower HDL cholesterol than the placebo subjects. There were no statistically significant differences in the measurements of progression of femoral atherosclerosis between treatment groups.

The findings of the PQRST trial have been interpreted as evidence against the lipid peroxidation -atherogenesis theory. This conclusion is not valid because of the multiple effects of probucol. The HDL reducing effect might have overcome any LDL oxidation reducing effects.

In the recently reported skin cancer prevention trial in 1312 patients, supplementation with 200 µg of selenium daily for 4.5 years reduced total cancer mortality by 52% (95% confidence interval 24-69%, p=0.001) and total mortality insignificantly by 21% (95% CI -2% to 39%, p=0.07, Clark et al. 1996).

On-Going Preventive Trials

The "Antioxidant Supplementation in Atherosclerosis Prevention" ("ASAP") study is a 2x2 factorial double-masked randomised placebo-controlled 3-year trial, testing the effect of 200 mg/d of d-α-tocopherol acetate and 500 mg/d of vitamin C on the progression of carotid atherosclerosis, blood pressure and cataract formation/progression in 500 men and women aged 50-69 years from eastern Finland. The progression of carotid atherosclerosis is assessed with B-mode ultrasonography. Although not primary study outcomes, also glucose status and the incidence of infections are assessed. Unlike other on-going trials, in the "ASAP" study also extensive measurements of lipid peroxidation are carried out to establish the mechanism through which antioxidants are hypothesised to affect atherogenesis. A cross-sectional analysis of a subsample of 48 male subjects has been published (Nyyssönen et al 1997b). ASAP is the first supplementation trial with measurements of in vivo lipid peroxidation.

The French "SUVIMAX" study is a double-masked, randomised, placebo-controlled-trial to test the effect of a combination of antioxidants on total mortality and incidence of cancers, myocardial infarction and cataracts. The supplementation consists of 15 mg of vitamin E, 120 mg of vitamin C, 6 mg of β-carotene, 100 μg of selenium and 20 mg of zinc daily. The sample size is 5,000 men aged 45-60 and 10,000 women aged 35-60 at entry.

The British "Heart Protection Study" is even a larger trial in 20,000 men and women aged 40-75 years, with a high risk of CHD. It has a randomised placebo-controlled 2x2 factorial design. One treatment is a LDL-lowering drug and the other a combination of 600 mg of vitamin E, 250 mg of vitamin C and 20 mg of β-carotene daily. The study outcomes for the vitamin part of the study are total CHD and fatal CHD.

In the U.S.A. two trials testing the preventive effect of β-carotene on cancer and CHD are underway: the "Physicians' Health Study" and the "CARET" study. A third primary prevention trial, the "Women's Health Study" concerns the effect of 600 mg of vitamin E or 30 mg of β-carotene every other day on cancer and CHD incidence in 40,000 women.

Conclusions

On the basis of the totality of currently existing evidence, antioxidants represent a promising, although yet unproven, means of reducing the risks of CHD and other atherosclerotic cardiovascular diseases. Another, yet untested approach, would be to reduce the availability of catalytic transition metals in the body. For smokers, quitting smoking may be the best way to reduce oxidative stress. Epidemiologic studies indicate that deficiencies of selenium and vitamin C are associated with an increased risk of CHD and suggest that high intakes of vitamin E, and possibly of carotenoids, vitamin C and flavonoids, are associated with reduced risk of CHD. Even if evidence from preventive trials would not be considered necessary for recommendations, the findings from epidemiologic studies have to be more consistent than they are at the moment to justify firm conclusions. For this reason, we need to wait for the results of the on-going preventive trials before recommendations about the role of antioxidants in atherosclerosis and in the prevention of CHD can be formulated.

References

Abbey M, Nestel PJ, Baghurst PA (1993) Antioxidant vitamins and low-density-lipoprotein oxidation. Am J Clin Nutr 58: 525-532

Ames BN, Shigenaga MK, Hagen TM (1993) Oxidants, antioxidants, and the degenerative disease of aging. Proc Natl Acad Sci 90: 7915-7922

Ascherio A, Willet WC, Rimm EB, Giovannucci E, Stampfer MJ (1994) Dietary iron intake and risk of coronary heart disease. Circulation 89: 969-974

Bellizzi MC, Franklin MF, Duthie GG, James WPT (1994) Vitamin E and coronary heart disease: the European paradox. Eur J Clin Nutr 48: 822-831

Benov LC, Benchev IC, Monovich O.H. (1990) Thiol antidotes effect on lipid peroxidation in mercury-poisoned rats. Chem Biol Interact 76: 321-332

Blot WJ, Li J-Y, Taylor PR Guo W, Dawsey S, Wang G-Q, Yang CS, Zheng S-F, Gail M, Li G-Y, Liu B-Q, Tangrea J, Sun Y-H, Liu F, Fraumeni Jr JF, Zhang Y-H, Li B (1993) Nutrition intervention trials in Linzian, China: Supplementation with specific vitamin/mineral combinations, cancer incidence, and disease-specific mortality in the general population. J Natl Cancer Int 85, 1483-1492

Clark LC, Combs GF, Turnbull BW, et al (1996) Effects of selenium supplementation for cancer prevention in patients with carcinoma of the skin. A randomized controlled trial. JAMA 276: 1957-1963

Cuvin-Aralar M.L., Furness R.W. (1991) Mercury and selenium interaction: a review. Ecotoxicol. Environ. Safety 21: 348-364

Eichholzer M, Stahelin HB, Gey KF (1992) Inverse correlation between essential antioxidants in plasma and subsequent risk to develop cancer, ischemic heart disease and stroke respectively: 12-year follow-up of the Basel Study. EXS 62: 398-410

Enström JE, Kanim LE, Klein MA (1992) Vitamin C intake and mortality among a sample of the United States population. Epidemiology 3: 194-202

Esterbauer H, Gebicki J, Puhl H, Günther J (1992) The role of lipid peroxidation and antioxidants in oxidative modification of LDL. Free Rad Med Biol 13: 341-390

Ganther HE (1980) Interactions of vitamin E and selenium with mercury and silver. Ann NY Acad Sci 355: 1-372

Gey KF (1986) On the antioxidant hypothesis with regard to arteriosclerosis. Bibl Nutr Dieta 37: 53-91

Gey KF (1993) Prospects for the prevention of free radical disease, regarding cancer and cardiovascular disease. In: Cheeseman KH, Slater TF (eds) Free Radicals in Medicine, Brit Med Bull 49: 679-699

Gey KF, Moser UK Jordan P, Stähelin HB, Eichholzer M, Lüdin E (1993) Increased risk of cardiovascular disease at suboptimal plasma concentrations of essential antioxidants: an epidemiologic update with special attention to carotene and vitamin C. Am J Clin Nutr 57: 787S-797S

Halliwell B (1994) Free radicals, antioxidants, and human disease: curiosity, cause, or consequence? Lancet 344: 721-724

Halliwell B, Gutteridge JMC (1989) Free radicals in biology and medicine. Clarendon Press, Oxford

Halliwell B, Gutteridge JMC (1990) Role of free radicals and catalytic metal ions in human disease: an overview. In: Packer L, Glazer AN (eds) Methods in Enzymology, Vol 186, Academic Press, San Diego, pp 1-85

Hense HW, Stender M, Bors W, Keil U (1993) Lack of an association between serum vitamin E and myocardial infarction in a population with high vitamin E levels. Atherosclerosis 103:21-28

Hensrud DD, Heimburger DC, Chen J, Parpia B (1994) Antioxidant status, erythrocyte fatty acids, and mortality from cardiovascular disese and Keshan disease in China. Eur J Clin Nutr 48: 455-464

Hertog MGL, Feskens EJM, Hollman PCH, Katan MB, Kromhaut D (1993) Dietary antioxidant flavonoids and risk of coronary heart disease: the Zutphen elderly study. Lancet 342: 1007-1011

International Programme on Chemical Safety (IPCS) (1990) Environmental health criteria 101. Methylmercury. World Health Organization, Geneva

Jansson G, Harms-Ringdahl M (1993) Stimulating effects of mercuric- and silver ions on the superoxide anion production in human polymorphonuclear leucocytes. Free Rad Res Comms 18: 87-98

Jialal I, Grundy SM (1993) Effect of combined supplementation with alpha-tocopherol, ascorbate, and beta carotene on low-density lipoprotein oxidation. Circulation 88: 2780-2786

Kardinaal AF, Kok FJ, Ringstad J, Gomez-Aracena J., Mazaev VP, Kohlmeier L, Martin BC, Aro A, Kark JD, Delgado-Rodrigues M, Riemersma P, van't Veer P, Huttunen JK

Martin-Moreno JM (1993) Antioxidants in adipose tissue and risk of myocardial infartion: the EURAMIC Study. Lancet 342: 1379-1384

Kiechl S, Aichner F, Gerstenbrand F, Egger G, Mair A, Rungger G, Spogler F, Jarosch E, Oberhollenzer F, Willeit J (1994) Body iron stores and presence of carotid atherosclerosis. results from the Bruneck Study. Arterioscler Thromb 14: 1625-1630

Knekt P, Reunanen A, Järvinen R, Seppänen R, Heliövaara M, Aromaa A (1994) Antioxidant vitamin intake and coronary mortality in a longitudinal population study. Am J Epidemiol 139: 1180-1189

Kok FJ, Hofman A, Witteman JCM, De Bruijn AM, Kruyssen DHCM, De Bruin M, Valkenburg HA (1989) Decreased blood selenium and risk of myocardial infarction. JAMA 261: 1161-1164

Linder MC (1991) Nutrition and metabolism of the trace elements. In: Linder MC (ed) Nutritional biochemistry and metabolism. 2nd edition. Elsevier, New York, pp 215-276

Lund BO, Miller DM, Woods JS (1993) Studies on Hg(II)-induced H_2O_2 formation and oxidative stress in vivo and in vitro in rat kidney mitochondria. Biochem Pharmacol 45: 2017-2024

Maggi E, Finardi G, Poli M, Bollati P, Filipponi M, Stefano PL, Paolini G, Grossi A, Clot P, Albano E et al (1993) Specificity of autoantibodies against oxidized LDL as an additional marker for atherosclerotic risk. Coron Artery Dis 4: 1119-1122

Miller OM, Lund BO, Woods JS (1991) Reactivity of Hg(II) with superoxide: evidence for the catalytic dismutation of superoxide by Hg(II). J Biochem Toxicol 6: 293-298

Morris DL, Kritchevsky SB, Davis CE (1994) Serum carotenoids and coronary heart disease. The lipid research clinics coronary primary prevention trial and follow-up study. JAMA 272: 1439-1441.

Morrison HI, Semenciw RM, Mao Y, Wigle DT(1994) Serum iron and risk of fatal acute myocardial infarction. Epidemiology 5: 243-246

Naganuma A, Koyama Y, Imura N (1980) Behavior of methylmercury in mammalian erythrocytes. Toxicol Appl Pharmacol 54: 405-410

Nyyssönen K, Salonen R, Korpela H, Salonen JT (1994a) Elevation of serum high density lipoprotein cholesterol concentration due to lowering of body iron stores by blood letting. Eur J Lab Med 2: 113-115

Nyyssönen K, Porkkala E, Salonen R, Korpela H, Salonen JT (1994b) Increase in oxidation resistance of atherogenic serum lipoproteins following antioxidant supplementation: a randomized double-blind placebo-controlled trial. Eur J Clin Nutr 48: 633-642

Nyyssönen K, Parviainen MT, Salonen R, Tuomilehto J, Salonen JT (1997a) Vitamin C deficiency and risk of myocardial infarction: prospective population study of men from eastern Finland. Brit Med J 314: 634-638

Nyyssönen K, Porkkala-Sarataho E, Kaikkonen J, Salonen JT (1997b) Ascorbate and urate are the strongest determinants of plasma antioxidative capacity and serum lipid resistance to oxidation in Finnish men. Atherosclerosis 130: 223-233

Parviainen MT, Salonen JT (1990) Vitamin C status of 54-year old Eastern Finnish men throughout the year. Int J Vit Nutr Res 60: 47-51
Parviainen MT, Nyyssönen K, Penttilä IM, Seppänen K, Rauramaa R, Salonen JT, Gref CG (1986) A method for routine assay of plasma ascorbic acid using high-performance liquid chromatography. J Liq Chromatogr 9: 2185-2197
Price JF, Fowkes FGR (1997) Antioxodant vitamins in the prevention of cardiovascular disease. Eur H J 18: 719-727
Riemersma RA, Wood DA, MacIntyre CCA, Elton RA, Gey KF Oliver MF (1991) Risk of angina pectoris and plasma concentrations of vitamins A, C, and E and carotene. Lancet 337: 1-5
Rimm EB, Stampfer MJ, Ascherio A, Giovannucci E, Colditz GA, Willet WC (1993) Vitamin E consumption and the risk of coronary heart disease in men. N Engl J Med 328: 1450-1456
Salonen JT (1985) Selenium in cardiovascular diseases and cancer - epidemiologic findings from Finland. In: Boström H and Ljungtedt N (eds) Trace Elements in Health and Disease. Scandia International Symposia, The Scandia Group, Almqvist & Wiksell International, Stockholm, pp 172-186
Salonen JT (1987) Selenium in ischaemic heart disease. Int J Epidemiol 16: 323-328
Salonen JT (1989) Antioxidants and platelets. Ann Med 21: 59-62
Salonen JT (1991) Dietary fats, antioxidants and blood pressure. Ann Med 23: 295-298
Salonen JT (1992) Statistical correction for measurement imprecision. Lancet 339: 1116
Salonen JT (1993) The role of iron as a cardiovascular risk factor. Curr Opinion Lipidol 4: 277-282
Salonen JT (1995) Epidemiological studies on LDL oxidation, pro- and antioxidants and atherosclerosis. In: Bellomo G, Finardi G, Maggi E, Rice-Evans C (eds) Free Radicals, Lipoprotein Oxidation and Atherosclerosis. Richelieu Press, London, pp 27-52
Salonen JT (1996) Body iron stores, lipid peroxidation and coronary heart disease. In: Hallberg L, Asp N-G (eds) Iron nutrition in health and disease. JL London Press, London
Salonen JT, Salonen R (1993) Ultrasound B-mode imaging in observational studies of atherosclerotic progression. Circulation 87 (suppl. II): 55-65
Salonen JT, Alfthan G, Huttunen JK, Pikkarainen J, Puska P (1982) Association between cardiovascular death and myocardial infarction and serum selenium in a matched-pair longitudinal study. Lancet II: 175-179
Salonen JT, Salonen R, Penttilä I, Herranen J, Jauhiainen M, Kantola M, Lappeteläinen R, Mäenpää PH, Alfthan G, Puska P (1985) Serum fatty acids, apolipoproteins, selenium and vitamin antioxidants and the risk of death from coronary artery disease. Am J Cardiol 56: 226-231
Salonen JT, Salonen R, Seppänen K, Kantola M, Parviainen M, Alfthan G, Mäenpää PH, Taskinen E, Rauramaa R (1988) Relationship of serum selenium and antioxidants to plasma lipoproteins, platelet aggregability and prevalent ischaemic heart disease in Eastern Finnish men. Atherosclerosis 70: 155-160
Salonen JT, Salonen R, Seppänen K, Rinta-Kiikka S, Kuukka M, Korpela H, Alfthan G, Kantola M, Schalch W (1991) Effects of antioxidant supplementation on platelet function: a randomized pair-matched placebo-controlled double-blind trial in men with low antioxidant status. Am J Clin Nutr 53: 1222-1229
Salonen JT, Ylä-Herttuala S, Yamamoto R, Butler S, Korpela H, Salonen R, Nyyssönen K, Palinski W, Witztum JL (1992a) Autoantibody against oxidised LDL and progression of carotid atherosclerosis. Lancet 339: 883-887
Salonen JT, Tatzber F, Salonen R, Nyyssönen K, Korpela H, Ehnholm C, Esterbauer H (1992b) Autoantibodies against oxidized LDL associated with accelerated atherosclerosis progression in hypercholesterolemic men. Eur Heart J 13: 391

Salonen JT, Salonen R, Seppänen K, Kantola M, Suntioinen S, Korpela H (1992c) Interactions of serum copper, selenium, and low density lipoprotein cholesterol in atherogenesis. Brit Med J 302: 756-760

Salonen JT, Nyyssönen K, Korpela H, Tuomilehto J, Seppänen R, Salonen R (1992d) High stored iron levels are associated with excess risk of myocardial infarction in eastern Finnish men. Circulation 86: 803-811

Salonen JT, Nyyssönen K, Salonen R (1994a) Body iron stores and the risk of coronary heart disease. N Engl J Med 331: 1159

Salonen R, Korpela H, Nyyssönen K, Porkkala E, Salonen JT (1994b) Reduction of blood pressure by antioxidant supplementation: A randomized double-blind clinical trial. Life Chem Rep 12: 65-68

Salonen JT, Seppänen K, Nyyssönen K, Korpela H, Kauhanen J, Kantola M, Tuomilehto J, Esterbauer H, Tatzber F, Salonen R (1995a) Intake of mercury from fish, lipid peroxidation and the risk of myocardial infarction and coronary, cardiovascular and any death in Eastern Finnish men. Circulation 91: 645-655

Salonen JT, Korpela H, Nyyssönen K, Porkkala E, Tuomainen T-P, Belcher JD, Jacobs DR Jr, Salonen R (1995b) Lowering of body iron stores by blood letting and oxidation resistance of serum lipoproteins: a randomized cross-over trial in smoking men. J Int Med 237: 161-168

Salonen JT, Nyyssönen K, Tuomainen T-P, Mäenpää PH, Korpela H, Kaplan GA, Lynch J, Helmrich SP, Salonen R (1995c) Increased risk of non-insulin dependent diabetes mellitus at low plasma vitamin E concentrations: a four year follow-up study in men. Brit Med J 311: 1124-1127

Salonen JT, Nyyssönen K, Salonen R, Porkkala-Sarataho E, Tuomainen T-P, Diczfalusy U, Björkhem I (1997) Lipoprotein oxidation and progression of carotid atherosclerosis. Circulation 95: 840-845.

Sempos CT, Looker AC, Gillum RF, Makuc DM (1994) Body irons stores and the risk of coronary heart disease. N Engl J Med 330: 1119-1124

Stampfer MJ, Hennekens CH, Manson JE, Coditz GA, Rosner B, Willett WC (1993) Vitamin E consumption and risk of coronary disease in women. N Engl J Med 328: 1444-1449

Steinberg D, Parthasarathy S, Carew TE, Khoo JC, Witztum JL (1989) Beyond cholesterol: Modifications of low-density lipoprotein that increase its atherogenicity. N Engl J Med 320: 915-924

Steinberg D and Workshop Participants (1992) Antioxidants in the prevention of human atherosclerosis. Circulation 85: 2337-2344

Street DA, Comstock GW, Salkeld RM, Schuep W, Lag MJ (1994) Serum antioxidants and myocardial infarction. Are low levels of carotenoids and alpha-tocopherol risk factors for myocardial infarction? Circulation 90: 1154-1161

Suadicani P, Hein OO, Gyntelberg F (1992) Serum selenium concentration and risk of ischaemic heart disease in a prospective cohort study of 3000 males. Atherosclerosis 96: 33-42

Sullivan JL (1989) The iron paradigm of ischaemic heart diseae. Am Heart J 117: 1177-1188

The Alpha-Tocopherol, Beta Carotene Cancer Prevention Group (1994) The effect of vitamin E and beta carotene on the incidence of lung cancer and other cancers in male smokers. N Eng J Med 330: 1029-1035

Tuomainen T-P, Nyyssönen K, Salonen R, Tervahauta A, Korpela H, Lakka T, Kaplan GA, Salonen JT (1997a) Body iron stores are associated with serum insulin and blood glucose concentrations. Population study in 1,013 eastern Finnish men. Diabetes Care 20: 426

Tuomainen T-P, Salonen R, Nyyssönen K, Salonen JT (1997b) Cohort study of relation between donating blood and risk or myocardial infarction 2682 men in eastern Finland. Brit Med J 314: 793-794

Walldius G, Erikson U, Olsson AG, Bergstrand L, Hådell K, Johansson J, Jaijser L, Lassvik C, Mölgaard J, Nilsson S, Schäfer-Elinder L, Stenport G, Holme I (1994) The effect of probucol on Femoral atherosclerosis: The probucol quantitative pregression Swedish trial (PQRST). Am J Cardiol 74: 875-883

Wayner DDM, Burton GW, Ingold KU, Barclay LRC, Locke SJ (1987) The relative contributions of vitamin E, urate, ascorbate and proteins to the total peroxyl radical-trapping antioxidant activity of human blood plasma. Biochim Biophys Acta 924: 408-419

Welsh SO (1979) The protective effect of vitamin E and N,N'-diphenyl-p-phenylenediamine (DPPD) against methyl mercury toxicity in the rat. J Nutr 109: 1673-1681

Witztum JL (1994) The oxidation hypothesis of atherosclerosis. Lancet 344: 793-795

Regulatory Immunotoxicology: The Scientist's Point of View

(Chair: I. Kimber, United Kingdom, and F. Højelse, Denmark)

Regulatory Immunotoxicology - The Scientist's Point of View : An Introduction

Ian Kimber
Zeneca Central Toxicology Laboratory, Alderley Park, Macclesfield, UK

Introduction and Definitions

Immunotoxicology describes adverse health effects that result from the interaction of xenobiotics with the immune system. In the context of consideration of some of the key regulatory issues, two main categories of adverse effects may be identified : immunotoxicity and allergy. The ability of chemicals to impair the functional integrity of one or more components of the immune system is defined as immunotoxicity. Here the concern is that compromised immune function will result in reduced host resistance and an increased susceptibility to infectious and malignant disease. Allergy describes the adverse effects that may be caused by the stimulation of specific immune responses; the most common manifestations of chemical-induced allergy being contact dermatitis and respiratory hypersensitivity associated with rhinitis and asthma.

For both immunotoxicity and allergy there exist questions regarding hazard identification and risk assessment and debate about how these issues may best be addressed.

Immunotoxicity

There is currently no consensus on the requirements for immunotoxicity testing. One difficulty is that, in principle at least, there are many and varied ways in which xenobiotics might influence immune function and a reflection of this is the fact that there has been described a wide range of test methods and endpoints. The approaches available can be summarized as follows:

a) Consideration of changes in the weight, cellular composition and histopathological appearance of lymphoid tissues, including blood.
b) Direct assessment of the functional integrity of the immune system by analyses of cell mediated immunity, antibody production and natural killer (NK) cell cytotoxic activity.
c) Evaluation of induced changes in host resistance by measurement of susceptibility to challenge with micro-organisms (viruses, bacteria or multicellular parasites) or transplantable tumour cells.

A question frequently posed is whether a detailed examination of the weight and histopathological characteristics of lymphoid tissues (thymus, spleen, lymph nodes and bone marrow), together with consideration of haematological parameters, are of sufficient sensitivity to permit the identification of potential immunotoxicants in the context of routine toxicity studies. This question has yet to be resolved. Should the view prevail that the apparently greater sensitivity provided by functional assays is necessary for hazard evaluation, then consideration must be given to which among the tests available are to be deployed and at what stage during toxicological investigations. A related issue is one of interpretation. It is not clear presently how small changes in immune function, or in the relative frequencies of lymphoid cell subpopulations, should be interpreted. The immune system undoubtedly has some functional reserve and there are known to exist complementary and compensatory mechanisms. As a consequence, a small change in a single immune parameter may be expected to have little or no impact on the overall integrity of immunological function and host resistance in the normal, healthy individual. It must be recognized, however, that the consequences of such a change may be of considerably greater significance for the infant, for the aged, for those who are chronically ill, or for those whose immune system is already partially compromised. Whereas, at an individual level, there may exist a clear threshold for alterations in immune function below which there is no apparent influence on host resistance or the ability to combat infectious disease, at a population level a more linear relationship between immunotoxicity and acquired health effects may pertain. It has been argued that as there exists a background level of infectious disease, then it is possible that modest perturbations of immune status may, if the exposed population is sufficiently large, translate into a significant increase in the prevalence of infectious disease.

A greater understanding of the cellular and molecular mechanisms through which chemicals may influence immune responses, and a more detailed appreciation of the relationships that exist between induced changes in immune function and health status, will provide the scientific foundations necessary to ensure that appropriate measures for immunotoxicity hazard and risk assessment are in place.

Allergy

A variety of chemicals has been shown to induce allergy in susceptible individuals and allergic disease is a widespread occupational health problem. Both contact (skin) and respiratory allergy develop in two phases. Following a first encounter with the chemical allergen, the susceptible individual mounts a specific immune response that results in sensitization. Subsequent encounter of the now sensitized subject with the same allergen provokes an accelerated and more aggressive secondary immune response that causes an inflammatory response (allergic contact dermatitis or respiratory hypersensitivity reaction) at the site of exposure. For both contact and respiratory allergy the issues are of accurate hazard identification, evaluation of relative potency, risk assessment and risk management.

Contact Sensitization: Allergic contact dermatitis is a form of delayed-type hypersensitivity reaction and as such is dependent upon the action of T lymphocytes and cell mediated immune responses. There is available a wide variety of methods for the identification of chemicals that have the potential to cause skin sensitization. The need now is to develop approaches that permit evaluation of relative potency as a first step towards an accurate assessment of risk. The methods available for skin sensitization testing are described elsewhere in this volume.

Sensitization of the Respiratory Tract: The situation here is rather different, insofar as there are currently no methods for the identification of chemical respiratory allergens that have gained wide recognition or acceptance. One difficulty is that chemicals may induce symptoms of respiratory hypersensitivity and asthma through both allergic and non-allergic mechanisms. Even within the context of allergic sensitization of the respiratory tract there is considerable debate regarding the important immunobiological processes and about the requirement for specific IgE antibody. Despite those uncertainties some progress has been made and approaches to testing using both guinea pigs and mice have been described. In the former species the emphasis has been on measurement of anti-hapten antibody responses and of challenge-induced respiratory reactions in previously sensitized animals. In mice, attention has focused instead upon either changes in the serum concentration of IgE, or induced patterns of cytokine production by draining lymph node cells, following topical exposure to the test chemical. Although some of these approaches show promise, none is yet confirmed to possess the standards of sensitivity and selectivity required for routine use as a predictive method.

Alternative Methods: It has proven difficult to develop purely in vitro methods for the measurement of either contact or respiratory sensitization potential. Approaches based upon the assumed protein-reactivity of sensitizing chemicals and their ability to stimulate cytokine production or other responses in isolated cells or tissues have been proposed but none has yet been evaluated fully or shown to have the reliability required for replacement of standard methods. Perhaps more realistic presently is the identification of sensitization hazard from consideration of chemical structure. Particularly in the context of contact allergy, important progress has been made in defining the relationship between structure and sensitizing activity and in developing computer-assisted models for screening purposes.

Concluding Comments

Immunotoxicity and chemical allergy pose a number of important and intriguing regulatory challenges. Development of a more detailed understanding of the ways in which xenobiotics interact with the immune system, alter immunological function and provoke specific immune responses will undoubtedly allow many of these issues to be addressed in an effective and rational way.

Contact and Respiratory Allergy: A Regulatory Perspective

Peter Evans
Health and Safety Executive, Stanley Precinct, Bootle, Merseyside, UK

Respiratory Allergy

Introduction

Over recent years the subject of the induction of asthma by chemicals has proved to be not only an important but also a difficult and controversial area of regulatory toxicology. The development of clear positions has been hampered by such factors as an incomplete understanding of underlying toxicological mechanisms, confusing terminology, the absence of internationally accepted experimental test systems, inconclusive clinical data, a lack of useful criteria for classification and labelling purposes and doubt concerning the impact on the risk of occupational asthma of different exposure patterns and routes.

These problem areas, together with some regulatory approaches which have been devised to enable progress to be made, are described below. They provide the background for the development of the specific criteria that have recently emerged within the European Union to aid in the classification and labelling of chemicals with respect to their potential to induce asthma.

Terminology and toxicological mechanisms

The varying and sometimes overlapping definitions available for key terms, such as hypersensitivity, respiratory sensitisation, allergy and asthma, constitute an important source of possible confusion. Medical, regulatory, industrial and academic scientists may each have their own understanding of the meaning of these terms. The terms "asthma" and "respiratory sensitisation" have been used synonymously and interchangeably by some in the occupational health field, but in other minds they are distinguished from each other. This lack of clarity surrounding definitions has been compounded by uncertainties regarding the toxicological mechanisms underlying the disease processes involved in asthma. Thus, possible meanings for "respiratory sensitisation" in relation to effects in the lung include:

1. asthma induced by a proven immunological mechanism,
2. asthma induced by an immunological mechanism which may be proven or simply presumed,
3. asthma induced by a mechanism specific to the substance in question, but which may be immunological or non-immunological in nature,
4. asthma induced by any means, or even

5. asthma "caused" by any means, with this definition encompassing the mere provocation of symptoms of asthma with or without the implication that the substance induced the condition in the first place.

A further possible refinement to these definitions is the differentiation of immunological mechanisms into those mediated by immunoglobulin E and those apparently not.

In these circumstances of potentially confusing terminology, one regulatory approach has been to focus on the occurrence of the disease of concern (asthma), and attempt to harmonise on the term "asthmagen", meaning an agent that *induces* asthma by any means.

There is for low-molecular-weight chemicals considerable uncertainty regarding the underlying toxicological mechanisms for the induction of asthma. Although for high-molecular-weight allergens such as proteins there is good evidence that asthma is mediated by immunoglobulin E, this is not often available for chemicals. It may well be that for some reactive chemicals other immunological processes, not directly mediated by antibody, are important. In addition, in recent years a number of alternative mechanisms have been proposed, including neurogenic inflammation (Meggs 1993), pharmacological reactions (Fabbri et al. 1993) and repeated low-level irritation (Kipen et al. 1994). It has also been suggested that the condition known as reactive airways dysfunction syndrome, which may be brought on by a single high exposure to an irritant chemical, is in fact a variant of occupational asthma (Brooks and Bernstein 1993).

Given the current rather rudimentary state of knowledge concerning the mechanisms underlying the production of asthma for many low-molecular-weight chemicals, it is reasonable that regulatory attention has focussed on the potential for production of the disease of concern (i.e. asthma), without imposing any absolute requirement to elucidate the underlying toxicological mechanism. This approach has been adopted in the EU classification criteria developed to reflect the hazard "respiratory sensitisation" (production of asthma), as described below. Nevertheless, consideration of the potential underlying mechanism is an important factor in determining the appropriate risk management option(s) for any confirmed "respiratory sensitiser"/ "asthmagen". For instance, different approaches to substance control and health protection may be taken for substances producing asthma via immunological as against non-immunological mechanisms.

Methods for identification of asthmagens

A factor that makes the identification of agents having the ability to cause asthma less straightforward than for most other toxicological endpoints is the lack of a fully validated predictive animal test. Although a number of methods using guinea pig and mouse show considerable promise, none has yet attained international regulatory recognition as a test guideline adopted by the Organisation for Economic Co-operation and Development (OECD). At the present time, the available methods are generally considered by regulators to be acceptable as screening tests, providing useful information for chemicals which give positive responses leading, for example, to classification and labelling. However, negative findings produced by these methods are not currently taken to be a reliable indicator of the absence of asthmagenic potential. Thus another aspect to the regulatory approach is to encourage the further validation and international acceptance of the most promising predictive tests. Similarly, no standard, validated *in vitro* method is available, although the potential for a chemical to interact with protein can be considered a prerequisite for immunogenic activity. Another potential source of information, structure-activity relationship modelling, is still at a relatively early stage of development. Clearly, however, simple examination of molecular structure for reactive groups and checking whether a particular chemical is of a type (isocyanates or anhydrides, for example) already associated with the induction of asthma is worthwhile. Overall, the regulatory approach needs to take a balanced view of all the available information.

The absence of routinely-used animal and *in vitro* test methods means that much of the information available for an evaluation of the potential of a substance to cause asthma is based on clinical and epidemiological findings in people exposed at the workplace or, occasionally, at home or elsewhere. In one sense such human data are ideal, in that they come directly from the species and biological system of interest. However, from the regulatory perspective this information can also suffer from a number of deficiencies, some deriving from the nature of the original purpose of the investigations. Many studies have been aimed at the clinical diagnosis of asthma in a patient without any particular need for stringent identification of the agent responsible for actually inducing the asthmatic state, as opposed to that simply provoking in a non-specific manner asthmatic symptoms in an individual already having hyperresponsive airways, for reasons known or unknown. Exposure data for the period leading up to the recognition of occupational asthma is rarely available, and in many cases unquantified, but possibly high, previous and/or concurrent exposures to substances other than the one under suspicion may serve to prevent a firm conclusion being drawn about which agent *induced* the asthmatic state.

The clinical investigations themselves may contribute further uncertainty by the very nature of their conduct and the interpretation of their findings, and it is necessary for the regulator to carefully assess the quality of key tests, particularly specific bronchial challenges. For such a test to be considered truly rigorous by regulatory standards, a series of conditions should be met, including use of a clearly sub-irritant concentration of the putative asthmagen, maintaining blind conditions for the subject (and preferably also for the investigator) to the nature of the exposure (i.e. whether to the test or control substance), and careful control of possible confounding factors, such as use of asthma medication, smoking habits and the existence of upper respiratory tract viral infection. For a positive result to be convincing for regulatory purposes, the response should be of an appropriate magnitude (e.g. a decrease in the forced expiratory volume in one second of 15% or greater) over and above any effect seen at the control challenge. Unfortunately, it is unusual for bronchial challenge tests reported in the scientific literature to meet all or even most of these conditions, reflecting the fact that the tests are normally carried out for reasons of medical diagnosis of a condition rather than regulatory identification of a hazardous property of a specified agent.

Thus, in attempting to form an opinion about the asthmagenic potential of a substance, the regulatory approach is again to take a balanced view of all the information available, in order to make the best scientific judgement possible.

The EU classification criteria for respiratory sensitisation

A crucial starting point within the EU framework for regulation of industrial chemicals is the identification of their hazardous properties. The classification system in place in the EU serves to identify the hazardous properties of chemicals which are supplied commercially, and the correct application of the system is a statutory requirement within each of the member states. Criteria used to derive the appropriate classification for a substance are available in Annex VI to the Dangerous Substances Directive, commonly referred to as the "labelling guide" (EEC 1993).

Determining that a substance warrants classification as a respiratory sensitiser results in the assignment of the classification category "sensitising" and justifies application of the warning phrase R42 (May cause sensitisation by inhalation). However, there has been a problem in that the guidance given in the current EU labelling guide with respect this phrase is not particularly helpful. To improve the situation, new and more extensive criteria for the assignment of R42 were developed and formally adopted by EU member states in May 1996, with an intention that they come into effect in national law by 31 May 1998.

The revised criteria and accompanying notes were officially published in September 1996 (EC 1996), and provide guidance on the nature of the human evidence (clinical history, data from lung function and bronchial challenge tests, immunological findings, structural and mechanistic considerations) and results from animal studies that lead to classification of a chemical as a cause of asthma. Critically, it is stated that "immunological mechanisms do not have to be demonstrated", and that it is important to make a distinction between agents that actually induce asthma and those that simply provoke symptoms.

It is also made clear that the classification decision needs to weigh all the available evidence against the size of the population exposed and the extent of the exposure. This is intended to produce a distinction between the possession by a substance of any degree of ability, however weak, to produce asthma and the identification of a substance as having significant asthmagenic potential. Only the latter type of substance warrants classification as sensitising and application of R42. This principle has been introduced in order to prevent a substance being classified on the basis of only one or two rare, idiosyncratic reactions, since there may be the possibility of this occurrence for very many substances. The key point is that the decision on classification needs to set the number of cases reported against the size of the population that has been exposed and the extent of the exposure that has occurred in that population. There needs to be a significant number of cases of asthma induced by a particular substance in relation to the total number of people exposed to it, before classification becomes appropriate. Thus, the conclusion could be that a high-production-volume chemical, used in large quantities in many workplaces throughout the world, would not warrant the R42 phrase if only a few cases of asthma associated with its use have been reported over the years. In contrast, 3 cases of asthma among a workforce of 20 in contact with a speciality chemical could well result in the conclusion that the substance warrants classification as a sensitiser. This sort of "clustering" of cases can provide strong evidence with respect to a particular substance, although the case reports would still need careful critical appraisal, and the possibility of shared exposure with another, unsuspected substance should also be considered. Regarding extent of exposure, if a substance is stringently controlled, perhaps due to concern for another toxicological endpoint such as carcinogenicity, it is likely that fewer cases of asthma will become apparent than for a substance of equivalent asthmagenic potential for which historically there has been no such concern and exposure has not been so well controlled. Thus this sort of consideration may also need to be taken into account when making the overall assessment.

An important and as yet unresolved issue concerns the significance of peak exposures in the induction of the hypersensitive state. Such peaks, consisting of brief periods (perhaps of less than a minute) of exposure to high concentrations of the agent, may be masked in 8-hour time-weighted average values for exposure derived by routine personal sampling, but in fact reflect the intermittent nature of exposures in many industrial processes. At the present time practical experience in several industries suggests that peak exposures are important in the induction of the hypersensitive state, although the scientific evidence remains inconclusive (Morris 1994).

For chemicals which cause asthma there is also some uncertainty regarding the relevant routes of exposure for the induction phase of the process (i.e. rendering the airways hypersensitive). Regarding the provocation phase (i.e. triggering the airway reaction), clearly the inhalation route is generally the only relevant one. For protein and other macromolecular asthmagens, it is likely that inhalation is also the only route involved at the induction phase, as skin penetration is unlikely. In the case of low molecular weight chemicals, however, there is some evidence from animal studies that an immune response sufficient to sensitise the respiratory tract may occur after dermal exposure (Kimber and Wilks 1995). The current regulatory view accepts that for a limited number of chemicals there is some indication from animal experiments that a hypersensitive state in the respiratory tract can be induced by skin contact. However, it may be that this is simply an experimental phenomenon rather than a reflection of a route that operates in exposed workers.

Contact Allergy

Introduction

Reports relating to contact allergy are submitted to EU regulatory authorities in relation to programmes concerned with the notification, classification and labelling and risk assessment of `new' and `existing' chemicals. For predictive testing in guinea pigs there are approved (`Annex V') methods and an OECD guideline (no. 406) available, but there has still been considerable discussion regarding preferences for and the validity of specific test protocols. This uncertainty has extended to issues related to the nature of the positive control substance, the number of guinea pigs needed, appropriate concentrations of test substance to be used at induction and challenge, the grading of skin reactions and the interpretation of results for classification purposes.

The aspects of these issues that are related to the successful identification of the skin sensitising hazard of chemicals are briefly discussed below. In addition, it is considered whether, given the increasing importance of the risk assessment process in the regulation of chemicals, it is now time for regulators to make more use of newer test methods such as the local lymph node assay, which has the potential to generate information on the potency of any sensitising effect.

Acceptable types of test

The EU has recently published an updated version of the Annex V method for skin sensitisation (EC 1996) which is largely consistent with the current OECD test guideline (OECD 1992). This updated method gives specific guidance on the maximisation test (the preferred adjuvant method) and the Buehler test (the preferred non-adjuvant method). It considers that in general adjuvant tests are preferred over non-adjuvant tests, because of their likely greater sensitivity and hence ability to predict an effect in humans. Experience has shown that it is technically possible to perform a maximisation test for the great majority of substances, and indeed more than 90% of the tests submitted to the EU on contact allergy of new substances are maximisation tests (Schlede and Eppler 1995). However, it must be recognised that testing for skin sensitisation is not as yet globally harmonised, so that reports of tests other than maximisation studies (for example, Buehler tests carried out in American laboratories) are occasionally received by EU regulatory authorities. In the interests of animal welfare, it is appropriate that the results of these tests are properly assessed and used whenever possible, and that a study is not rejected unless there are good scientific reasons.

The OECD guideline published in 1992 introduced the possibility of using positive results obtained with mice (in the mouse ear swelling test or the local lymph node assay) to identify a substance as a potential skin sensitiser, and thus avoid the need to test in guinea pigs. However, if a negative result is obtained in the mouse assay, this is not currently acceptable for classification purposes, and a guinea pig test must be conducted subsequently in order to confirm that the substance is not a sensitiser. The challenge that has faced the mouse assays is to achieve full validation, so that negative findings (which do not lead to classification and labelling) also become acceptable to regulatory authorities. Over the last few years a considerable body of evidence has accumulated to indicate that this point has now been reached in regard to the local lymph node assay (Basketter et al. 1996). This issue is discussed further below.

Although no method even approaching regulatory status is currently available, the development of an *in vitro* screen for skin sensitisation is a goal that remains desirable and in theory appears reasonable, given that individual components of the sensitisation reaction (such as the ability to penetrate skin and to bind to protein) can be assessed using *in vitro* techniques.

Positive control substances

Although it is not necessary to run concurrent positive controls with every contact sensitisation test, the sensitivity and reliability of the method being used should be assessed about every 6 months using known sensitisers. The revised Annex V method and the OECD guideline are now agreed that this check should use substances with mild-to-moderate skin sensitising properties, with the preferred substances being hexylcinnamic aldehyde, mercaptobenzothiazole and benzocaine.

The introduction of the use of mild-to-moderate sensitisers to define the sensitivity of the assay system used has important implications for the selection of the test method in the first place. If the method consistently gives adequate sensitisation response rates with one or more of the recommended positive control substances, it would not be reasonable to reject that method on the grounds that it lacks sensitivity. This is obviously more likely to be relevant to methods other than the maximisation test, and would apply, for example, to the 'other methods' referred to in the updated Annex V guidance as requiring validation and scientific justification. Conversely, an assay previously acceptable to a regulatory authority because it responded well to a strong sensitiser such as dinitrochlorobenzene, but now found to give an inadequate sensitisation response with mild-to-moderate allergens, would require refinement to increase its sensitivity.

Number of animals in the maximisation test

Although early protocols for the maximisation test stipulated group sizes of at least 20 treated and 10 control guinea pigs, more recently there have been demonstrations of the feasibility of using only 10 treated and 5 control animals, but obtaining the same conclusions regarding the need to classify the substance as a skin sensitiser as when the full '20 and 10' method is followed (Shillaker et al. 1989). In the light of these findings, the most recent versions of both the Annex V method and the OECD guideline now state that a minimum of 10 animals is used for the treated group and at least 5 for the control group. However, there is also the proviso that "when it is not possible to conclude that the test substance is a sensitiser, testing in additional animals to give a total of at least 20 test and 10 control animals is strongly recommended". The EU scheme requires classification of a substance as a skin sensitiser when it causes a response in at least 30% of the animals in an adjuvant test (EEC 1993). Thus, this qualification means that since a proportion of '10 and 5' tests (those with 1 or 2 /10 test guinea pigs showing sensitisation reactions) will need to be repeated to bring them up to '20 and 10', it will in some cases appear more practical to conduct a '20 and 10' study from the outset. For animal welfare reasons, however, it is strongly recommended that a '10 and 5' study is initially performed, then repeated if necessary. In the event of a clear positive response (i e 3-10/10 sensitisation reactions), or completely negative findings in this initial test, no further animals need to be tested. If the result of the initial test is equivocal in classification terms (i e there are 1 or 2/10 sensitised guinea pigs in the test group), the test will need to be repeated. However, there will still be a benefit in that over a period of time it will become possible to assess the value of this repetition in determining the eventual classifications of the test substances.

Interpretation of results for classification purposes

According to the EU classification scheme, if a substance produces sensitisation reactions in 30% or more of the guinea pigs tested in an adjuvant assay, or 15% or more in a non-adjuvant assay, it will normally be classified as 'sensitising' and the associated phrase R43 ('May cause sensitisation by skin contact') will appear on the label of its package (EEC 1993). When no skin reactions are seen in the control guinea pigs, the interpretation of the results for classification purposes is straightforward when clear-cut skin reactions are obtained in the test animals at challenge. However, weak challenge reactions in test animals may be evidence of either sensitization or irritation, so that false-positive results are possible, according to a recent assessment of experience with the maximisation test (Kligman and Basketter 1995). The mechanism proposed for these false-positive irritant reactions is that irritating induction doses of test substance cause a general state of hyperirritability in the test animals. To distinguish between sensitisation and false-positive irritation reactions obtained at challenge, it is therefore advisable to perform a second challenge after 2 to 3 weeks. A true allergic state normally persists over this period (and often for much longer), while the hyperirritable state tends to decrease or disappear. Rapid fading of a challenge reaction would also suggest irritation rather than sensitisation (Kligman and Basketter 1995).

Further uncertainty over interpretation of challenge reactions can arise when effects are also seen in the control animals. Since these must be irritant effects, they may cast doubt on the nature of the reactions apparent in the test animals, particularly if they are of the same grade. In this situation, one approach is to consider as evidence for sensitisation only skin reactions in test animals that are of a higher grade than the maximum observed in any of the control animals. Thus an assay in which 4/10 test guinea pigs showed grade 2 and 6/10 grade 1 reactions would lead to classification of the test substance as a skin sensitiser even if 5/5 control animals also gave grade 1 reactions. However, there are potential pitfalls to this approach. For example, 9/10 grade 1 reactions seen in the test group would be countered by just one similar reaction found among the controls, so that no classification would follow.

An alternative approach to defining the percentage of test animals showing contact allergy is to subtract from the percentage of test animals showing skin reactions the percentage of control animals giving these reactions at challenge. Thus if 6/10 test and 1/5 control animals gave grade 1 effects, the response taken to be sensitisation (60 - 20 = 40%) would be sufficient to classify. Similarly, if there were 1/10 grade 2 and 9/10 grade 1 reactions in the test group, and 4/5 grade1 effects in the control group, the sensitisation response (10 + [90 - 80] = 20%) would be enough to classify in a non-adjuvant but not in an adjuvant test.

Overall, there is probably no single method of evaluating challenge results for classification purposes that is appropriate for all circumstances. The approach taken will largely depend on the scientific judgement of those conducting the study, and the regulator can do nothing more than appraise that judgement on a case-by-case basis.

The regulatory status of the local lymph node assay

Regulatory methods for predicting the skin sensitising potential of chemicals have historically used guinea pigs, and over the years these methods have provided positive results for a relatively high proportion of chemicals tested. During the period 1993-1996, for example, fully 31% (98/314) of the new substances notified in the UK received the R43 warning phrase, compared with 10% in total for the 3 phrases relating to skin corrosion and irritation (R34, R35 and R38). Indeed, for the same period, the labels of no less than 60% of all UK new substances that received any classification at all for acute toxicity, corrosivity/irritancy or sensitisation endpoints included R43. Furthermore, in half these cases skin sensitisation was the only such human health endpoint identified. All of these values include only chemicals producing a sensitisation rate above the classification threshold and so exclude, for example, those maximisation tests with response rates of less than 30%.

By the very nature of their design, the guinea pig tests (particularly those using adjuvant) do not provide robust information on the potency of any sensitising effect that they identify. Such information is becoming increasingly important as toxicologists need it to refine their risk assessments for sensitising chemicals.

Overall, therefore, it seems appropriate that regulators should now reassess how useful the guinea pig assays continue to be for their purposes, particularly as an alternative method has been proposed in the form of the local lymph node assay. This method appears to have a number of advantages over the maximisation test, including straightforward and objective interpretation of results, the potential to obtain potency data that are more directly applicable to risk assessment, and animal welfare benefits. The method has also been extensively and rigorously validated against both animal and human test data. From the regulatory viewpoint, there does indeed seem to be good scientific justification for this assay to be formally adopted by the OECD and accepted by the EU as a suitable method for classification purposes. At the very least, the time is right for regulatory authorities to develop a clear and consistent view regarding the utility of the method for hazard identification, classification and risk assessment of contact sensitisers.

References

Basketter DA, Gerberick GF, Kimber I, Loveless SE (1996) The local lymph node assay: A viable alternative to currently accepted skin sensitization tests. Food Chem Toxicol 34:985-997

Brooks SM, Bernstein IL (1993) Reactive airways dysfunction syndrome or irritant-induced asthma. In: Bernstein IL, Chan-Yeung M, Malo J-L, Bernstein DI (eds) Asthma in the workplace. Dekker, New York, pp 533-549

EC (1996) Annex to Commission Directive 96/54/EC. Off J Eur Comm L248:1-230

EEC (1993) Annex to Commission Directive 93/21/EEC. Off J Eur Comm L110A:45-86

Fabbri LM, Ciaccia A, Maestrelli P, Saetta M, Mapp CE (1993) Pathophysiology of occupational asthma. In: Bernstein IL, Chan-Yeung M, Malo J-L, Bernstein DI (eds) Asthma in the workplace. Dekker, New York, pp 61-92

Kimber I, Wilks MF (1995) Chemical respiratory allergy. Toxicological and occupational health issues. Hum Exp Toxicol 14:735-736

Kipen HM, Blume R, Hutt D (1994) Asthma experience in an occupational and environmental medicine clinic. Low-dose reactive airways dysfunction syndrome. J Occup Med 36:1133-1137

Kligman AM, Basketter DA (1995) A critical commentary and updating of the Guinea-pig maximization test. Cont Derm 32: 129-134

Meggs WJ (1993) Neurogenic inflammation and sensitivity to environmental chemicals. Environ Health Perspect 101:234-238

Morris L (1994) Respiratory sensitisers. Controlling peak exposures. Tox Sub Bull 24:10

OECD (1992) Guidelines for testing of chemicals, No.406. Skin sensitisation

Schlede E, Eppler R (1995) Testing for skin sensitization according to the notification procedure for new chemicals: The Magnusson and Kligman test. Cont Derm 32:1-4

Shillaker RO, Bell GB, Hodgson JT, Padgham MDJ (1989) Guinea pig maximisation test for skin sensitisation: The use of fewer test animals. Arch Toxicol 63:283-288

Immunotoxicology: Extrapolation from Animal to Man - Estimation of the Immunotoxicologic Risk Associated with TBTO Exposure

H. Van Loveren[1], W. Slob[1], R.J. Vandebriel[1], B.N. Hudspith[2], C. Meredith[2], J. Garssen[1]

[1] National Institute of Public Health and the Environment, PO Box 1, 3720 BA Bilthoven, the Netherlands, and

[2] BIBRA, Carshalton, United Kingdom.

Abstract

Bis(tri-*n*-butyltin)oxide (TBTO) has been shown to be immunotoxic in rodents, resulting in decreased resistance to infections. The no-effect level assessed by estimating effects on host resistance in rats has been found to lie between 0.5 and 5.0 mg TBTO/kg food (0.025 and 0.25 mg/kg body weight). For risk assessment such animal data need to be extrapolated to the human situation. In risk assessment procedures uncertainty factors are used to account for interspecies variation (extrapolation from animal to man) and for variation within the human species. For both factors a value of 10 is often used, based on international guidelines. Hence, exposures below 0.00025 mg/kg body weight should not pose a risk for the human population.

In the present study we have taken an alternative approach. We have produced dose-response curves for the effect of TBTO exposure on resistance to *Trichinella spiralis*. To extrapolate this curve to the human situation, we produced additional dose response data concerning *in vitro* effects of TBTO exposure on the mitogen responsiveness of both rat lymphoid cells and human blood cells. Using regression analyses of these dose-response data, we calculated a factor that accounts for interspecies variation (IEV) and a factor that accounts for intraspecies variation (IAV) within the human samples.

Using these factors, we estimated the dose that decreases resistance in man to an infection. We choose a 10% increase of the infectious load as a reference point which in our view is of biological significance. Based on these considerations, we estimated the dose that may affect resistance in adult humans at 0.04 mg/kg body weight. Pre- and postnatal exposure will probably result in effects at lower concentrations, due to the vulnerability of the developing immune system.

Introduction

Trialkyltin compounds are widely used as biocides (e.g. in wood preservatives and as an antifouling component in paints). Controlled-release formulations of bis(tri-*n*-butyltin)-oxide (TBTO) and -fluoride have been proposed as molluscicides for the control of snails serving as vectors for Schistosoma infections in man. Several studies have indicated that oral TBTO exposure in rodents can lead to immunomodulation, and, as a consequence, to a decrease in resistance to infections, such as bacterial infection (*Listeria monocytogenes*), parasitic infection (*Trichinella spiralis*) (Vos et al 1984, 1990) and viral infection (cytomegalovirus) (Garssen et al 1995). The effects of TBTO on the immune system were mainly due to effects of TBTO on thymus-dependent immunity (cellular immunity), although non-specific immune functions such as natural killer cell function and macrophage function were also affected by TBTO exposure (Vos et al 1984, 1990).

Since TBTO is a potent and well characterized immunosuppressive compound, it is suitable as a model compound for new developments within the field of immunotoxicology, such as the development of risk assessment procedures, as presented in this report.

The effects of TBTO-like compounds on the immune system, and particularly on host resistance, have proven to be the most sensitive parameter of toxicity in the rat (which is the most sensitive species of all species tested). The no-observed adverse effect level (NOAEL), using the *Trichinella spiralis* host(rat)-resistance model, lies between 0.5 and 5.0 mg/kg diet (0.025-0.25 mg/kg body weight), while in immune function tests the NOAEL is 0.6 mg/kg body weight. Based on these host resistance studies the safety margin for exposure of humans was calculated and published in an IPCS/WHO report (IPCS 1990).

In the present study a new approach was developed for the estimation of risk of exposure to TBTO regarding deleterious effects on human health, i.e. lowered resistance to infectious diseases. The relative sensitivity of the immune system for TBTO exposure was studied in *in vitro* test systems of human lymphocytes and rat lymphocytes. Lymphocyte proliferation induced by mitogens was used as the immune parameter in this comparison. Dose-response studies and regression analyses were performed to estimate the dose producing 50% (Effective Dose 50, ED_{50}), or 75 % (ED_{25}) or 90% (ED_{10}), respectively. Thus, instead of threshold doses (LOAEL or NOAEL), non-threshold doses were determined. Similar dose-response analyses were performed for host resistance to *Trichinella spiralis* and *Listeria monocytogenes* infections in rats, orally exposed to TBTO.

The results from the sensitivity comparison studies and the dose-response studies for resistance in rats were combined to estimate the dose that may affect host resistance in humans.

Methods

***In vivo* studies:** Effects of TBTO exposure on resistance to *Trichinella spiralis* and *Listeria monocytogenes* in Wistar rats have been published by Vos et al (1984, 1990). The original data from these experiments were used in the present study to evaluate the relationship between dose and effect.

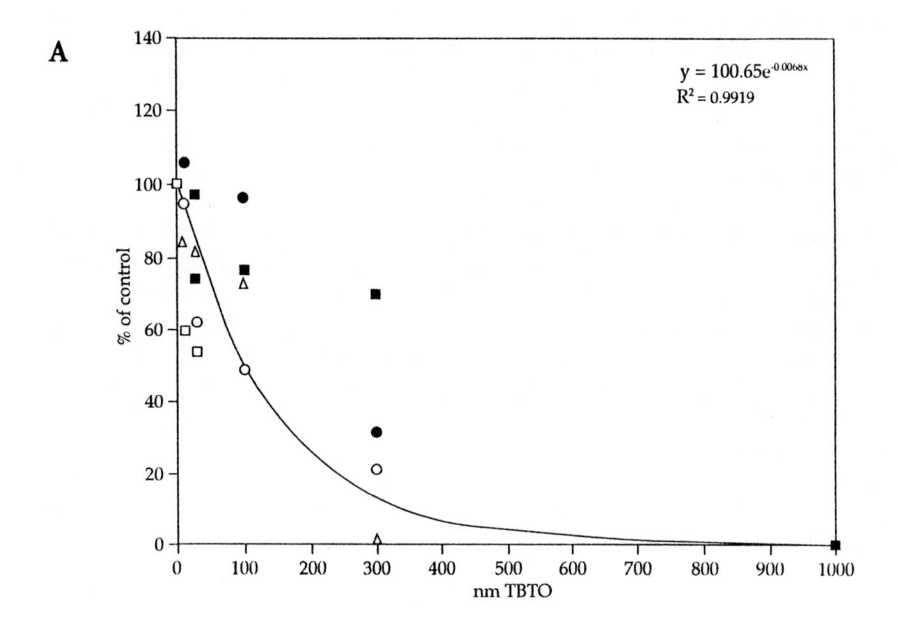

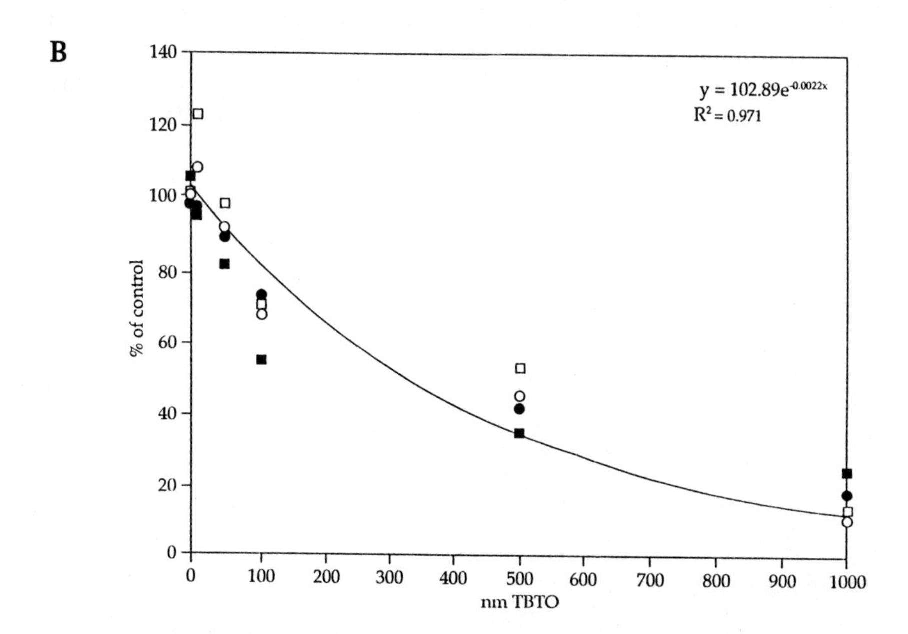

Figure 1. Fitted dose response curve of effect of *in vitro* exposure of human (A) and rat (B) lymphocytes to TBTO.

***In vitro* studies:** Spleen cells from 8 week old male Wistar rats, and peripheral blood cells from healthy adult humans were exposed *in vitro* to TBTO. Subsequently, the proliferative response of the lymphocytes to ConcanavalinA (ConA) was assessed by measuring tritium-thymidine uptake. Dose-response relationships were evaluated at concentrations that did not affect viability (judged by trypan blue dye exclusion).

Quantitative Methods: Linear and non-linear regression methods were applied to evaluate the dose response relationships in the *in vivo* and *in vitro* studies.

Based on the curves fitted through the host resistance data, dose levels were calculated that resulted in an increase of 10 % or 100 % of the pathogen load as compared to unexposed controls. Based on the curves fitted through the *in vitro* data, the dose that produced 10, 25, or 50 % (ED_{10}, ED_{25}, ED_{50}) of the maximal effect were calculated. The interspecies variation (IEV) was calculated by dividing the human ED by the ED of rat cells.

The intraspecies variability among the human samples was calculated as a range between the ED_{50} of the most sensitive individual divided by the mean ED_{50} and the ED_{50} of the least sensitive individual divided by the mean ED_{50}.

Results

TBTO inhibits ConA induced lymphocyte proliferation of both rat and human lymphocytes *in vitro* in a dose dependent fashion. The fitted curves derived from these data are shown in Fig. 1. IEV values based on either ED_{10}, ED_{25}, or ED_{50} showed no major differences. They indicated that rat lymphocytes were approximately 3 times more sensitive than human lymphocytes (Table 1).

Table 1. ED and IEV values for TBTO effects on lymphocyte responsiveness to Con A.

In vitro LST (Con A)	TBTO human (nM)	TBTO rat (nM)	IEV
ED_{50}	328	102	3.18
ED_{25}	643	204	3.14
ED_{10}	1059	339	3.12

IAV values were based on the ED_{50} value of individual regression analyses within the human samples tested. They showed that the most sensitive individual needed 0.64 times the amount of TBTO and the least sensitive individual 1.7 times the amount of TBTO compared to the mean of the whole group tested.

Based on the dose response curves of the host resistance assays (Fig. 2), the IEV values, and the IAV range, the dose of TBTO that results in an increment of 10 % or 100 % of pathogens in the most and least sensitive individuals was calculated (Table 2). The effect of exposure on *Trichinella spiralis* load was more pronounced than that on *Listeria monocytogenes* load.

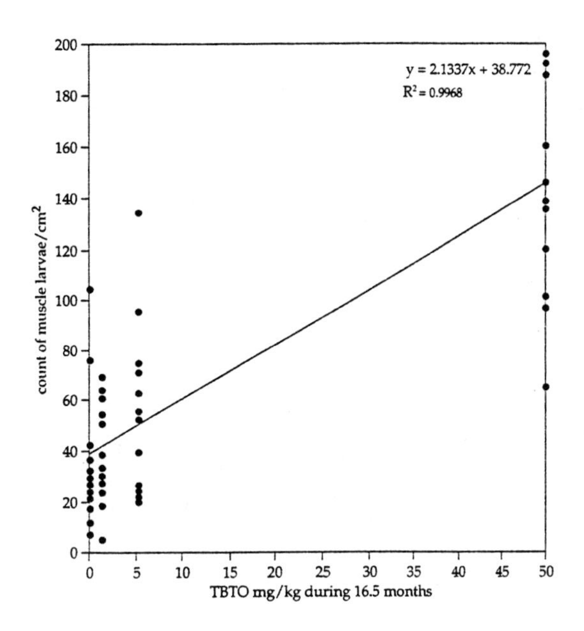

Figure 2. Fitted dose response curve of effect of TBTO exposure on resistance to *Trichinella spiralis.*

Table 2. Effect of TBTO on pathogen load after infection

pathogen load	TBTO dose (mg/kg) resulting in increased *Trichinella spiralis* larve in tissue		TBTO dose (mg/kg) resulting in increased *Listeria monocytogenes* load in spleen	
	most sensitive individual	least	most sensitive individual	least
110%	0.04	0.11	0.89	2.36
200%	1.58	4.21	2.54	6.75

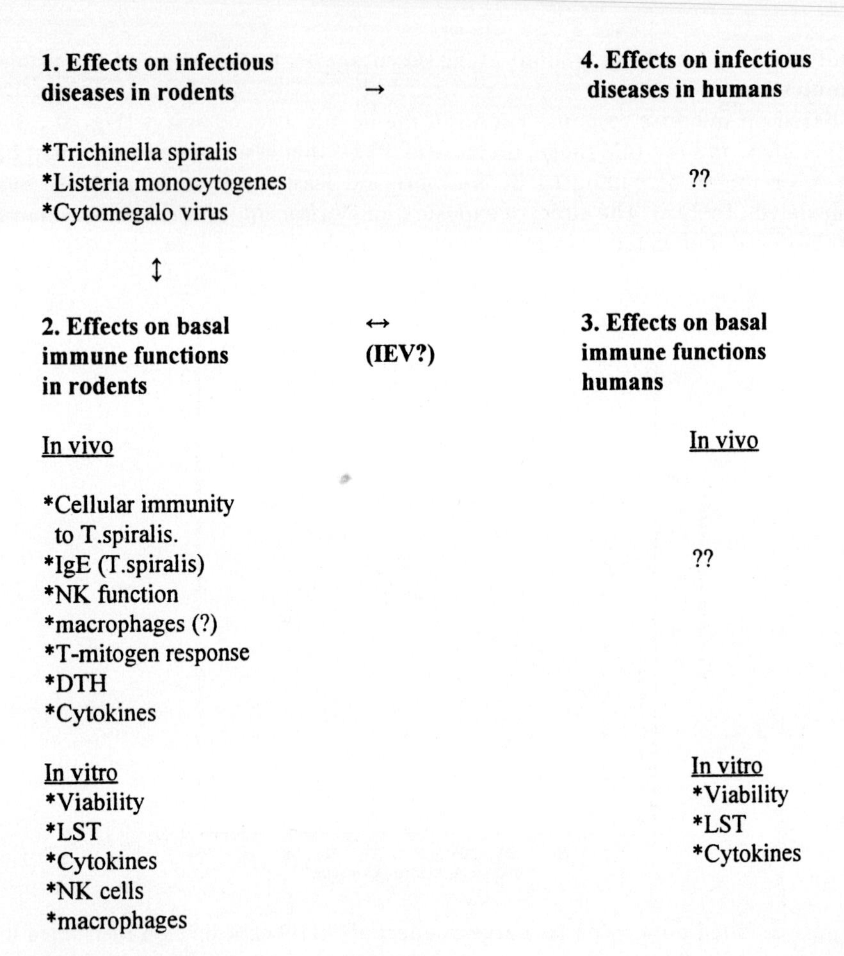

Figure 3. Parallellogram approach for risk assessment of TBTO immunotoxicity

Discussion

Bis(tri-*n*-butyltin)oxide (TBTO), an antifouling agent used in paints of ship hauls, is a well known immunotoxicant. It interacts with -SH groups in cytoskeleton structures, and its prime targets are immature T lymphocytes and macrophages. Exposure of rats to TBTO leads to deficient immune responses, especially cellular immune responses, and reduced resistance to infections, such as to the parasite *Trichinella spiralis*, the bacterium *Listeria monocytogenes*, and rat cytomegalovirus. The effect of TBTO on host resistance appears to be one of the most sensitive parameters tested, as in rats exposed for a prolonged time (16.5 months) to 5 mg TBTO/kg food, no effects of TBTO on general toxicologic or basal immunotoxicologic parameters were observed, whereas decreased resistance to *Trichinella spiralis*, as evidenced by increased numbers of muscle larvae, could already be detected. Internationally, this parameter has now been accepted to determine the no-adverse-effect-level for exposure to TBTO (IPCS 1990).

The assessment of maximal tolerable doses in humans by IPCS/WHO was based on a NOAEL obtained from animal studies published in the literature during the last 20 years. To elucidate the relevance of the observations in experimental animals, we have used the so-called parallellogram approach (see Fig. 3). According to this approach, the exposure effect on host resistance in rats was extrapolated to the human situation, using a quantitative species comparison of *in vitro* effects in peripheral lymphocytes of rat and human origin. Since TBTO affects cellular immune responses, and has a direct effect on immature T lymphocytes in the thymus, these cells would have been the best choice for interspecies comparison. For obvious reasons, it is not easy to use such cells from human origin. Therefore, we chose human peripheral lymphocytes and rat peripheral (spleen) lymphocytes, and used ConA responsiveness as a measure of functionality. As ConA is also mitogenic for immature T-lymphocytes, esponsiveness to this mitogen forms the best surrogate for the effects of TBTO on T lymphocytes accessible both in rats and humans.

In vitro exposure to TBTO was carried out at concentrations that did not affect the viability of the cells. Dose-response curves were calculated, and the dose leading to a reduction of the functionality by 50 % (Effective Dose 50, ED_{50}) was calculated. Dividing the ED_{50} of human cells by that of rats resulted in an Interspecies Variability (IEV) factor of 3.18. If the IEV was calculated with the ED_{20} or ED_{10}, the resulting IEV was 3.14 or 3.12, respectively, indicating that the selection of the Effective Dose on which the IEV was based was not critical.

In addition to interspecies variability, intraspecies variability was also accounted for. For this purpose the ED_{50} of the most sensitive sample of human peripheral lymphocytes was calculated and divided by the mean ED_{50} of all human samples, leading to a factor of 0.64. Similarly, the ED_{50} of the least sensitive sample divided by the mean ED_{50} was calculated to be 1.7. These figures were used as an indication for the upper and lower margin of the risk estimate of TBTO exposure.

Using dose-response curves on numbers of muscle larvae after infection with *Trichinella spiralis* of TBTO-exposed rats, we calculated the dose of TBTO required to increase the larvae burden by 10 % (expressed as numbers of muscle larvae per m^2 of striated (tongue) tissue). In our view this percentage represents a clinically significant increase. This increase was found to occur at 0.0195 mg TBTO/kg body weight. Hence, the calculated dose of TBTO to induce an increase in parasitic load in humans after infection with *Trichinella spiralis* ranged from approximately 0.04 mg (= 0.0195 X 3.18 X 0.64) to approximately 0.11 mg (= 0.0195 X 3.18 X 1.7) TBTO/kg body weight. Since the developing immune system is likely to be more sensitive than the fully developed system in adult animals and man, it is likely that the risk-producing dose in pre- and postnatal exposure will be lower than in adults.

Dose-response curves for resistance to *Listeria monocytogenes* in TBTO-treated rats showed that a considerably higher TBTO dose was required to provide an increase of 10 % in bacterial load in the spleen, i.e. 0.4 mg TBTO/kg body weight. These different outcomes for the ED_{50}s in rats concerning an increased infectious burden after TBTO exposure represents the difference in sensitivity of resistance to *Trichinella spiralis* and *Listeria monocytogens*, and indicates that for TBTO *Trichinella* may be the preferential host resistance model on which to base risk analysis.

The classical risk assessment of (immuno)toxicity of chemicals is heavily dependent on arbitrarily chosen uncertainty factors. The alternative approach for estimating risk may serve to further understand in quantitative terms the risk of exposure to immunotoxic chemicals with respect to resistance to infectious diseases in humans, and may diminish the need to use arbitrarily chosen uncertainty factors for inter- and intraspecies extrapolation.

References

Garssen, J., Van der Vliet, H., De Klerk, A., Goettsch, W., Dormans, J.A.M.A., Bruggeman, C.A., Osterhaus, A.D.M.E., Van Loveren, H.(1995)A rat cytomegalovirus infection model as a tool for immunotoxicity testing. Eur. J. Pharmacol., 292, 223-231

IPCS. Environmental Health Criteria 116 (1990) Tributyltin Compounds. World Health Organization, Geneva, Switzerland, pp 273

Vos, J.G., De Klerk, A.,. Krajnc, E.I., Kruizinga, W., Van Ommen, B., Rozing, J. (1984) Toxicity of bis(tri-*n*-butyltin)oxide in the rat. II. Suppression of immune responses and of parameters of non specific resistance after short-term exposure. Toxicol. Appl. Pharmacol., 75, 387-408

Vos, J.G., De Klerk, A., Krajnc, E.I., Van Loveren, H., Rozing, J.(1990) Immunotoxicity of bis(tri-*n*-butyltin)oxide in the rat: Effects on thymus-dependent immunity and non-Appl. Pharmacol., 105, 144-155

Regulating Immunotoxicity Evaluation: Issues and Needs

J. Descotes
Department of Pharmacology, Medical Toxicology and Environmental Medicine
& INSERM U 80, Faculté de Médecine Lyon-R.T.H. Laennec, 69008 Lyon, France

Introduction

The potential immunotoxicity of xenobiotics has been a matter of growing concern during the past two decades. A large number of research papers showed that many industrial and environmental chemicals as well as medicinal products or food additives, can adversely influence the immune competence of mammals, including man, but can also induce hypersensitivity and autoimmune reactions. In 1982, the US Environmental Protection Agency issued guidelines for the immunotoxicity evaluation of pesticides and this was followed by a recommendation from the Council of the European Communities to pay attention to immunotoxicity during the safety evaluation of new medicinal products in 1983. Presumably because these regulatory texts were either premature (EPA) or grossly general (CEC), they have never been implemented. Surprisingly, no guidelines on immunotoxicity evaluation have been published until very recently and therefore, immunotoxicity evaluation remains largely unregulated today.

When considering whether regulating immunotoxicity evaluation is necessary, a number of issues and needs arise. This paper reflects the author's personal views on some of these issues and needs.

What are the Issues ?

At least, four major issues can be addressed :

1. Human health hazards associated with immunotoxicity

The first issue to be addressed when one considers regulating the immunotoxicity evaluation of new medicinal products and chemicals, is whether hazards related to immunotoxic effects have been adequately identified and may pose a threat to human health.

Based on the long experience gained with the therapeutic use of immunosuppressive agents, particularly in transplant patients (Vial and Descotes, 1996), immunosuppression has been shown to result in impaired resistance to microbial pathogens with more frequent, severe, relapsing and often atypical infectious complications, and in lymphomas. Similarly, the expanding therapeutic use of recombinant cytokines indicates that immunostimulation (or immune activation) can result in sometimes severe flu-like reactions ("acute cytokine syndromes"), more frequent autoimmune diseases, such as autoimmune thyroiditis, more frequent allergic reactions to unrelated allergens, and inhibition of hepatic drug metabolism potentially resulting in drug interactions (Vial and Descotes, 1996). All these adverse events can be considered as direct consequences of pharmacological or toxic effects on the immune system.

Hypersensitivity reactions are the commonest immunotoxic complications of exposures to xenobiotics, particularly medicinal products and industrial chemicals (Vos et al., 1996). Although mechanisms are ill-understood in many instances, it appears that immune-mediated as well as nonimmune-mediated (i.e. "pseudo-allergic") mechanisms can be involved.

Finally, drugs and chemicals can induce a variety of autoimmune reactions (Bigazzi, 1997) in particular systemic reactions, such as lupus syndromes and scleroderma-like diseases.

Thus, the possible human health consequences of immunotoxicity associated with xenobiotics are reasonably well identified and some of these consequences can be severe, even life-threatening.

2. Likelihood and magnitude of associated risks

Whereas immunotoxicological hazards are reasonably well identified, it remains to assess related risks to man in realistic conditions of exposure. Several risks have been adequately characterized, but others remain excessively uncertain, essentially because few human studies have been conducted, so that the issue of risks associated with immunotoxicity is at best partly settled.

Immunosuppression, when severe, can result in infectious complications and lymphomas in a substantial fraction of treated patients as shown by the clinical experience gained with transplant patients (Vial and Descotes, 1996). Fewer data suggest nevertheless that less severe immunosuppression (or immunodepression) resulting from low-dose immunosuppressive regimens in rheumatoid patients or accidental exposure to biphenyls for instance, is potentially associated with the same adverse events, which are however less frequent and less severe. Unfortunately, no widely accepted definition of immunosuppression is available, although the impact on human health of a 10% decrease in antibody levels is certainly different from that of a total suppression of the antibody response. The concept of functional reserve capacity of the immune system has often been used, but no operational definition has been proposed.

Immunostimulation is associated with various adverse events as described above. Interestingly, these adverse events were suspected long before recombinant cytokines with potent immunopharmacological properties were introduced in the clinic (Descotes, 1985). To some extent, a parallelism can be noted between the immunostimulating potency of medicinal products and the frequency and severity of adverse events associated with these products. Therefore, any new immunoenhancing compound can be suspected to induce these adverse events.

Hypersensitivity reactions are quite common with chemicals acting as direct immunogens (large-molecular-weight compounds) or haptens (chemicals with marked chemical reactivity). Although these reactions can be severe, even life-threatening in some instances, the majority is fortunately mild and self-limiting. Immunological mechanisms, such as IgE-mediated anaphylaxis, IgM-mediated cytotoxicity or T lymphocyte-mediated contact sensitization, can be involved, but do not account for all hypersensitivity reactions among which many are presumably non immune-mediated (Vos et al., 1996; Wedner, 1987).

Autoimmune reactions associated with drug therapy are seemingly very rare (Vial et al., 1997b) and probably still more so in association with chemical exposures. However, when they do develop, these reactions are usually severe. A major problem today is the nearly total lack of clear understanding of the mechanisms involved.

3. Acceptability of risks associated with immunotoxicity

The results of risk assessment, but also the perception of risks must be taken into account when addressing this issue. There is no definition of what an acceptable risk is and it is unsure that definitions based on probabilistic estimates (Freudenburg, 1988) can be easily accepted by the public.

In any case, several immunotoxic effects, such as infectious complications and lymphomas associated with immunosuppression, acute cytokine syndromes, anaphylactic shocks and systemic autoimmune reactions, can be severe, sometimes life-threatening, suggesting that at least some imunotoxic risks may not be acceptable.

4. Status of available methods to predict immunotoxicological hazards

If risks associated with immunotoxicological hazards are judged unacceptable, another set of issues arise.

In the past decade, extensive efforts have been put into the design and evaluation of methodologies for the purpose of predicting the unexpected immunosuppressive effects of xenobiotics. Based on the results of Luster et al. (1992; 1993) or White et al. (1994) as well as those of others, a battery of validated tests can now be included in two- or three-tier rodent protocols.

In contrast, other aspects of immunotoxicity received only limited attention. For instance, no validated protocol can be proposed to predict unexpected immunostimulation. Similarly, predicting hypersensitivity reactions is extremely difficult, if not impossible, in most instances, particularly as far as medicinal drugs or food additives are concerned, because they usually are not directly immunogenic or lack chemical reactivity to act as haptens. Finally, only systemic autoimmune reactions can be expected to be identified in the future using the popliteal lymph node assay (Vial et al., 1997a), provided further validation of this assay is conducted.

Despite obvious limitations in available methodologies, various approaches can be used to evaluate immunotoxicity. The same tiered protocols as those used for detecting unexpected immunosuppression can supposedly be useful to predict immunostimulation and studies are underway to confirm this assumption. Assays can be used as in vitro screen to assess the cytokine releasing properties of novel compounds (Revillard et al., 1995). Although much remains to be done as regards the allergic potential of xenobiotics, the local lymph node assay in mice, a validated and accepted tool to identify contact sensitizers (Kimber et al., 1994), was suggested to help predict respiratory sensitizers as well, and the value of the mouse IgE test is currently being investigated (Hilton et al., 1996).

What are the needs ?

Several unmet needs might explain why adequately standardized and validated methods are not available to evaluate every potential immunotoxic effect.

1. Immunotoxicity should not be restricted to immunosuppression
Despite the fact that hypersensitivity is the most common immunotoxic risk of drug treatment and occupational exposure to industrial chemicals, immunosuppression was, and to some extent still is, the primary focus of immunotoxicological investigations. This can explain why methods to predict immunosuppression have been extensively standardized and validated. Time has come to shift focus on other critical aspects of immunotoxicity, such as allergenicity, immunostimulation, but also mild to moderate immunosuppression.

2. Efforts should be undertaken to conduct less science-oriented immuno-toxicological studies
Experimental studies in immunotoxicology have been largely science-oriented, which resulted in obvious limitations: for instance, the mechanism of dioxin immunotoxicity, especially in rodents, has been extensively investigated (Kerkvliet, 1994), but it still remains to be known whether and if so, how potently immunosuppressive dioxin actually is in man.

Validation studies consistently used prototypic compounds and it remains to be established to what extent results obtained with the immunosuppressant cyclosporin, the contact sensitizer dinitrochlorobenzene or the respiratory allergen isothiocyanate, for instance, are actually helpful to predict the effects of other compounds with lesser immunotoxic potential. Science-oriented studies are absolutely essential to understand the mechanism of immunotoxic effects, but additional studies are needed to delineate the reliability and predictive value of available methods and determine the magnitude and likelihood of immunotoxic effects in animals and man. As is true in any other area of regulatory toxicology, regulatory immunotoxicology cannot be entirely based on the results of science-oriented studies.

3. Development of clinical immunotoxicology
A major limitation of current immunotoxicology is the near total lack of a human database. Much information is available in rodents, but it is often unknown how immunotoxic findings in rodents compare with immunotoxic effects in humans. The development of biomarkers of immunotoxicity for use in

the course of the clinical trials of novel medicinal products, as well as field and epidemiological studies of industrial and environmental chemicals, is essential. The value of available biomarkers of immunotoxicity is debatable in most instances and efforts should be undertaken to designing better endpoints (Vandebriel et al., 1996).

4. Models specifically designed to address immunotoxicity issues
The majority of available methods and assays for use in immunotoxicity evaluation have been derived directly from immunology. Some of these assays actually proved useful as illustrated by the plaque assay which was shown to be a very reliable indicator of immunosuppression in rodents (Luster et al., 1992). However, models specifically designed to address immunotoxicity issues are needed.

In vitro assays can be expected to fill the gap, but a more thorough understanding of immunotoxic effects is required to develop reliable in vitro screens to be used at an early stage of immunotoxicity evaluation. The popliteal lymph node assay which has so far been essentially used for the prediction of systemic autoimmune reactions, could also be developed to predict other aspects of immunotoxicity as an early screen for immunotoxicity as recently suggested (Bloksma et al., 1995).

5. Renewed immunotoxicological research can emerge from implementing regulations
Previous research work in immunotoxicology provided a fair amount of useful information as briefly described above. Regulations in immunotoxicology can be expected to have a major impetus on the development of immunotoxicology as many laboratories throughout the world, particularly in the industry, would have to generate data and in doing so, would gain experience and design new investigative methologies.

The issue of possible immunotoxicity is vividly present in the minds of toxicologists, be they from the industry, the Academia or regulatory agencies. Therefore, immunotoxicity is readily considered as a possible explanation to unexpected findings noted during conventional toxicity testing. As no systematic immunotoxicity evaluation is performed, it is often impossible to compare findings observed with one compound to findings observed with similar or dissimilar compounds, which is a major drawback in the interpretation of these findings. Therefore, the industry could take benefit from a more systematic immunotoxicity evaluation.

Conclusion

The potential immunotoxic effects of a number of xenobiotics have been evidenced. Although these effects were mainly investigated in laboratory animals, there is a sufficient body of evidence to show they can result in adverse consequences to the human health. Efforts have been paid to standardise and validate predictive methods. When no such methods are available, alternative assays can be proposed on a case-by-case approach.

However, when considering the need for regulating immunotoxicity evaluation, it is essential to recall that immunotoxicologists cannot address all issues, but then nobody would claim that genotoxicity evaluation should not be performed because of the on-going debate on which battery of tests is the most appropriate, or that carcinogenicity evaluation should not be performed because of the debated predictive value of mouse carcinogenicity assays. It could be helpful to catalogue those issues which can be reasonably addressed, and a classification of chemicals according to their established or suspected immunotoxicity or lack of toxicity could also be instrumental.

Based on what is known and what is unknown regarding immunotoxicity, realistic (i.e. both relevant and cost-effective) regulations can be proposed to be implemented. Despite all previous efforts, regulatory perspectives in immunotoxicology have not significantly changed during the last ten years (Descotes, 1986; De Waal et al. 1996) and unmet needs cannot be entirely blamed for this situation.

References

Bigazzi P (1997) Autoimmunity caused by xenobiotics. Toxicology 119, 1-21.

Bloksma N, Kubicka-Muranyi M, Schupe HC, Gleichmann E, Gleichmann H (1995) Predictive immunotoxicological test systems: suitability of the popliteal lymph node assay in mice and rats. Crit. Rev. Toxicol. 25, 369-396.

Descotes J (1985) Adverse consequences of chemical immunomodulation. Clin. Res. Pract. Drug Regul. Affairs 3, 45-52.

Descotes J (1986) Immunotoxicology : health aspects and regulatory issues. Trends Pharmacol. Sci. 7, 1-4.

De Waal EJ, Van der Laan JW, Van Loveren H (1996) Immunotoxicity of pharmaceuticals: a regulatory perspective. Toxicol. Environ. Newslet. 3, 165-172.

Freudenburg WR (1988) Perceived risk, real risk : social science and the art of probabilistic risk assessment. Science 242, 44-49.

Hilton J, Dearman RJ, Boylett MS, Fielding I, Basketter DA, Kimber I (1996) The mouse IgE test for the identification of potential chemical respiratory allergens: considerations of stability and controls. J. Appl. Toxicol. 16, 165-170.

Kerkvliet NI (1994) Immunotoxicology of dioxins and related chemicals. In: Dioxins and Health, A.Schecter ed. New York: Plenum Press, p.199-225.

Kimber I, Dearman RJ, Scholes EW, Basketter DA (1994) The local lymph node assay: developments and applications. Toxicology 93, 13-31.

Luster MI, Portier C, Pait DG, White KL, Gennings C, Munson AE, Rosenthal GJ (1992) Risk assessment in immunotoxicology. I. Sensitivity and applicability of immune tests. Fund. Appl. Toxicol. 18, 200-210.

Luster MI, Portier C, Pait DG, Rosenthal GJ, Germolec DR, Corsini E, Blaylock BL, Pollock P, Kouchi Y, Craig W, White KL, Munson AE, Comment CE (1993) Risk assessment in immunotoxicology. II. Relationships between immune and host resistance tests. Fund. Appl. Toxicol. 21, 71-82.

Mathews KP (1984) Clinical spectrum of allergic and pseudoallergic drug reactions. J. Allergy Clin. Immunol. 74, 558-566.

Revillard JP, Robinet E, Goldman M, Bazin H, Latinne D, Chatenoud L (1995) In vitro correlates of the acute toxic syndrome nduced by some monoclonal antibodies: a rationale for the design of predictive tests. Toxicology 96, 51-58.

Vial T, Descotes J (1996) Drugs affecting the immune system. In: Meyler's Side-Effects of Drugs. 13th edition. MNG Dukes ed. Amsterdam: Elsevier Science, p.1090-1167.

Vial T, Legrain B, Carleer J, Verdier F, Descotes J (1997a) The popliteal lymph node assay: results of a preliminary interlaboratory validation. Toxicology, in press.

Vial T, Nicolas B, Descotes J (1997b) Drug-induced autoimmunity. Experience of the French Pharmacovigilance system. Toxicology 119, 23-27.

Vos JG, Younes M, Smith E (1996) Allergic hypersensitivities induced by chemicals. Boca Raton: CRC Press.

Vandebriel RJ, Meredith C, Scott MP, Gleichmann E, Bloksma N, Vant'erve EHM, Descotes J, Van Loveren H (1996) Early indicators of immunotoxicity: development of molecular biological test batteries. Hum. Exp. Toxicol. 15, Suppl.1, 2-9.

Wedner HJ (1987) Allergic reactions to drugs. Primary Care, 14, 523-545.

White KL, Gennings C, Murray MJ, Dean JH (1994) Summary of an international methods validation study, carried out in nine laboratories, on the immunological assessment of cyclosporin A in the Fischer 344 rat. Toxicol. in vitro, 8, 957-961.

Novel Issues of Risk Assessment of Chemical Carcinogenesis

(Chair: H. Vainio, France, and T. Malmfors, Sweden)

Biomarkers in Metabolic Subtyping - Relevance for Environmental Cancer Control

Harri Vainio
International Agency for Research on Cancer, 150 cours Albert Thomas, F-69372 Lyon France

Abstract

People differ in their susceptibility to particular cancers and in their sensitivity to certain carcinogens. Differences in sensitivity to environmental carcinogens are determined by variations in genetic background - genetic polymorphism. Susceptibility or sensitivity factors can act at any stage in the multistage process of carcinogenesis, from exposure to carcinogens to the clinical appearance of cancer. This paper addresses the use and limitations of studies on human polymorphism for carcinogen metabolizing enzymes and their relevance for cancer control. The practical use of susceptibility and sensitivity markers in cancer control is not yet clear. Some of the potential dangers of their use are job discrimination and genetic exculpation - the 'blame the victim' attitude. Furthermore, an imprudent focus on genetic predisposition could shift the attention from carcinogens in the environment to biological 'defects' in the individual. Although carcinogens act in predisposed subjects, this should not overshadow the fact that most cancers are due to environmental factors, in both susceptible and unsusceptible individuals, and are therefore preventable. Even if individuals differ in their sensitivity to carcinogens, the primary option in cancer control must be to reduce exposure in order to include and protect the most sensitive fraction of the population.

Introduction

Controversy about the relative importance of nature and nurture, i.e. whether cancer is caused principally by genetic or environmental factors, has been raging for years. The fact that humans show various metabolic responses to environmental mutagens and carcinogens undermines any facile distinction between a 'purely' genetic or environmental etiology of cancer. Only a small fraction of cancers, however, are genetic in the sense that they are heritable, being passed from generation to generation via the DNA contained in the fertilized egg that gives rise to an embryo. These 'germ-line' cancers are sometimes passed from parent to offspring regardless of the external environment within which the individual develops. Cancers with such high penetrance are fairly rare; they include childhood cancers such as retinoblastoma, familial forms of colon and breast cancer, and cancers associated with certain rare genetic syndromes.

People have long been aware that cancer sometimes clusters in families. The classical problem with studies of hereditary cancers is to distinguish whether familial patterns are due to a shared genetic background or to shared environmental exposures. Thus, Henry Earle, a grandson of Percival Pott, conceded in 1823 that while chimney soot was doubtless the cause of the 'chimney sweepers' cancer, the disease also tended to run in families, suggesting a 'constitution predisposition' that 'renders the individual susceptible of the action of the soot'.

While the debate about 'nature or nurture' continues there is now a good body of evidence to suggest that the genetic and environmental theories are not incompatible. Soot, coal-tar, and tobacco tar, for example, induce cancer, but individuals are differently sensitive to the agents in these mixtures which cause cancer. One of the most pervasive dogmas in cancer research is that carcinogenesis is a multistage process (Vainio et al. 1992), which implies that it involves a number of distinct events that may be separated in time. It is likely that one or more of the factors the contribute to increased risks for most malignancies will be shown to be the enzymes that activate and inactivate the exogenous carcinogens. In this contribution, I shall not review each polymorphic enzyme that has been used as a marker of susceptiblity (for recent reviews, see e.g. Raunio et al. 1995; Nebert et al. 1996; Puga et al. 1997) but shall discuss selected enzymes briefly and address the practical and ethical aspects in more detail.

Genetic Susceptibility and Sensitivity

It is becoming increasingly clear that people are not homogeneous in their responses to carcinogens. For example, individuals who are heterozygous for ataxia telangiectasia appear to be not only *susceptible*, i.e. more likely to develop cancer 'spontaneously' in the absence of an obvious external factor, but are also more *sensitive* to radiation-induced cancer (Swift et al. 1991). For the purposes of cancer control and risk assessment, it is important to distinguish between interindividual susceptibility and sensitivity. Genetic susceptibility to cancer is indicated by an increase in the background frequency of a specific tumour type or of any tumour. Susceptible individuals carry a genetic alteration, such as a germ-line mutation in the tumour suppressor gene *p53*, that predisposes them to developing cancer in the absence of a readily identifiable exposure (Li 1995). Germ-line mutations in *p53* predispose to various childhood cancers and breast cancer in young women (Malkin et al. 1992; Kleihues et al. 1997). In contrast, sensitivity to a carcinogen predisposes individuals to a specific type of tumour or any tumour in response to exposure to e.g. carcinogenic chemicals or radiation, as is the case for patients with xeroderma pigmentosum, who are sensitive to solar radiation and have an increased risk for cancer (Bootsma and Hoeijmakers 1991). A single genetic alteration can confer either or both susceptibility and sensitivity. For some cancers, population subgroups are both susceptible and sensitive; for others, subgroups are either susceptible and not sensitive to a particular agent or sensitive but not susceptible (Preston, 1996).

Metabolic Subtyping for Increased Sensitivity: Theoretical Basis and Potential Limitations

Several dozen genetic polymorphisms for drug-metabolizing genes have been characterized, and some have been reported to be associated with enhanced risks for cancer (Gonzalez and Idle 1994; Nebert et al. 1996). A wide range of enzymes known or assumed to be toxicologically important have been shown to have substantially different levels of activity within the population; these include N-acetyltransferases, several cytochromes P450, and glutathione-S-transferases (Raunio et al. 1995; Nebert et al. 1996). Each of these enzymes has a potential role in the activation or inactivation of one or more potent carcinogens or other chemicals. Variations in the expression of these enzymes could, theoretically, account for different sensitivity to the effects of carcinogens. For instance, individual differences in the inducibility of aryl hydrocarbon hydroxylase activity in response to exposure to cigarette smoke may contribute to individual differences in the risk for lung cancer (Kellermann et al. 1973; Kouri et al. 1982), although convincing evidence is still lacking. Common variations in gene sequences can result in altered proteins and they are termed "polymorphisms". As the genetic loci of these and other metabolic enzymes have been determined, it is possible to identify polymorphisms and, therefore, genotypic and phenotypic differences in populations. All the consistent differences in cancer risk predicted theoretically have not yet been explained, but newer methods will allow the study of a wider array of polymorphisms. One limitation is the identification of individuals with well-documented exposure to carcinogens of interest and with established health outcomes, such as cancer; however, even when study groups are available, the differences between genotypes with a high and a low risk are typically threefold or less (Raunio et al. 1995; Hirvonen et al. 1996).

Oxidative metabolism of carcinogens: The superfamily of cytochrome P450 enzymes catalyses the oxidation of a large number of endogenous and exogenous chemicals. The P450 *CYP* genes in the families *CYP1*, *CYP2*, and *CYP3* code for enzymes that are primarily responsible for the metabolism of most procarcinogens and promutagens (Nebert 1991). The dual role of these enzymes in detoxification and metabolic activation has prompted a vigilant survey of *CYP* genes for polymorphisms likely to result in variable enzyme activities.

The human CYP1A subfamily comprises two enzymes, CYP1A1 and CYP1A2. Inducibility of CYP1A-related phenotypic activity can be regarded as an indirect quantitative measure of effects related to both exposure and host factors. Individuals vary widely in the extent to which CYP1A1 is induced, e.g. by tobacco smoking: 10% of Caucasians have been demonstrated to be highly inducible

(Nebert 1991). The inducibility of the CYP1A1 phenotype has been shown to be higher in lung cancer patients than in controls (Kellermann et al. 1973; Kouri et al. 1982). The *CYP1A1* genotype and its association with lung cancer risk has been studied extensively. In the Japanese population, a structural polymorphism affecting the size of *Msp* I fragments of the human *CYP1A1* gene appears to be associated with a higher incidence of lung cancer (Kawajiri et al. 1990) and with an isoleucine to valine mutation (I462V) in the haem-binding region of *CYP1A1* (Hayashi et al. 1991). Hayashi and coworkers (1992) subsequently demonstrated that this mutation, combined with a glutathione-S-transferase M1 null mutation, is associated with a nine-fold increased risk for lung cancer among cigarette smokers. These studies indicate that the human *CYP1A1* structural gene, or a region near this gene, might be correlated with the inducibility phenotype and with an increased risk for lung cancer. Studies of *CYP1A1* in other ethnic groups do not, however, support the Japanese findings: Norwegians, Finns, American Caucasians and Blacks, for instance, lack the association between the *Msp* I fragments, the I462V mutation, and lung cancer (Tefre et al. 1991; Hirvonen et al. 1992; Shields et al. 1992). Thus, although the *Msp* I fragments may explain some of the genetic predisposition to increased risk among Japanese individuals for particular types of cigarette smoke-induced cancer, it does not hold true for other ethnic groups.

N-Acetylation polymorphisms and cancer risk: The phase II enzymes, such as glutathione-S-transferases, N-acetyltransferases, and UDP-glucuronosyl-transferases, are involved mainly in the detoxification of various foreign substances. As they appear to have evolved as a defense system against toxic environmental challenges, genetic variation in their activity is likely to be of considerable toxicological significance (Raunio et al. 1995).

The N-acetyltransferases (NAT) will serve as an exemple of the action of phase II enzymes, since this class of enzymes has recently received a lot of scientific attention. N-Acetylation is a major conjugative metabolic step for many drugs and carcinogens, and genetic variation in acetylation processes is well characterized in humans. Furthermore, the human N-acetyltransferase functional genes (*NAT1* and *NAT2*) have been cloned, the *NAT2* gene encoding for the NAT2 enzymes, which are mainly responsible for the rapid- and slow-acetylator phenotypes. Worldwide, the frequency of the slow-acetylator phenotype ranges from 10% in Japanese to over 90% in some Mediterranean populations. NAT2 was known previously as polymorphic NAT, and NAT1 as monomorphic NAT. Recent evidence suggests, however, that allelic variation also exists in the *NAT1* gene (Bell et al. 1995); therefore, both genes should be taken into account in future studies when assessing the risk associated with acetylator phenotype.

Slow acetylators have an increased risk for urinary bladder cancer but a lower risk for colorectal carcinoma (Cartwright et al 1982; Gonzalez and Idle 1994). Cigarette smoking and occupational exposure to aromatic amines are important factors for the development of urinary bladder cancer. In attempts to correlate exposure with risk by identifying arylamine adducts in the haemoglobin of slow and fast acetylators, the risk of slow acetylators appears to be about twice that for fast acetylators exposed to low levels of tobacco smoke (Vineis et al. 1994).

There is evidence to suggest that the genetically determined acetylator phenotype of an individual also modulates the risk for colorectal cancer associated with meat intake (Vineis and McMichael 1996). A recent study from Australia indicates that the rapid-acetylator phenotype is associated with odds ratios of 1.1 for colorectal adenoma and 1.8 for colorectal cancer, and the risk for adenoma or cancer increased with increasing intake of meat in rapid but not in slow acetylators (Roberts-Thomson et al. 1996). Rapid acetylators may convert the heterocyclic amines in meat to reactive aryl nitrinium ions. In colonic mucosa, the N-hydroxy derivatives of arylamines are good substrates for O-acetylation by NAT2. Another recent study, based on hierarchical modelling, showed, however, that the risk for colorectal polyps increases with increasing consumption of red meat among genotypically slow acetylators, whereas the pattern was not as clear among genotypically fast acetylators (Aragaki et al. 1997). The authors concluded that "given the imprecision of both studies, the conflict could be due to random errors".

In a study in which caffeine metabolites were used to determine CYP1A2 and NAT2 phenotypes in patients with colorectal cancer and in controls, rapid CYP1A2 and NAT2 phenotypes were only slightly higher in cases than in controls, but the combined rapid CYP1A2 and rapid NAT2 phenotypes resulted in an odds ratio of 2.8 for the incidence of colorectal cancer or nonfamilial polyps (Lang et al. 1994).

Cigarette-smoking postmenopausal women with the slow-acetylator NAT2 genotype have been shown to be as much as four times more likely to develop breast cancer than those with the rapid-acetylator NAT2 allele (Ambrosone et al. 1996). No difference from controls was observed for premenopausal women with breast cancer.

Use of polymorphic markers in cancer control: Even if certain carcinogens could be shown to affect human risk as a direct function of measurable enzyme activity, the implications for disease control might still be limited. The two genes involved in acetyltransferase activity illustrate the complexities involved in epidemiological studies of polymorphisms in metabolic genes. Only when a small segment of the population has a marked excess risk could such knowledge be used either ethically or in the interests of public health to exclude people from work or to implement comparable host-based control strategies. A search for individuals at particularly high risk, such as those with xeroderma pigmentosum, who constitutively express certain oncogenes or harbour defective tumour suppressor genes closely associated with important cancer types, may be fruitful but would probably be relevant to only a small segment of the potentially exposed population. In contrast, there would be little practical benefit from using a marker of genetic polymorphism that is too widespread in the population. Factors associated with a small difference in risk, however theoretically or mechanistically important, are also of little practical use. In all of these situations, the primary control option must be to reduce exposure in order to include and protect the most sensitive fractions of the population.

Markers for Metabolic Subtyping and Risk Management

The recent fascination with genetic predisposition distracts attention from the fact that the incidence of cancer varies according to diet, occupation, socioeconomic status, and personal habits such as smoking (Tomatis et al. 1997). Genetics may eventually tell us who is more likely to develop certain kinds of cancer - everything else being equal, which it rarely is - but it will never tell us anything interesting about large-scale shifts in cancer patterns with changing exposures to carcinogens. Lung cancer rates, for example, have risen dramatically over the course of the century; genetic propensity can have little to do with this increase, or with the slight decline observed among males in recent years. Genetic propensity will not tell us why stomach cancer was more common during the first half of this century than the second. Barring diagnostic artefacts, such dramatic changes point suspicion at environmental changes as the primary culprit. Although carcinogens work with predisposition, this should not overshadow the fact that most cancers are environmentally induced and therefore preventable. Genetics will not explain why asbestos workers have higher rates of mesothelioma than people who work in air-conditioned offices, or why people who work in underground mines where there is radon seepage are more likely to contract lung cancer. Asbestosis is obviously not a genetic disease, but whether it progresses to lung cancer might be shaped, at least in part, by genetic factors that affect how well the damage caused by asbestos fibres can be repaired (Hirvonen et al. 1996). Even if individuals vary in their sensitivity to different agents, it is probably wishful thinking to imagine that physicians may one day be able to use profiles of genetic biomarkers to inform people whether they are at risk for common diseases such as cancer.

Increasing knowledge about individual differences may have a darker side. For example, an employer might use knowledge about the genotypes of the workers to hire only those at presumed lower risk for toxic or malignant effects due to exposure to particular agents in the workplace. Insurance companies might refuse to insure 'high-risk' persons. For these reasons, it is important to prevent abuse of information about individuals' genotype and phenotype.

Acknowledgements

I would like to thank Ms Elisabeth Heseltine for editing this manuscript and Mrs Agnès Meneghel for her help in its preparation.

References

Ambrosone CB, Freudenheim JL, Graham S, Marshall JR, Vena JE, Brasure JR, Michalek AM, Laughlin R, Nemoto T, Gillenwater KA, Harrington AM, Shields PG (1996) Cigarette smoking, N-acetyltransferase 2 genetic polymorphisms, and breast cancer risk. JAMA 276:1494-1501

Aragaki CC, Greenland S, Probst-Hensch N, Haile RW (1997) Hieararchical modeling of gene-environment interactions: estimating *NAT2** genotype-specific dietary effects on adenomatous polyps. Cancer Epidemiol Biomarkers Prev 6:307-314

Bell DA, Badawi AF, Lang NP, Ilett KF, Kadlubar FF, Hirvonen A (1995) Polymorphism in the N-acetyltransferase 1 (NAT1) polyadenylation signal: association of NAT1*10 allele with higher N-acetylation activity in bladder and colon tissue. Cancer Res 55:5226-5229

Bootsma D, Hoeijmakers JH (1991) The genetic basis of xeroderma pigmentosum. Ann Genet 34:143-150

Cartwright RA, Glashan RW, Rogers HJ, Ahmad RA, Barham Hall D, Higgins E, Kahn MA (1982) Role of N-acetyltransferase phenotypes in bladder carcinogenesis: a pharmacogenetic epidemiological approach to bladder cancer. Lancet 2:842-845

Gonzalez FJ, Idle JR (1994) Pharmacogenetic phenotyping and genotyping. Present status and future potential. Clin Pharmacokinet 26:59-70

Hayashi S, Watanabe J, Nakachi K, Kawajiri K (1991) Genetic linkage of lung cancer-associated MspI polymorphisms with amino acid replacement in the heme binding region of the human cytochrome P450IA1 gene. J Biochem Tokyo 110:407-411

Hayashi S, Watanabe J, Kawajiri K (1992) High susceptibility to lung cancer analyzed in terms of combined genotypes of P450IA1 and Mu-class glutathione S-transferase genes. Jpn J Cancer Res 83:866-870

Hirvonen A, Husgafvel Pursiainen K, Karjalainen A, Anttila S, Vainio H (1992) Point-mutational MspI and Ile-Val polymorphisms closely linked in the CYP1A1 gene: lack of association with susceptibility to lung cancer in a Finnish study population. Cancer Epidemiol Biomarkers Prev 1:485-489

Hirvonen A, Saarikoski ST, Linnainmaa K, Koskinen K, Husgafvel Pursiainen K, Mattson K, Vainio H (1996) Glutathione S-transferase and N-acetyltransferase genotypes and asbestos-associated pulmonary disorders. J Natl Cancer Inst 88:1853-1856

Kawajiri K, Nakachi K, Imai K, Yoshii A, Shinoda N, Watanabe J (1990) Identification of genetically high risk individuals to lung cancer by DNA polymorphisms of the cytochrome P450IA1 gene. FEBS Lett 263:131-133

Kellermann G, Shaw CR, Luyten Kellerman M (1973) Aryl hydrocarbon hydroxylase inducibility and bronchogenic carcinoma. N Engl J Med 289:934-937

Kleihues P, Schauble B, zur Hausen A, Esteve J, Ohgaki H (1997) Tumors associated with p53 germline mutations: a synopsis of 91 families. Am J Pathol 150:1-13

Kouri RE, McKinney CE, Slomiany DJ, Snodgrass DR, Wray NP, McLemore TL (1982) Positive correlation between high aryl hydrocarbon hydroxylase activity and primary lung cancer as analyzed in cryopreserved lymphocytes. Cancer Res 42:5030-5037

Lang NP, Butler MA, Massengill J, Lawson M, Stotts RC, Hauer Jensen M, Kadlubar FF (1994) Rapid metabolic phenotypes for acetyltransferase and cytochrome P4501A2 and putative exposure to food-borne heterocyclic amines increase the risk for colorectal cancer or polyps. Cancer Epidemiol Biomarkers Prev 3:675-682

Li FP (1995) The 4th American Cancer Society Award for Research Excellence in Cancer Epidemiology and Prevention. Phenotypes, Genotypes, and Interventions for Hereditary Cancers. Cancer Epidemiol Biomarkers Prev 4:579-582

Malkin D, Jolly KW, Barbier N, Look AT, Friend SH, Gebhardt MC, Andersen TI, Borresen AL, Li FP, Garber J, et al (1992) Germline mutations of the p53 tumor-suppressor gene in children and young adults with second malignant neoplasms. N Engl J Med 326:1309-1315

Nebert DW (1991) Role of genetics and drug metabolism in human cancer risk. Mutat Res 247:267-81

Nebert DW, McKinnon RA, Puga A (1996) Human drug-metabolizing enzyme polymorphisms: effects on risk of toxicity and cancer. DNA Cell Biol 15:273-280

Preston RJ (1996) Interindividual variations in susceptibility and sensitivity: linking risk assessment and risk management. Toxicology 111:331-341

Puga A, Nebert DW, McKinnon RA, Manon AG (1997) Genetic polymorphisms in human drug-metabolizing enzymes: potential uses of reverse genetics to identify genes of toxicological relevance. Crit Rev Toxicol 27:199-222

Raunio H, Husgafvel-Pursiainen K, Anttila S, Hietanen E, Hirvonen A, Pelkonen O (1995) Diagnosis of polymorphisms in carcinogen-activating and inactivating enzymes and cancer susceptibility--a review. Gene 159:113-121

Roberts-Thomson IC, Ryan P, Khoo KK, Hart WJ, McMichael AJ, Butler RN (1996) Diet, acetylator phenotype, and risk of colorectal neoplasia. Lancet 347:1372-1374

Shields PG, Sugimura H, Caporaso NE, Petruzzelli SF, Bowman ED, Trump BF, Weston A, Harris CC (1992) Polycyclic aromatic hydrocarbon-DNA adducts and the CYP1A1 restriction fragment length polymorphism. Environ Health Perspect 98:191-194

Swift M, Morrell D, Massey RB, Chase CL (1991) Incidence of cancer in 161 families affected by ataxia-telangiectasia. N Engl J Med 325:1831-1836

Tefre T, Ryberg D, Haugen A, Nebert DW, Skaug V, Brogger A, Borresen AL (1991) Human CYP1A1 (cytochrome P(1)450) gene: lack of association between the Msp I restriction fragment length polymorphism and incidence of lung cancer in a Norwegian population. Pharmacogenetics 1:20-25

Tomatis L, Huff J, Hertz Picciotto I, Sandler DP, Bucher J, Boffetta P, Axelson O, Blair A, Taylor J, Stayner L, Barrett JC (1997) Avoided and avoidable risks of cancer. Carcinogenesis 18:97-105

Vainio H, Magee P, McGregor DB, McMichael AJ (eds) (1992) Mechanisms of Carcinogenesis in Risk Identification. IARC Scientific Publication No. 116, International Agency for Research on Cancer, Lyon

Vineis P, Bartsch H, Caporaso N, Harrington AM, Kadlubar FF, Landi MT, Malaveille C, Shields PG, Skipper P, Talaska G, et al (1994) Genetically based N-acetyltransferase metabolic polymorphism and low-level environmental exposure to carcinogens. Nature 369:154-156

Vineis P, McMichael AJ (1996) Interplay between heterocyclic amines in cooked meat and metabolic phenotype in the etiology of colon cancer. Cancer Causes and Control 7:479-486

Assessment of Animal Tumour Promotion Data for the Human Situation

Lars Wärngård[1,2], Marie Haag-Grönlund[1], and Yvonne Bager[1].
[1] Institute of Environmental Medicine, Karolinska Institutet, Box 210, S-171 77 Stockholm, Sweden
[2] Astra AB, S-151 85 Södertälje, Sweden.

Introduction

There is considerable evidence that cancer development in humans and experimental animals includes different stages that are results of interactions between target cells and various endogenous/exogenous factors. In order to protect man from chemically induced cancer, the development of suitable test systems for the detection of carcinogenic potency has been and still is an important task of applied toxicological research. Conventional cancer bioassays in which rodents are exposed to a certain chemical over two years is commonly used to investigate possible carcinogens. This test procedure is costly with respect to time and money and makes use of a large number of animals.

To improve the precision in human risk assessment of carcinogenic substances, there is a need to integrate results from short term assays on molecules and cells with those from whole animal systems (Figure 1). In addition, results from such short term assays provide a better understanding of the mechanisms of carcinogenesis.

We have tried to apply this integrated approach to the risk assessment of chlorinated environmental pollutants. Three polychlorinated biphenyls with principally different structures have been chosen as model compounds; these are the dioxin-like 3,4,5,3',4'-pentachlorobiphenyl (PCB 126), the non dioxin-like 2,4,5,2',4',5-hexachlorobiphenyl (PCB 153) and the mixed inducing 2,4,5,3',4'-pentachlorobiphenyl (PCB 118) (Figure 2). For this purpose we have investigated the effect of these substances on intercellular communication, enzyme induction and the growth of enzyme altered hepatic foci in rat liver. In addition we have initiated a study with the specific aim to estimate the interactive effects of these chemicals when administered together.

Abbreviations: PCB, polychlorinated biphenyl; PCB 126, 3,4,5,3',4'-pentachlorobiphenyl; PCB 118, 2,4,5,3',4'-pentachlorobiphenyl; PCB 153, 2,4,5,2',4',5-hexachlorobiphenyl; CYP, cytochrome P-450; PH, partial hepatectomy; AHF, altered hepatic foci; EROD, 7-ethoxyresorufin-O-dealkylase; PROD, 7-pentoxyresorufin-O-dealkylase; cx, connexin; Ah, aryl hydrocarbon; GST-P, glutathione-S-transferase P; GJIC, gap junctional intercellular communication.

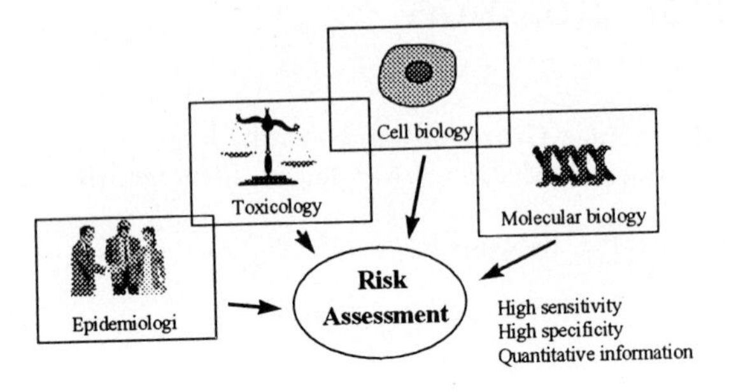

Fig. 1. The integration of various types of data.

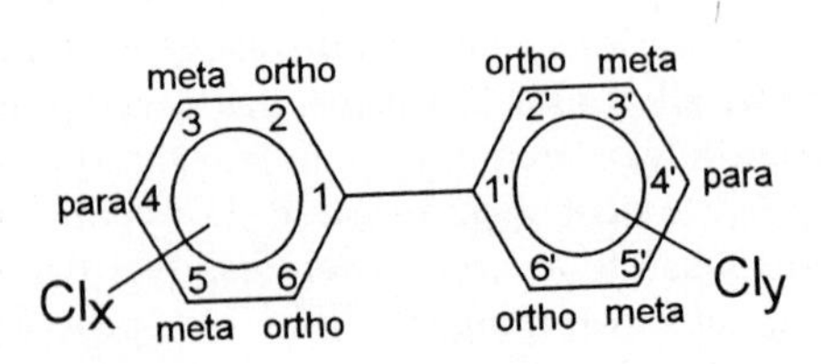

Fig. 2. The structure of the PCB molecule.

Materials and Methods

Test Compounds and Studied Parameters: Three PCB congeners were selected; the dioxin-like 3,4,5,3',4'-pentachlorobiphenyl (PCB 126), the non dioxin-like 2,4,5,2',4',5-hexachlorobiphenyl (PCB 153) and the mixed inducing 2,4,5,3',4'-pentachlorobiphenyl (PCB 118).

The parameters studied after long term exposure *in vivo* with these three congeners were growth of altered hepatic foci and the induction of CYP1A1 in rat liver (Hemming et al., 1993; 1995; Bager et al., 1995; Haag-Grönlund et al., 1997). In addition, these PCBs were also studied for their ability to alter gap junction proteins in rat liver (Bager et al., 1994; 1997a). Gap junctional intercellular communication in the non tumourigenic rat liver epithelial cell line IAR 20 was also studied (Bager et al., 1997b).

Enzyme Altered Hepatic Foci: The initiation/promotion assay described by Pitot (1978) was used to study the tumour promotive capacity of the PCBs in this study. The medium-term, two-stage, initiation/promotion protocols were originally designed to study the mechanisms of hepatic promoting agents and have been suggested as potential tools for the detection of non-genotoxic carcinogens. These assays are based on the assumption of multistage carcinogenesis, and that clonal expansion is a critical step in tumour promotion (Figure 3).

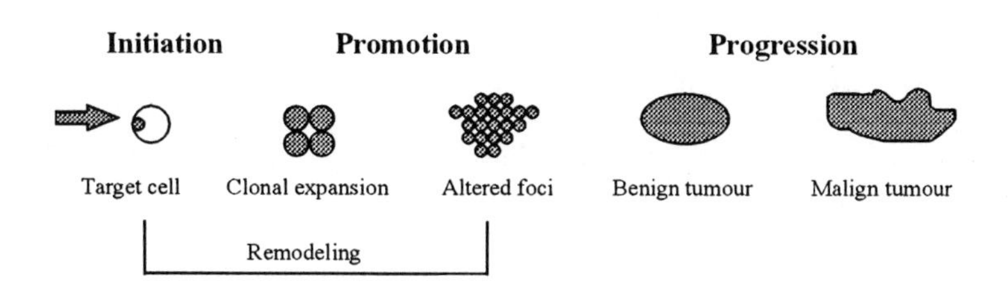

Fig. 3. A schematic illustration of the different steps in multistage carcinogenesis.

In these assays enzyme altered areas of the liver (altered hepatic foci, AHF) are visualized and quantified after treatment with chemicals. Chemically induced enhancement of such AHF is believed to indicate a tumour promoting ability of the chemical. An outline of the experimental protocol is shown in Figure 4. Female Sprague-Dawley rats were given a full initiation treatment, i.e., 2/3 partial hepatectomy (PH) followed by 30 mg N-nitrosodiethylamine/kg body weight (ip) 24 hours later. After 5 weeks recovery the promotion periods started by weekly subcutaneous injections with the test substance for 20 weeks (some groups of animals given PCB 118 and PCB 126 were administered for 52 weeks). The experimental details are presented in Hemming et al., 1993; 1995; Bager et al., 1995; Haag-Grönlund et al., 1997. Sections of rat liver were immunohistochemically stained for glutathione-S-transferase P (GST-P) as a marker of AHF (Haag-Grönlund et al., 1997) and evaluated for size and number of foci (Flodström et al., 1988).

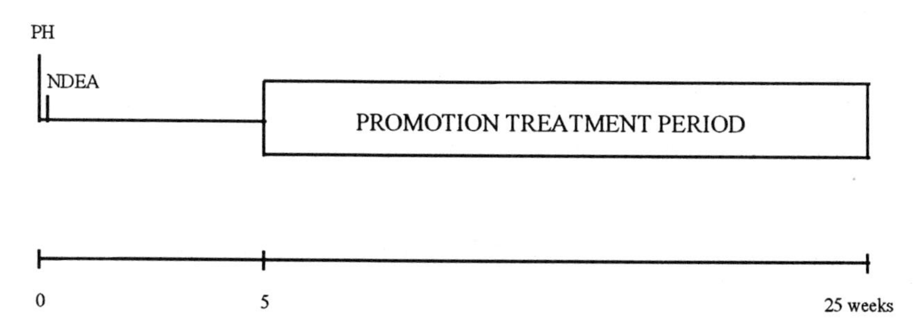

Fig. 4. Outline of the experimental protocol.

Biochemical Assays: Livers of all animals were analysed for induction of O-dealkylation of 7-ethoxyresorufin (EROD) as a marker of CYP1A1 activity essentially according to Hemming et al., 1993.

Gap Junctional Intercellular Communication: Gap junctional intercellular communication is a unique type of cell communication (Loewenstein, 1966). Intercellular communication is believed to be involved in many biological processes such as cell growth, differentiation and maintenance of cell homeostasis (Yamasaki, 1990). During carcinogenesis, growth control mechanisms are impaired, and several lines of evidence support that inhibition of gap junctional intercellular communication plays an important role.

Gap junctions consist of channels aggregated together forming plaques in the plasma membranes. Six gap junction proteins (connexin, cx) form a hemi-channel (connexon) that couples to a connexon in the neighbouring cell (Bennet et al., 1991). In the liver, the two major connexins expressed are cx 26 and cx 32.

Gap junctional intercellular communication can be examined by the spreading of a dye after microinjection. The dye most frequently used is Lucifer Yellow, a fluorescent dye, which is detected by a fluorescence microscope. The number of dye-coupled cells per injection is determined by counting the fluorescent cells surrounding the injected donor cell, allowing quantitative determination of intercellular communication (Figure 5). In liver sections, gap junctions were detected by using immunofluorescence staining (Bager et al., 1994; 1997a).

Results and Discussion

Promotion Related Effects Induced by PCB 126, PCB 153 and PCB 118: The studies demonstrate that PCB 126, PCB 153 and PCB 118 enhance the formation of altered hepatic foci and can thus probably act as tumour promoters in rat liver. However, the ability of these compounds to induce foci is highly variable. PCB 126 is the most potent tumour promoter of the compounds tested, whereas PCB 118 and PCB 153 have less potent tumour promoting capacity in rat liver (Hemming et al., 1993; 1995; Bager et al., 1995; Haag-Grönlund et al., 1997). In addition, administration of PCB 118 and PCB 126 for 52 weeks induced hepatocellular adenomas and carcinomas, respectively (Haag-Grönlund et al., 1997). This is in agreement with other studies, which have shown that mixtures of PCBs produce tumours in rat liver (Ahlborg et al., 1992).

An understanding of why the dioxin-like PCBs are much more potent tumour promoters than the non dioxin-like PCBs is essential for the risk assessment of the PCBs. It has been suggested that the Ah receptor mediates the biological effects, for example the induction of CYP1A1, induced by the dioxin-like PCBs (Safe, 1990). However, hepatic CYP1A1 activity, which is highly induced after PCB 126 exposure, is not directly correlated to enhancement of altered hepatic foci when data from individual animals are compared (Hemming et al., 1995). Additionally, PCB 118 treatment for 20 weeks clearly shows that foci development is not directly correlated to induction of CYP1A1, since after 20 weeks of treatment, EROD was increased 130-180 times without any significant increase in the percentage of the liver occupied by foci (Haag-Grönlund et al., 1997). These results support that neoplastic development is caused by multifactorial mechanisms and that enzyme induction *per se* is not sufficient for the formation of foci.

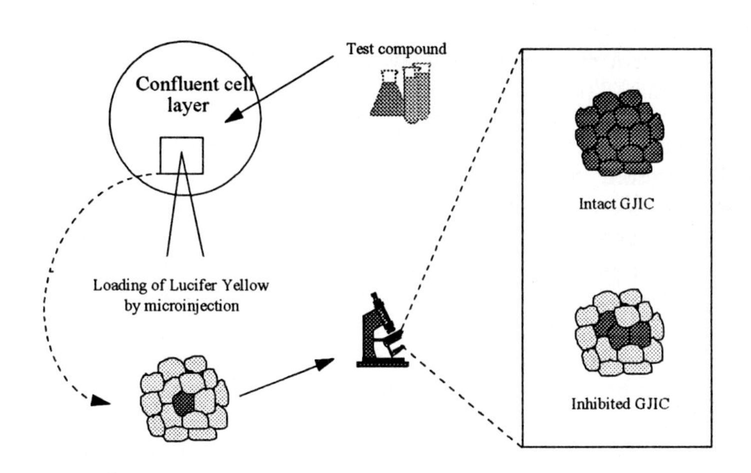

Fig. 5. A schematic illustration of the microinjection/dye transfer assay.

Interaction of PCBs: Since human exposure to PCBs is via a mixture of different PCB-congeners, it is also important to study the interaction of congeners after co-administration. We have initiated an interaction study with a non-*ortho* PCB (CYP1A1 inducing), a mono-*ortho* PCB (mixed-inducing) and a di-*ortho* PCB (CYP2B1 inducing). In this study, interactive effects will be investigated regarding i.e. the development of altered hepatic foci in rats. In order to effectively analyse the interactive effects, the study was designed for a multivariate analysis based on CCF (Central Composite Face centred analysis). Using multivariate analysis, a large amount of data can be evaluated at the same time and be summarized in a more understandable way. Covariation between variables, that are not immediately obvious from the data, can efficiently be detected with this method.

Gap Junctions *in vivo* and *in vitro*: Rat livers from studies using the initiation-promotion protocol previously described were investigated for possible differences in the ability of PCB-congeners (PCB 126, PCB 153 and PCB 118) to alter gap junctions. In rats, only exposure to the non-*ortho* substituted PCB 126 resulted in a reduction of gap junction proteins (cx 26 and cx 32) in the hepatocyte plasma membranes outside altered hepatic foci, and visible amounts of cx 43 began to appear after 20 weeks of promotion treatment. This increase in the cx 43 protein may have been a compensatory up-regulation in response to a decrease of other connexins in the liver (Bager et al., 1994; 1997a). Within foci, connexin expression was decreased after exposure to all tested substances. Furthermore, PCB 126 treatment of rats resulted in a general decrease of gap junction protein expression in the livers, preceding the development of altered hepatic foci (Bager et al., 1994).

All PCB-congeners tested down-regulated gap junctional intercellular communication in the rat liver epithelial cell line, IAR 20 (Bager et al., 1997b). However, the decrease in communication occurs after different exposure periods and different regulatory pathways of gap junctions are involved for the different PCB-congeners tested.

Structure-Activity Relationships of PCBs in Their Ability to Alter Gap Junctions Compared with Tumour Promotion Potency: The results are summarized in Table 1. In order to better understand the harmful effects of the different PCB-congeners, it is important that both the variation in tumour promotion potency of these compounds, and the mechanisms responsible for these variations are investigated. The ability of a compound to reduce gap junctions in rat liver may be important for its liver tumour promoting potency. This is demonstrated by results from the present study, which show that only the most potent tumour promoter, the non-*ortho* chlorine substituted PCB 126 (Hemming et al., 1995), reduces the amount of connexins in the parenchymal cell plasma membranes outside altered hepatic foci (Bager et al., 1994; 1997a). The less potent tumour promoters, the di-*ortho* chlorine substituted PCB 153 (Hemming et al., 1993; Bager et al., 1995) and the mono-*ortho* chlorine substituted PCB 118 (Haag-Grönlund et al., 1997), do not alter the gap junctions outside GST-P positive foci (Bager et al., 1997a). Thus, these results demonstrate that only the non-ortho chlorine substituted PCB-congener, PCB 126, cause a general decrease of the gap junction proteins in rat liver, which may in turn contribute to their high tumour promoting capacity.

All PCB-congeners tested down-regulate intercellular communication *in vitro*, but after different exposure periods and by different mechanisms (Bager et al., 1997b). These *in vitro* results support a promoting activity of these compounds *in vivo*, which is in agreement with previously performed studies.

Assessment of Animal Tumour Promotion Data for the Human Situation: Risk assessment of tumour promoters is a critical issue for environmental protection as well as for the chemical industry. In the future, there will certainly be a need for identification of the most important health risks and the most cost effective control strategies. The continuing change of focus from fairly high-level exposure of defined groups of people to low level exposure of almost the entire population, will have an impact on the risk assessment and management of environmental pollutants. Established dose-response relationships for environmental pollutants are mainly based on observations in high dose exposed rodents. The present knowledge about adverse effects that may occur at low level exposure is scarce, since classical methods for toxicity testing are not sensitive enough to provide accurate dose-response data in the low dose range with acceptable certainty. Reliable extrapolation from high to low doses requires knowledge about mechanisms over the whole dose-response range. Consequently, there is a need for mechanistically based short-term assays to bridge the gap between the mechanisms of responses at high and low doses.

Table 1. A brief summary of the ability of PCB-congeners to alter gap junctions compared to their tumour promotion potency.

Treatment	Down-regulate GJIC[a] *in vitro*	Act as tumour promoter in rat liver[b]	Alter gap junction expression in rat liver[c]	Tumour promotion potency in rat liver
PCB 126 non-*ortho*[d]	Yes	Yes	Yes	High
PCB 153 di-*ortho*[d]	Yes	Yes	No	Weak
PCB 118 mono-*ortho*[d]	Yes	Yes	No	Weak

[a] GJIC; gap junctional intercellular communication.
[b] Tumour promoter; shown by enhancement of the volume fraction of AHF.
[c] Outside GST-P positive foci after 20 or 52 weeks of promotion treatment.
[d] Chlorine substituted.

In the present investigation, we have used two different short-term assays for detecting potential tumour promoters representing both *in vivo* and *in vitro* test systems. It is important to stress that experiments *in vivo* are considered more reliable predictors of tumour promotion activity than are studies *in vitro*, which can never completely predict tumour promoting activity. If the molecular mechanisms involved in tumour promotion *in vivo* were characterized, *in vitro* tests could be developed to assess the ability of a certain chemical to induce one or more critical changes important for tumour promotion. Since such a complete characterization is still lacking, we have to rely on rather unspecific tests for tumour promotion. However, an intelligent use of short-term assays in combination with additional mechanistical data such as receptor interaction, induction of cell-proliferation, programmed cell death, change in gene or protein expression etc. could provide a platform for strategic decisions concerning the relative risk and potential thresholds.

We have investigated three polychlorinated biphenyls with principally different structures for their effect on intercellular communication and their potential to promote preneoplastic lesions in rat liver. The correlation found between inhibition of intercellular communication, induction of preneoplastic foci and induction of hepatocellular tumours were good. However, there was no strict correlation between the activation of the Ah-receptor and the enhancement of altered hepatic foci. The overall conclusion from these experiments is that the short term tests in combination with available mechanistic data could provide a good estimation of the relative risk of various PCBs to promote tumour growth in rat liver. Since humans are exposed to mixtures of PCBs, it is important to elucidate the interactive effects between various PCB-congeners. If the interactions are purely additive, potency data from studies on individual PCBs can be used for estimation of the adverse effects of defined mixtures of PCBs.

Our ability to extrapolate cancer risks from animal experiments conducted at high doses to humans, typically exposed to doses that are orders of magnitude lower, requires an understanding of the details of the metabolism of chemicals and their molecular mechanism of interaction at the target site. For obvious reasons, tumour promotion cannot be studied directly in humans. However, demonstration of the same mechanisms of toxicity for rodents and humans would support extrapolation of rodent data to humans, although the presence of additional mechanisms can never be excluded. A step forward in our understanding of the carcinogenicity of PCBs to humans could be achieved by using primary cell cultures for comparisons of mechanistically related parameters, such as inhibition of gap junctional intercellular communication, between rodents and humans. In this way, the existence of thresholds in animals and humans could also be studied. Such mechanistic comparisons between species at the cellular level could provide a necessary bridge between animal and human risk assessment of carcinogenicity of PCBs.

An ultimate confirmation of the relevance of animal carcinogenicity data to humans can only be achieved with epidemiologic studies of large groups of humans, but also these types of studies are impaired by a high degree of uncertainty.

References

Ahlborg UG, Hanberg A, Kenne K (1992) Risk assessment of polychlorinated biphenyls (PCBs). Nord 1992:26, Nordic Council of Ministers, ISBN 9291200751. pp 1-99

Bager Y, Kenne K, Krutovskikh V, Mesnil M, Traub O, Wärngård L (1994) Alteration in expression of gap junction proteins in rat liver after treatment with the tumour promoter 3,4,5,3',4'-pentachlorobiphenyl. Carcinogenesis 15:2439-2443

Bager Y, Hemming H, Flodström S, Ahlborg UG, Wärngård L (1995). Interaction of 3,4,5,3',4,'-pentachlorobiphenyl and 2,4,5,2',4',5'-hexachlorobiphenyl in promotion of altered hepatic foci in rats. Pharmacol Toxicol 77:149-154

Bager Y, Kato Y, Kenne K, Wärngård L (1997a) The ability to alter the gap junction protein expression outside GST-P positive foci in liver of rats was associated to the tumour promotion potency of different polychlorinated biphenyls. Chem-Biol Interact 103:199-212

Bager Y, Lindebro M-C, Martel M, Chaumontet C, Wärngård L (1997b) Altered function, localization and phosphorylation of gap junction intercellular proteins in rat liver epithelial, IAR 20, cells after treatment with PCBs or TCDD. Environ Toxicol Pharmacol, in press

Bennet MVL, Barrio LC, Bargiello TA, Spray DC, Hertzberg E, Saez JC (1991) Gap junctions: New tools, new answers, new questions. Neuron 6:305-320

Flodström S, Wärngård L, Ljungquist S, Ahlborg UG (1988) Inhibition of metabolic cooperation *in vitro* and enhancement of enzyme altered foci incidence in rat liver by the pyrethroid insecticide fenvalerate. Arch Toxicol 61: 218-223

Haag-Grönlund M,Wärngård L, Flodström S, Scheu G, Kronevi T, Ahlborg UG, Fransson-Steen R (1997) Promotion of altered hepatic foci by 2,3',4,4',5-pentachlorobiphenyl in Sprague-Dawley female rats. Fund Appl Toxicol 35:120-130

Hemming H, Flodström S, Wärngård L, Bergman Å, Kronevi T, Nordgren I, Ahlborg UG (1993) Relative tumour promoting activity of three polychlorinated biphenyls in rat liver. Eur J Pharmacol 248:163-174

Hemming H, Bager Y, Flodström S, Nordgren I, Kronevi T, Ahlborg UG, Wärngård L (1995) Liver tumour promoting activity of 3,4,5,3',4'-pentachlorobiphenyl and its interaction with 2,3,7,8-tetrachlorodibenzo-p-dioxin. Eur J Pharmacol 292:241-249

Loewenstein WR (1966) Permeability of membrane junctions. Ann NY Acad Sci 137:441-472

Pitot HC, Barsness L, Goldsworthy T, Kitigawa T (1978) Biochemical characterization of stages of hepatocarcinogenesis after a single dose of diethylnitrosoamine. Nature 271:456-457

Safe S (1990) Polychlorinated biphenyls (PCBs), dibenzo-p-dioxins (PCDDs), dibenzofurans (PCDFs) and related compounds: environmental and mechanistic considerations which supports the development of toxic equivalency factors (TEFs). Crit Rev Toxicol 21:51-88

Yamasaki H (1990) Gap junctional intercellular communication and carcinogenesis. Carcinogenesis 11:1051-1058

Use of Transgenic Mutational Test Systems in Risk Assessment of Carcinogens

Peter Schmezer, Claudia Eckert, Ute M. Liegibel, Reinhold G. Klein, and Helmut Bartsch
Division of Toxicology and Cancer Risk Factors, German Cancer Research Centre, Im Neuenheimer Feld 280, 69120 Heidelberg, Germany.

Abstract

Two transgenic in vivo mutation assays are described which are based on LacZ (Muta™ Mouse) and LacI (Big Blue™) shuttle vector systems. Their utility has already been explored by a number of investigators including our laboratory. The evaluation of data derived from these assays confirm that they offer a practical method for studying mutagenic activity and mechanism in a wide range of tissues including those of the respiratory and gastrointestinal tract. Therefore, these transgenic mutation assays are valuable tools to assess the organotropic effects of genotoxic carcinogens.

Introduction

The majority of human carcinogens act through mutational mechanisms (Shelby 1988; Bartsch and Malaveille 1989; Ashby 1996). Thus, in vivo mutagenicity assays play a critical role in the prediction of risk posed by genotoxic chemicals. Recent developments in transgenic technology have led to the increasing use of transgenic rodent mutation assays (Sullivan et al. 1993; Mirsalis et al. 1994; Forster 1995). While other systems are still in various stages of development, two commercially available mouse gene mutation assays based on lacZ and lacI lambda shuttle vector systems have been explored by a considerable number of investigators including our laboratory (Morrison and Ashby 1994; Schmezer et al. 1994). The following review summarises the methodology of the two most frequently used systems and discusses their major advantage, namely to allow in vivo mutagenicity studies in a wide range of rodent tissues.

Methods

LacI system (Big Blue™)

Two transgenic mouse strains (C57Bl/6 and B6C3F1; Kohler et al. 1991) and a transgenic rat (Fischer 344; Dycaico et al. 1994) have been developed carrying the lacI gene as mutational target for the in vivo mutation analysis. The constructed λLIZα shuttle vector is shown in Fig. 1.

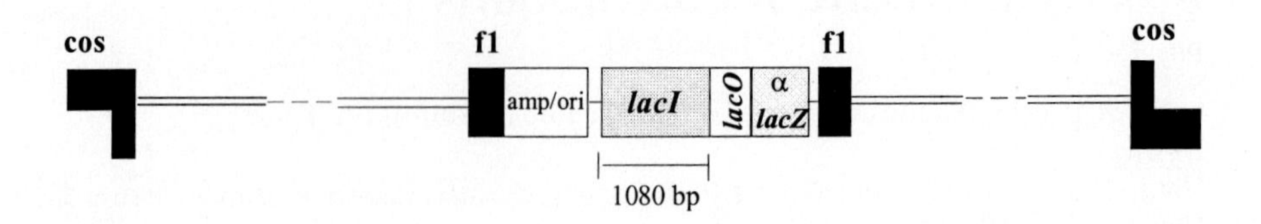

Fig. 1 λLIZα shuttle vector

The construct comprises the lacI gene, the lacI promotor, the operator sequence (lacO) and a lacZ fragment (αlacZ). The whole vector has a length of 45.6 kb, while the lacI gene measures 1080 bp. About 30-40 copies of the transgene construct are stably integrated in head-to-tail arrangement at a single locus on chromosome 4. After treatment of the transgenic rodents with the test agent and a subsequent observation (expression) time which allows fixation of mutations, genomic DNA is isolated from the relevant tissues/organs. The λ phage system serves as vehicle between the eucaryotic and the procaryotic cells.

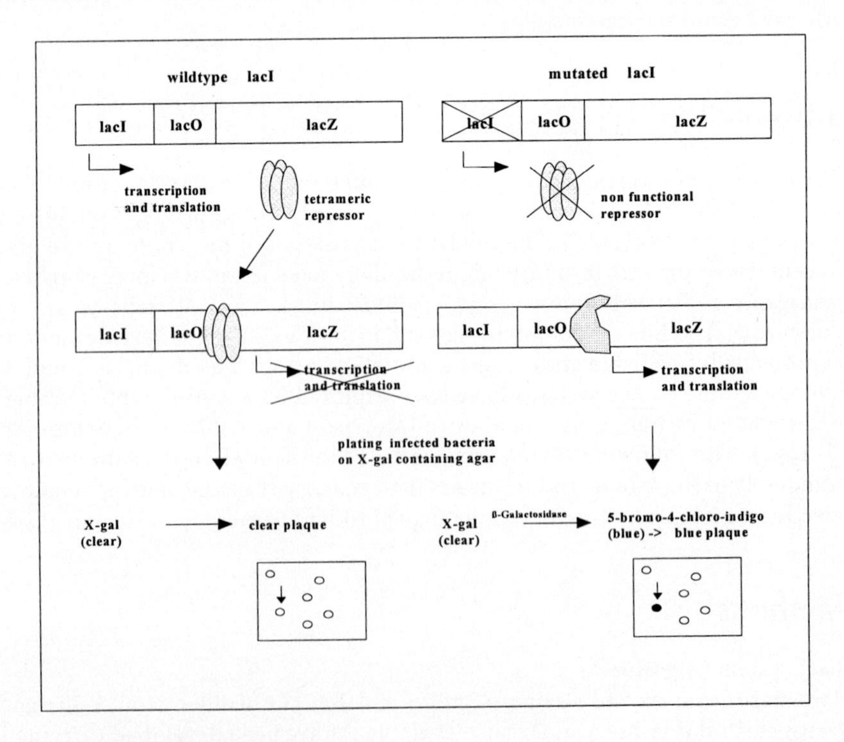

Fig. 2 Detection of lacI mutants

When the genomic DNA is mixed with λ packaging extract, a terminase recognizes and cleaves the *cos* sites flanking the vector on both sides, and single copies of the mutational target are packaged into infectious phage particles. This phage stock is subsequently allowed to adsorb to restriction-deficient *E. coli SCS8*, then mixed with top agar containing a chromogenic substrate (X-gal) and incubated overnight. Mutations in the lacI gene can be detected by using lacZ as reporter gene. The amino-terminal αlacZ of the vector complements the carboxy-terminal part of the *E. coli* host resulting in a functional β-galactosidase. LacI mutants are detected as illustrated in Fig. 2. Binding of the tetrameric non-mutated lacI repressor to the lacO inhibits the interaction of RNA polymerase with the promotor and thus the transcription of the lacZ gene (see left part of Fig. 2).

In the case of a mutation in the lacI gene, a non-functional repressor protein can result which does not bind to the promotor sequence and thus allows the expression of the β-galactosidase gene (see right part of Fig. 2). The produced enzyme cleaves X-gal and gives rise to a blue plaque. The mutation frequency is determined by counting the number of blue plaques per total plaque-forming units. This system is able to detect base substitutions, frameshift mutations, insertions and small deletions up to approximately 7.5 kb (Dycaico et al. 1994).

LacZ system (Muta™ Mouse)

Transgenic CD2F1 (BALB/c x DBA/2) mice have been developed carrying the lacZ gene as mutational target (Gossen et al. 1989). Homozygous animals contain about 80 copies of the λgt10lacZ shuttle vector (Fig. 3), integrated on chromosome 3. The whole vector has a length of about 47 kb and is flanked by *cos* sites. The lacZ target gene comprises 3.1 kb.

The colorimetric detection is following the same principle as shown above for the lacI system but, as the lacZ gene serves as mutational target, the wildtype plaques are blue and the mutants white. An improvement for the practical use of the lacZ system is the development of a rapid and convenient mutant selection system (Dean and Myhr 1994).

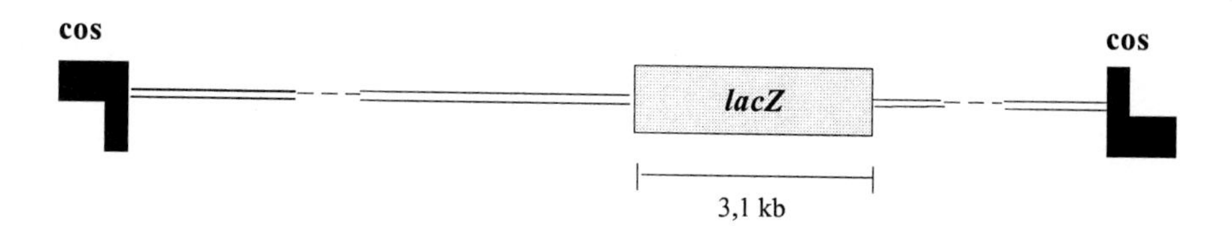

Fig. 3 λgt10lacZ shuttle vector

This selective system is based on the toxicity of galactose for *E. coli C galE⁻* bacteria which carry a mutation in the galE gene. These bacteria do not replicate in the presence of galactose because they contain an inactive UDP-epimerase which leads to the accumulation of toxic UDP-galactose (see left part of Fig. 4).

The wildtype lacZ expresses a functional β-galactosidase which accepts phenylgalactose (P-gal) in the medium as substrate to produce galactose. Host and viral replication are stopped before visible plaques are formed. In the case of a mutated lacZ leading to a non functional β-galactosidase, P-gal is not converted into the toxic metabolite and plaques can be produced on the bacterial plates (see right part of Fig. 4). Concurrent non-selective titer plates (absence of P-gal) are prepared in order to count the total number of phage particles and to determine the resulting mutation frequency.

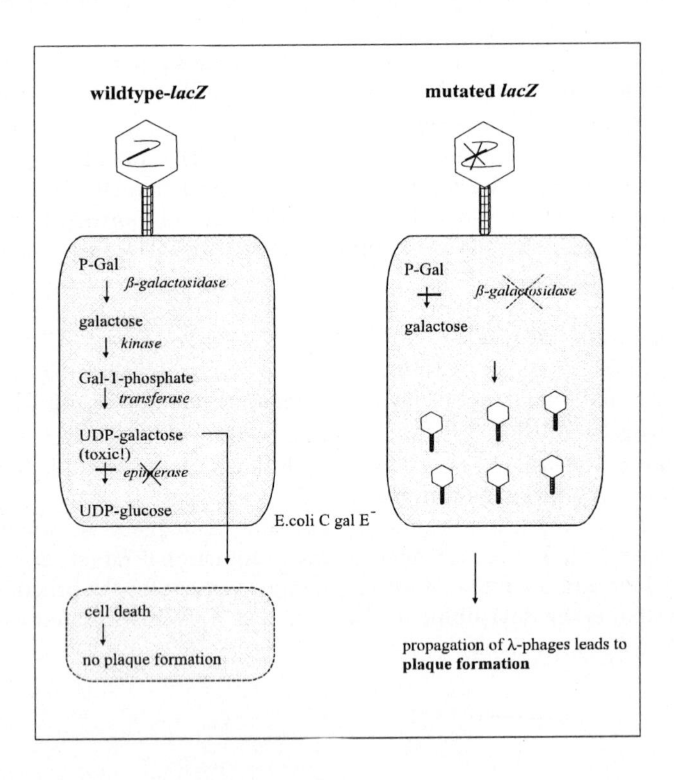

Fig. 4 Positive selection for lacZ mutants

Results and Discussion

It is obvious that in vivo assays are better models of the human situation than are in vitro assays because they include important factors such as uptake, distribution and metabolic activation of a chemical compound, its detoxification and excretion, as well as cell proliferation, sex and species differences, all of which can influence the induction of mutations in different tissues. Although it is not yet clear how good in vivo mutation assays are in predicting rodent or even human carcinogenesis (Heddle 1995), the detection of genetic toxicity of a test chemical in rodents is an important step to identify carcinogens (Ashby 1996).

The most frequent endpoints currently used in in vivo assays comprise the formation of micronuclei, chromosomal aberrations and unscheduled DNA synthesis. These endpoints are routinely evaluated only in bone marrow cells, peripheral blood and liver while other important target tissues e.g. of the gastrointestinal or the respiratory tract are generally not analyzed. In contrast, the newly developed transgenic rodent mutation assays can provide, as a major advantage, mutation data in a wide range of tissues/organs. They play therefore a valuable role in elucidating the mechanisms of tissue-specific genotoxicity and/or carcinogenicity (Gorelick and Mirsalis 1996). Another important aspect of these assays with regard to carcinogenesis is that heritable mutations are scored in viable cells one to several weeks after exposure, i.e. cells that may eventually die as a consequence of acute genotoxic effects are not likely to be assessed (Ashby 1996).

Tissue-Specific Induction of Mutations

Using C57Bl/6 mice bearing the lacI gene, we have investigated the organ-specific induction of mutations in vivo by the rodent carcinogen streptozotocin (Schmezer et al. 1994). Following parenteral administration, this compound is carcinogenic in the mouse kidney but not in the mouse liver, although DNA adducts have been formed in the latter organ. We confirmed the in vivo genotoxicity of streptozotocin in both organs by alkaline single cell microgel electrophoresis (comet assay). The mutation analysis, however, revealed the kidney to be more sensitive than the liver. At the same dose level, a 2-fold increase of the background mutation frequency was observed in the liver while in the target organ of carcinogenicity, the kidney, the induction was about 6-fold. Thus, the organotropic carcinogenesis induced by streptozotocin is better correlated with its tissue-specific mutagenicity than with its overall DNA damage.

In another series of experiments (Liegibel et al. 1995), we used the lacZ system to study the tissue-specific mutagenesis induced by N-nitrosodimethylamine (NDMA). This compound is a well known rodent liver carcinogen following oral application. In contrast, following chronic inhalative exposure of rats, hepatic tumors appear only in rare cases while a high incidence of nasal cavity tumors was induced (Klein et al. 1991). Lung tumors were not observed in the inhalation experiments. Therefore, we investigated whether this route-dependent shift in target organs correlates with the tissue-specific induction of mutations. The results of the inhalation experiments are summarized in Table 1.

There was a strong dose-dependent increase of the mutation frequency in the nasal mucosa with a 10-fold induction at 8.7 mg NDMA / kg bw. At the same dose level, only a doubling of the spontaneous rate in the liver was observed but no mutation induction in lung tissue. In contrast, after oral application (gavage) of 10 mg/kg bw. following the same protocol for mutation analysis as in the inhalation experiments, a 4-fold increase of the spontaneous mutation frequency was induced in the liver while there was no response in the nasal mucosa (Fig. 5).

Taken together, the experiments with NDMA demonstrate that the in vivo mutagenicity of this compound highly correlates with its tissue-specific carcinogenicity in rodents. Similar results have been reported by Suzuki et al. (1996) using the lacI system following intraperitoneal application of NDMA. They observed mutation induction in target organs (liver, kidney, lung) but not in non-target organs (bone marrow, bladder, testis) of NDMA carcinogenesis.

Table 1 Induction of mutations in the lacZ system (Muta™ Mouse) following NDMA exposure by inhalation

Tissue/ Dose[a] (mg/kg)	No. of Animals	Plaque-Forming Units	No. of Mutants	Mutation Frequency (x 10^{-5})
Nasal Mucosa				
control	9[b]	1,441,760[c]	122	8.5 ± 2.8
5.1	9[b]	1,531,330[c]	1030	67.3 ± 29.7
8.7	4[b]	358,154[c]	303	84.6
Lung				
control	9	2,848,200	218	7.7 ± 3.5
5.1	9	3,333,661	301	9 ± 4.2
8.7	4	1,355,388	91	6.7 ± 1.5
Liver				
control	9	3,023,667	254	8.4 ± 3.1
5.1	9	2,860,223	376	13.1 ± 2.4
8.7	4	1,403,113	251	17.9 ± 6

[a] tissues were analyzed from 8-weeks old male mice after a single 2 hr inhalation with the given dose for total uptake; the expression time was 14 days

[b] only small tissue samples can be explanted from the mouse nasal cavity; therefore, pooled tissue samples from 2-4 animals were analyzed

[c] a lower number of plaque-forming units was examined due to small tissue samples available

Other mechanistic studies in which these transgenic rodent mutagenesis assays have been successfully applied to detect tissue- and species-specific actions of carcinogens are summarized by Gorelick and Mirsalis (1996). o-Anisidine, a potent urinary bladder carcinogen in mice and rats, gave negative results in standard rodent genotoxicity assays (Ashby et al.1991). Evaluation of the mutagenicity of this compound to the mouse bladder (lacI system) gave consistent evidence for a positive response (Ashby et al. 1994). The rat liver carcinogen tamoxifen has also been identified as a rat liver mutagen in lacI F344 rats (Davies et al. 1996). The species-specific liver carcinogen aflatoxin B_1 induces

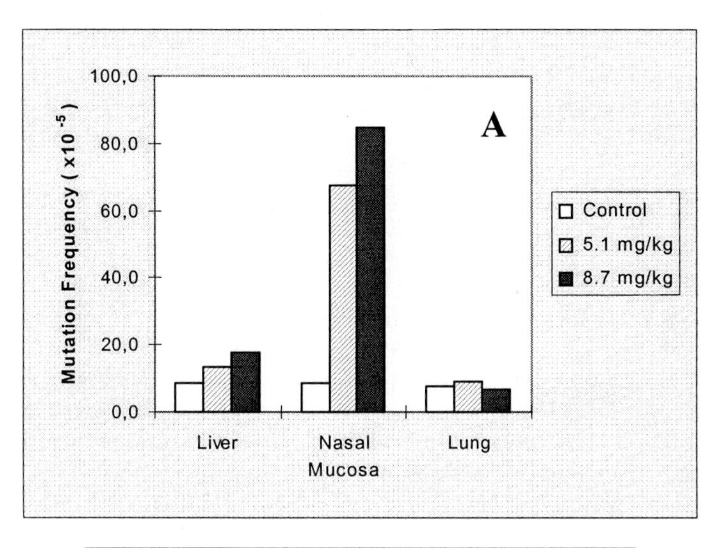

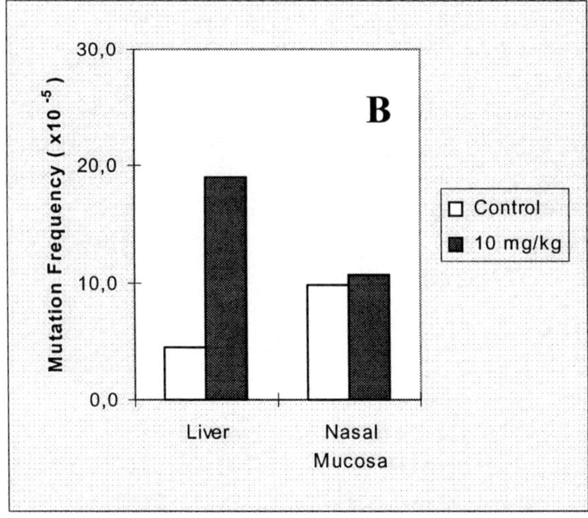

Fig. 5 Tissue-specific induction of mutations by NDMA, A: inhalative exposure, B: oral application (gavage)

hepatic tumors in rats but is non-tumorigenic in the mouse liver. Using the lacI system, Dycaico et al. (1996) found hepatic mutagenesis in rats but not in mice. The direct-acting, alkylating agents N-methyl-N'-nitro-N-nitrosoguanidine and β-propiolactone are known to act at the site of administration and therefore induce tumors in stomach following oral treatment of rodents. Brault et al. (1996) used the lacZ system to demonstrate the gastric specificity of these genotoxic carcinogens.

These examples as well as others reported in the current literature demonstrate that both the lacI (Big Blue™) and the lacZ (Muta™ Mouse) system are suitable for a wider use as in vivo rodent mutagenicity assays. This is further supported by the good qualitative agreement resulting from published interlaboratory comparisons. These include a large collaborative evaluation of transgenic mouse germ cell mutation assays (Ashby et al. 1997; Liegibel and Schmezer 1997), and an *EU Environment Programme*-sponsored interlaboratory study on mouse liver in which we have also been participating (Tinwell et al 1995). In an EU-sponsored project (in collaboration with J. Ashby, Alderley Park, UK, and V. Thybaud, Vitry-sur-Seine, France), we are further evaluating and applying these transgenic systems to identify presumptive human carcinogens by analyzing selected chemicals in a wide range of rodent tissues. Hereby it is assumed that a hierarchy of events leads to chemically-induced carcinogenesis. This hierarchy distinguishes the ability of a chemical to damage DNA, measured by the comet assay, from its ability also to induce mutations, measured in rodent mutation assays, and tumors, evaluated in hemizygous p53 mice.

In conclusion, our summary report demonstrates already that the two transgenic rodent mutagenicity assays have opened new possibilities for performing tissue-specific mechanistic studies. As valuable tools they should play an important role in the prediction of risks posed by genotoxic carcinogens, although they are still undergoing development. Important issues such as optimal dose selection and treatment regimen for detecting tissue-specific mutagenicity have to be further defined.

Acknowledgement

The financial support of the EU Environment Programme (Contract #EV5V-CT92-0212 and #ENV4-CT96-0200) is gratefully acknowledged.

References

Ashby J (1996) Alternatives to the 2-species bioassay for the identification of potential human carcinogens. Human Exp Toxicol 15:183-202

Ashby J, Lefevre PA, Tinwell H, Brunborg G, Schmezer P, Pool-Zobel BL, Shanu-Wilson R, Holme JA, Soderlund EJ, Gulati D, and Wojciechowski JP (1991) The non-genotoxicity to rodents of the potent rodent bladder carcinogens o-anisidine and p-cresidine, Mutation Res 250:115-133

Ashby J, Short JM, Jones NJ, Lefevre PA, Provost GS, Rogers BJ, Martin EA, Parry JM, Burnette K, Glickman BW, and Tinwell H (1994) Mutagenicity of o-anisidine to the bladder of lacI transgenic B6C3F1 mice: absence of ^{14}C or ^{32}P bladder DNA adduction. Carcinogenesis 15:2291-2296

Ashby J, Gorelick NJ, and Shelby MD (1997) Mutation assays in male germ cells from transgenic mice: overview of study and conclusions. Mutation Res 388:111-122

Bartsch H, and Malaveille C (1989) Prevalence of genotoxic chemicals among animal and human carcinogens evaluated in the IARC monograph series. Cell Biol Toxicol 5:115-127

Brault D, Bouilly C, Renault D, Thybaud V (1996) Tissue-specific induction of mutations by acute oral administration of N-methyl-N'-nitro-N-nitrosoguanidine and β-propiolactone to the Muta™ Mouse: preliminary data on stomach, liver and bone marrow. Mutation Res 360:83-87

Davies R, Martin EA, White INH, Smith LL, and Styles JA (1996) Tamoxifen causes gene mutations, DNA adducts and hyperplasia in the livers of Big Blue™ transgenic F344 rats. Environ Molec Mutagen 27 (Suppl 27):16

Dean SW and Myhr B (1994) Measurement of gene mutation in vivo using Muta™ Mouse and positive selection for lacZ- phage. Mutagenesis 9:183-185

Dycaico MJ, Provost GS, Kretz PL, Ransom SL, Moores JC and Short JM (1994) The use of shuttle vectors for mutation analysis in transgenic mice and rats. Mutation Res 307:461-478

Dycaico MJ, Stuart GR, Tobal GM, de Boer JG, Glickman BW, and Provost GS (1996) Species-specific differences in hepatic mutant frequency and mutational spectrum among lambda/lacI transgenic rats and mice following exposure to aflatoxin B_1.. Carcinogenesis 17:2347-2356.

Forster R (1995) Measuring genetic events in transgenic animals. In: Phillips DH, Venitt S (eds) Environmental Mutagenesis. ßios Scientific Publishers, Oxford, pp 291-314

Gorelick NJ and Mirsalis JC (1996) A strategy for the application of transgenic rodent mutagenesis assays. Environ Molec Mutagen 28:434-442

Gossen JA, de Leeuw WJF, Tan CHT, Zwarthoff EC, Berends F, Lohman PHM, Knook DL, and Vijg J (1989) Efficient rescue of integrated shuttle vectors from transgenic mice: a model for studying mutations in vivo. Proc Natl Acad Sci 86:7971-7975

Heddle JA (1995) In vivo assays for mutagenicity. In: Phillips DH, Venitt S (eds) Environmental Mutagenesis. ßios Scientific Publishers, Oxford, pp 141-154

Klein RG, Janowsky I, Pool-Zobel BL, Schmezer P, Hermann R, Amelung F, Spiegelhalder B, and Zeller WJ (1991) Effects of long-term inhalation of N-nitrosodimethylamine in rats. In: O'Neill IK, Chen J, and Bartsch H (eds) Relevance to Human Cancer of N-Nitroso Compounds, Tobacco Smoke and Mycotoxins, International Agency for Research on Cancer, Lyon, Scientific Publication No. 105, pp 322-328

Kohler SW, Provost GS, Fieck A, Kretz PL, Bullock WO, Putman DL, Sorge JA, and Short JM (1991) Analysis of spontaneous and induced mutations in transgenic mice using a lambda ZAP/lacI shuttle vector. Environ Mol Mutagen 18:316-321

Liegibel UM, Schmezer P, Klein RG, Kuchenmeister F, and Bartsch H (1995) Tissue-specific induction of mutations in transgenic mice after dimethyl nitrosamine inhalation. Environ Mol Mutagen 25 (Suppl 25):30

Liegibel UM and Schmezer P (1997) Detection of the two germ cell mutagens ENU and iPMS using the lacZ/transgenic mouse mutation assay. Mutation Res 388:213-218

Mirsalis JC, Monforte JA, and Winegar RA (1994) Transgenic animal models for measuring mutations in vivo. Crit Rev Toxicol 24:255-280

Morrison V, and Ashby J (1994) A preliminary evaluation of the performance of the Muta™ Mouse (lac Z) and Big Blue™ (lac I) transgenic mouse mutation assays. Mutagenesis 9:367-375

Schmezer P, Eckert C, and Liegibel UM (1994) Tissue-specific induction of mutations by streptozotocin in vivo. Mutation Res. 307:495-499

Shelby MD (1988) The genetic toxicity of human carcinogens and its implications. Mutation Res 204:3-15

Sullivan N, Gatehouse D, and Tweats D (1993) Mutation, cancer and transgenic models: relevance to the toxicology industry. Mutagenesis 8:167-174

Suzuki T, Itoh T, Hayashi M, Nishikawa Y, Ikezaki S, Furukawa F, Takahashi M, and Sofuni T. (1996) Organ variation in the mutagenicity of dimethylnitrosamine in Big Blue™ mice. Environ Mol Mutagen 28:348-353

Tinwell H, Liegibel U, Krebs O, Schmezer P, Favor J, and Ashby J (1995) Comparison of lacI and lacZ transgenic mouse mutation assays: an EU-sponsored interlaboratory study. Mutation Res 335:185-190

Assessing Toxicological Impacts in Life Cycle Assessment

Stig Irving Olsen and Michael Z. Hauschild
Life Cycle Engineering Group, Technical University of Denmark, Building 403, DK-2800 Lyngby (Denmark)

Introduction

The focus of the traditional approach of regulating toxic emissions to the environment has been on single processes and plants. In Denmark, this approach has been very successful in decreasing the amounts of toxics released from industrial facilities (Danish EPA, 1996). However, it is a common perception that the overall environmental exposure of humans to toxics is not decreasing. It is therefore necessary to look into the more diffuse sources of toxics in the life cycle of products. Life Cycle Assessment (LCA) is an environmental management tool that covers the entire life cycle of a product from the exploitation of raw materials to the final disposal of the product. LCA aims to include all relevant environmental impacts which can be attributed to that product, primarily as a tool for resource conservation and pollution prevention. The LCA methodology has been subject of intensive development in the last decade[1].The present paper introduces the framework for assessing toxic impacts in LCA and presents a methodology developed for this purpose.

The Life-Cycle Assessment Framework

One primary objective of LCA is to provide a realistic presentation of the potential environmental impacts of a service throughout the whole life-cycle of that service. Impacts are caused by the inputs and outputs, hereafter called the environmental interventions, from the chain of processes connected with the service. Most often this service is provided by a product (see Figure 1). To be able to compare the services provided by different products, the concept of a

[1] The Society of Environmental Toxicology and Chemistry (SETAC) is one of the most active organizations in the development (see e.g. Consoli et al., 1993, Fava et al., 1991, 1993, Guinée et al., 1993, Udo de Haes et al., 1994, Udo de Haes, 1996) but development goes on in other fora as well (e.g. Lindfors et al., 1995, Hauschild & Wenzel, 1997). Furthermore, the framework and methodologies of LCA are presently being standardized in the International Standardization Organisation, ISO, in the ISO 14000-series (ISO, 1997).

functional unit is introduced. This functional unit could be e.g. the packaging of 1000 l of milk, which would be equally fulfilled by the use of 1000 disposable polyethylene bottles or of 100 reuseable glass bottles, assuming a 10 times reuse. For each process in the system, the environmental interventions are quantified and allocated to one functional unit. Next, the interventions from all processes are aggregated to an inventory of the whole system. This inventory forms the starting point for the assessment of the product's environmental impacts.

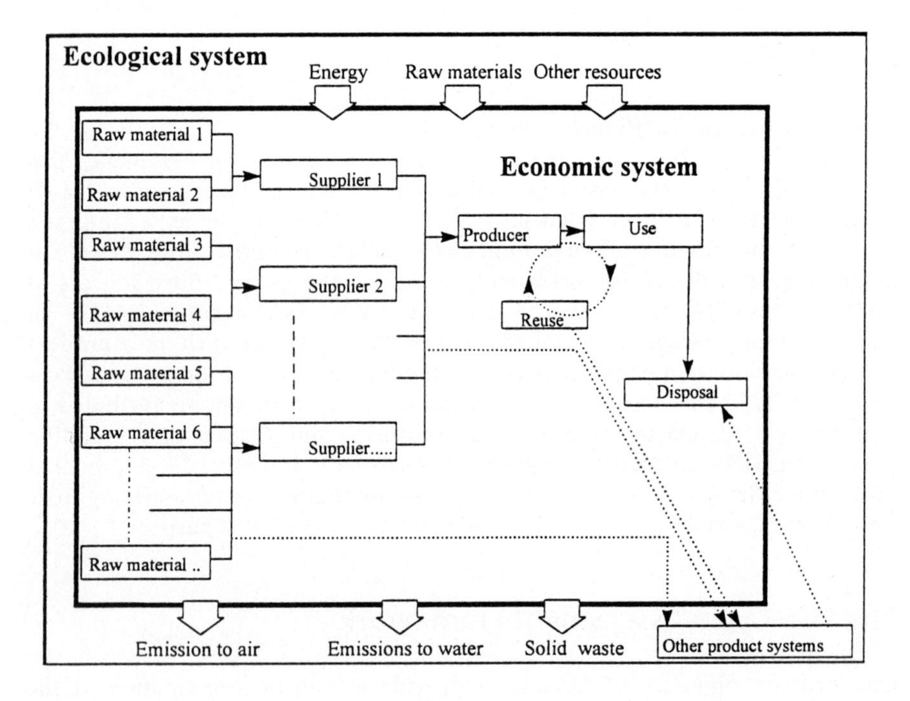

Figure 1 Schematic overview of a product system

The environmental relevance of the system's environmental interventions are assessed in terms of their impact on (Consoli et al., 1993):

- Resources
- Ecological health
- Human health

These very broad categories of environmental impacts have to be broken down into different impact categories because it is difficult, if not impossible, to establish cause-effect relationships between the environmental interventions and the three endpoints mentioned[2]. It is commonly accepted methodology to relate the inventory data to a surveyable number of "environmental themes" or impact categories (see Table 1).

Ideally, impact assessment should list and quantify all potential effects caused by the life-cycle of the studied system. Some general criteria to be applied in life-cycle impact assessment are (Klöpffer, 1994):

- Objectivity (based on science)
- Feasibility within the framework of an LCA
- Transparency and repeatability

Table 1: Examples of impact categories considered in LCA. The categories can be further divided into subcategories. Modified from Udo de Haes (1996)

Impact categories	
Abiotic resources	Photo-oxidant formation
Biotic resources	Acidification
Resources - Land	Eutrophication (inkl. BOD and heat)
Global warming	Odour
Depletion of stratospheric ozone	Noise
Human Toxicological impacts	Radiation
Ecotoxicological impacts	Casualties

Several environmental impacts considered in LCA have a common effect mechanism within each of the impact categories, e.g. ozone depletion or acidification. Furthermore, the number of substances contributing to the impact is limited and a list of these substances can be made. For these impact categories, the contribution of a particular substance to the effect can also be expressed in terms of an equivalent contribution from a reference substance (e.g. CFC-11 equivalents for ozone-depletion or CO_2-equivalents for global warming). For the global impact categories “global warming” and “stratospheric ozone-depletion”, equivalency factors for different compounds have been assessed and developed by international bodies (e.g. IPCC, 1995). The equivalency factor expresses the potential contribution to these impacts based on fate and effect considerations, and corresponding equivalency factors for other impact categories have been developed as well (e.g Guinee et al. (1993), Hauschild & Wenzel (1997)). Figure 2 shows an example of the application of equivalency factors in LCA.

In principle, all substances are toxic and the substances exhibit a wide range of effect mechanisms even within the same type of effect e.g. carcinogenesis or

---

[2] All environmental interventions forms part of an impact web where each intervention between the environment and the process (e.g. a substance emitted to air) is assigned to an effect. One substance may have several effects and each effect may be caused by several substance, hence the web.

Figure 2 Example of the application of equivalency factors

$$\textit{Impact score}_{cat} = \sum_{subs} \textit{equivalency factor}_{cat,subs} \times \textit{emission}_{subs}$$

To give an example: The total emission of CH_4, from a product's life-cycle is 30 g ($emission_{CH_4}$ = 30 g). CH_4 contributes to Global Warming, and the *equivalence factor*$_{GW,\ CH_4}$ is 25 kg CO_2/kg CH_4 i.e., each kg of CH_4 contributes as much to global warming as 25 kg of CO_2. So the impact score of CH_4 for Global Warming is 0.030 kg CH_4 x 25 kg CO_2/kg CH_4 = 0.75 kg CO_2. This can be done for all substances contributing to Global Warming and all the CO_2-equivalencies can be aggregated/added to yield the system total potential contribution to Global Warming.

teratogenesis. It is therefore not possible to make a list of contributors to toxic effects and it is not possible to relate the effect of all substances to one reference substance having the same effect mechanism as is done for other impact categories.

Condition for Assessing Toxicological Impacts in LCA

A methodology for assessing toxicological impacts in the LCA-framework must of course be feasible within this framework. The LCA-framework imposes several requirements to the methodology:

1. *The potential contribution to toxicological impact of each single emission in the inventory must be expressed independently of site-specific and temporal parameters.* When assessing toxicological risks, knowledge of the exposure (i.e. amount released, fate and partitioning in the environment etc.) and the effect (toxicokinetics, effect mechanism(s), dose-response relation) is paramount to the evaluation. NRC (1994) deals with the importance of these issues and the uncertainties involved in a comprehensive way. With respect to these parameters, the most important limitations to toxicological evaluations imposed by the LCA-framework described above are:
    - The inventory is related to a functional unit which generally is chosen arbitrarily. Thus, the magnitude of the impact is a relative value not related to actually occurring impacts at the given sites of the processes considered.
    - The actual impacts at a given site are related to the nature and number of processes at that site. Many other products will generally be produced by these processes. Therefore, only some of the interventions at the site will be allocated to the product

under study.

- The inventory is generally aggregated over the entire life-cycle thus abstracting from temporal and spatial characteristics of the emissions.

These limitations are primarily concerned with the exposure part of the assessment and restricts the evaluations of toxicological impacts to be based on the inherent properties of the substances in combination with some standard scenarios, thus representing the hazard potential of the substance.

2. *The potential contribution to toxicological impact must, for each single emission in the inventory, be expressed as an additive figure.* To give a representation of the total potential toxicological impact of a products system it must be possible to add all contributions to yield a single or a few figure(s). The most prominent problems associated with this requirement are:

- that the impacts are added irrespective of their mechanism of effect and
- impacts of different severity are added, implicitly giving the same weight to all effects

There have been proposals to subdivide the impact category "Toxicological impacts" into several subcategories, which would represent effect mechanisms and severity of effects more homogeneously (Burke et al., 1996, Jolliet, 1996).

3. *It must be operational on a limited database.* The estimated number of chemicals being used in industry is over 100,000. All of these chemicals will participate in the life cycle of some product, but only a few thousand are thoroughly investigated toxicologically. Apart from pharmaceuticals and pesticides, which are heavily regulated, High Production Volume (HPV) chemicals (>1000 t/year) are expected to be the best investigated ones. It has been estimated by the European Chemical Bureau that for the 2000-2500 European HPV chemicals, acute toxicity data are available for approximately 90%, whereas toxicity data for other effects are available for only 20-60% of the chemicals, depending of the type of effect (Bro-Rasmussen et al., 1996). In general, the physical and chemical properties necessary for generic exposure assessment are better available or easier to estimate.

Methodologies for Assessing Toxicological Impacts in LCA

So far, methodologies for assessing toxicological impact have primarily been developed for exposure via the environment, i.e. in-door and occupational exposures have not been included. At least three methodologies have been made operational in a quantitative way. They differ primarily by their method of estimating the fate of compounds in the environment. For the estimation of the toxicological effects of these compounds, the initially occurring adverse effects (the effect occurring at the lowest dosage/concentration level, i.e. NOAEL or LOAEL) are used as basis.

The CML-methodology has been developed by Guinée et al. (1996). The exposure estimation is built on the USES model (RIVM, 1994) which has been developed for generic risk assessment of chemical compounds and is now being applied as a tool for this purpose in the European Union. Based on the inherent properties of a compound, modeling of the fate of this compound in the environment is performed using a fugacity model with a standard scenario (see e.g. Mackay (1991) for further explanation). For the estimation of indirect exposure of humans (i.e. through food, beverages etc.) an intake model using bioconcentration or biotransfer factors has been employed, estimating the concentration of a substance in drinking water, fish, crops, beef, and milk (De Nijs and Vermeire, 1990). These factors mostly rely on the octanol-water partition coefficient (P_{ow}) as does the fugacity model. This kind of modeling is only valid for non-ionic organic substances, but for other kinds of substances, e.g. metals, default values have been used as input for the model. To meet the requirement of being additive, all equivalency factors derived from the model are related to the equivalency factor for 1,4-dichlorobenzene emitted to air, thus presenting equivalency factors as 1,4-dichlorobenzene-equivalents (Guinée et al., 1996).

The Critical Surface-Time methodology developed by Jolliet (1994) determines empirically the overall response of the environmental system and to calculates a fate factor as the ratio of the measured ambient concentration to the corresponding total emission flow for Switzerland. Jolliet and Crettaz (1996) have generalised this methodology to a world level. This approach is most suitable for well-known pollutants, where concentration measurements are available for large areas (e.g. NOx, SOx, VOC etc.). Orders of magnitude of the fate factors for other pollutants (presently 100) have been extrapolated as a function of their life times. They can also be combined with models to include inter-media transfer. To meet the requirement of being additive, all equivalency factors derived from the model are related to the equivalency factor for lead emitted to air, thus presenting equivalency factors as lead-equivalents.

The EDIP (Environmental Design of Industrial Products) methodology developed by the authors and described in detail below uses a critical volume concept, i.e. the volume of the environmental compartment (air, soil, or water) polluted to the No Effect Concentration for that compartment is calculated.

The EDIP Methodology for Assessing Toxicological Impacts in LCA

The EDIP methodology includes fate considerations and effect consideration based on the inherent properties of the substances emitted. An overview of the parameters included is given in Figure 3, which shows the general principles of the methodology applied to each environmental compartment (indicated by subscript *c*).

The emitted amount (in g) of a given substance can be directly multiplied with the equivalency factor (unit: m^3/g) for that substance to give the impact potential (unit: m^3) of the emission

$$EP_{c,i} = EF_{c,i} * Q_i$$

where $EP_{c,i}$ is the environmental impact potential in environmental compartment c of an amount Q of substance i, and $EF_{c,i}$ is the equivalency factor for substance i in compartment c. In words, EP_c could be explained as the volume of compartment c polluted up to the level with no adverse toxicological effect in humans.

To calculate the total potential toxicity in compartment c, the impact potentials of all the substances are added:

$$EP_c = \sum_{i=1}^{n}(EF_{c,i} * Q_i)$$

Figure 3 Determination of equivalency factors for a compound involves fate and effect considerations (Hauschild et al., 1997)

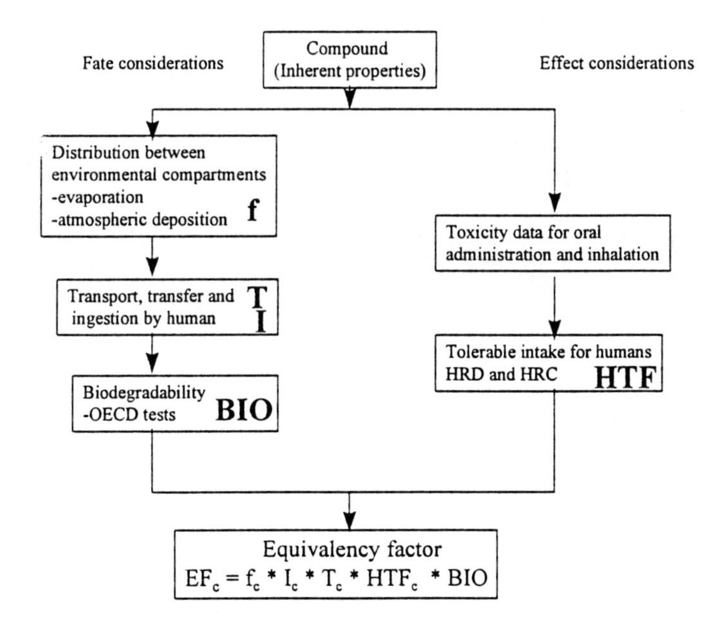

f_c is the distribution factor, expressing distribution of the substance emitted to environmental compartment c between other environmental compartments

T_c is a transfer factor expressing transfer of a substance into the foodchain of humans. For direct exposure it is 1

I_c expresses the average daily intake of air, food, water etc.

BIO expresses the biodegradability of the substance

HTF is an expression for the toxicity of the substance

Human exposures to substances through the following routes are considered(see also Figure 4):

- Air (by inhalation)
- Surface water (through ingestion of fish and shellfish)
- Soil (direct ingestion, ingestion of vegetables, meat or milk-products from organisms exposed to the soil)
- Ground water (direct ingestion)

Below a short explanation is given of the considerations that are taken into account when calculating the individual parameters composing the equivalency factor $EF_{c,i}$. Subscripts c and i, are environmental compartment c and substance i, respectively.

The distribution factor $f_{c,i}$ is introduced into the calculations because substances emitted to one compartment may contribute to toxicity in other compartments depending on the inherent properties of the substance and the environmental processes involved, e.g. a volatile substance emitted to water will also contribute to toxicity through air. The numerical value of $f_{c,i}$ lies between 0 and 1 and is based on information on the substance half-life in air, τ, Henry's law constant, H, and the relative percentage of water and soil surface in the area considered.

The intake factor I_c gives a Danish average daily intake (Danish National Food Agency, 1990) of the seven media shown in Figure 4 per kg body weight (average 70 kg assumed). The intake factors are shown in Table 2.

Figure 4 The exposure routes included in the model for humans in the environment (Hauschild et al., 1997)

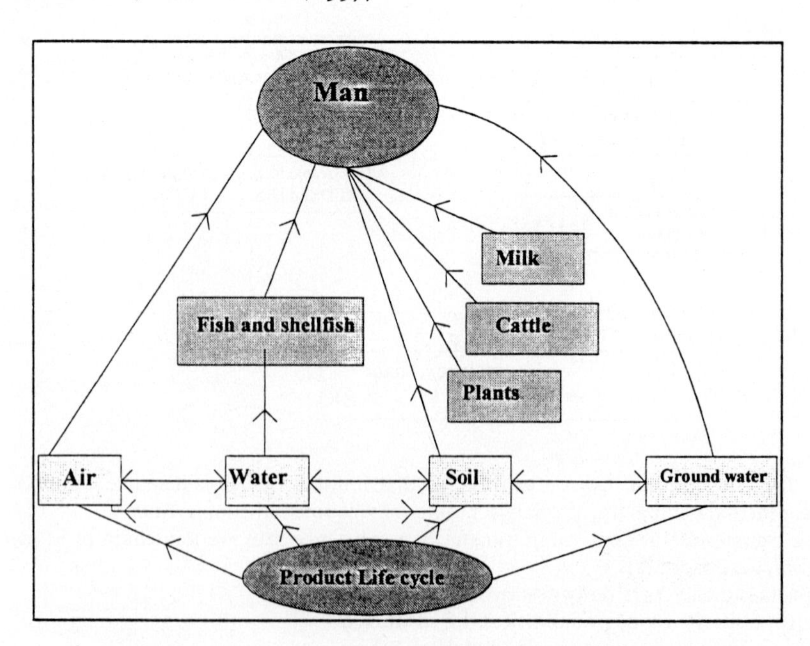

The transfer factor $T_{c,i}$ is introduced to account for concentration or dilution of the substance in the media taken in, in relation to the environmental compartment. For example, a substance for which the final environmental compartment is surface water may be taken up by fish and shellfish (which are eaten by man) in an amount characterised by $T_{c,i}$. The expressions used for calculating $T_{c,i}$ are show in Table 3.

The bioconcentration factor BCF can be found directly from experimental studies or it may be estimated from the octanol-water partition coefficient.

The biodegradability factor, BIO, can take the values of 0.2, 0.5, or 1, corresponding to readyly biodegradable, inherently biodegradable, and not biodegradable, respectively, as determined by OECD- or EEC-test-guidelines.

The stem concentration factor, SCF, and the transfer factors from soil or plant to meat (B_b) and milk (B_m) are based on the work by De Nijs and Vermeire (1990) and the literature their studies were based on. For non-ionic compounds the following relationships were found, primarily from an empirical basis:

$$SCF = (0.82 + 10^{(0.95 \log P_{ow} - 2.05)}) * (0.748\, e^{-[(\log P_{ow} - 1.78)^2 / 2.44]})$$

$B_b = 2.5 * 10^{-8} * P_{ow}$, and

$B_m = 7.9 * 10^{-9} * P_{ow}$

For other substances either literature values or default values are used.

The adsorption coefficient, K_d, can be found from experimental studies or be estimated from P_{ow}.

Table 2: Determination of I_c (Hauschild et al., 1997)

Final environmental compartment	Exposure through	I_c	Unit
Air	direct	1 or 0.286[1]	dimensionless or m^3 air/kg bw/d
Surface water	fish/ shellfish	$3.71*10^{-4}$	kg fish/kg bw/d
Soil	direct plants meat milkprod.	$2.86*10^{-6}$ $9.30*10^{-3}$ $1.53*10^{-3}$ $1.32*10^{-3}$	kg soil/kg bw/d kg plant/kg bw/d kg meat/kg bw/d kg milkprod./kg bw/d
Ground water	direct	$2.86*10^{-2}$	l water/kg bw/d

[1] If $HTF_{air,i}$ is based on air-concentrations from inhalation studies, then I_{air} = 1 (dimensionless), while I_{air} = 0,286 m^3/kg bw/d, if $HTF_{air,i}$ is based on oral studies.

Table 3 Determination of $T_{c,i}$ (Hauschild et al., 1997)

Final environmental compartment	Via	$T_{c,i}$	Unit
Air	direct	1	-
Surface water	fish/ shellfish	BCF	kg water/ kg fish
Soil	direct	1	-
	plants	$\frac{SCF}{K_d + \frac{f_w^*}{\rho_{st}}}$	kg soil/ kg plant
	milkprod.	$\frac{B_m * BI_p * SCF}{K_d + \frac{f_w^*}{\rho_{st}}} + B_m * BI_s$	kg soil/ kg milk
	meat	$\frac{B_b * BI_p * SCF}{K_d + \frac{f_w^*}{\rho_{st}}} + B_b * BI_s$	kg soil/ kg meat
Ground water	direct	1	-

BCF is the substance's bioconcentration factor from surface water to fish/shellfish
BIO expresses the biodegradability of the substance
SCF is the substance's bioconcentration factor from soil liquid to plant shoots (stem concentration factor)
K_d is the substance's adsorption coefficient in soil
f_w^* is the average water content of soil based on volume - default value 0,4 l soil liquid/l soil
ρ_{st} is the average density of soil - default value 1,5 kg dry matter soil/l soil
B_b is the substance's transfer factor from soil or plant to meat in cattle
B_m is the substance's transfer factor from soil or plant to milk from cows
BI_s or BI_p is the average intake of soil or plants by cattle; 0,41 and 16,9 kg/day, respectively.

The toxicity factor, *HTF*, expresses the substance's toxicity to humans following exposure via one of the routes shown in Figure 4. The No Effect Level in man is estimated from NOAEL, LOAEL, LD_{50}, or other toxicological parameters found in toxicological databases, primarily in RTECS (1997), HSDB (1997) and IRIS (1997), using assessment factors as shown in Table 4. The resulting value is called Human Reference Dose/Concentration (HRD/HRC) and is parallel to ADI-values assessed by JECFA (FAO/WHO) and Reference Dose (RfD) assessed by the US EPA, although not based on as thorough an evaluation as these values are. HTF is defined as the reciprocal of HRD/HRC.

$$HTF_{c,i} = \frac{1}{HRD_i \text{ or } HRC_i}$$

The assessment factors, except for those extrapolated from acute studies, are similar to those used by the US EPA when setting RfD-values (Barnes and Dourson, 1988). The reason for including assessment factors even for acute studies, thus allowing determination of HRD from acute data, is that for many substances occurring in the inventory of an LCA only limited toxicological data are available, and the methodology must be able to include these substances as well. For a detailed description and guidance for using the methodology reference is given to Hauschild et al. (1997), where equivalency factors for more than 100 substances have been calculated.

Finally, it could be mentioned that the methodology includes procedures for normalisation of the toxicity effect potential, i.e. relating the *EP* of a product system to the average toxicity potential emitted by one person per year, thus expressing the toxicity potential in person equivalents, which can be compared between impact categories. Also included is a procedure for weighing different impact categories against each other based on political reduction targets or environmental carrying capacity (Hauschild et al, 1997).

Discussion

The most important simplification and assumptions related to the toxicological aspects of the methodology are:

- Assumption of additivity of all toxic effects irrespective of the effect's mechanism and severity of the effect. This simplification is being met by a trend of developing subcategories for toxic impacts (e.g. irreversible, possibly reversible, and reversible effects as proposed by Burke et al. (1996)). The limited data availability for toxicological parameters may hamper the application of such an approach and a more pragmatic approach could be to subcategorise according to the data availability, and thus the certainty of the effect.
- All contributions to toxic impact are included even if they may be below a threshold for effect, implicitly assuming linearity of the toxic response. It has been argued, that probably all substances in an LCA inventory will occur at concentrations below threshold levels, and that linearity therefore is justified (Potting and Hauschild, 1997). Additionally, because of the NOAEL approach, no

Table 4 Determination of assessment factors based on the quality and relevance of the toxicity data available (Hauschild et al., 1997)

Criteria	Assessment factor, AF
Extrapolation from LC_{50} or LD_{50} from experimental studies in animals	10^5
Extrapolation from LC_{Lo} or LD_{Lo} from experimental studies in animals	5*104
Extrapolation from LOAEL from subchronic experimental studies in animals	10^4
Extrapolation from LC_{Lo} or LD_{Lo} from acute observations in human	$5*10^3$
Extrapolation from subchronic experimental studies in animals (duration < 1 year) or extrapolation from LOAEL in chronic experimental studies in animals (duration> 1 year)	10^3
Extrapolation from validated chronic experimental studies in animals or extrapolation from LOAEL from observations in humans or extrapolation from lowest irritative concentration in humans	10^2
Extrapolation from NOAEL in validated long-term observations in humans	10

consideration is given to the slope of the dose-response curve. Approaches have been forwarded proposing an only above-threshold approach (White et al. (1995), Hogan et al. (1996)), but such an approach is not applicable in the LCA-framework due to the limitations mentioned previously.

• The use of assessment factors for acute toxicological studies does assume that there is a relation between the effect mechanism in acute and chronic toxic responses. In addition, this procedure may give too much weight to sparsely tested substances in comparison to those substances which have been tested thoroughly. Furthermore, in risk assessment, or when setting limit values, assessment factors are used to provide a virtually safe dose. In LCA the aim is merely to estimate a realistic no-effect-level. Consequently, the assessment factors may be overly conservative.

Furthermore, all emitted substances are given the same weight even though there may be site-specific differences in their potential effect. There is a trend towards developing methodologies to include some site-dependency in a generic way (Dokkum et al. (1997), Potting et al. (1997)) in order to reach a more fair presentation of the potential effect.

Conclusion

The outlined methodology indeed represents a simplistic view of reality. Scientists in the fields of environmental chemistry, ecotoxicology and toxicology would all agree that the equivalency potentials calculated for each substance does not give a satisfactory presentation of the actual toxicological impact imposed by that substance at a given locality. But the methodology must be feasible in an LCA framework, for which one of the limitations is that it is only possible to characterise the hazard potential of a substance. Through the fate considerations, the methodology does estimate a potential exposure of humans to the emitted substance, and the toxicity factor does give an estimate of the doses/concentrations, which may have adverse effects on humans. The rationale for applying the developed methodology is that a simple estimation of the potential toxic impacts is better than neglecting toxic impacts in LCA, because such negligence could lead to decisions which would not improve the environmental performance of the product system.

References

Barnes DG and Dourson M (1988) Reference Dose (RfD): Description and Use in Health Risk Assessments. Regulatory Toxicology and Pharmacology 8: pp. 471-486

Bro-Rasmussen F et al. (1996) The non-assessed chemicals in EU. Report and recommendations from an interdiciplinary group of Danish experts. Report from the Danish Board of Technology 1996/5

Burke TA et al. (1996) Human Health Impact Assessment in Life Cycle Assessment: Analysis by an Expert Panel. International Life Sciences Institute, Health and Environmental Sciences Institute, Life Cycle Assessment technical Committee. Draft July 1. 1996

Consoli, F et al. (1993) Guidelines for Life-Cycle Assessment: A "Code of Practice". Society of Environmental Toxicology and Chemistry. Edition 1

Danish EPA (1996a) A strenghtened product oriented environmental policy. In Danish (En styrket produktorienteret miljøpolitik, Et debatoplæg. Miljøstyrelsen)

Danish EPA (1996b) Status and perspectives for the area of chemical. In Danish (Status og perspektiver for kemikalieområdet, et debatoplæg. Miljøstyrelsen)

Danish National Food Agency (1990) National monitoring of foods, Nutrients and pollutants 1983-1987. Danish National food Agency, Soeborg, Denmark

De Nijs ACM and Vermeire TG (1990) Soil-plant and plant mammal transfer factors. Report no. 670203001, National Institute of Public Health and Environmental Protection (RIVM), Bilthoven, The Netherlands

Dokkum HP et al (1997) Location specific assessment within LCA - a pilot study. Presented at the Seventh Annual Meeting of SETAC in Amsterdam, April 1997

Fava, J et al. (1993) A conceptual framework for Life-cycle Impact Assessment. Workshop report from the Sandestin, Florida, USA, workshop held February 1992. SETAC, March, 1993

Fava, J et al. (1991) A Technical Framework for life-cycle Assessment. SETAC workshop report from a workshop in Smugglers Notch, Vermont, August 18-23, 1990

Guinée, JB et al. (1993(a)) Quantitative Life Cycle Assessment of Products: 1. Goal definition and Inventory. Journal of Cleaner Production Vol. 1, No. 1: pp. 3-13

Guinée, JB et al. (1993(b)) Quantitative Life Cycle Assessment of Products: 2. Classification, valuation and improvement analysis. Journal of Cleaner Production Vol. 1, No. 2: pp. 81-91

Guinée JB et al. (1996) LCA Impact Assessment of Toxic Releases, Generic Modeling of fate, exposure, and effect for ecosystems and human beings with data for about 100 chemicals. Report 1996/21, Ministry of Housing, Spatial Planning and Environment. The Hague, The Netherlands

Hauschild M, Olsen SI, Wenzel H (1997) Human toxicity as a criterion in the environmental assessment of products. Chapter 7 in Hauschild M and Wenzel H (Eds.) Environmental assessment of products. Vol 2. Scientific background. Chapman & Hall, United Kingdom. In press

Hauschild MH and Wenzel, H(1997) Environmental Assessment of Products. Volume 2: Scientific background. Chapman & Hall, United Kingdom. In press

Hogan LM, Beal RT and Hunt RG (1996) Threshold Inventory Interpretation Methodology - A case Study of Three Juice Containers systems. Int. J. LCA Vol. 1(3) 1996: pp. 159-167

HSDB (1997) Hazardous Substance Data Bank. Published by National Library of Medicine, USA. Available on CD-rom

IPCC (1995) Technical summary forming part of the contribution of working group I to the IPCC second assessment report (submitted by the chairman of the IPCC for acceptance by the panel). Eleventh session, Rome 11-15 December, 1995. Intergovernmental Panel on Climate Change, Geneva, 1995.

IRIS (1997) Integrated Risk Information System. Elaborated by the US EPA. Available on CD-rom

ISO, 1997 Environmental Management - Life Cycle Assessment - Principles and Framework. Final Draft International Standard ISO/FDIS 14040

Jolliet O (1994a) Critical Surface Time: A Valuation method for Life-cycle Assessment from emission to concentration. In Udo de Haes HA, Jensen AA, Klöpfer W,and Lindfors L-G (Eds.): Integrating Impact Assessment into LCA, SETAC-Europe, Brussels, Belgium

Jolliet O (1994b) Impact Assessment of Ecotoxicity and Human Toxicity in Life-Cycle Assessment, including exposure and fate: Critical Surface Time II. Paper prepared for the SETAC-Europe Working group on Toxicity Assessment on LCA, November 25, 1994

Jolliet O (1996) Impact assessment of human and eco-toxicity in Life Cycle Assessment. Part IV in Towards a Methodology for Life Cycle Impact Assessment. Ed. H.A. Udo de Haes. SETAC-Europe, 1996.

Jolliet O and Crettaz P (1996) Critical Surface-Time 95 (CST 95), a Life Cycle Impacts Assessment methodology including exposure and fate. Paper 1/96, prepared for the workshop on impact assessment of the concerted action on LCA in agriculture, Revised version 2, November 1, 1996. Swiss Federal Institute of Technology (EPFL), Institute of Soil and Water management, Lausanne Switzerland

Klöpfer W (1994) Review of life-cycle impact assessment. In Udo de Haes, H.A., Jensen, A.A., Klöpffer, W. and Lindfors, L.-G. (Eds.): Integrating Impact Assessment into LCA, SETAC-Europe, Brussels, Belgium: pp. 11-16

Lindfors, L-G et al. (1995) Nordic Guidelines on Life-Cycle Assessment. Nord 1995:20, Nordic Council of Ministers

NRC (1994) Science and Judgment in Risk Assessment. National research Council. National Academy Press, Washington DC

Mackay D (1991) Multimedia Environmental Models, the Fugacity Approach. Lewis Publishers, Chelsea, MI Potting J, Hauschild M, Blok, K and Wenzel H (1997) Spatial aspects of human toxicity in life-cycle assessment. Submitted to Int J LCA.

Potting J and Hauschild M (1997) The linear nature of environmental impact from emissions in life-cycle assessment. Accepted for publication in Int J LCA

RIVM, VROM, WWC (1994) Uniform System for the Evaluation of Substances (USES). Version 1.0 National Institute of Public Health and Environmental Protection (RIVM), Ministry of Housing, Physical Planning and Environment (VROM), Ministry of Welfare, Health, and Cultural Affairs (WWC), The Hague, RIVM Publication distribution No. 11144/150.

RTECS: Registry of Toxic Effects of Chemical Substances. Published by the National Institute for Occupational safety and Health, USA. Available on CD-rom

Udo de Haes, HA, Jensen, AA, Klöpffer, W and Lindfors, L-G (Eds.) (1994) Integrating Impact Assessment into LCA SETAC-Europe, Brussels, Belgium

Udo de Haes, HA (Ed.) (1996) Towards a Methodology for Life Cycle Impact Assessment. SETAC-Europe, Brussels.

White P, de Smet B, Udo de Haes HA and Heijungs R (1995) LCA back on the track, but is it one or two? LCA News, Vol. 5, Nr. 3: pp.2-4

Complex Mixtures

(Chair: J. Gray, Norway, and
V. Feron, Netherlands)

Risk Assessment for Complex Chemical Exposure in Aquatic Systems: The Problem of Estimating Interactive Effects

John S. Gray
Biological Institute, University of Oslo, P.O.B. 1064 Blindern, 0316 Oslo, Norway

Abstract

The traditional, but little used, way of assessing effects of the interaction between known chemicals is to use factorial experimental designs. Such designs allow one to test for less than additive (antagonistic) and greater than additive (synergistic) effects. Whilst synergism can be demonstrated in such experiments the concentrations at which synergistic effects occur are extremely high and are unlikely to occur in nature.

Recently developed techniques allow one to measure directly the effects of combined stressors in the field. These biological effect techniques range from tests on individual organisms to tests on communities. At the biochemical level the tests can indicate that the organism has been exposed to certain groups of chemicals (for example cytochrome P-450 enzymes responding to PAHs or metallothioneins responding to heavy metals). At the community level of organisation there are highly sensitive statistical techniques that indicate clearly the combined effect of stressors. The effects of oil exploration and production on benthic communities in the North Sea can be linked to concentrations of chemicals. However, such relationships are correlative and do not necessarily indicate cause and effect. Experiments are needed to test the hypotheses generated concerning the interactive effects of chemicals on the benthic species. The statistical analyses do, however, show which species have been affected and their relative sensitivity to chemical and physical disturbances. Such species are preferable to the traditional "laboratory weeds" usually utilised.

A strategy for risk assessment is needed that combines an experimental protocol for making predictions, from laboratory experiments, of likely effects to be found in the field. This should be combined with field monitoring that allows one to detect changes that were not predicted. At present most monitoring designs cannot adequately detect trends. This is due to concentration on Type-I statistical errors rather than properly considering Type-II errors. By concentrating on Type-II errors one can design monitoring programmes that are able to detect trends with a given degree of precision. There are also strong ethical grounds for a change to giving more emphasis to Type-II errors.

Laboratory Tests of Effects of Interacting Chemicals

One of the main aims of risk assessment in relation to the effects of environmental chemicals is to reduce the uncertainties. Yet, making predictions from laboratory tests is still rather primitive. Most acute toxicity tests are concerned with the effects of a single chemical on a given species and over a limited time period, usually 24 or 96 h, (Swanson & Lloyd, 1994). Long term impacts are assessed by chronic toxicity tests, but the length of these rarely covers the life-span of the species used and usually high doses are used for a small part of the life-span. From these tests one makes predictions of the likely effects that can be expected in the field. Monitoring then takes place to check that the predicted levels are not exceeded. If there are mixtures of chemicals being discharged the usual assumption is that the chemicals act together in an additive manner. Regulatory authorities limit the discharges so that if there are 3 chemicals then one third of each LC_{50} value is acceptable to discharge. Yet it is often claimed that there are dangers in this assumption, since chemicals may act in a synergistic manner. Whilst there are standard techniques for assessing the effects of mixtures of chemicals (i.e. factorial experimental designs) these are rarely used.

I have tested the effect of interacting heavy metals on the growth rate of a marine ciliate protozoan, (Gray & Ventilla 1974) and of interacting environmental factors and mercury on the larvae of a polychaete worm (Gray 1976). I will use the former to illustrate the technique. Table 1 shows the experimental design. This is orthogonal, i.e. there is a similar increase in concentration for each of the tested chemicals, and there are 3 concentrations of 3 chemicals: a 3 x 3 factorial design. Each concentration is replicated. This design allows one to test whether or not the effects are linear or quadratic for the chemicals alone and for their interactions. Statistically significant *interactions* are defined as greater than, or less than, additive expectation. For example at the concentration pair Zn 0.125 ppm/Hg 0.0025 ppm the growth rate reductions are 8.3 and 9.5% for the two chemicals alone, giving a 17.8% reduction in growth rate on additive expectation. However, the data show that mortality was only 13.9%. Similar effects are found at Zn 0.25 ppm/Hg 0.0025 ppm: additive expectation was 23.7%, while 18.8% were observed. These data show *antagonism*. At higher concentrations,however, greater than additive expectations are found. For example at the triple combination Hg 0.005 ppm/Zn 0.25 ppm/Pb 0.3 ppm growth rate reductions for the single chemicals were 12.1%, 14.2%, and 11.8%, respectively, giving an additive expectation of 38.1%, whereas a 67.8% reduction was observed, showing *synergism*. These results show clearly the great advantage of the factorial design. Yet, it is time-consuming, and testing of more than 3 chemicals would prove logistically difficult. Here 3 x 3 x 3 = 27 experiments were performed, each replicated, giving 54 separate experiments.

Whilst these results show that synergism can occur in nature, I doubt that anywhere in the marine domain there are concentrations that are as high as those used here. Most heavy metals occur at much lower levels and the most likely effects are thus expected to be antagonistic. Why this should be so is unclear but it has been shown many times that at low doses toxic chemicals often stimulate higher growth rates, a process that Stebbing (1985) has called *hormesis*. Clearly much more work needs to be done, particularly on effects of interacting organic compounds since these are known to be the most likely sources of environmental effects.

Table 1 Reduction of growth rate of a ciliate (% cp. control, mean of two replicates) caused by interacting heavy metals (from Gray & Ventilla, 1973)

Concentrations (ppm)									
$ZnSO_4$	0	0	0	0.125	0.125	0.125	0.25	0.25	0.25
$Pb(NO_3)_2$	0	0.15	0.3	0	0.15	0.3	0	0.15	0.3
$Hg(Cl)_2$									
0	0	8.5	11.8	8.3	14.4	18.8	14.2	18.3	25.9
0.0025	9.5	10.7	14.5	13.9	16.2	23.6	18.8	29.0	51.3
0.005	12.1	18.7	21.8	18.9	21.3	23.2	35.5	36.5	67.8

Table 2. Contaminant levels for total hydrocarbons and copper in the field and mesocosms used by Widdows and Johnson (1988)

Mesocosm Basin	THC (μg /l)	Cu (μg /l)	Field site	Total PAH (mg/kg)
C	3.0	0.5	1	2.24
L	6.4	0.8	2	5.97
M	31.5	5.0	3	11.43
H	124.5	20.0	4	15.45

Field Tests of Effects of Interacting Chemicals

Today there are new techniques which allow one to assess the effects of chemicals on organisms in the field. These are based on biochemical and physiological responses of individual organisms to stress, so-called biomarkers (see Depledge & Fossi, 1994; Peakall, 1994; McCarthy & Shugart 1990; GESAMP, 1995). Some biomarkers can indicate metal contamination (e.g. metallothionein response, see Hogstrand & Haux, 1991; Chan, 1995), or organophosphorus pesticide contamination (acetylcholinesterase inhibition, Galgani et al. 1992), or effects of PAHs (cytochrome P-450 system, Stagg & Addison, 1995). Such techniques indicate whether individual organisms have been exposed to contaminants or are showing symptoms of stress from the exposure.

Fig 1 shows the application of two biomarkers, a variant of the cytochrome P-450 test (EROD) and acetylcholinesterase, applied in a decreasing gradient of contamination in the North Sea. The EROD data show a decrease in the induced biomarker along the gradient, indicating a decrease in effect of inducers such as PAHs or/and PCBs, whereas the acetylcholinesterase biomarker shows an increase with increased stress. Both data sets suggest that at the Dogger Bank there is unexpected evidence of effect of stress, so that this station is more affected than Station 8. Subsequent chemical analyses have shown that the Dogger Bank area is indeed impacted and studies are being done to establish likely causes. There are other tests that are more general and do not responed to specific chemicals but rather to the totality of chemicals. One test is the "scope for growth" of blue mussel *Mytilus edulis* (Widdows & Johnson, 1988). In this test the energy demand of the mussel under different conditions of stress is measured and that remaining, the scope for growth, is used as an indicator of stress: High stress leads to low scope for growth.

The biomarker tests are on individuals of a given species and make predictions of likely effects on the population. Analyses of changes in community patterns in space and/or time in suspected impact areas has been shown to be an effective way of assessing impacts (Gray et al 1990; Clarke 1993). Olsgard and Gray (1995) show clearly how, using a simple sampling design, these statistical techniques can extract clear patterns of impact of oil activities. From these analyses it is then possible to establish correlations between the faunal data and chemical data, and to generate hypotheses of what variables might cause the effects on the fauna found.

Fig 2 shows the development of pollution effects around the Gyda oil field, North Sea, from 1987, the year of the baseline survey when little drilling waste had been discharged, to 1993. The area of altered benthic community structure is clearly shown. Table 3 shows an analysis of the environmental data showing that Cu and Sr were significantly correlated with the fauna in the baseline survey, whereas after 6 years of oil activities total hydrocarbon concentrations (THC), Zn and Ba show the highest correlations. Clearly, oil-related activities led to these effects. These data suggest that one can assess with high accuracy the effects of disturbances on benthic communities, in order to test the risk assessments made. Such monitoring and analysis techniques should, therefore, be part of modern environmental management practices.

Fig 2 and Table 3 show clearly that in 1987 the year of the baseline survey there was no pattern in the faunal distribution. The CANOCO analysis showed that the correlation with environmental variables was low in 1987 but that in 1990 there were clear influences of hydrocarbons (THC) and barium, both of which derive from oil industry discharges. By 1993 the patterns of effect cover large areas (ca 30 km^2) and there are very clear correlations with discharges. When permission was given by the Norwegian State Pollution Board for oil activities the companies did environmental impact assessments (EIAs) and predicted that

Figure 1 Biomarker techniques used on flat fish in the North Sea

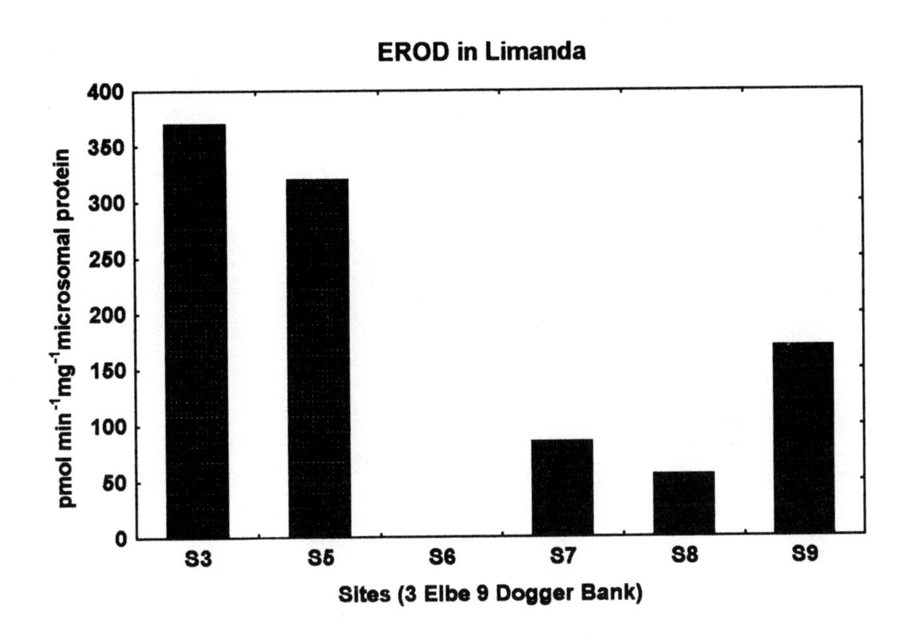

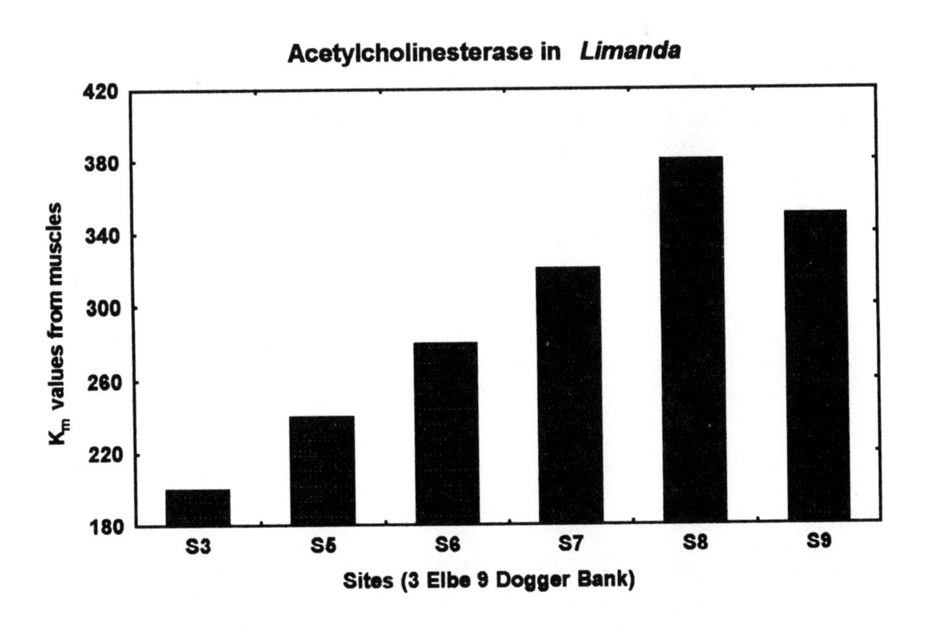

Figure 2 Effects of oil on benthic communities. Disturbance state of the different zones: a = undisturbed, b = slightly disturbed, c = moderately disturbed, d = grossly disturbed (from Olsgard and Gray 1995)

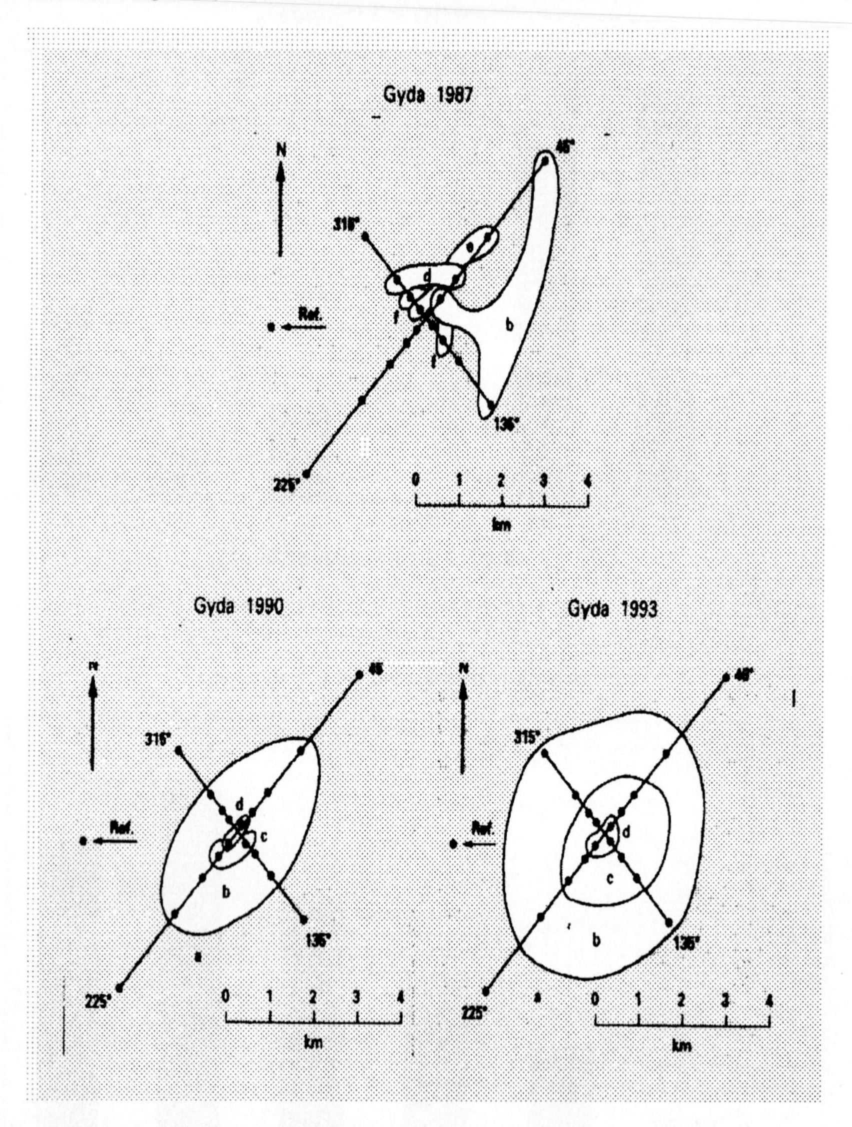

they would affect the sea bed to a 1 km radius, i.e. 3 km^2 . The effects on many fields within the Norwegian sector have been shown to be to much greater areas (up to 50 km^2 at some fields, Olsgard and Gray 1995). Since these effects were not predicted by the companies, the Norwegian State Pollution Board has acted to reduce discharges of oil-based drilling cuttings. Technological solutions were found within 2 years: now, water-based drilling muds are used, and effects of pollution have been reduced. The lessons are that the predictions of the EIA must be testable if one is to protect the marine environment.

Table 3 Relationship between environmental variables and fauna (using the forward selection procedure in CANOCO; ter Braak, 1988, 1990).
Mdφ is the median particle diameter in phi units and Sorting is the grain size sorting, a measure of heterogeneity. The percentage explanation of each environmental variable to the total faunal variance is given together with the level of significance: ** p = 0.01, * p = 0.05, n.s. = not significant.

1987		1990		1993	
Cu	14.7%**	Cu	38.1%**	THC	39.7%**
Sr	9.4%**	THC	37.7%**	Ba	39.2%**
Mdφ	7.3% n.s.	Zn3	7.2%**	Sr	39.2%**
TOC	6.3% n.s	Sorting	34.5%**	Zn	37.5%**
		Ba	33.6%*	Cu	37.0%**

Risk Assessment for Effects of Interacting Chemicals

From the above results it is possible to develop a strategy for risk assessment. Firstly, assess the risk of exposure to the chemicals discharged into the environment. The precautionary principle lists three characteristics of chemicals that are considered as important: toxicity, persistence and bioaccumulatability. Gray et al. (1991) list seven characteristics *viz*:

- Quantities: the production methods, loads and sources and discharge patterns
- Distribution: physicochemical characteristics, affinity for environmental compartments, sinks
- Persistence: kinetics of hydrolysis, photolysis, biodegradation
- Bioaccumulation: n-octanol/water partition coefficients, metabolic pathways in different organisms
- Toxicity: measures of biological activity of the substance (ideally from cells to ecosystems)
- Biological system typologies: biotic and abiotic characteristics of the structure and function of the biological communities
- Targets of exposure: consideration about the vulnerability and stability of (a) potential target communities and (b) life cycle stages of potential target species

Fig 3 shows the outline of a comprehensive approach to risk assessment and risk management in a marine context (adapted from Gray et al 1991).

Figure 3 Risk Assessment Procedure

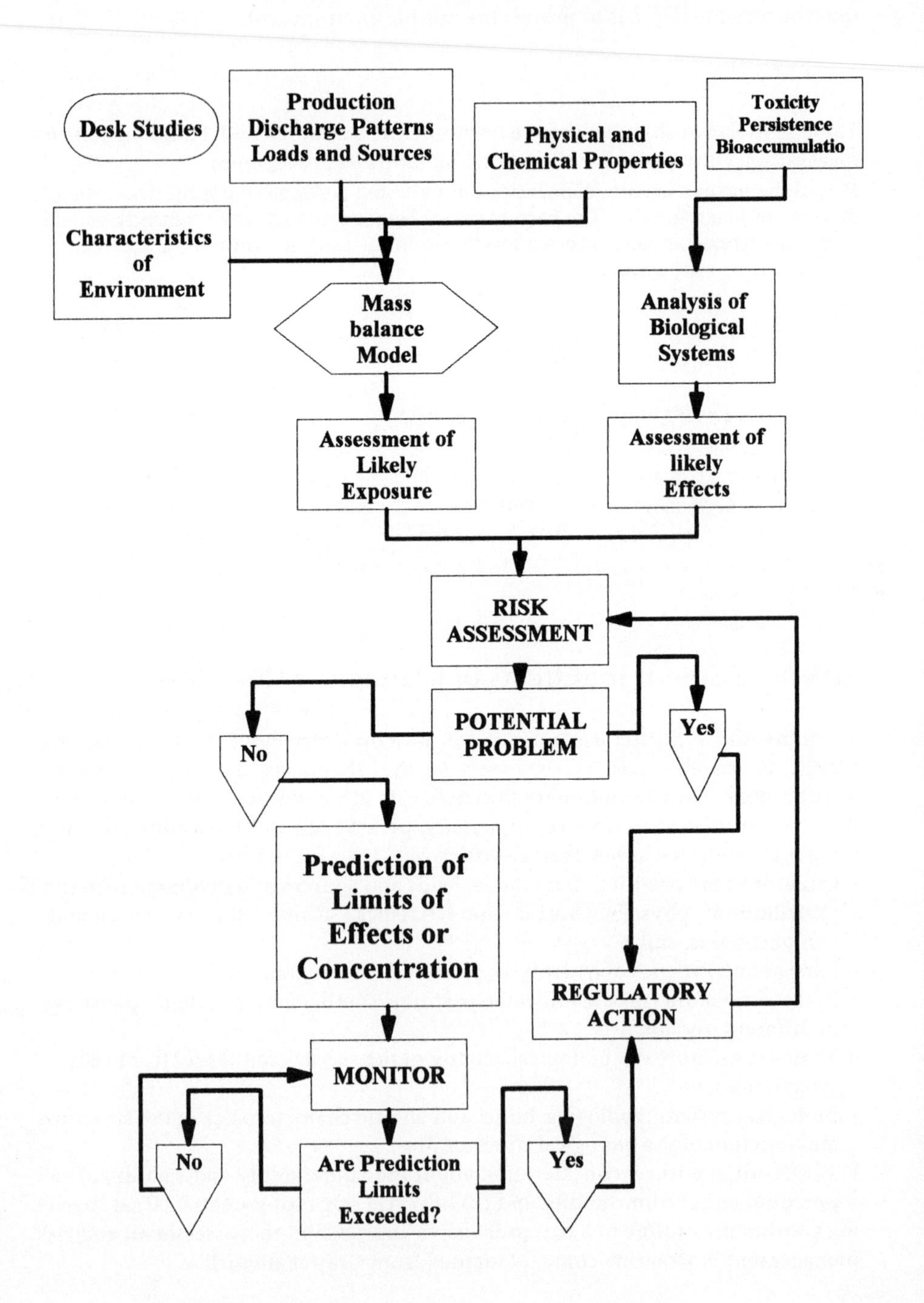

There have been few attempts to quantify the loads and sources of chemicals reaching the marine environment. The North Sea Task Force (1993) has reviewed sources of chemicals entering the North Sea from land, via rivers and via the atmosphere. Recently Neal et al. (1997) have reviewed the loads input to the North Sea and have shown that there are large discrepancies between countries in the manner in which the loads are measured. A study by GESAMP (1990) shows that on a global scale, atmospheric inputs dominate over riverine inputs for lead, cadmium, zinc, and for most high molecular weight hydrocarbons, whereas riverine inputs dominate for copper, nickel, arsenic, iron, nitrogen and phosphorus. In the environment chemicals are subjected to physical, chemical and biological processes which determine where and in what concentration they will occur. From knowledge of the physical forms of the chemicals and from kowledge of the physico-chemical conditions of the environment in which they are to be discharged, it is possible to predict in which environmental compartment they will be found (e.g. water or sediment). Paustenbach (1994) gives a comprehensive review of the properties that can be measured, such as water solubility, photodegradation either directly or indirectly, biodegradation, dissociation constant, sorption and desorption, and partition coefficient (to predict bioaccumulation and bioconcentration). Fugacity models have been developed for terrestrial systems, where, from physico-chemical properties, it is possible to predict the relative amount of a substance in each environmental compartment (Scheunert et al 1994). This has not yet been done in the marine environment.

Most chemicals adsorb onto particles and are sedimented out of the water column, so that marine sediments are the sinks for contaminants. In areas subjected to periodic erosion and resuspension, sediments provide a source of contaminants to the water column (reviewed by Harland et al. 1993). In particular, the sorption of hydrophobic organic chemicals by partitioning in or onto organic matter is a key process (Lyman 1983; Elzerman & Coates 1987; EPA 1989). Oxidation/reduction (the redox potential) of sediments also plays a major role in the chemistry of sediments, particularly for iron and manganese.

Data on the chemicals are combined with knowledge of physical, chemical and biological characteristics of the environment to produce a mass balance model that allows one to check that all sources and sinks are known. The importance of the biological systems should not be underestimated. Failings of mass balance models to predict, for example, eutrophication events may be due to the fact that the significant roles played by biological organisms in biogeochemical cycling were not known. Predictions can now be made of the concentration and physicochemical form of the chemical in a given environmental compartment at a point in space and time, giving the likely exposure of organisms to the chemical.

From knowledge of the physical attributes of the chemicals it is often possible to predict the effects on biological systems. Quantitative structure activity relationships (QSARs) have been developed for several classes of chemicals and within these classes reliable predictions can be made of the toxicity (Connell 1995). Use is made of toxicity data. It is usually assumed that no-effect-concentrations (NOEC) can be established (see Forbes and Forbes, 1994, for a review). Information must be obtained on the biological systems in the areas that are likely to be affected by the chemical. The following biological characteristics are recommended by a US National Research Council panel (1989)

1. a characterisation of major habitat types
2. a catalogue of representative species (or major species groups) present in the area
3. seasonal patterns of distribution and abundance
4. basic ecological information on feeding behaviour and reproduction
5. basic information on factors determining the vulnerability of various species

The panel added that "where unique habitats or endangered or rare species exist, more extensive characterisation of the biota .....may be required."

Once the exposure concentrations have been predicted, a risk assessment can be made. The usual procedure is to adopt a one-hundred-fold safety factor, i.e. to divide the NOEC by 100 to arrive at a "safe discharge" value. Paustenbach (1994) gives a clear and full account of how these factors are applied in relation to human health and the problems associated with the assessment of risk factors. Yet although the procedures are established they have rarely been applied in a comprehensive manner in the marine environment.

One important but highly neglected aspect is the design of monitoring prgrammes to check whether the predictions of the risk assessment procedure are upheld or not. Gray et al. (1991) suggest that in the marine environment there are two different basic strategies to environmental risk assessment, dependent on whether the discharge is a point-source and/or the effect of local disturbance, or a diffuse discharge and/or the effect of chronic disturbance. Monitoring is performed for point sources where one checks the predictions made by the risk assessment procedure. With diffuse sources and /or chronic disturbance it is suggested that surveillance should be undertaken in which levels of contaminants or degrees of disturbances are compared with control sites that are not thought to be contaminated. If the level of one or more contaminants or the state of a physically impacted site are shown to be different to controls then a risk assessment should be carried out.

Sampling designs, both for monitoring and surveillance, are needed in order to detect changes in populations and communities in the highly variable natural environment (Underwood 1991, 1993 and 1996). These new designs require a series of control sites, not just one. Furthermore, if effects are expected over time, sampling should be random within seasons. Neal et al. (1997) point out that the frequency and periodicity of sampling within the North Sea monitoring programme varies greatly between countries and that as a consequence "the data cannot be used for comparative purposes in an international policy-making and legislative context."

Another important aspect related to interpretation of surveillance and monitoring data is the plea by Peterman & M'Gonigle (1992) and Fairweather (1991) for statements of statistical power when presenting results. If the mean of a variable from a suspected impacted site is compared with the corresponding mean from a control site and no statistically significant difference is found, this may mean that either there is no difference or, alternatively, that the sampling methods were too poor to detect even a very large difference with statistical significance. Power analyses take into account variance and number of samples and allow a simulation of how many samples for a given variance would be needed to detect a significant difference (see Cohen, 1988, for a thorough description of methods). Power analyses are slowly being incorporated into surveillance and monitoring programmes. As an example, a baseline study was done in Denmark using various biological variables to measure the condition of an eelgrass bed. The environmental authorities agreed that a 25% reduction in these variables constituted an acceptable impact. A power analysis was then performed to check whether such a difference could be detected by the use of these variables. The generally accepted criterion is that a change can be detected with 80% power. However, none of the measured variables met this criterion! Some only had a 20% power for the statistically significant detection of a change. Further analyses were then performed in order to look for means of improving power. Increasing the sampling effort to 5-fold after the baseline survey made little difference, but a 4-fold increase in sampling effort before and after gave the necessary power to detect a change. This example illustrates that the application of these techniques would greatly improve the chances of detecting trends in monitoring programmes. However, the decision of whether or not to use such techniques rests not with scientists but with managers who fund such monitoring programmes. A four-fold increase in sample number may not be economically acceptable, yet the consequence of not accepting this would be that a really existing change may not be detected in a statistically significant way.

In summary, risk assessment procedures have been developed for predicting and testing the effects of mixtures of chemicals in the marine environment. However, the procedures are not yet incorporated into national and international legislation, and are therefore seldom applied. Science has produced the tools, but the managers have not taken them up; thus it is science that is being wrongly blamed for not taking proper care of the marine environment. Let us hope that this situation will be changed in the very near future.

References

Chan, KM (1995) Metallothionein: Potential biomarker for monitoring heavy metal pollution in fish around Hong Kong. Marine Pollution Bulletin 31, 411-415.

Clarke, KR (1993). Nonparametric multivariate analyses of changes in community structure. Australian Journal of Ecology 18, 117-143.

Cohen, J (1988) Statistical Power Analysis for the Behavioural Sciences. 2nd edn. L. Erlbaum Associates, Hillsdale, New Jersey.

Connell, DW (1995) Predictions of bioconcentrations and related lethal and sublethal effects with aquatic organisms. Marine Pollution Bulletin 31, 201-208.

Depledge, MH and Fossi, MC (1994) The role of biomarkers in environmental assessment (2). Invertebrates. Ecotoxicology 3 (3): 173-179.

Elzerman, AW, and Coates, JT (1987) Hydrophobic organic compounds on sediments. Equilibria and kinetics of sorption. in: Sources and Fates of Aquatic Pollutants (RA Hites and SJ Eisenreich, eds.), American Chemical Society, Washington DC. pp. 263-317

EPA (1989) Briefing report to the EPA Science Advisory Board on the Equilibrium Partitioning Approach to Generating Sediment Quality Criteria. EPA 440/5-89-002.

Fairweather, PG (1991) Statistical power and design requirements for environmental monitoring. Australian Journal of Marine and Freshwater Research 42, 555-568.

Forbes, VE , and Forbes, TL (1994) Ecotoxicology in Theory and Practice. Chapman & Hall, London. .

Galgani, F, Bocquené, G, and Cadiou, Y (1992) Evidence of variation in cholinesterase activity in fish along a pollution gradient in the North Sea. Marine Ecology Progress Series 91, 77-82.

GESAMP (1990) The State of the Marine Environment. UNEP Regional Seas Reports and Studies GESAMP 115, 1-111.

GESAMP (1995). Biological indicators and their use in the measurement of the condition of the marine environment. Reports and Studies GESAMP 55, 1-56.

Gray, JS (1976) Effects of salinity, temperature and mercury on mortality of the trocophore larvae of *Serpula vermicularis* L., Annelida, Polychaeta. J. exp. mar. Biol. Ecol. 23, 127-134.

Gray, JS, Calamari, D, Duce, R, Portmann, JE Wells, PG, and Windom, HL (1991) Scientifically based strategies for marine environmental protection and management. Marine Pollution Bulletin 22, 432-440.

Gray, JS, Clarke, KR, Warwick, RM, and Hobbs, G (1990) Detection of initial effects of pollution on marine benthos: an example from the Ekofisk and Eldfisk oilfields, North Sea. Marine Ecology Progress Series 66, 285-299.

Gray, JS, and Ventilla, RJ (1973) Growth Rates of a Sediment-living Protozoan as a Toxicity Indicator for Heavy Metals. Ambio 2, pp. 118-121.

Harland, BJ, Matthiesen, P, and Murray-Smith, R (1993) The prediction of fate and exposure of organic chemicals in aquatic sediments. In: Chemical Exposure Predictions. (Calamari, D, ed.), Lewis, Bacon Raton, pp. 115- 136

Hogstrand, C, and Haux C (1991) Binding and tetoxification of heavy metals in lower vertebrates with special reference to metallothionein. Comparative Biochemical Physiology 100C, 137-141.

Lyman, W (1983) Adsorption coefficients for soils and sediments. in: Handbook of Chemical Property Estimation Methods. McGraw-Hill, New York.

McCarthy, JF, and Shugart, LR (1990) Biological Markers of environmental Contamination. Lewis Publishers Inc., Bacon Raton, Florida.

National Research Council (1989). The Adequacy of Environmental Information for Outer Continental Shelf Environment Studies Program II. Ecology. National Academy Press, Washington D.C.

Neal, CW, House, WA, Jarvie, HP, Leeks, GJL, and Marker, AH (1997) Cocnlusions to the special volume of Science of the Total Environment concerning UK fluxes to the North Sea, Land Ocean Interaction Study River basins research, the first two years. Science of the Total Environment 194/195 467-477.

North Sea Task Force (1993). North Sea Quality Status Report 1993. Oslo and Paris Commissions, London. Olsen & Olsen, Fredensborg, Denmark 132 pp.

Olsgard, F, and Gray, JS (1995) A comprehensive analysis of the effects of offshore oil and gas exploration and production on the benthic communities of the Norwegian continental shelf. Marine Ecology Progress Series 122, 277-306.

Paustenbach, DJ (1994) A survey of environmntal risk assessment. In: Paustenbach, DJ, The Risk Assessment of Environmental and Human Health Hazards: A Textbook of case studies.. Wiley, New York, pp. 27- 124

Peakall, D (1994) The role of biomarkers in environmental assessment (1). Introduction. Ecotoxicology 3, 157-160.

Peterman, RM, and M'Gonigle, M (1992) Statistical power analysis and precautionary principle. Marine Pollution Bulletin 24, 231-234.

Scheunert, I, Schroll, R, and Trapp, S (1994) Prediction of plant uptake of organic xenobiotics vy chemical substance and plant properties. In: Chemical exposure Predictions (D Calamari, ed.), Lewis Publishers, Boca Raton, pp. 137-146.

Stagg, RM, and Addison, R (1995) An inter-laboratory comparison of measurements of ethoxyresorfin O-de-ethylase activity in Dab (*Limanda limanda*) liver. Marine Environmental Research 40, 93-108.

Stebbing, ARD (1985) The use of hyroids in toxicology. In: The Effects of Stress and Pollution in Marine Animals (Bayne, B.L. et al., eds.), Praeger, New York, pp 267-300.

Swanson, T.W. & Lloyd, R. (1994). The regulation of chemicals in agricultural production. In: Environmental Toxicology, Economics and Institutions (Bergman, L, and Pugh, DM, eds.), Kluwer Academic Publishers, pp 15-38.

ter Braak, CJF (1988) CANOCO - a Fortran program for canonical community ordination by (partial) (detrended) (canonical) correspondence analysis, principal components analysis and redundance ananlysis (version 3.10), TNO Institute of Applied Computing Science. Statatistics Department, pp 1-95.

ter Braak, CJF (1990) Update notes: CANOCO version 3.10. Agricultural Mathematics Group, Wageningen, Ntherlands.

Underwood AJ (1991) Beyond BACI: experimental designs for detecting human environmental impacts on temporal variations in natural populations. Australian Journal cf Marine and Freshwater Research 42, 569-587.

Underwood AJ (1993) The mechanics of spatially replicated sampling programmes to detect environmental impacts in a variable world. Australian Journal of Ecology 18, 99-117.

Underwood, AJ (1996) Ecological Experiments: Their Logic, Design, Analysis and Interpretation Using Analysis of Variance. Cambridge U. P., Cambridge.

Widdows, J, and Johnson, D (1988) Physiological energetics of *Mytilus edulis*: Scope for Growth. Marine Ecology Progress Series 46, 113-121.

Exposure of Humans to Complex Chemical Mixtures: Hazard Identification and Risk Assessment

Victor J. Feron, John P. Groten and Peter J. van Bladeren
TNO-Nutrition and Food Research Institute, Toxicology Division, P.O. Box 360, 3700 AJ Zeist, The Netherlands

Abstract

A *complex* chemical mixture is defined as a mixture that consists of tens, hundreds or thousands of chemicals, and of which the composition is qualitatively and quantitatively not fully known. In contrast, a *simple* mixture consists of a relatively small number of chemicals, say ten or less, and the composition of which is fully known. In the present paper a number of options for hazard identification and risk assessment of complex chemical mixtures is discussed, and a scheme aimed at selecting the most appropriate approach for each (type of) complex mixture is presented. A conspicuous element of this scheme is the dichotomy of complex mixtures into mixtures that are readily available and mixtures that are virtually unavailable for testing in their entirety. Another characteristic aspect of the scheme is the inclusion of the "top-ten" and "pseudo top-ten" approaches, which in essence are ways to select the, say ten, most risky chemicals or pseudocomponents to be dealt with as a simple chemical mixture.

Introduction

Historically, exposure assessment, hazard identification and risk assessment of chemicals have dealt with single substances and have often been confined to a single route of exposure. However, in reality humans are exposed to a complex and ever-changing mixture of compounds in the air they breath, the water they drink, the food and beverages they consume, the surfaces they touch, and the consumer products they use (Sexton et al., 1995). Fortunately, the interests of toxicologists and regulators in chemical mixtures is increasing; three years ago an impressive book dealing with the toxicology of mixtures was published by Yang (1994); two years ago the proceedings of a very successful symposium on quantitative risk assessment of chemical mixtures appeared (Simmons, 1995), and last year proceedings were published of a European conference on combination toxicology held in The Netherlands (Feron and Bolt, 1996). Later this year a 3-day international conference on current issues on chemical mixtures will take place in Fort Collins, Colorado, USA. Moreover, symposia, workshops or round table discussions on chemical mixtures are on the programme of meetings of societies of toxicology and related fields. Clearly, it has been widely recognized that evaluating the potential health effects caused by simultaneous or sequential exposure to large numbers of chemicals is a reality of considerable importance.

To successfully study the toxicology of mixtures and to properly assess their potential health risks it is essential to understand the basic concepts of combined action and interaction of chemicals (Mumtaz et al., 1994), and to distinguish between whole mixture analysis (top-down approach) and component interaction analysis (bottom-up approach) (Yang et al., 1995). In our view it is equally important to make a clear distinction between simple and complex chemical mixtures (Feron et al.,1995a; 1995b; Ray and Feron, 1996). A *simple* mixture is defined as a mixture that consists of a relatively small number of chemicals, say ten or less, and the composition of which is qualitatively and quantitatively known; examples are a cocktail of pesticides, a combination of medicines, a group of irritating aldehydes. A *complex* mixture is defined as a mixture that consists of tens, hundreds or thousands of chemicals, and of which the composition is qualitatively and quantitatively not fully known; examples are welding fumes, a workplace atmosphere, environmental tobacco smoke, drinking water, a new food product. It is self-evident that a bottom-up approach for studying the toxicology of complex mixtures is virtually impossible, implying that the top-down approach is the option left. For some complex mixtures the top-down approach may be appropriate, for others such as for instance the atmosphere at a hazardous waste site or a workplace atmosphere, whole mixture analysis seems to be impracticable and meaningless.

Recently, Fay and Feron (1996) reported the results of a Working Group discussion on complex chemical mixtures, held during the European conference on combination toxicology (Feron and Bolt, 1996). The Working Group discussed the strengths and limitations of (a) whole mixture analysis, (b) fractionation of mixtures, and (c) identification of the "top-ten" chemicals to be treated as a simple mixture. The present paper briefly discusses a number of options for hazard identification and risk assessment of complex mixtures, and proposes a framework for these options, aimed at selecting for each (type of) complex chemical mixture the most appropriate approach.

Top-down Approach for Complex Chemical Mixtures

Mixtures Studied in Their Entirety: There are a number of difficulties with characterizing the toxicity of complex mixtures by studying the mixture *in toto*. First, the complex mixture may not be available in its entirety for toxicity testing e.g. a workplace atmosphere. Second, often adequate testing is impossible because a sufficiently high dose cannot be applied: for example, in case of a new food product a high dose incorporated in the diet may lead to an imbalanced diet; or, in case of cigarette smoke, which contains highly acutely toxic

substances such as carbon monoxide, the use of a high dose in long-term carcinogenicity studies is impossible. Third, low-dose extrapolation may not be scientifically sound; at high concentrations the contribution of different chemicals to the adverse effect may be proportionally different from that at low concentrations; this may be particularly true when the adverse effect is an increased tumour incidence: the carcinogenic mixture may contain cocarcinogenic and/or antagonistic factors which are less active or inactive at lower exposure levels (Feron et al., 1986; 1995a).

On the other hand, testing of complex mixtures in their entirety has produced relevant toxicological information. Examples are: safety evaluation of single cell proteins (De Groot et al., 1971), and long-term toxicity studies with recycled drinking water (Lauer et al., 1994).

Mixtures Studied by Breaking them into Fractions Acording to Toxicity: Testing of all known individual chemicals in a complex mixture is virtually impossible and, even if we were able to produce the toxicity data and to set an exposure limit for all constituents, we would still be facing the diffilculty of a complex mixture with its numerous possible combined actions or interactions. Breaking the complex mixture into fractions according to toxicity might be an appropriate approach, for instance, for identification of toxic, mutagenic or carcinogenic compounds or group of compounds, and for substitution of those components to create a less toxic mixture (Fay and Feron, 1996). This fractionation approach has been successfully used to study the toxicity of diesel exhaust which consists of three major fractions *viz.* the gas phase, the fine particulate carbon soot phase, and the organic phase adsorbed on the carbon soot. The studies suggested that the particulate phase is the fraction of concern, and that high lung burdens of carbon soot particles as such are sufficient to induce the neoplastic and nonneoplastic lung lesions induced by inhaled total diesel exhaust (Henderson and Mauderly, 1994). On the other hand, fractionation may lead to other (artefactual) complex mixtures, the toxicity of which may be different from that of both the individual compounded chemicals and the mixture as a whole (Fay and Feron, 1996). Moreover, fractionation according to toxicity may just not work, the toxicity being more or less evenly distributed over the different fractions.

Selection of the "Top-ten" Chemicals to be Approached as a Simple Mixture (Two-step Procedure): When confronted with the request to develop a method for safety evaluation of mixtures of (hundreds of) chemicals occurring at workplaces, we suggested to focus on a limited number of compounds, e.g. the ten chemicals that rank highest on the priority list for chemicals with high health risk potential at a particular workplace. We also recommended to assess the potential health risk of these "top-ten" chemicals not only compound by compound but also as a group of chemicals to which the workers are exposed simultaneously (Feron et al., 1993; 1995a; 1995b). A comparable approach has been successfully used in the United States to select the (top-ten) priority substances that are released from hazardous waste sites and that pose the greatest hazard to human health. Once the top-ten chemicals have been identified, their potential health risk is assessed as a mixture, using a narrative weight-of-evidence approach incorporating mechanistic insights on possible chemical interactions (Johnson and DeRosa, 1995).

Figure 1 shows the essential elements of this two-step procedure we developed for the safety evaluation of (certain) complex mixtures *viz.* selection of the "top-ten" chemicals (step one) followed by hazard identification and risk assessment of the selected chemicals to be approached as a simple mixture (step two). The selection of the "top-ten" chemicals is based on their risk quotient (Figure 1); it is a measure for the probability of adverse health effects. The higher the value, the higher the risk of health effects in humans. Hazard identification and risk assessment of the "mixture" of selected chemicals (the "top-ten" chemicals), the second step, should be based on toxicity data and on data on mechanism of action of the individual chemicals and on prediction of presence or absence of additive effects or potentiating interactive effects. A major practical problem is often the lack of this type of information. However, where no data are available experts should fill in the gaps, using both computer-based strucuture-toxicity relationships and their expertise and experience (Feron et al., 1995a). Actually, this approach is essentially similar to the weight-of-evidence scheme for assessment of chemical interactions as developed by Mumtaz and Durkin (1992), and applied for risk assessment purposes to mixtures of selected chemicals released from hazardous waste sites (Johnson and DeRosa ,1995; Mumtaz et al., 1994).

In brief, a strategy for hazard identification and risk assessment of complex mixtures could be to identify the, say, ten most risky chemicals in the mixture, and to approach these ten chemicals as a simple mixture, assuming that hazard and potential risk of this simple mixture are representative for hazard and risk of the entire complex mixture. This two-step procedure is, in our view, particularly suitable for analysis of complex mixtures such as workplace atmospheres, other indoor atmospheres and local atmospheres around hazardous waste sites that are difficult to characterize both qualitatively and quantitatively, and more importantly, are virtually unavailable for testing in their entirety.

Figure 1 shows that we added to step 2 of the procedure, as a kind of an appendix, a rather pragmatic approach that could be used for analysis of complex mixtures which are not available for testing *in toto* and which are not thought to be notoriously hazardous to humans e.g. an office atmosphere. The idea is to test the (ten) priority chemicals *as a mixture* for general toxicity in a 4-week repeated exposure study in rats and for genotoxicity in one of the standard screening assays, using exposure concentrations of the individual chemicals which are e.g. 3 or 10 times higher than those occurring in practice. When these studies do not produce evidence of toxic and/or mutagenic effects of the entire simple mixture, the entire complex mixture can be considered relatively harmless. When, however, toxic and/or mutagenic effects are found, concern about possible health effects in humans is justifiable and further studies on the group of selected chemicals are warranted, using full or fractionated designs suitable for testing simple, defined mixtures (Eide, 1996; Gennings, 1996; Groten

STEP 1

IDENTIFICATION OF PRIORITY CHEMICALS

Select a limited number of chemicals (e.g. ten) with the highest health risk potential, using the risk quotient (RQ):

$$RQ = \frac{\text{level of exposure}}{\text{level of toxicity}}$$

In brief, identify the "top ten" chemicals

STEP 2

HAZARD IDENTIFICATION AND RISK ASSESSMENT

Identify the hazard and assess the health risk of the defined mixture of the ("ten") priority chemicals, using approaches appropriate for simple mixtures of chemicals

A pragmatic approach: carry out limited toxicity studies e.g. one 4-week rat study and one screening assay for genotoxicity with the defined mixture of the ("ten") priority chemicals, using exposure concentrations e.g. 3 to 10 times higher than those occurring in the complex mixture

Fig.1 Original two-step procedure for the safety evaluation of complex chemical mixtures (modified from Feron et al., 1995a)

et al., 1996, 1997; Schoen, 1996). Clearly, studies along these lines are neither simple from a management point of view nor cheap, but they are most certainly informative and relevant to health risk assessment of certain types of complex mixtures (Feron et al., 1995a)

Scheme for Hazard Identification and Risk Assessment of Complex Chemical Mixtures

As appears from the previous sections there are several ways to identify the hazard and to assess the potential health risk of complex chemical mixtures. Clearly, different (types of) complex mixtures call for different approaches. Moreover, the usefulness of a certain approach depends on the context in which one is confronted with the mixture, and on the amount, type and quality of the information that is available on the chemistry and the toxicity of the mixture (Fay and Feron, 1996).

Recently, Sexton et al (1995 presented a framework for describing major aspects of mixtures and mixture-related research. This framework illustrates the complexity of mixture toxicology, and offers a matrix for a structured approach of research on mixtures. However, this framework does not distinguish between simple and complex chemical mixtures, which hampers its application.

The report of a Working Group on complex mixtures (Fay and Feron, 1996) defines simple and complex mixtures, and recommends the development of a decision tree for risk evaluation of complex mixtures. Challenged by this report, we present a scheme for hazard identification and risk assessment of complex mixtures. The various elements of the scheme are depicted in Figure 2. After showing that chemical mixtures should be divided into simple and complex mixtures (for definitions see one of the previous sections), the scheme continues with a dichotomy of complex mixtures into mixtures readily available and mixtures virtually unavailable for testing in their entirety. Examples of the former are drinking water, diesel exhaust, welding fumes, tobacco smoke and new food products; examples of the latter are workplace atmospheres, coke oven emissions and atmospheres at a waste sites.

Complex Mixtures Virtually Unavailable for Testing in Their Entirety: For complex mixtures that are virtually unavailable for testing as a whole, we suggest the "top-ten" approach described in detail in the previous section of this paper[1].

Complex Mixtures Readily Available for Testing in Their Entirety: For mixtures readily available for testing in their entirety, we see three possible approaches: (a) testing as a whole, (b) identification of the "top-ten" chemicals to be treated as a simple mixture ("top-ten" approach), and (c) identification of the "top-ten" classes of chemicals to be lumped per class to the "top-ten" chemicals to be treated as a simple mixture ("pseudo top-ten" approach).

Testing as a whole. A reason for testing a complex mixture in its entirety is the verification of its (presumed) safety. An example is drinking water. Apart from the difficulty – inherent to each type of complex mixture – of changes in chemical composition over time, relevant studies with drinking water require concentration of the sample to be tested, entailing all kinds of problems such as

[1] The "pseudo top-ten" approach (see below) might also be considered for this type of mixtures. However, its use may unnecessarily complicate the evaluation.

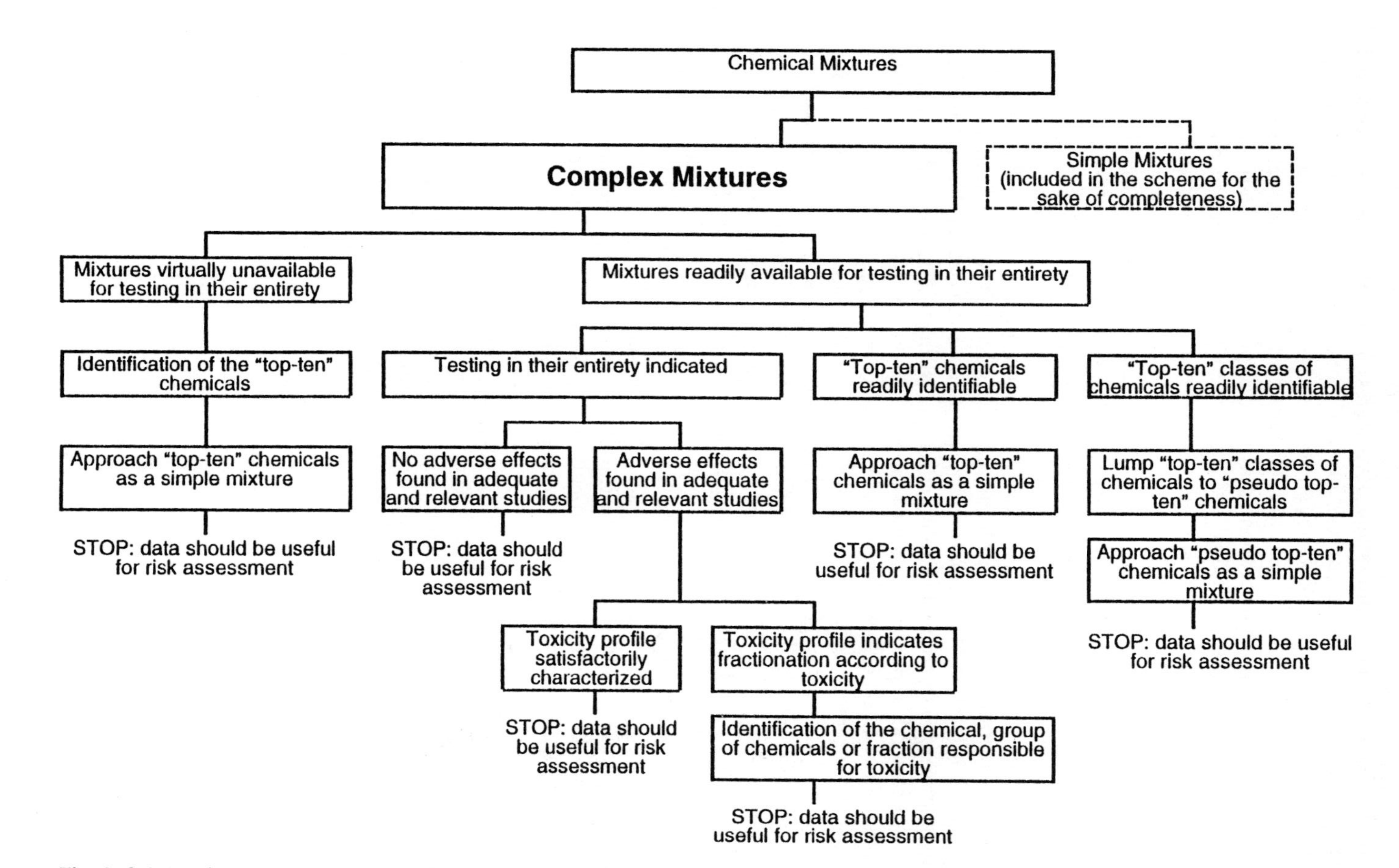

Fig. 2. Scheme for safety evaluation of complex chemical mixtures

loss of volatile organic compounds. Nevertheless, relevant studies with recycled drinking water have been successfully performed, resulting in scientifically sound conclusions (Lauer et al., 1994). New food products are another example of readily available complex mixtures, testing of which in their entirety seems to be the first option. A major problem may be incorporation of the test material in the diet at a sufficiently high dose, avoiding an unbalanced diet and nutritional deficiencies. However,there are numerous examples of successful safety evaluation of new food products.

Another reason for testing a complex mixture in its entirety is the characterization of its toxicity profile. Once the toxicity profile is satisfactorily characterized (e.g. critical adverse effect and dose-effect/response relationship, including a dose without effect/response, have been established) the data might be useful for risk assessment. On the other hand, the toxicity profile might be such that further studies are indicated e.g. to find out what chemical, group of chemicals or fraction of the mixture is responible for the critical effect. Breaking the mixture into fractions and subfractions according to toxicity might prove to be a successful approach as it has appeared to be in the case of the safety evaluation of diesel exhaust (Henderson and Mauderly, 1994).

"Top-ten" approach. Also for complex mixtures that are easily available for testing as a whole, it might be worthwhile to consider the "top-ten" approach as the primary option, thereby getting around the disadvantages of testing the mixture in its entirety. The "top-ten" approach may be particularly useful when the available knowledge about the chemical composition and the (defective) information on the toxicity of the mixture seem to designate a small number of constituents as presumably responsible for the toxicity of the whole mixture. Welding fumes might be good candidates for such an approach.

"Pseudo top-ten" approach. Identification of the "top-ten" chemicals of a complex mixture might be very difficult, if not impossible, for mixtures which consist of a large number of widely varying chemicals with no obvious ranking of individual constituents according to potential health risks. However, in view of all the limitations inherent to testing complex mixtures in their entirety (for details see one of the previous sections of this paper), an approach that deserves serious consideration is identification of the "top-ten" classes of chemicals to be lumped together per class to the "top-ten" chemicals. The lumping technique is based on grouping chemicals with relevant similarity such as for instance the same target organ or similar mode of action. The selected "top-ten" chemicals are either representative chemicals of each class or pseudocomponents representing a fictional average of a certain class. The lumping technique has elegantly been described in detail in a recent paper by Verhaar et al (1997) who proposed the lumping technique combined with QSARs (quantitative structure-activity relationship) and PBPK/PD (physiologically based pharmacokinetic/pharmacodynamic) modeling as an integrated method to predict the toxicity of complex mixtures of petroleum products such as e.g. JP-5, a navy jet fuel.

Composite Standards

It has been propsed to supplement, or preferably to gradually replace, single chemical-oriented standard setting by mixture-oriented standard setting (Willems, 1988; Feron et al., 1995). Indeed, there is a need to move forward with instituting standards for mixtures, particularly at the workplace (Fay and Feron, 1996). A first step could be identification of multiple exposure situations of presumed health concern followed by accurate determination of the profile of such multiple exposures (Fay and Mumtaz, 1996). Once a priority multiple-exposure situation has been identified and characterized, its health hazard and potential health risk should be assessed, using approaches as mentioned in Figure 1. The challenge is to substitute gradually mixture-oriented (real life-oriented) standard setting for (unrealistic) single chemical-oriented standard setting.

Concluding Remarks

To study successfully the toxicology of mixtures and to properly assess their potential health risks, it is essential (a) to understand the basic concepts of combined action and interaction of chemicals, (b) to distinguish between whole mixture analysis (top-down approach) and component interaction analysis (bottom-up approach), and (c) to distinguish between simple and complex mixtures.

There are various ways to deal with hazard identification and risk assessment of complex chemical mixtures. Different (types of) complex mixtures call for different approaches to evaluate their safety, and the usefulness of a certain approach depends on the context in which one is confronted with the mixture and also on the amount, type and quality of the available data on the chemistry and toxicity of the mixture. A conspicuous element of the scheme presented for the safety evaluation of complex mixtures is the dichotomy of complex mixtures into mixtures that are virtually unavailable and mixtures that are readily available for testing in their entirety. Moreover, the inclusion in the scheme of the "top-ten" and "pseudo top-ten" approaches [in essence selection of the (say, ten) most risky chemicals or pseudocomponents to be dealt with as a simple mixture] is regarded as another characteristic aspect of the scheme. The scheme aims at stimulating progress with hazard identification and risk assessment of complex chemical mixtures, bringing together and effectively using all available relevant information, methods, technologies, expertise and experience.

References

De Groot AP, Til HP, Feron VJ (1971) Safety evaluation of yeast grown on hydrocarbons. III. Two-year feeding and multigeneration study in rats with yeast grown on gas oil. Fd Cosmet Toxicol 9: 787-800

Eide I (1996) Strategies for toxicological evaluation of mixtures. Fd Chem Toxicol 34: 1147-1149

Fay RM, Feron VJ (1996) Complex mixtures:hazard identification and risk assessment. Fd Chem Toxicol 34:1175-1176

Fay RM, Mumtaz MM (1996) Development of a priority list of chemical mixtures occurring at 1188 hazardous waste sites, using the HazDat database. Fd Chem Toxicol 34:1163-1165.

Feron VJ, Bolt HM (1996) Combination toxicology: proceedings of a European conference. Fd Chem Toxicol 34: I-IV, 1025-1185

Feron VJ, Griesemer RA, Nesnow S (1986) Testing of complex chemical mixtures. In: Montesano R, Bartsch H, Vainio H, Wilbourn J, Yamasaki H (eds) Long-term and short-term assays for carcinogens: a critical appraisal. IARC Scientific Publications, No 83, Lyon, pp 483-494

Feron VJ, Groten JP, Jonker D, Cassee FR, van Bladeren PJ (1995a) Toxicology of chemical mixtures: challenges for today and the future. Toxicol 105: 415-427

Feron VJ, Jonker D, Groten JP, Horbach GJMJ, Cassee FR, Schoen ED, Opdam JJG (1993) Combination toxicology: from challenge to reality. Toxicol Trib 14: 1-3

Feron VJ, Woutersen RA, Arts JHE, Cassee FR, de Vrijer F, van Bladeren PJ (1995b) Safety evaluation of the mixture of chemicals at a specific workplace: theoretical considerations and a suggested two-step procedure. Toxicol Lett 76: 47-55

Gennings C (1996) Economical designs for detecting and characterizing departure from additivity in mixtures of many chemicals. Fd Chem Toxicol 34: 1053-1058

Groten PJ, Schoen ED, van Bladeren PJ, Kuper CF, van Zorge J, Feron VJ (1997) Subacute toxicity of a mixture of nine chemicals in rats: detecting interactive effects with a fractionated two-level factorial design. Fund Appl Toxicol 36: 15-19

Groten PJ, Schoen ED, Feron VJ (1996) Use of factorial designs in combination toxicity studies. Fd Chem Toxicol 34: 1083-1089

Henderson RF, Mauderly JL (1994) Diesel exhaust: approach for the sudy of the toxicity of chemical mixtures. In: Yang RSH (ed) Toxicology of chemical mixtures. Academic Press, San Diego, pp 119-133

Johnson BL, DeRosa CT (1995) Chemical mixtures released from waste sites: implications for health risk assessment. Toxicol 105: 145-156

Lauer WC, Wolfe GW, Condie LW (1994) Health effect studies on recycled water from secondary wastewater. In: Yang RSH (ed) Toxicology of chemical mixtures. Academic Press, San Diego, pp. 63-81

Mumtaz MM, DeRosa CT, Durkin PR (1994) Approaches and challenges in risk assessments of chemical mixtures. In: Yang RSH (ed) Toxicology of chemical mixtures. Academic Press, San Diego, pp. 565-597

Mumtaz MM, Durkin PR (1992) A weight-of-evidence scheme for assessing interactions in chemical mixtures. Toxicol Ind Health 8: 377-406

Sexton K, Beck BD, Bingham E, Brain JD, DeMarini DM, Hertzberg RC, O`Flaherty EJ, Pounds JG (1995) Chemical mixtures from a public health perspective: the importance of research for informed decision making. Toxicol 105: 429-441

Simmons JE (1995) Chemical mixtures: challenge for toxicology and risk assessment. Toxicol 105: 111-119

Verhaar HJM, Morroni JR, Reardon KF, Hays SM, Gaver DP Jr, Carpenter RL, Yang RSH (1997) A proposed approach to study the toxicology of complex mixtures of petroleum products: the integrated use of QSAR, lumping analysis and PBPK/PD modeling. Environm Health Persp 105: 179-195

Willems JHBM (1988) Feasibility of composite workplace standards for chemical and/or physical workplace factors. In: Notten WRF, Herber RFM, Hunter WJ, Monster AC, Zielhuis RL (eds) Health surveillance of individual workers exposed to chemical agents. Springer, Berlin, pp. 128-134

Yang RSH (1994) Toxicology of chemical mixtures. Academic press, San Diego, p. 720

Yang RSH, El-Masri HA, Constan AA, Tessari, JD (1995) The application of physiologically-based pharmacokinetic/pharmacodynamic (PBPK/PD) modelling for exploring risk assessment approaches of chemical mixtures. Toxicol Lett 79: 193-200

Receptor Mediated Toxic Responses

(Chair: J.D. Tugwood, United Kingdom,
and F. Fonnum, Norway)

Peroxisome Proliferator-Activated Receptor-alpha and the Pleiotropic Responses to Peroxisome Proliferators

Jonathan D. Tugwood[1], Thomas C. Aldridge[1], Kevin G. Lambe[2], Neil Macdonald[1] and Nicola J. Woodyatt[1].
[1]Zeneca Central Toxicology Laboratory, Alderley Park, Macclesfield. SK10 4TJ. U.K.
[2]Zeneca Pharmaceuticals, Alderley Park, Macclesfield. SK10 4TG. U.K.

Introduction

Peroxisome proliferators (PPs) form a discrete class of rodent non-genotoxic hepatocarcinogens (see Stringer, 1992 for review). These chemicals, which include industrially and pharmaceutically important compounds such as chlorinated solvents, plasticisers and hypolipidaemic drugs, have been the subject of intensive study over a number of years. Generally, following administration of PPs rats and mice undergo a range of responses including proliferation of peroxisomes in the liver and kidney, induction of certain hepatic enzymes, and liver enlargement. Although the vast majority of PPs are non-genotoxic by in vivo and in vitro criteria (reviewed in Ashby et al., 1994), these compounds cause the development of hepatocellular carcinomas when administered continually to rats and mice.

Although rats and mice are acutely sensitive to the toxic and carcinogenic effects of PPs, it is clear that there are marked species differences in response to PP administration (reviewed in Stringer, 1992). In particular, there is a substantial body of evidence to suggest that humans are relatively resistant to the hepatotoxic effects of PPs. For example, cultured human hepatocytes do not show signs of peroxisome proliferation or increased enzyme activity when treated with PPs (Blaauboer et al., 1990), and follow-up studies of patients that have received fibrate drug therapy for hyperlipidaemia and hypertriglyceridaemia show no evidence of an increased incidence of liver tumours (Frick et al., 1987). To substantiate these observations, and to provide definitive evidence that there is no increased carcinogenic risk to humans from therapeutic or environmental exposure to PPs, it is necessary to establish the mechanisms by which these compounds exert their effects. Such mechanistic information should provide an answer to the apparent conundrum that while humans clearly respond to PPs in the sense of lowered circulating lipids following fibrate therapy, there is no overt toxicity as seen in the mouse and the rat.

Peroxisome Proliferator-Activated Receptors

Studies on the biochemical effects of PPs in rodents have shown that the activities of certain classes of enzymes are strongly and rapidly upregulated. These enzymes include those involved in the peroxisomal β-oxidation of fatty acids, such as the fatty acyl CoA oxidase and the bifunctional enoyl-CoA hydratase/3-hydroxyl-CoA dehydrogenase (Reddy et al., 1986), and the cytochrome P450 4A enzymes, involved in fatty acid ω-oxidation (Hardwick et al., 1987). The upregulation of the activity of these enzymes is due to increased rates of transcription (Bars et al., 1993). Also, the ability of PPs to interact with a liver cytosolic protein was demonstrated (Lalwani et al., 1987). These properties of binding to an intracellular factor, and regulating specific gene transcription, are very reminiscent of the way that steroid hormones act via interaction with ligand-activated transcription factors (Evans, 1988). Consistent with this notion, a novel member of this transcription factor family that can be activated by PPs was cloned from mouse liver (Issemann and Green, 1990) - this was termed the Peroxisome Proliferator-Activated Receptor, or PPAR. It became apparent shortly afterwards that this receptor is one of a family of closely related receptors that have now been isolated from several species. Four such receptor subtypes have been identified to date (see Table 1).

The original PPAR isolated from mouse liver was designated the alpha subtype, PPARα. Subsequent characterisation of PPARα has indicated very strongly that it is this receptor which mediates the peroxisome proliferation phenomenon and associated effects in rodents, and the hypolipidaemic effects of these compounds in humans. The key observations that have allowed this inference are as follows.

- The PPARα is highly expressed in those rodent tissues that show peroxisome proliferation in response to PP treatment (mainly liver and kidney), and the efficacy with which a compound activates PPARα in an in vitro assay is broadly consistent with its potency as a rodent peroxisome proliferator and hepatocarcinogen (Issemann and Green, 1990).
- The PPARα forms a heterodimeric complex with another nuclear receptor, the Retinoid X Receptor (RXR), and this complex binds to DNA in a sequence-specific fashion (Kliewer et al., 1992). The sequence motif that is recognised by the PPARα:RXR complex is based on a direct repeat of the hexanucleotide sequence TGACCT, the so-called DR-1 motif (Tugwood et al., 1992), and these motifs have been identified in the regulatory regions of a number of PP-responsive genes (see Schoonjans et al., 1996 for review).
- A transgenic mouse line that lacks PPARα shows no peroxisome proliferation, enzyme induction or liver enlargement in response to PP treatment (Lee et al., 1995).
- There is evidence that at least one potent peroxisome proliferator, Wy-14,643, can activate PPARα through direct binding (Devchand et al., 1996), and recent data suggests that structurally diverse PPs can interact with PPARα to cause a conformational change (Dowell et al., 1997).

Table 1. Peroxisome Proliferator-Activated Receptors.

Receptor	Species in which identified	Function	References
PPARα	Mouse, Rat, *Xenopus*, Human	Regulation of lipid metabolism, peroxisome proliferation (rodents)	Issemann and Green, 1990; Dreyer et al., 1992; Göttlicher et al., 1992; Sher et al., 1993
PPARβ	*Xenopus*	Unknown	Dreyer et al., 1992
PPARγ	Mouse, *Xenopus*, Human, Hamster	Adipocyte differentiation	Dreyer et al., 1992; Tontonoz et al., 1994; Aperlo et al., 1995; Lambe and Tugwood, 1996
PPARδ	Rat, Human	Unknown	Schmidt et al., 1992; Xing et al., 1995

Based on these key observations, a schematic model of peroxisome proliferator action can be constructed (Fig.1). By analogy with other well studied receptor systems, the binding of ligand by PPARα and by RXR (the natural ligand of which is 9-*cis* retinoic acid) allows the formation of the heterodimeric complex, which binds to the DR-1 element, or peroxisome proliferator response element (PPRE - see Mangelsdorf and Evans, 1995 for review). The bound heterodimer is presumed to influence the rate of transcription by interacting with the RNA polymerase II complex and other components of the 'basal' transcription machinery, either directly or via 'bridging' factors. Hepatic peroxisome proliferation and associated effects are brought about by the regulation of key genes involved in the peroxisomal β-oxidation of fatty acids, including acyl CoA oxidase, acyl CoA synthetase, and enoyl-CoA hydratase/3-hydroxyl-CoA dehydrogenase. Similarly, transcriptional regulation of other hepatic genes contributes to the lipid homeostatic effects of PPs which occur both in rodents and humans. The important genes in these pathways include those encoding certain apolipoproteins and hepatic lipases (reviewed in Schoonjans et al., 1996).

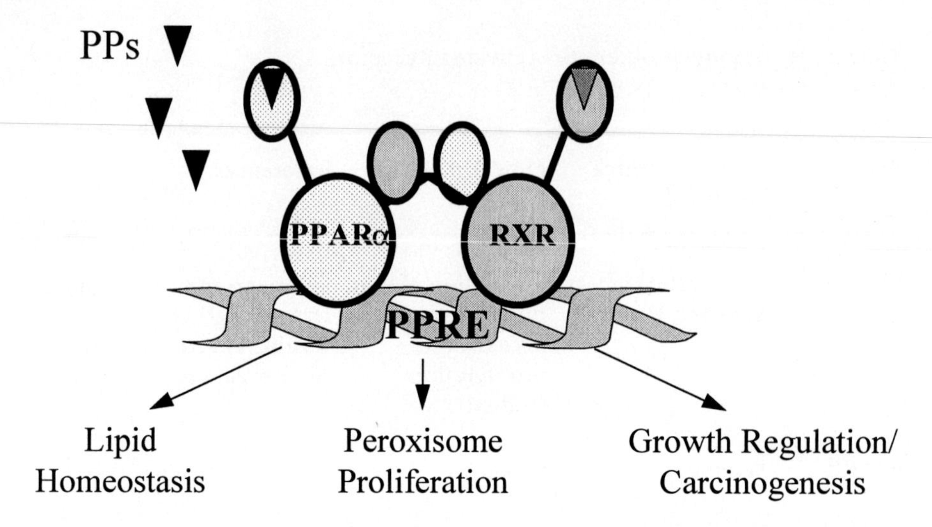

Fig.1. Schematic representation of peroxisome proliferator action.
Following ligand activation, the PPARα:RXR complex binds specifically to DNA at peroxisome proliferator response elements (PPREs), thereby regulating gene expression. Genes regulated in this fashion by PPs include those involved in lipid homeostasis, peroxisome proliferation and, it is hypothesised, liver growth regulation and hepatocarcinogenesis in rodents.

Human PPARα s

A substantial amount of evidence supports the contention that the response of rodents and humans to PPs differs in several important respects. Since the pleiotropic effects of PPs in rodents are mediated by PPARα, it follows that inter-species variations in PPARα abundance, distribution or function may be responsible for such differences. With this in mind, we have been studying PPARαs in humans to try to understand more fully the basis of species differences in PP responses.

Whereas PPARα mRNAs can be detected in human liver by hybridisation experiments, the abundance of the transcripts appears to be substantially less than in rodent liver (Tugwood et al., 1996). This simple analysis also reveals inter-individual variations in PPARα transcript abundance. The lower overall PPARα levels and the inter-individual variation are evident with protein analysis as well as with RNA (F.J. Gonzalez, personal communication).

The human PPARα cDNA sequences that have previously been reported in the literature indicate that the predicted amino acid sequences are greater than 80% conserved between mouse and human, and that human PPARαs can be activated by PPs in *in vitro* assays (Sher et al., 1993; Mukherjee et al., 1994). We have isolated from a human liver biopsy a PPARα cDNA which shows subtle variations when compared to those sequences previously identified, and some of these changes are in important functional domains of the PPARα molecule. This variant cDNA we have designated hPPARα6/29, and the amino acid changes therein are shown schematically in Fig.2.

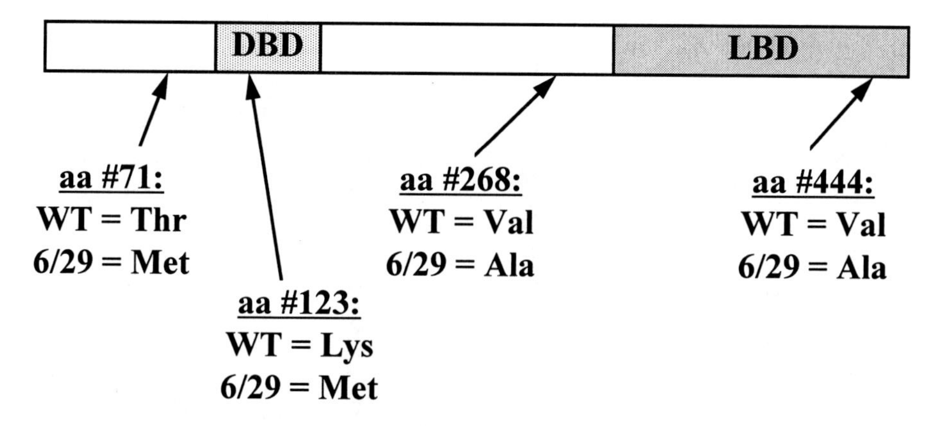

Fig.2. Diagrammatic representation of human PPARα cDNA, showing the regions encoding the DNA binding domain (DBD) and the putative ligand binding domain (LBD). The amino acid variations between hPPARα6/29 (6/29) and previous published hPPARα sequences (WT) are indicated, together with numbered amino acid locations. The full length hPPARα protein comprises 468 amino acids.

A non-conservative lysine to methionine change was observed in the DNA binding domain of hPPARα6/29, which was expected to be functionally significant as this amino acid is conserved not just throughout the PPAR family, but also throughout other nuclear receptor families (Umesono and Evans, 1989). However, the variant receptor was able to bind specifically to a PPRE-containing oligonucleotide in the presence of RXR in an *in vitro* DNA binding assay (Tugwood et al., 1996). As the DNA binding activity of hPPARα6/29 was uncompromised, we were able to use a PPRE-containing luciferase reporter plasmid to analyse the ability of the receptor to be activated by PPs. Transient transfection assays using rat hepatoma cells (Bardot et al., 1993) were used to compare the ability of mouse and human PPARαs to be activated by PPs (Fig.3).

Whereas the mouse PPARα and the human PPARαWT receptors could be activated by PPs *in vitro*, the variant receptor hPPARα6/29 showed no such activity. Since this receptor is capable of binding to DNA, the loss of activity could be due either to loss of the ability to activate transcription, or to loss of the ability to interact with ligand.

The lack of responsiveness of hPPARα6/29 to PPs *in vitro*, coupled with the ability to interact with RXR to bind specifically to PPREs, raises the possibility

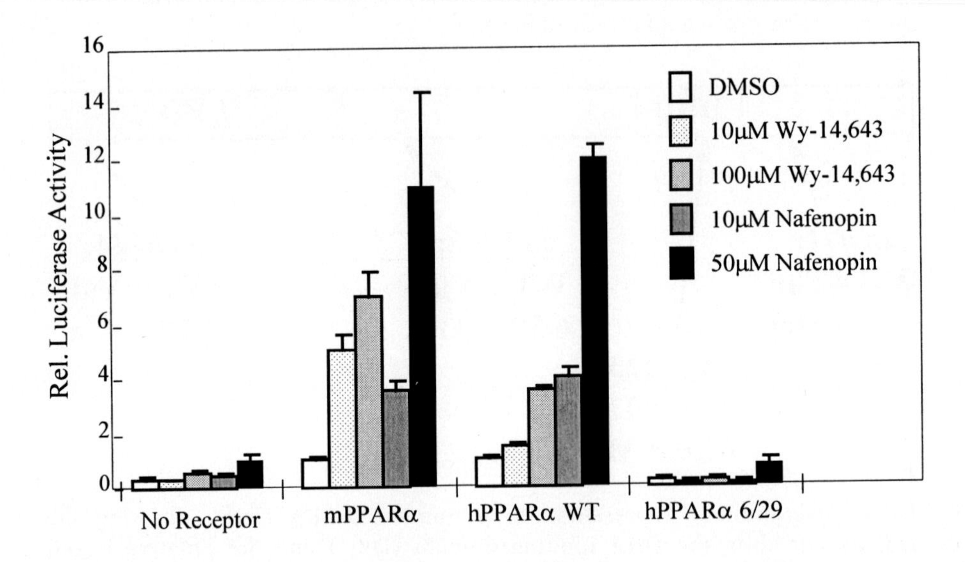

Fig.3. Analysis of PPARα activity using transient transfection analysis. A reporter plasmid containing the luciferase reporter gene under the control of a PPRE-containing promoter DNA fragment was introduced into rat hepatoma cells, together with expression vectors containing PPARα cDNAs as indicated. Cell cultures were treated either with dimethyl sulphoxide vehicle (DMSO), or with the PPs Wy-14,643 or nafenopin as indicated. Cell extracts were assayed for luciferase activity.

that this receptor can behave as a specific 'dominant negative' repressor (reviewed in Yen and Chin, 1994). In this scenario, transcriptionally inactive heterodimers would be formed between RXR and hPPARα6/29, which would compete with active PPARα:RXR heterodimers for PPRE regulatory elements in the promoters of PP target genes. Therefore, if present in excess, hPPARα6/29 could ameliorate the activity of PPARs and, thereby, PP-regulated genes. The ability of hPPARα6/29 to inhibit the activity of mouse PPARα was tested using an adaptation of the transient transfection assay (Fig.4). Increasing amounts of hPPARα6/29 expression vector were introduced into rat hepatoma cells along with a standard amount of mPPARα, in the presence of the PP Wy-14,643. The hPPARa6/29 receptor was able to downregulate the activity of mPPARα in a concentration-dependent fashion, and adding hPPARα6/29 in a six-fold excess over mPPARα resulted in a complete abolition of activity (Fig.4, column 9).

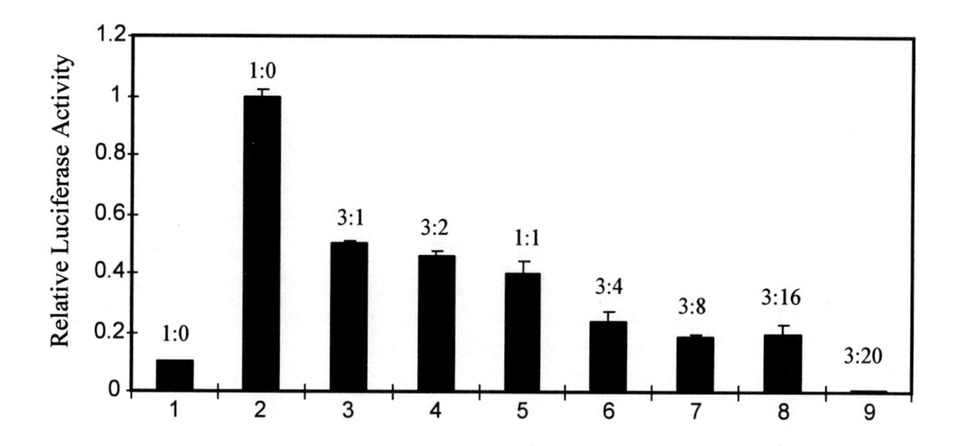

Fig.4. Dominant negative activity of hPPARα6/29. Separate rat hepatoma cell cultures were transfected with a PPRE-containing luciferase reporter plasmid, together with varying ratios of mPPARα to hPPARα6/29 as indicated by the figures above the columns. Cultures were maintained in the absence (column 1) or the presence (columns 2-9) of 100 μM Wy-14,643. Cell extracts were assayed for luciferase activity.

These results indicate that the variant receptor hPPARα6/29 can indeed behave as a dominant negative repressor of PPAR activity, and as such will be a useful tool with which to investigate PPAR function.

PPARα and Human Responses to Peroxisome Proliferators

Analyses on the relatively few human PPARα cDNAs isolated so far provides preliminary evidence for inter-individual variation in receptor structure and function. This is consistent with clinical observations that there are variations in responses of patients to fibrate hypolipidaemic drugs. Furthermore, the observation in some studies that a small proportion of patients are complete non-responders to these drugs (J. Auwerx, personal communication) is also consistent with the possibility that some individuals may possess aberrant PPARαs like hPPARα6/29. Further investigation will be necessary to identify more human PPARαs, and to relate their structure and function to patient status and response to fibrates.

Thus, the existence of variant human PPARαs such as hPPARα6/29 may explain the lack of response to PPs in some individuals, but a more general explanation is required for the differential nature of human and rodent responses. Perhaps the simplest hypothesis is that the lower levels of PPARαs in human livers allows only a subset of the responses to PPs to take place, whereas in rodent livers the greater abundance of receptors allows the full range of pleiotropic responses to occur. This hypothesis requires that different subsets of PP-responsive genes have different 'thresholds' for activation by PPARα, which is conceptually valid but for which evidence is lacking at present.

Another possible explanation of species specific responses to PPs is the existence of PPARα co-activators or repressors, which may be expressed in a species-specific manner. In addition to RXR, there are several other nuclear receptors which recognise similar DR-1 - type response elements and which have been shown to modulate the activity of PPARα, for example COUP-TFII (Marcus et al., 1996) and HNF-4 (Hertz et al., 1995).

Whereas both rodent and human 'PP-responsive' genes have been shown to contain PPREs (reviewed in Schoonjans et al., 1996), the ability of cognate human and rodent genes to respond to PPs can be very different. For example, whereas the rat apo-AII gene is downregulated by fibrates (Staels et al., 1992), the human apo-AII is upregulated by the same stimulus (Vu-Dac et al., 1995). This suggests that gene promoter 'context' may be important in species-specific gene expression, and that PPREs may bind a variety of factors dependent on their location within the promoter, and on the nature of adjacent DNA sequences. This may provide an explanation of why the human acyl CoA oxidase gene is not induced in human hepatocytes by PPs, even though a consensus PPRE element has been identified within the promoter (Varanasi et al., 1994, 1996).

It now seems clear that the pleiotropic effects of PPs in rodents are mediated by PPARα, and that this receptor plays an important natural role in fatty acid metabolism. There is an emerging picture of key differences in PPARα status between rodents and humans, including receptor abundance, structure and responsiveness to PPs. A more complete understanding of these inter-species differences is required before it can be stated definitively that PPs pose no toxic or carcinogenic risk to humans.

References

Aperlo C, Pognonec P, Saladin R, Auwerx J, Boulukos KE (1995) cDNA cloning and characterization of the transcriptional activities of the hamster peroxisome proliferator-activated receptor haPPARgamma. Gene 162:297-302

Ashby J, Brady A, Elcombe CR, Elliott BM, Ishmael J, Odum J, Tugwood JD, Kettle S, Purchase IFH (1994) Mechanistically-based human hazard assessment of peroxisome proliferator-induced hepatocarcinogenesis. Hum Exp Tox 13:S1-S117

Bardot O, Aldridge TC, Latruffe N, Green S (1993) PPAR-RXR heterodimer activates a peroxisome proliferator response element upstream of the bifunctional enzyme gene. Biochem Biophys Res Comm 192:37-45

Bars RG, Bell DR, Elcombe CR (1993) Induction of cytochrome P450 and peroxisomal enzymes by clofibric acid in vivo and in vitro. Biochem Pharmacol 45:2045-2053

Blaauboer BJ, van Holstein CW, Bleumink R, Mennes WC, van Peit FN, Yap SH, van Pelt JF, van Iersel AA, Timmerman A, Schmidt BP (1990) The effects of beclobric acid and clofibric acid on peroxisomal β-oxidation and peroxisome proliferation in primary cultures of rat, monkey and human hepatocytes. Biochem Pharmacol 40:521-528

Devchand PR, Keller H, Peters JM, Vazquez M, Gonzalez FJ, Wahli W (1996) The PPARα-leukotriene B4 pathway to inflammation control. Nature 384:39-43

Dowell P, Peterson VJ, Zabriskie TM, Leid M (1997) Ligand-induced peroxisome proliferator-activated receptor alpha conformational change. J Biol Chem 272:2013-2020

Dreyer C, Krey G, Keller H, Givel F, Helftenbein G, Wahli W (1992) Control of the peroxisomal β-oxidation pathway by a novel family of nuclear hormone receptors. Cell 68:879-887

Evans RM (1988) The steroid and thyroid hormone receptor superfamily. Science 240:889-895

Frick H, Elo, Haapa K, Heinonen OP, et al. (1987) Helsinki heart study: Primary-prevention trial with gemfibrozil in middle-aged men with dislipidemia. New Eng J Med 317:1235-1247

Göttlicher M, Widmark E, Li Q, Gustafsson J-Å (1992) Fatty acids activate a chimera of the clofibric acid-activated receptor and the glucocorticoid receptor. Proc Natl Acad Sci 89:4653-4657

Hardwick JP, Song B-J, Huberman E, Gonzalez FJ (1987) Isolation, complementary DNA sequence, and regulation of rat hepatic lauric acid omega-hydroxylase (cytochrome P-450LA(). J Biol Chem 262:801-810

Hertz R, Bishara-Shieban J, Bar-Tana J (1995) Mode of action of peroxisome proliferators as hypolipidemic drugs. J Biol Chem 270:13470-13475

Issemann I, Green S (1990) Activation of a member of the steroid hormone receptor superfamily by peroxisome proliferators. Nature 347:645-649

Kliewer SA, Umesono K, Noonan DJ, Heyman RA, Evans RM (1992) Convergence of 9-cis retinoic acid and peroxisome proliferator signalling pathways through heterodimer formation of their receptors. Nature 358:771-774

Lalwani ND, Alvares K, Reddy MK, Reddy MN, Parikh I, Reddy JK (1987) Peroxisome proliferator-binding protein: identification and partial characterization of nafenopin-clofibric acid-, and ciprofibrate-binding proteins from rat liver. Proc Natl Acad Sci 84:5242-5246

Lambe KG, Tugwood JD (1996) A human peroxisome-proliferator-activated receptor-gamma is activated by inducers of adipogenesis, including thiazolidinedione drugs. Eur J Biochem 239:1-7

Lee SST, Pineau T, Drago J, Lee EJ, Owens JW, Kroetz DL, Fernandez-Salguero PM, Westphal H, Gonzalez FJ (1995) Targeted disruption of the Ω isoform of the peroxisome proliferator-activated receptor gene in mice results in abolishment of the pleiotropic effects of peroxisome proliferators. Mol Cell Biol 15:3012-3022

Mangelsdorf DJ, Evans RM (1995) The RXR heterodimers and orphan receptors. Cell 83:841-850

Marcus SL, Capone JP, Rachubinski RA (1996) Identification of COUP-TFII as a peroxisome proliferator response element binding factor using genetic selection in yeast: COUP-TFII activates transcription in yeast but antagonizes PPAR signaling in mammalian cells. Mol Cell Endocrinol 120:31-39

Mukherjee R, Jow L, Noonan D, McDonnell DP (1994) Human and rat peroxisome proliferator activated receptors (PPARs) demonstrate similar tissue distribution but different responsiveness to PPAR activators. J Ster Biochem Mol Biol 51:157-166

Reddy JK, Goel SK, Nemali MR, Carrino JJ, Laffler TG, Reddy MK, Sperbeck SJ, Osumi T, Hashimoto T, Lalwani ND, Rao MS (1986) Transcriptional regulation of peroxisomal fatty acyl-CoA oxidase and enoyl-CoA hydratase/3-hydroxyacyl-CoA dehydrogenase in rat liver by peroxisome proliferators. Proc Natl Acad Sci 83:1747-1751

Schmidt A, Endo N, Rutledge SJ, Vogel R, Shinar D, Rodan GA (1992) Identification of a new member of the steroid hormone receptor superfamily that is activated by a peroxisome proliferator and fatty acids. Mol Endocrinol 6:1634-1641

Schoonjans K, Staels B, Auwerx J (1996) Role of the peroxisome proliferator activated receptor (PPAR) in mediating the effects of fibrates and fatty acids on gene expression. J Lipid Res 37:905-925

Sher T, Yi H-F, McBride OW, Gonzalez FJ (1993) cDNA cloning, chromosomal mapping, and functional characterisation of the human peroxisome proliferator activated receptor. Biochemistry 32:5598-5604

Staels B, Van Tol A, Andreu T, Auwerx J (1992) Fibrates influence the expression of genes involved in lipoprotein metabolism in a tissue selective manner in the rat. Arterioscler & Thromb 12:286-294

Stringer DA (ed) (1992) ECETOC Monograph No.17 Hepatic Peroxisome Proliferation. Brussels, Belgium

Tontonoz P, Hu E, Graves RA, Budavari AI, Spiegelman BM (1994) mPPAR-gamma2: tissue-specific regulator of an adipocyte enhancer. Genes & Devel 8:1224-1234

Tugwood JD, Issemann I, Anderson RG, Bundell KR, McPheat WL, Green S (1992) The mouse peroxisome proliferator activated receptor recognizes a response element in the 5' flanking region of the rat acyl CoA oxidase gene. EMBO J 11:433-439

Tugwood JD, Aldridge TC, Lambe KG, Macdonald N, Woodyatt NJ (1996) Peroxisome proliferator-activated receptors: structures and function. In: Reddy JK, Suga T, Mannaerts GP, Lazarow PB, Subramani S (eds) Peroxisomes: Biology and role in toxicology and disease. Annals of the New York Academy of Sciences, Vol.804, New York, pp. 252-265

Umesono K, Evans RM (1989) Determinants of target gene specificity for steroid/thyroid hormone receptors. Cell 57:1139-1146

Varanasi U, Chu R, Chu S, Espinosa R, LeBeau MM, Reddy JK (1994) Isolation of the human peroxisomal acyl-CoA oxidase gene: organization, promoter analysis, and chromosomal localization. Proc Natl Acad Sci 91:3107-3111

Varanasi U, Chu R, Huang Q, Castellon R, Yeldandi AV, Reddy JK (1996) Identification of a peroxisome proliferator-responsive element upstream of the human peroxisomal fatty acyl coenzyme A oxidase gene. J Biol Chem 271:2147-2155

Vu-Dac N, Schoonjans K, Kosyth V, Dallongeville J, Fruchart JC, Staels B, Auwerx J (1995) Fibrates increase human apolipoprotein A-II expression through activation of the peroxisome proliferator-activated receptor. J Clin Invest 96:741-750

Xing G, Zhang L, Zhang L, Heynen T, Yoshikawa T, Smith M, Weiss S, Detera-Wadleigh S (1995) Rat PPAR-delta contains a CGG triplet repeat and is prominently expressed in the thalamic nuclei. Biochem Biophys Res Comm 217:1015-1025

Yen PM, Chin WW (1994) Molecular mechanisms of dominant negative activity by nuclear hormone receptors. Mol Endocrinol 8:1450-1454

Excitotoxicity in the Brain

Frode Fonnum
VISTA, Norwegian Defence Research Establishment, Division for Environmental Toxicology, 2007 Kjeller (Norway)

Excitotoxins

Excitotoxins are a special group of neurotoxic substances that excite somatic and dendritic receptors in such a way that the neurons may die. All excitotoxins are in principle agonists of glutamate receptors in the brain and are structurally related to glutamate. The excitotoxins can be administered systemically before the blood brain barrier is fully developed (at day 10 in rats) or they can be injected locally into the brain. In the latter case, they produce the so called axon-sparing/dendritosomatic lesions.

Hayashi (1954) was the first to show that glutamate added to the motor cortex produced excitation and convulsions. Subsequently, Lucas and Newhouse (1957) showed degeneration of GABAergic neurons in the retina when they administered glutamate systemically to young rats. In 1960, the fundamental study by Curtis and Watkins appeared, showing that a large number of amino acids, structurally related to glutamate, caused excitation of neurons. It was, however, Olney (1969) who showed a correlation between excitatory potential and neurotoxicity of such compounds.

When glutamate (0.1 to 1 g) is given systemically to rodents, the most sensitive regions of the brain were the retina, the mediobasal hypothalamus (nucleus arcuatus), the area postrema and the so called circumventricular nuclei. The lesions were first described histologically by Olney and coworkers in a series of papers (Olney, 1969; Olney et al, 1990). In our laboratory we saw this as a means of identifying neutrotransmitters in specific pathways. In the retina we showed that there was not only a loss of GABAergic fibres, but also a loss of cholinergic and dopaminergic fibres together with the loss of horizontal and amacrine cells (Karlsen and Fonnum, 1976). From lesions produced in the nucleus arcuatus and median eminence, we showed the existence of cholinergic and GABAergic elements in the tuberoinfundural tract and also indicated that dopaminergic cells in the nucleus arcuatus were less sensitive to glutamate, possibly because they contain less glutamate receptors.

Later a series of more effective compounds such as kainate, ibotenate, quinolinate and NMDA were described. They could be separated into two groups according to their mode of toxic action when injected into adult rat amygdala. N–methyl-D-aspartate (NMDA) and quinolinate together with L-glutamate, L-aspartate and homocysteate caused local neuronal necrosis without other toxic effects. On the other hand kainate, quisqualate, ibotenate and AMPA caused, in addition to local neuronal necrosis, sustained limbic seizures and a pattern of distant lesions which included hippocampus (Köhler, 1982). The effect of these four excitotoxins has been compared by local injection into different brain regions. In the hippocampus the order of vulnerability to kainate is CA3 pyramidal cells>CA1 pyramidal cells>granular cells>CA2 pyramidal cells. In contrast ibotenate causes degeneration of pyramidal and granular cells to the same extent. Quinolinate is more toxic to CA4 and CA1 pyramical cells than CA3 cells and granular cells. In all cases the cholinergic and serotonergic fibres originating from medial septum and raphe nuclei, respectively, are not lesioned (Köhler, 1982).

Table 1 Excitotoxins

Endogenous candidates	Exogenous candidates
L-glutamate	kainate
L-aspartate	quisqualate
L-cysteine sulphonate	ibotenate
L-cysteine sulphinate	domoate
L-homocysteate	N-methyl-D-aspartate (NMDA)
folates	α-amino-3-hydroxy-5-methyl-
quinolinate	isoxazole-4-propinate (AMPA)
cysteine	β-N-oxylamino-L-alanine

Excitotoxins are also found in food. Domoate is believed to be responsible for the cognitive deficit and also death after mussel poisoning in Canada. Domoate is an excitotoxic analogue of glutamate that selectively interacts with the kainate receptor (Steward et al, 1990). Likewise the poisonous ingredient in *Lathyrus sativus* is β-N-oxalylamino-L-alanine which has powerful excitotoxic properties mediated by the AMPA receptor. β-methylaminoalanine, a weak excitotoxin, has been suggested as the causative agent of the triactive complex (amyotrophic lateral scelerosis/Parkinson/dementia) in Guam (Spencer et al, 1987). Other compounds like MAM which is also isolated from the same plant may participate in the development of the disease.

A compound which does not resemble glutamate but produces excitotoxic effects is cysteine (Olney et al, 1990; Fonnum et al, 1992).

The excitotoxins have been used to (1) make specific lesions in the brain and thereby allowing detailed identification of neurotransmitters belonging to specific pathways, (2) produce lesions simulating certain neurodegenerative diseases like Huntington's disease (3) investigate the mechanism of the excitotoxicity as a model for the degeneration of nerve cells during disease and during ageing.

Glutamate Receptors

Excitotoxicity in the brain may occur when an endogenous or exogenous compound activates a glutamate receptor. Since glutamate is quantitatively the most important neurotransmitter in the brain (Fonnum, 1984), there is a wide distribution of glutamate receptors. The receptors can be classified into ionotropic receptors (NMDA and non-NMDA receptors) and metabotropic receptors (Table 2). The non-NMDA receptors are termed the AMPA and kainate receptors.

Table 2 Glutamate receptors in the brain

Receptor type	NMDA	Non-NMDA		Metabotropic
Agonist	NMDA	AMPA	Kainate	
Subtypes	NR1	GLuR1	GLuR5	mGLuR1 and 5
	NR2A	GLuR2	GLuR6	mGLuR2 and 3
	NR2B	GLuR3	GLuR7	GLuR4, 6, 7, 8
	NR2C	GLuR4	KA1	
	NR2D		KA2	

The NMDA receptor is particularly important for excitotoxicity. It has binding sites for glutamate/NMDA and also for glycine. Both binding sites are necessary for activating the receptor. The ion channel of the NMDA receptor is permeable for Ca^{2+}, K^{+} and Na^{+}. The ion channel has a binding site for phencyclidine and is normally blocked by Mg^{2+}. When the membrane potential is lowered from +90 to +60 - +30 mV, Mg^{2+} is expelled from the ion channel and ion current can take place (Hollmann and Heinemann, 1994). The heteromeric NMDA receptor is apparently modulated by sulphydryl redox reagents and this could be a contributing factor to explain cysteine toxicity (Mathisen et al, 1996). The NMDA receptor can be blocked by NMDA antagonists, glycine antagonists like HA-966 and 7-chlorokynurinine or ion channel blockers such as MK801 (Hollmann and Heinemann, 1994).

The AMPA receptors consist of 4 subunits which are heterooligomers. They have a high permeability for Na^{+} and all except GluR2 have permeability for Ca^{2+}. All the subunits can occur in a "flip or flop" mode. Kainate has a moderate affinity for the AMPA receptors.

There are 5 kainate receptor subunits. Subgroup GluR5, 6 and 7 have weak affinity for AMPA. Domoate isolated from sea weed *Chondria armata* or acromelate isolated from the poisonous mushroom *Clitocybe acromelalga* are alternative agonists. The quinoxalinediones DNQX and CNQX antagonise both AMPA and kainate. NBQX clearly antagonises AMPA better than kainate (Hollmann and Heinemann, 1994).

The metabotropic receptors are all coupled to G-proteins. mGlu1 and 5 are linked to PI turnover, mGlu2 and 3 are linked to adenyl cyclase and mGluR4, 6 and 7 are linked to phospholipase C. Activation of mGluR1, mGluR2, mGluR3 and mGluR5 during exposure to NMDA in hippocampal slices prevented the toxic effects (Pizzi et al, 1966).

Mechanism of Excitotoxicity

Excitotoxicity is not due to a single event, but to a cascade of events. It starts when there is an excessive exposure to an extracellular excitatory compound. Exogenous compounds can be locally injected or administered systemically when the blood brain barrier is weak. Endogenous compounds like aspartate, glutamate and quinolinate can be found in high concentrations extracellularly e.g. during ischemia or when their transport is inhibited (Rothstein et al, 1996). The normal extracellular glutamate concentration is estimated from CSF to be 1 to 5 μM, whereas the concentration within a glutamatergic terminal is 30-50 mM (Fonnum, 1991).

There are two phases of excitotoxicity, one fast and one slow (Choi, 1988). The fast phase is due to swelling caused by an increase of Na^+ influx via the glutamate receptors. This swelling leads to a further increase of water and Cl^- influx. The swelling is a reversible process if the agonist is removed. The slow phase is believed to be more important for neurodegenerative disease. The increase of intracellular Na^+ leads to a depolarization of the neuron, subsequently to a release of Mg^{2+} from the NMDA receptor and an increase of intracellular Ca^{2+}. The increase of intracellular Na^+ can also increase intracellular Ca^{2+} through the Na^+/Ca^{2+} exchanger (Choi, 1988).

The high intracellular Ca^{2+} concentration is a major event in excitotoxicity and has severe consequences. It can activate membrane destructive enzymes such as calpain, endonucleases and phospholipase A_2. It is sequestered into mitochondria and this Ca^{2+} current induces depolarization of the mitochondrial inner membrane. This is consistent with many findings that inhibition of mitochondrial function and the reduced energy levels in neurons make the cells more vulnerable to exitotoxicity (Novelli et al, 1988). The depolarization allows the release of reactive oxygen species (Schinder et al, 1966). Such free radicals can also be produced by calcium activation from xanthine oxidase, NO synthase and arachidonate. The oxygen radicals will attack proteins, lipids and DNA and may be the last step in the excitotoxic destruction of neurons.

Mitochondrial Function During Excitotoxicity

Mitochondria are involved in excitotoxicity either by their ability to take up cytoplasmic Ca^{2+} or by their ability to provide energy (ATP) to the cells. Mitochondria are the only subcellular structure with high enough capacity to deal with the increase in cytosolic Ca^{2+} during excitotoxicity. Ca^{2+} is sequestered into the mitochondrial matrix driven by the proton electrochemical gradient generated by the electron transport chain. This reduction in the electrochemical gradient decreases ATP synthesis at a time with great demand for the Ca-pump (White and Reynolds, 1995). The reduction of the electrochemical gradient induces the opening of the permeability transition pore of the mitochondrial inner membrane thereby allowing ionic diffusion and collapse of the membrane potential (Schinder et al, 1966). The drop in ATP levels in concert with the generation of free oxygen radicals is the prime cause of cell death. The boundary between cell survival and cell death lies in the subtle balance between mitochondrial function and dysfunction centered on membrane potential (Ψ).

This is in agreement with the many findings that inhibition of the TCA cycle or the electron transport chain will also lead to a decrease in ATP and thereby increase the vulnerability to excitotoxins. Novelli et al (1988) showed that addition of 1mM KCN or omission of glucose increase the toxicity of glutamate to cerebellar granular cells substantially. Also preincubation with 1 μM ouabain, an inhibitor of the Na^+/K^+ pump, potentiates the toxicity of glutamate and NMDA. Similarly, the toxicity of NMDA is increased when the cells were treated with aminooxyacetate, an inhibitor of the malate-aspartate cycle (McDonald and Schoepp, 1993), or nitropropionate, an irreversible inhibitor of succinic dehydrogenase (Brouillet et al, 1993). The excitotoxic potential of NMDA, AMPA and glutamate was increased when administered together with malonate, a reversible succinic dehydrogenase inhibitor (Greene and Greenamyre, 1995). The most likely explanation for the results is the inability of the neuron to maintain energy production in the absence of glucose or in the presence of metabolic inhibitors. This leads to a reduction in activity of the ion pumps involved in maintaining the resting potential. We assume that partial depolarization occurs when the energy levels are reduced relieving the voltage dependent Mg^{2+} block of the NMDA receptor channel. The effect of ouabain, an inhibitor of ion pumps responsible for maintaining the resting potential, also shows that reduced activity of these pumps permits glutamate to become more toxic.

Free Radicals in Excitotoxicity

Oxidative stress is being increasingly implicated as a mediator of excitotoxic cell death (Coyle and Putfarken, 1993). NMDA activation of cultured granular cells showed free radical formation by electron spin resonance (Lafor-Cazal et al, 1993). Kainate treatment also leads to free radical formation (Coyle and Putfarken, 1993). When all cell cultures were deprived of oxygen for 2-24 hours after an excitotoxic insult, there was no further neuronal death beyond those caused by oxygen depletion. Therefore oxygen-based reactions are an essential component of calcium-mediated cell death (Dubinsky et al, 1995). Also overexpression of Bcl-2 in neurons protects them from both free radicals and excitotoxicity showing a similar reaction (Lawrence et al, 1996).

Several sources have been proposed for glutamate induced reactive oxygen species. Mitochondria are a principal source of reactive oxygen species (ROS) in the brain. The leakage of high energy electrons along the electron transport chain causes the production of O^- and H_2O_2. This is important during a decreased membrane potential. FCCP which blocks the ROS production in mitochondria, also blocks the glutamate response in cultured forebrain membranes (Reynolds and Hastings, 1995). Dykens (1994) demonstrated that intact mitochondria generated $OH^\cdot$ when exposed to pathological concentrations of Ca^{2+} and ADP.

There are several other alternative ways of producing ROS. Although xanthin oxidase activity is normally low in the brain, Dykens (1994) have shown a conversion of xanthin dehydrogenase to xanthine oxidase by Ca^{2+} activated proteases during excitotoxic action. In cultured cerebellar granule cells, Lafon-Cazal et al (1993) showed that activation of phospholipase A_2 leads to arachidonate which gives ROS on further metabolism. A third source of ROS may come from the conversion of NO to peroxynitrite by Ca^{2+} activated nitric oxide synthease. Inhibition of this enzyme has in few cases limited excitotoxicity (Clementi et al, 1966). $NO^\cdot$ may play a dual role having both detrimental and protective effects since it removes $O^{\cdot-}$ on production of peroxynitrite.

Neuroprotective Strategies

Different pharmacological approaches to attenuate excitotoxic lesions may contribute to our understanding of the toxic mechanism. Blockers of glutamate release can in principle protect against excitotoxins. MK801, HA-966 and dextromethorphan, acting on different sites of the NMDA receptor, selectively antagonize the receptor and ameliorate lesions. Also NBQX and CNQX which antagonize the AMPA receptor attenuate lesions (Ikonomidou and Turski (1995). Specific Calcium channel blockers are of less use with the exception of dantrolene which reduces Ca^{2+} release from intracellular stores (Rothstein & Kuncl, 1995). Enhancers of mitochondrial function like coenzyme Q_{10} and nicotinamide can block lesions and reduce ATP depletion (Schultz et al, 1966). Blockage of the permeability transition pore in mitochondria by cyclosporine or other agents should in principle attenuate lesions (Brockmeier and Pfeiffer, 1995). Antioxidants such as tBPN are useful (Schultz et al, 1966). ATX, an inhibitor of calcium endonuclease has been shown to block glutamate-mediated neurotoxicity (Roberto-Lewis et al, 1993).

There is substantial evidence that growth factors can protect neurons from excitotoxicity. In primary cultures of hippocampal pyramidal neurons, bFGF (fibroblast growth factor) increased the level of glutamate required for toxicity and reduced the increase in intracellular Ca^{2+} caused by glutamate. Protection of cultured cells from glutamate toxicity has now been reported by the use of several growth factors such as for BDNF, TGFβ, TNFs, APP and EGF. In most cases pretreatment of 12-24 h is required for maximum protection of excitotoxicity by growth factors suggesting an action at the gene level. The same factors also offer protection against cerebral ischemia (Barger and Mattson, 1995).

Role of Excitotoxicity in Neurodegenerative Disorders

Excitotoxicity has been indicated in several neurodegenerative disorders such as stroke, trauma, hypoglycaemia, Huntington's disease, amylotrophic lateral sclerosis (ALS) and to lesser extent in Parkinson's syndrome and Alzheimer's disease. There is strong evidence for the participation of exitotoxicity in both focal and global ischemia and hypoglycaemia (Review: Ikonomidou and Turski, 1966). Head trauma like concussion, high cerebral pressure and acute subdural hematoma cause high extracellular level of glutamate (Engelsen et al, 1985) and the primary damage is ameliorated by NMDA receptor antagonists and NBQX (Lipton, 1993; Faden et al, 1989).

There has been considerable progress in our knowledge on the toxic mechanisms involved in ALS. Both excitotoxicity partly due to the loss of glutamate transport (Rothstein et al, 1996) and ROS due to mutations in Cu/Zn superoxide dismutase (Bowling et al, 1993) have been forwarded as causal agents in some forms of the illness. In Huntington's disease the neurons degenerated in the striatum are similar to those destroyed by NMDA agonists (Brouillet et al, 1993). Also the neurotoxic effects of the Parkinson inducing compound 1-methyl-4-phenylpyridinium could be blocked by NMDA antagonists (Turski et al, 1991).

The difficulty in applying the excitotoxic theory to the neurological disorders other than stroke and trauma, is the fact that the disorders develop over several years and that glutamate, the main endogenous candidate, is actively removed from the extracellular space by active transport. The finding that disturbances of the mitochondrial energy metabolism make the neurons more sensitive to excitotoxicity has increased the interest in excitotoxicity. The mitochondrial DNA is 10 times more exposed to ROS and has less efficient repair systems than nuclear DNA (Mecocci et al, 1993). This makes brain mitochondria less active during ageing and could thus increase the vulnerability of neurons during ageing. Neurological disorders such as Huntington's disease can therefore be a product of both a decreased energy level in neurons either caused by an exogenous or endogenous factor decreasing the mitochondrial function and an increased susceptibility to glutamate toxicity. The effect of nitropropionate which inhibits succinic dehydrogenase, but kills by excitotoxicity in aged animals is an example of such a mechanism (Brouillet et al, 1993).

References

Barger SW, Mattson MP (1995) Excitatory amino acids and growth factors: biological and molecular interactions regulating neuronal survival and plasticity. In: Stone TW (ed) CNS neurotransmitters and neuromodulators: Glutamate. CRC Press, pp 273-294

Bowling AC, Schulz JB, Brown RH, Beal MF (1993) Superoxide dismutase activity, oxidative damage, and mitochondrial energy metabolism in familial and sporadic amyotrophic lateral sclerosis. J Neurochem 61:2322-2325

Brouillet E, Jenkins BG, Hyman T, Ferrante J, Kowall NW, Srivastava R, Roy D, Rosen BR, Beal M (1993). Age-dependent vulnerability of the striatum to the mitochondrial toxin 3-nitropropionic acid. J Neurochem 60:356-360

Choi DW (1988) Glutamate neurotoxicity and diseases of the nervous system. Neuron 1:623-634

Clementi E, Racchetti G, Melino G, Meldolesi J (1996). Cytosolic Ca^{2+} buffering, a cell property that in some neurons markedly decreases during aging, has a protective effect against NMDA/nitric oxide-induced excitotoxicity. Life Science 59 (5-6):389-397

Coyle JT, Puttfarcken P (1993) Oxidative stress, glutamate and neurodegenerative disorders. Science 262:689-695

Curtis DR and Watkins JC (1960) The excitation and depression of spinal neurons by structurally related amino acids. J Neurochem 6:117-127

Dubinsky, Kristal S, Fournier ME (1995) Obligate role for oxygen in the early stages of glutamate-induced, delayed neuronal Death. J Neuroscience 15, 7071-7078

Dykens JA (1994) Isolated cerebral and cerebellar mitochondria produce free radicals when exposed to elevated Ca^{2+} and Na^{+}: implications for neurodegeneration. J Neurochem 63:584-591

Engelsen BA, Fosse VM, Myrseth E, Fonnum F (1985) Elevated concentrations of glutamate and aspartate in human ventricular cerebrospinal fluid (vCSF) during episodes of increased CSF pressure and clinical signs of impaired brain circulation. Neurosci-Lett 20; 62(1):97-102

Faden AI, Demeduck P, Scott Panter S, Vink R (1989) The role of excitatory amino acids and NMDA receptors in traumatic brain injury. Science 244:798-800

Fonnum F (1984) Glutamate a neurotransmitter in mammalian brain. J Neurochem 42: 1-10

Fonnum F (1991) Neurochemical studies on glutamate mediated neurotransmission, in: Meldrum BS, Moroni F, Simen RP and Woods J H (eds) Excitatory Amino Acids. Raven Press New York 15-25

Fonnum F, Malthe-Sorenssen D, Lund-Karlsen R, Oddan E (1992) Change in neurotransmitter parameters in the brain induced by L-cysteine injections in the young rat. Brain Res 579:74-84

Greene JG, Greenamyre JT (1995) Exacerbation of NMDA, AMDA and L-glutamate excitotoxicity by the succinate dehydrogenase inhibition malonate. J Neurochem 64:2332-2338

Hayashi T (1954) Effect of sodium glutamate on the nervous system. Keio J Med 3:183-192

Hollman M and Heinemann S (1994) Cloned glutamate receptors. Ann Rev Neurosci 17:31-108

Ikonomidou C, Turski L (1995) In: Stone T W (ed) CNS neurotransmitters and neuromodulator: Glutamate. CRC Press, pp 253-272

Karlsen RL, Fonnum F (1976) The toxic effect of sodium glutamate on retina: neurotransmitter changes. J Neurochem 27:1437-1441

Köhler C (1982) Neuronal degeneration after intracerebral injections of excitotoxin. In: Fuxe K, Roberts P, Schwartz R (eds) Excitotoxin, MaxMillan pp 99-111

Lafon-Cazal M, Pletri S, Culcasi M, Backaert J (1993) NMDA-dependent superoxide production and neurotoxicity. Nature 364:535-537

Lawrence MS, Ho DY, Sun GH, Steinberg GK, Sapolsky RM (1996) Overexpression of Bcl-2 with herpes simplex virus vectors protects CNS neurons against neurological insults in vitro and in vivo. J Neurosci 16:486-496

Lipton SA (1993) Molecular mechanisms of trauma-induced neuronal degeneration. Curr Opinon Neurol Neurosurg 6:588-596

Lucas DR, Newhouse (1957) The toxic effect of sodium L-glutamate on the inner layers of the retina. Arch Ophthalmol 58:193-201

Mathisen GA, Fonnum F, Paulsen RE (1996) Contributing mechanisms for cysteine excitotoxicity in cultured cerebellar granule cells. Neurochem Res 21, 293-298

McDonald JW, Schoepp DD (1993) Aminoxyacetic acid produces excitotoxic brain injury in neonatal rats. Brain Res 624:239-244

Mecocci P, MacGarvey V, Kaufman AE, Koontz D, Shoffner JM, Wallace D, Beal MF (1993) Oxidative damage to mitochondrial DNA shows marked age-dependent increases in human brain. Ann Neurol 34:609-616

Novelli A, Reilly JA, Lysko PG, Henneberry RC (1988) Glutamate becomes neurotoxic via the NMDA receptor when intracellular energy levels are reduced. Brain Res 451:205-212

Olney JW (1969) Brain lesions, obesity and other disturbances in mice treated with monosodium glutamate. Science 164:719-721

Olney JW, Zorumski C, Price MT, Labryere J (1990) Cysteine, a bicarbonate-sensitive endogenous excitotoxin. Science 248:596-599

Pizzi O, Consolandi M, Spano PF (1996) Activation of multiple metabotropic glutamate receptor subtypes prevents NMDA-induced excitotoxicity in rat hippocampal slices. J Europ Neuroscience 8:1516-1521

Reynolds J, Hastings G (1995) Glutamate induces the production of reactive oxygen species in cultured forebrain neurons following NMDA receptor activation. J Neurochem 15:3318-3327

Roberto-Lewis JM, Marcy VR, Zhao Y, Vaught JL, Seman R, Lewis ME (1993) Aurintricarboxylic acid protects hippocampal neurons from NMDA- and ischemia-induced toxicity in vivo. J Neurochem 61:378-381

Rothstein, JD, Kuncl RW (1995) Neuroprotective strategies in a model of chronic glutamate-mediated motor neuron toxicity. J. Neurochem 65:643-652

Rothstein, JD, Dykes-Hoberg M, Pardo CA, Bristol LA, Jin L, Kuncl RW, Kanai Y, Hediger MA, Wang Y, Schielke JP, Welty DF (1996) Knockout of glutamate transporters reveals a major role for astroglial transport in excitotoxicity and clearance of glutamate. Neuron 16:675-86

Schinder AF, Olsen EC, Spitzer NC, Montal M (1966) Mitochondrial dysfunction is a primary event in glutamate neurotoxicity. J Neuroscience 16:6125-6133

Schulz JB, Matthews RT, Henshaw DR, Beal MP (1996) Neuroprotective strategies for treatment of lesions produced by mitochondrial toxins: implications for neurogenerative diseases. Neuroscience 71:1043-1048

Spencer PS, Nunn PB, Hugon J, Ludolph AC, Ross SM Roy DN, Robertson RC (1987) Guam amyotrophic lateral sclerosis-Parkinsonism-dementia linked to a plant excitant neurotoxin. Science 237:517-522

Stewart GR, Zorumski CF, Price MT, Olney JW (1990) Domoic acid: A dementia-inducing excitotoxic food poison with kainic acid receptor specificity. Exp Neurol 110:127-138

Turski L, Bressler K, Rettig K-J, Löschmann PA, Wachtel H (1991) Protection of substantia nigra from MPP^+ neurotoxicity by N-methyl-D-aspartate antagonists. Nature 349:414-417

White RJ, Reynolds IJ (1995) Mitochondria and Na^+/Ca^{2+} exchange buffer glutamate-induced calcium loads in cultured cortical neurons. J Neurosci 15:1318-1328

New Frontiers in Human and Ecological Toxicology: Determining Genetic *vs* Environmental Causes of Susceptibility

(Chair:D. Baird, United Kingdom, and V. Forbes, Denmark)

Variability in the Response of *Daphnia* Clones to Toxic Substances: Are Safety Margins Being Compromised?

Donald J Baird and Carlos Barata
Environment Group, Institute of Aquaculture, University of Stirling, Stirling, Scotland. FK9 4LA

Introduction

Daphnia magna continues to be the most widely used organism in aquatic hazard assessment, for reasons relating to its ease of culture in the laboratory and the ability to maintain clonal lines of female individuals. However, despite its widespread use for estimation of the acute and chronic toxicity of new and existing substances, doubts still remain concerning its use as a representative of aquatic fauna, with interlaboratory variation in response being recognised as a serious problem. Experimental studies which have attempted to identify and quantify the components of interlaboratory variation in response (Baird et al., 1989, 1991; Soares et al., 1992) have indicated that both the genotype of the clone used in tests and the environmental conditions prior to and during the tests play significant roles in determining this variability. Moreover, it is clear that the relative response of clones does not remain consistent across the substances tested, not even for similar classes of substance, indicating that genotype-environment interactions also play an important role in governing response.

One question which has not yet received consideration with regard to variability in the response of laboratory clones is: do such clones yield consistent performance over time, both in terms of their life-history characteristics and in their response to toxic substances? This is clearly an important issue, since it is possible that the level of between-clone variability reported in previous studies may itself vary, thus compounding the problem of interpretation of test data. Here we address this problem by presenting new data on a set of standard laboratory clones which have been maintained in culture at Stirling for over 100 generations. These results, together with results from other studies, are discussed with reference to the processes involved in establishing and maintaining laboratory cultures. In doing this, our aim is to raise important issues regarding the stability in clone response over time, and whether the ongoing move towards increasing test standardisation is likely to yield any benefits in test repeatability within and between laboratories.

The Use of *Daphnia* Data in Risk Assessment

Current procedures for hazard assessment within the EU routinely require the performance of standard toxicity tests using *Daphnia*. These may be either acute EC_{50} estimations or chronic life-cycle tests. Given the nature of the standard risk assessment process, it is quite possible that *Daphnia* tests may yield the most sensitive value - indeed this is quite common, given its sensitivity. Thus it is important that if PNEC values are to be calculated using *Daphnia* data, we should be assured that such data are representative of sensitive organisms within the target environment. In situations where a *Daphnia* EC_{50} value forms the basis of a PNEC calculation, a factor of 100 is recommended as a margin of safety; similarly, if a *Daphnia* chronic NOEC proves the most sensitive endpoint, a safety margin of division by 10 is recommended (Bates et al., 1996). This situation is illustrated in Fig. 1, where the sensitivity distribution of a putative group of target organisms is superimposed to illustrate the fraction of the target group which could be considered 'at risk' from exposure.

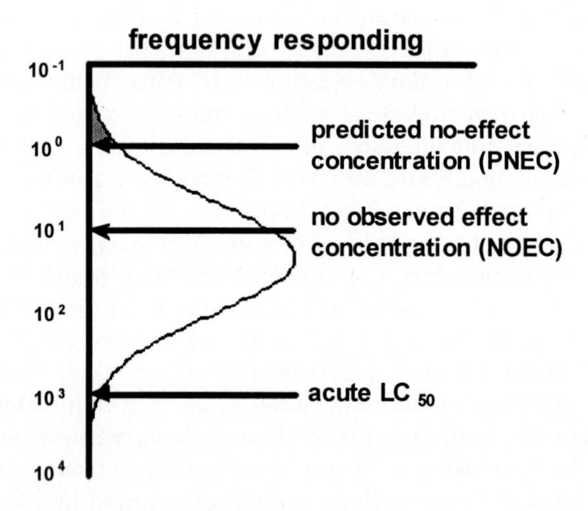

Fig 1. The relationship between laboratory toxicity endpoints derived from *Daphnia magna* tests and the derived predicted no-effect concentration (PNEC), indicated by arrows, and the underlying distribution of sensitivity within a putative target group of individuals along an arbitrary concentration scale. The grey area indicates those individuals which remain at risk from exposure. For further explanation, see text.

Safety margins are used to account for a variety of uncertainty factors which may result in the laboratory tests being underprotective, which include variations in the bioavailability of the test substance, test errors, and the fact that species more sensitive may exist within the target species assemblage. Of course, a further source of underestimation may result from the fact that the laboratory clone of *Daphnia* used in the PNEC derivation may not adequately represent the true range in genetically-based sensitivity to the substance of interest. In addition, it is possible that the calculated EC_{50} or NOEC values were underestimated due to environmental modification of the genotype response within the test itself. Such plasticity in response has not received much attention in ecotoxicological studies, yet it is potentially as important as genetically-based variation, as we shall show in examples below.

Constancy in Laboratory Clone Performance: Life-History Variation

The problem of constancy in performance of laboratory animals used in toxicity tests and bioassays has been identified as an important issue (Calow, 1992), although it has received little attention in aquatic toxicology in general, and in *Daphnia magna* in particular. While laboratories culturing *D. magna* do screen stocks for 'baseline' performance data, the compounds employed tend to be general stressors, and thus minimise the potential variability in response which can be observed, since general stressors are associated with low response variability (for a discussion of this point, see Baird et al., 1990). Moreover, there is an assumption that with increasing standardisation of laboratory culture systems and test designs, such problems can be avoided. Recently, work in our laboratory has revealed that there are limits to the degree of standardisation which can be achieved in test animal performance, relating to the mechanisms underlying maturation in Cladocera.

In *D. magna*, maturation can be considered as a size-dependent phenomenon. Maturation is thus a function of several interacting variables within a culture system: food level, egg size and genotype. Since it is not possible, we believe, to specify food level with absolute precision, even in flow-through systems, egg size will vary within a genotype, depending on micro-environmental differences in food density within the culture system. Since egg size within the normal culture environment varies across a range which encompasses maturation in the fifth or sixth instars, then genetically identical animals (even from the same clutch of eggs) will show heterogeneity in maturation instar, as can be seen in Fig 2a. Moreover, the proportion maturing within a given instar varies among clones, with some genotypes exhibiting more variability than others. This is apparently due to an interaction between egg size differences and food level (expressed as mg carbon/ml) which is complex (for a detailed explanation, see Barata and Baird, in press). The consequence is that time to maturity varies over time within clones, as can be seen in Fig 2b.

Surprisingly, the results shown in Fig 2a indicate that late maturity is more likely at high food levels, but this can be explained by the fact that animals at high food levels produce large clutches of eggs which are relatively smaller than those produced at low food levels. The consequence is that individuals mature more slowly at high food levels, since even at the low food levels used here, juvenile growth is not food-limited.

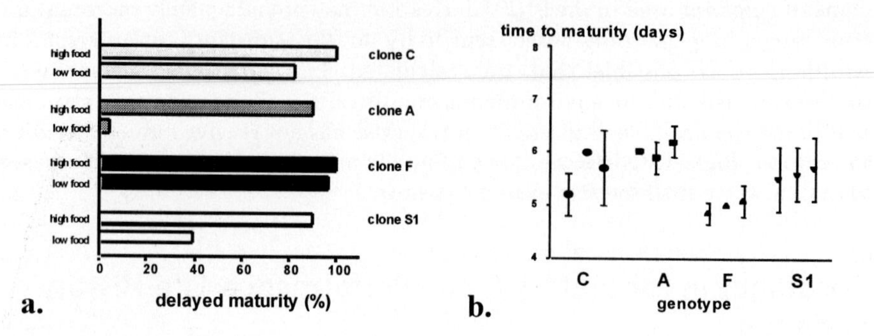

Fig 2. Variability in maturation in clones of *D. magna*.
a. The proportion of individuals with delayed maturity (i.e. maturing in the 6th rather than the 5th instar) maintained at high (1.5 mg C/l) and low (0.3 mg C/l) food levels.
b. Time to maturity (mean ± SD) in animals held at food levels of 1.5 mg C/l in repeat trials over a three-month period.

The implications of variability in maturation instar are considerable for life-history studies, since time to maturity influences other factors, including size at maturity and reproductive allocation patterns, and thus egg size. This is true also for ecotoxicological studies, since egg size has been identified as a major factor determining response of *Daphnia* in acute toxicity tests (Enserink et al. 1990).

Another key variable in *Daphnia* which has the potential to influence both life-history responses and toxicological responses is feeding rate.

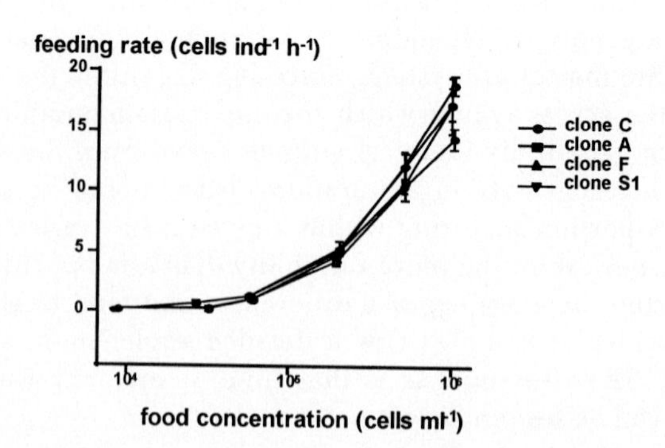

Fig 3. Feeding rate of four clones of *D. magna* as a function of food concentration.
Chlorella vulgaris was the food source (for comparison, 10^5 cells/ ml = 0.3 mg C/l) .

Feeding rate has been identified as a key variable in modelling the response of *Daphnia magna* to stress (Nogueira, Baird and Soares, in prep.). However, studies have revealed that feeding, although highly sensitive to environmental conditions, does not seem to vary among clones (Fig 3), after correcting for body size differences (ANCOVA ns; $p>0.05$). This is encouraging, as it has been hypothesised that toxic effects on feeding are the underlying mechanism behind chronic exposure responses (Allen et al., 1995) and the lack of between-clone-variation in feeding behaviour is thus consistent with the observation that chronic responses are less genetically variable than acute responses (Soares et al. 1992).

Constancy in Clone Performance: Toxicological Studies

It has been demonstrated that environmentally-induced variation in maturation time can influence other life-history traits, such as egg size, and that this tendency for maturation time to vary differs among genotypes. Thus we might expect that one consequence of this would be that within-clone response in exposure to toxic substances would be variable in time. If so, then the question arises: how important is this source of variation relative to between-clone variation, which has already been demonstrated to be important for some compounds (Baird et al. 1991).

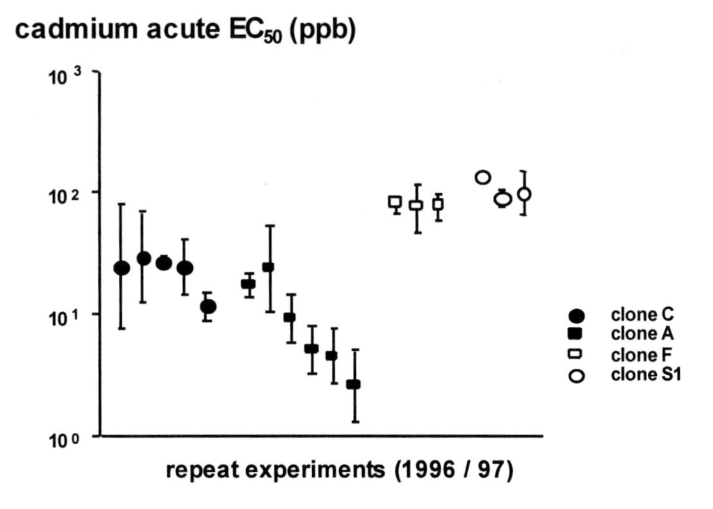

Fig 4. Constancy in acute sensitivity to cadmium in four clones in experiments carried out under standard, controlled conditions over a six month period. Points are the $EC_{50} \pm 95\%$ CL, calculated using a maximum-likelihood estimation procedure following Finney (1971).

Acute sensitivity: Using the same four laboratory genotypes, which were known to differ in response to cadmium (see Baird et al. 1990, 1991), we examined sensitivity to cadmium using a standard 48h acute immobilisation test.

As expected, genotypes exhibited considerable variation in response over time, although in general, rank order in sensitivity remained consistent (Fig 4). As found in previous studies, variation in acute sensitivity to cadmium varied by two orders of magnitude. As predicted, significant variation in acute sensitivity also occurred within clones. Most notably, the clone found to be the most variable in response was the clone which has been recommended as a standard genotype for laboratory testing (OECD, 1997), clone A, which is the most widely-cultured genotype within testing labs in the EU. This finding is consistent with experience in our laboratory, where clone A is recognised as a genotype which is prone to culturing problems (unpublished observations).

Chronic sensitivity: Although we have not yet attempted to measure consistency in chronic test performance over time, we would expect to see less variation in response within clones over time than for acute exposure. The general lack of variation in chronic sensitivity among clones can be illustrated by examining the effects of toxic substances on feeding rate , as a cost-effective alternative to conducting 21-day life-cycle tests. This can be justified by noting that the inhibition of feeding rate, or toxic anorexia, integrated over time, is the underlying mechanism behind the chronic test endpoint, reproductive inhibition. Fig 5 illustrates that for the same four clones exposed to a series of sublethal non-photoactivated fluoranthene concentrations, response is consistent across clones for each concentration.

Although it appears from Fig 5 that there could be differences among clones in control feeding rates, this was found to be due to body size differences (which are environmentally determined). After removing the effect of body size, no significant between-genotype variation in response was observed (ANCOVA ns $p=0.05$). We have observed similar patterns in response for other compounds , and to date we have failed to find significant between-clone variation in chronic response to toxic substances. For this reason, we do not consider that within-clone variation response in this factor is likely to be a significant problem, except possibly in the case where response to sublethal exposure is based on specific response mechanisms i.e. where the response is determined by major genes .

Conclusions

From these preliminary investigations, it appears that significant within-clone variation in life-history performance in cultures and, more importantly, in acute sensitivity to toxic substances exists in *D. magna*. It is very likely that these two phenomena are linked by a common underlying mechanism: the inability to precisely control micro-environmental conditions during culture. At present, this remains a hypothesis, but if true, undermines the current move towards increasing test standardisation with this species. It is somewhat ironic to consider that one of the major criticisms regarding the relevance of *D. magna* tests relates to the high food levels generally given to animals in culture, and in chronic tests, which is around 5-10 times higher than animals normally experience in the field. We can see here that in fact high food, rather than improving performance of test animals, actually impairs it, a situation not dissimilar to that noted for laboratory rats, which succumb to high tumour incidence when fed *ad libitum*. Of course, it could be argued that high food levels are required to ensure that enough *Daphnia* are generated for use in tests. Clearly a trade-off between test repeatability and reliability exists here, and more thought needs to be given to how test designs can be liberated from the strictures

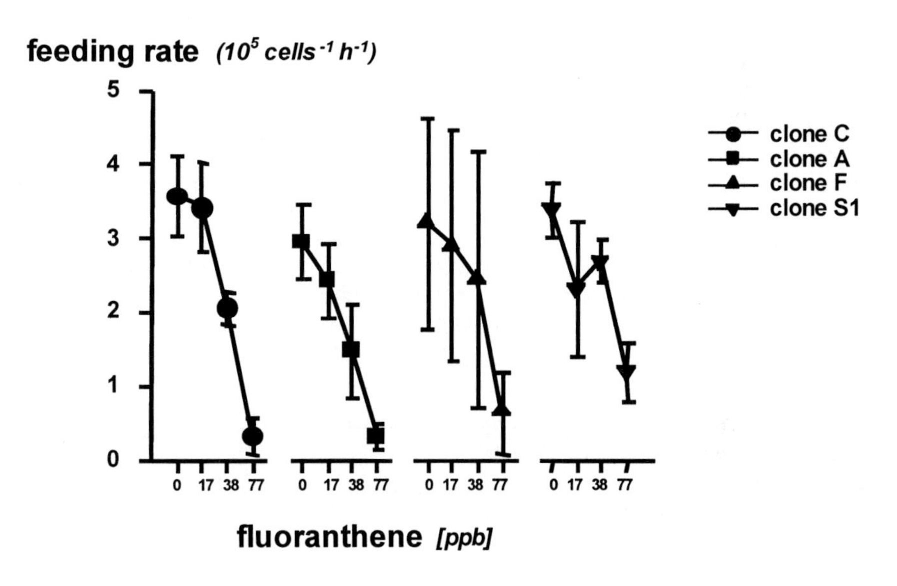

Fig 5. Effect of fluoranthene exposure on feeding rate in four clones of *Daphnia magna*. All concentrations are based on actual measurements. Points are means ± SD. For comparison, a 48h acute EC_{50} for fluoranthene is ca. 120ppb.

of standardisation without compromising their value as benchmarks of toxicity for aquatic life. Clearly, a simple message could be given to those interested in *Daphnia* test design - they should get out more...!

Finally, it should be noted that in addition to the problem of genetic variability in acute response we must also now add problems of lack of constancy in performance. Every additional problem of this type erodes the safety margins which are used in risk assessment. Given a two orders of magnitude genetically-based component (Baird et al. 1990), and an order of magnitude 'constancy' component, the maximum three order of magnitude safety margin (based on a direct extrapolation from acute to PNEC, see Fig 1) seems rather compromised.

When we consider that bioavailability problems, particularly for so-called difficult substances, are likely to further tax the credibility of this method, it is clear that a new, more flexible approach to risk assessment of chemicals is urgently needed.

References

Allen Y, Baird DJ and Calow P (1995) A mechanistic model of contaminant-induced feeding inhibition in *Daphnia magna*. Environmental Toxicology & Chemistry 14:1625-1630.

Baird DJ, Barber I, Bradley MC, Calow P, Soares AMVM (1989) The *Daphnia* bioassay: a critique. Hydrobiologia 188/189:403-406

Baird DJ, Barber I and Calow P (1990) Clonal variation in general responses of *Daphnia magna* Straus to toxic stress. I. Chronic life-history effects. Functional Ecology 4:399-407

Baird DJ, Barber I, Bradley MC, Soares AMVM, Calow P (1991) A comparative study of genotype sensitivity to acute toxic stress using clones of *Daphnia magna* Straus. Ecotoxicology and Environmental Safety 21:257-265

Barata C, Baird DJ The influence of newborn length on age at maturity in *Daphnia magna* Straus. Functional Ecology, in press.

Bates K, Hedgecott S, Grimwood M, Sims I, Fawell JK (1996) Ecotoxicological Risk Assessment Manual for Chemicals in the Aquatic Environment. Scottish and Northern Ireland Forum for Environmental Research Report SR 3846/1

Calow P (1992) The three Rs of ecotoxicology. Functional Ecology 6:617-619

Enserink L, Luttmer W and Maas-Diepeveen H (1990) Reproductive strategy of *Daphnia* affects the sensitivity of its progeny in acute toxicity tests. Aquatic Toxicology 17:15-25

Finney DJ (1971) Probit Analysis. 3rd Edition. Cambridge University Press, Cambridge, UK.

OECD (1997) Guideline for Testing of Chemicals no. 202. *Daphnia* sp. Acute immobilisation test and reproduction test. Organization for Economic Co-operation and Development, Paris.

Soares AMVM, Baird DJ, Calow P (1992) Interclonal variation in the performance of *Daphnia magna* Straus in chronic bioassays. Environmental Toxicology and Chemistry 11:1477-1483

Sources and Implications of Variability in Sensitivity to Chemicals for Ecotoxicological Risk Assessment

Valery E. Forbes
Department of Life Sciences and Chemistry, P.O. Box 260, Roskilde University, DK-4000 Roskilde, Denmark

Abstract

Variability among individuals in their responses to toxic chemicals arises from several sources, the most important of which are genetic differences, environmental influences (including maternal effects and historical factors) and measurement error. Effective risk assessment requires that estimates of toxicant response (e.g., LD_{50}, EC_{50}, LOEC, NOEC) are precise - that is, have narrow confidence limits -, repeatable - that is, different laboratories must obtain the same or very similar result –, and accurate - that is, they must provide a reasonable approximation of the effects of toxicants on real ecological systems. Determining which of the above-mentioned sources of variability has the greatest influence on toxicant response has implications for both the design and interpretation of ecotoxicological tests. If, for example, genetic influences are of overriding importance, controlling genotype (by using clones or inbred strains) can lead to greater precision but at the expense of accuracy when the objective is to estimate toxicant response for the species as a whole. Likewise, if environmental influences are of primary importance in controlling the response to toxicants, performing experiments under a standard temperature, light, and food regime may provide highly repeatable test results that have little relevance to the responses of populations in nature. Although there is little doubt that the development of standard ecotoxicological test guidelines (e.g., by the OECD), that control genetic and environmental sources of variability, has led to improvements in the practice of risk assessment, further advances will require a more sophisticated approach for dealing with these sources of uncertainty. There is a need for more systematic approaches for quantifying the sources of variability in toxicant response and for formally combining the error associated with each source in key risk assessment endpoints.

Variability is a Key Feature of Population Tolerance

That individuals in a population vary in response to toxicant exposure is a key feature of toxicological and ecotoxicological theory and explains why the response of a test population to a given toxicant is typically described in terms of a distribution of tolerances (Forbes and Forbes 1994). According to toxicological theory the shape of the distribution of responses to toxicant exposure is normal when plotted against the logarithm of dose (or concentration in ecotoxicology). Like all normal distributions, population tolerance distributions are defined by two key parameters - their central tendency and their spread about the central tendency - in other words their mean (or median) and standard deviation. To facilitate estimation of these key parameters, the tolerance distribution is first transformed to a cumulative scale (giving an s-shaped curve), and the resulting cumulative percentages transformed to standard deviation units (with the addition of 5 to give the so-called probit curve) (Finney 1978). The resulting straight line is analytically convenient, e.g., the median of the tolerance distribution is obtained by dropping a vertical line to the concentration corresponding to a probit value of 5, and the standard deviation of the tolerance distribution is simply one over the slope. Additionally, transforming tolerance distributions to straight lines facilitates statistical comparisons among them (allowing the response of the same population to different chemicals or the response of different populations to the same chemical to be compared) by analysis of covariance (ANCOVA). Traditionally the median of the tolerance distribution (i.e., LD_{50} or LC_{50}) has been the parameter of primary comparative interest, with less attention focused on the standard deviation (or slope on a probit scale). In order to maximize the chance of detecting real differences in median values among different toxicants or test populations, it is desirable that the confidence limits about the median be as small as possible. In other words, in order to detect differences among groups in LD(or LC)$_{50}$s, it is desirable that the test populations respond as uniformly as possible to the exposure. However, there is always some degree of variability in the response of test populations to toxicant exposure, and toxicologists have long been concerned with dealing with this variability. Hence Gaddum (1933) writes:

"The conception of a minimal lethal dose implies that a dose can be found which is just sufficient to kill all animals of a given species, while a slightly smaller dose would kill none. In practice this conception has no value, because of the wide variations in the sensitivity of individual animals ... Among the variables which may be controlled, in order to increase the homogeneity of the animals, are their genetic composition, weight, age, sex, diet, and environmental temperature".

Thus, identifying the sources of response variability among individuals in defined test populations has long provided a focus for study in toxicology (and subsequently ecotoxicology). Such insights as have been obtained have generally been used to remove the offending sources of variability from toxicological test designs in order to obtain as precise a measure of the populations' median response as possible. I will demonstrate that identification of the sources of variability in organisms' responses to toxicant exposure can contribute much more to the fields of toxicology and ecotoxicology than improved precision of population median responses. In addition to contributing to our understanding of the underlying mechanisms of toxicant impact, identification of the sources of variability in the responses of organisms to toxicant exposure can potentially lead to improvements in the practice of environmental risk assessment to the extent that general rules for incorporating variability into risk assessment models can be determined.

Sources of Variability

Variability among individuals in their responses to toxic chemicals arises from several sources, the most important of which are genetic differences, environmental influences (including maternal effects and historical factors) and measurement error. Genetic differences in tolerance can arise because populations have been selected for resistance during exposure to particular toxicants in the field, or they may be byproducts of selection by agents unrelated to toxicant stress. Environmental factors include conditions under which toxicity tests are conducted (e.g., temperature, food regime, light, etc.) as well as various factors that have operated on an organism prior to its being employed in a toxicity test. Following are examples from the published literature demonstrating the potential importance of each of these sources of variability in influencing population tolerance distributions.

Genotype: Figure 1 shows selected results from Baird, et al. (1990) demonstrating the potential importance of genetic sources of variability in toxic responses. They studied the tolerance of different clones of the common test species, *Daphnia magna*, to a number of chemicals. In Fig. 1 are shown acute tolerance distributions for 8 clones in response to cadmium (top) and dichloroaniline (bottom). Several points are worth noting from these data. First, as indicated by their position along the x-axis, different clones vary in their sensitivity to each of the chemicals, and the difference can be up to several orders of magnitude. Second, the ranking of the clones for the two chemicals is not consistent. In other words, the clones that are the most tolerant to cadmium exposure are not necessarily to most tolerance to dichloroaniline, and *vice versa*. Third, clones varied substantially in terms of their response uniformity as indicated by differences in the width of the tolerance distributions shown in figure 1. These data demonstrate very clearly the concept of genotype times environment interaction in that the relative responses of the different genotypes (LC_{50}s) are significantly dependent on the environment (toxicant) in which they are reared. That the ranking of genotypes was not consistent across chemicals (this was further supported by Baird et al. (1991) in which data were reported for a total of 9 chemicals) is inconsistent with the prediction of an inverse relationship between maintenance metabolic costs and tolerance to a broad range of environmental stresses (Koehn and Bayne 1989).

Environment: Figure 2, from Møller et al. (1994), demonstrates the influence of environmental factors on the variability among individuals in toxic response. In this study, the acute lethal response of a single clone of the parthenogenetic gastropod, Potamopyrgus antipodarum, exposed to cadmium was examined. Both exposure temperature (i.e., the temperature at which the toxicity test was

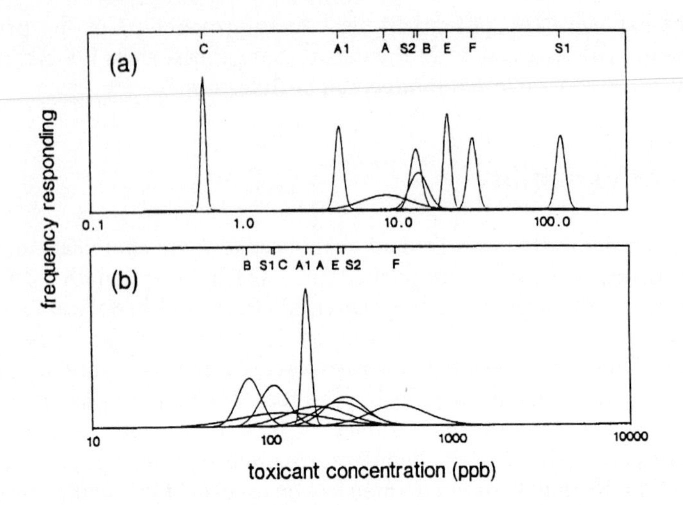

Fig. 1. Population tolerance distributions for 8 genotypes of Daphnia magna in response to acute exposure to a) cadmium and b) 3,4-dichloroaniline. Clone designations are given across the top of each figure. From Baird et al. (1990), reproduced by permission of the publisher.

conducted) and acclimation temperature (i.e., the temperature at which the snails had been cultured previous to the toxicity test) were investigated. LC_{50}s varied by a factor of four among the different temperature treatments. Increasing exposure temperature significantly increased the sensitivity of gastropods to cadmium exposure (Fig. 2a), which is a typical pattern observed for many species and toxicants. What is most interesting about these results is that not only the temperature at which the experiment was carried out, but also the temperature to which snails had been previously acclimated, influenced tolerance. In fact, the latter was the more important determinant of toxicity and explained 57% of the total variance in mortality, compared to exposure temperature which explained 40% of the variance (the remaining 3% remaining was error variance). Somewhat unexpectedly, snails acclimated to the intermediate temperature of 15oC, which is the most favorable temperature for growth of the three temperatures tested, were the most sensitive to acute cadmium exposure. Although differences in the total amount of cadmium taken up by the different treatment groups was ruled out as an explanation for the results, it is possible that differences in the internal distribution or handling of cadmium acted to increase the susceptibility of the (presumably) fastest - growing snails. These results are important for two reasons: 1) they show that historical environmental factors can potentially play an even larger role than present environmental factors in controlling the response of organisms to toxicants, and 2) they suggest that culturing test organisms under optimal conditions for growth does not necessarily increase their tolerance to subsequent toxicant exposure.

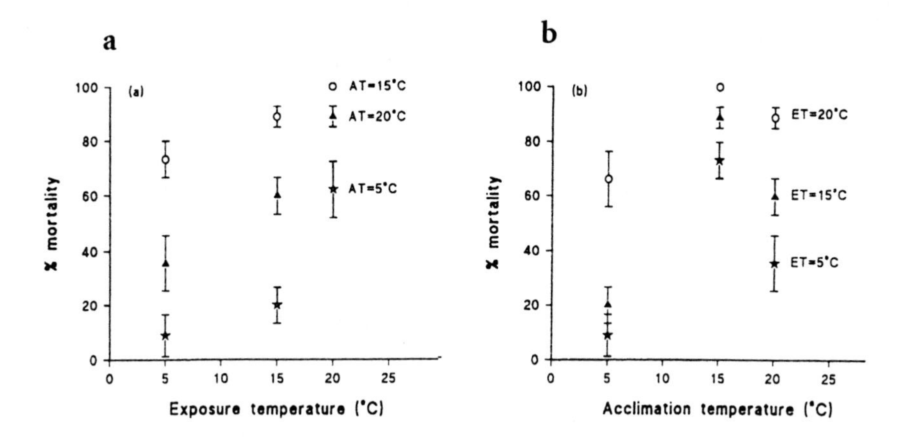

Fig. 2. Percent mortality at 2 mg Cd/l plotted against a) exposure temperature (ET) and b) acclimation temperature (AT). Error bars represent standard error of the mean. From Møller et al. (1994), reproduced by permission of the publisher.

Maternal and Common Environmental Effects: Figure 3 shows data from Lam (1996) examining the mortality response of two populations of the gastropod, *Brotia hainanensis*, to cadmium exposure. The upstream population was collected from a clean habitat and the downstream population from a metal polluted site. Field collected snails were allowed to acclimate to laboratory conditions for one week and then tested for their acute tolerance to cadmium exposure. Field-collected snails from the metal polluted site showed a higher acute tolerance to cadmium compared to snails from the clean site (Fig. 3a). In order to determine whether such observed differences in tolerance are genetically based, the typical approach is to breed field-collected individuals under clean conditions in the laboratory and test the tolerance of their offspring. The extent to which the offspring differ in tolerance is an indicator of the extent to which the tolerance is genetically determined. In this case the offspring of snails from the polluted site initially showed a higher tolerance to cadmium, suggesting that the tolerance had a large genetic component (Fig. 3b). However, when a subset of the same group of offspring was tested at a slightly later time, the difference in tolerance between populations had disappeared (Fig. 3c). Quantitative genetic analysis indicated that the variance in survival among offspring from a single mother was significantly less than the variance among snails from different mothers. Furthermore, much of the difference in tolerance among newly hatched offspring could be attributed to differences in body size at hatching. When allowed to feed in the laboratory for several days prior to testing for cadmium tolerance, the body size differences (and hence differences in cadmium tolerance) disappeared. Thus these data provide a good example of what may be termed maternal effects or common micro-environmental factors shared among siblings. In addition, they demonstrate the power of quantitative

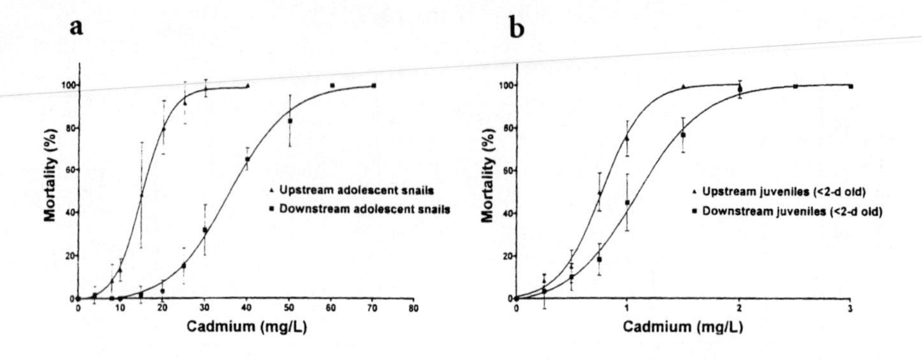

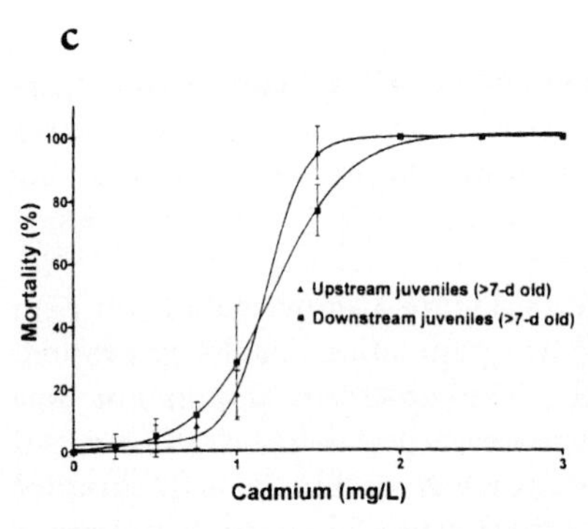

Fig. 3 Percent mortality in response to 96 h cadmium exposure for the gastropod *Brotia hainanensis* from upstream (clean) and downstream (metal-contaminated) field sites. Error bars are ± 1 S.D., number of replicates = 6. Data shown are for a) freshly-collected adolescents, b) newly-hatched (< 2 d), laboratory-reared, first generation offspring, and c) older (> 7 d), laboratory-reared, first generation offspring. From Lam (1996), reproduced by permission of the publisher.

genetical analyses for distinguishing sources of variability in responses totoxicant exposure and emphasize the care that is required when preforming laboratory breeding designs for such analyses.

Measurement Error: An additional potentially important source of variance in toxicant responses is what is most often referred to as measurement error. That is, after genetic differences have been eliminated and environmental factors have been controlled, the variability in response remaining among individuals is referred to as measurement error. Measurement error includes human error as well as limitations of the machinery and methods used. Although it could be expected that if the same subject was measured with the same equipment every day for a month we would probably observe a small degree of variability as a result of human error, this source of measurement error is generally small because it is easy to identify and control. In terms of methodological error the key limitations are in terms of sensitivity and detection limits. For example, every GC/MS has a certain minimum detection limit (that can be altered to some extent by choosing the appropriate column, program, etc.). The closer we make measurements to the detection limit, the more important become the small random variations among replicate samples. Again, this source of error is generally easy to quantify (e.g., by measuring replicate standards) and to control (e.g., by adjusting sample volume to be well above the detection limit). Both human and methodological sources of measurement error clearly belong in the category of true error, and minimizing them should be an important goal of any experiment.

Microenvironmental Heterogeneity: There is an additional source of variability which, because it is difficult both to measure and to control, is usually placed (somewhat misleadingly) under the category of measurement error. This variability arises from the response of biological systems to uncontrollable small-scale environmental factors - so-called microenvironmental heterogeneity. Despite the fact that great effort may be spent to control the environmental conditions for a test, for example, by performing the test using a standard temperature, test medium, food, light level, etc., a population of test organisms may experience their environment on very different spatial and temporal scales, and the actual environment experienced from the organisms' point of view may be much more heterogeneous than is apparent. Such differences have been used to explain the phenotypic variability that is often observed within single clones reared in presumably uniform environmental conditions. What makes this source of variability particularly interesting is that there are both theoretical and empirical grounds for expecting that it is a function of both genotype and the type and degree of toxicant exposure.

Referring again to Fig 1., it is apparent that variations in the width of the tolerance distributions indicate the phenotypic variability in response within single clones reared under uniform laboratory conditions. Notice that the width of the tolerance distribution - or the phenotypic variability within clones - varies both among clones for a single toxicant as well as among toxicants for a single clone. One interpretation of these data is that different genotypes differ in their sensitivity to microenvironmental heterogeneity and that there are interactions between a genotype's degree of sensitivity and exposure to particular toxicants. Understanding why some genotypes are more phenotypically uniform than others could yield insight into the genetic basis of tolerance, and the observation that some genotypes appear more sensitive to uncontrollable environmental heterogeneity than others has practical implications with regard to selecting genotypes for toxicity testing.

Chemical Exposure May Enhance Variability

Whereas problems associated with variability among individuals in their response to toxicants have long been recognized, more recently evidence has been accumulating that stress associated with chemical exposure may itself increase variability in a wide variety of biological traits (Forbes and Depledge 1996). Although the mechanisms behind stress-associated increases in variability remain unclear (and may be varied), their occurrence has clear implications for the ecological and evolutionary consequences for populations exposed to chemicals as well as practical implications for toxicity test design.

Table 1, modified from Forbes, et al. (1995), shows mean growth rate and its coefficient of variation (CV) for two clones of the gastropod, *Potamopyrgus antipodarum* cultured under clean and cadmium-exposed conditions. The CV provides a measure of variation that is independent of the mean and allows comparison of the variability among populations whose means vary widely. The first point to note from Table 1 is that the CVs - that is the phenotypic differences among individuals of the same genotype reared under "identical" laboratory conditions - differ between clones, with clone B being more variable than clone A. For both clones (and for several sexually reproducing populations not shown here) the CVs increase when snails are cultured under conditions of cadmium exposure. Furthermore, the degree of increase in variability in response to cadmium is genotype-dependent.

Table 1. Growth of *Potamopyrgus antipodarum* exposed to 200 µg Cd/l for 3 weeks. Growth rates are given in units of mm shell length/3 weeks. Data shown are means (standard deviations) and coefficients of variation (CVs). Data are from Forbes et al. (1995).

Genotype	Control		Cadmium	
	Mean (sd)	CV	Mean (sd)	CV
Type A	0.54 (0.11)	0.21	0.11 (0.03)	0.25
Type B	0.24 (0.09)	0.32	0.06 (0.04)	0.70

Studies of detoxification enzyme activity in weevil populations transferred to novel food sources have detected significant increases in the variance of enzyme activity despite little difference in mean activity. The largest increases in variance were observed for populations transferred to the more toxic of the food sources tested (Holloway et al., 1997).

Implications of Variability for Environmental Risk Assessment

Variability in the responses of organisms to toxicants can come from a variety of sources, some of which are easier to control than others. This variability has practical implications for the way in which the effects of chemicals on ecological systems can be evaluated. Effective risk assessment requires that estimates of toxicant response (e.g., LD_{50}, EC_{50}, LOEC, NOEC) are precise – that is, have narrow confidence limits – repeatable – that is, different laboratories must obtain the same or very similar result –, ad accurate – that is, they must provide a reasonable approximation to the effects of toxicants on real ecological systems (Calow 1992). Determining which of the above-mentioned sources of variability has the greatest influence on toxicant response has implications for both the design and interpretation of ecotoxicological tests. Ideally we would like to develop tests that are both accurate and precise, but in practice this has proven to be difficult.

An important difference between ecotoxicology and toxicology is that in the latter we are usually trying to predict the response of a single species (i.e., humans) from one or a few test species, whereas in ecotoxicology we are usually trying to predict the response of many species – of whole ecological systems – from the responses of one or a few test species. If we want the most accurate measure of the response of all species in an ecosystem, we should probably test as many species as possible. However, there are two problems with this approach. The first is that it is highly impractical. The second is that measuring the response of a truly random sample of species present in a particular ecosystem would very likely result in an estimated response that was very accurate but very imprecise.

The other extreme, which has its roots in the 1930s during the early development of drug testing, is to focus on maximizing precision. It was recognized very early on that different genetic strains of test animals sometimes showed markedly different toxicant responses (Gaddum 1933), and likewise that the environmental conditions during the test could be of paramount importance for obtaining precise and repeatable results. Hence there has been a great deal of effort in toxicology, and subsequently ecotoxicology, for minimizing sources of genetic and environmental variability in test results. There is a problem with this approach, and that is that it achieves precision at the expense of accuracy. If, for example, genetic influences are of overriding importance, controlling genotype (by using clones or inbred strains) can lead to greater precision but at the expense of accuracy when the objective is to estimate toxicant response for the species as a whole (e.g., see Baird, this volume). If the susceptibility of developing cancer has an identifiable genetic basis, then setting appropriate exposure limits for suspected carcinogens needs to take into account differences among genotypes. Likewise, if environmental influences are of primary importance in controlling the response to toxicants, performing experiments under a standard temperature, light, and food regime may provide highly repeatable test results that have little relevance to the responses of populations in nature where all of these variables will vary across many temporal and spatial scales.

Options for Incorporating Variability into Environmental Risk Assessment

The first option for addressing the problems posed by variability is to standardize tests to maximize precision and then apply application factors to minimize inaccuracy. This is indeed the present approach. Test guidelines, such as those developed by the OECD Chemicals Programme, include recommendations for standardizing test populations as well as environmental conditions. When the results of such tests are used to estimate effect concentrations for environmental risk assessment, they are generally divided by an application factor (typically 10, 100 or 1000), the size of which is inversely related to how many test results are available for consideration.

A second option has been proposed in recent years in response to the recognized arbitrariness of the application factor approach. This option involves the application of species distribution models which will protect a defined fraction of species (Aldenberg and Slob 1991; Wagner and Løkke 1991; Van Leeuwen et al. 1992). These models presume to increase accuracy by fitting the tolerances of selected 'representative' species to a distribution function which is then used to define toxicant concentrations that will have no effect on say 95% of the species in an ecosystem. There are a number of potentially serious problems with these models (Forbes and Forbes 1993). First, the distribution of species sensitivities in any real ecosystem has never been tested. Second, the species presently used for testing are in no way a random sample of species from any known ecosystem. Third, the models are more complex, but the results do not appear to differ substantially from the more crude application factor approach.

A third, and I believe most promising option is to attempt a more sophisticated approach for incorporating variability into risk assessment models. The first step in this approach would be to quantify sources of variability. This could be done, for example using traditional methods of quantitative genetics (e.g., Soares et al. 1992; Lam 1996). In the past, quantifying variability has been performed for the purpose of selecting variables for standardization. However, as argued above, this approach is not ideal in that it is likely to increase precision at the expense of accuracy. A more effective way to use knowledge of the sources of variability is to attempt to develop general rules that will allow the influence of the key sources to be incorporated into our estimates of risk. Classification according to chemical type and /or category of response are two obvious avenues to explore. With regard to the former, Baird et al. (1991) found much greater variability among genotypes in their response to heavy metals than to organics. With regard to the latter, it appears that single genotypes are more uniform in acute lethal responses to toxicants (Møller et al. 1996) than in chronic sublethal responses (Forbes et al. 1995) compared to genetically mixed test populations. Likewise, single genotypes of *Daphnia magna* appear to differ more from each other in acute than in chronic responses (Baird et al. 1990).

Once some general rules have been identified , it will be necessary to combine the variability associated with each factor to provide appropriate confidence limits around the effect concentration which is used as an input variable in risk calculations. Fortunately a number of methods for combining variance terms are presently available and have even already been incorporated into other steps in the risk assessment process (Slob 1994; Van Leeuwen and Hermens 1995). In some cases assumptions about the shape of the response distribution may be appropriate and can simplify calculation of confidence limits (Slob 1994). If no assumptions can be made about the way in which sources of variability may combine to influence the shape of the resulting response distribution then more sophisticated and computer-intensive resampling methods (e.g., bootstrapping or Monte Carlo estimation) may be required. The end result should be a more precise and accurate estimate of the variability in organism response to toxicant exposure.

Conclusions

In conclusion, there is little doubt that the development of standard ecotoxicological test guidelines (e.g., by the OECD), that control genetic and environmental sources of variability, has led to improvements in the practice of risk assessment. However, further advances will require a more sophisticated approach for dealing with these sources of uncertainty. There is a need for more systematic approaches for quantifying the sources of variability in toxicant response and for formally combining the errors associated with each source in key risk assessment endpoints.

References

Aldenberg T, Slob W (1991) Confidence limits for hazardous concentrations based on logistically distributed NOEC toxicity data. National Institute of Public Health and Environmental Protection (RIVM), report no. 719192992, The Netherlands

Baird DJ, Barber I, Calow P (1990) Clonal variation in general responses of *Daphnia magna* Straus to toxic stress. I. Chronic life-history effects. Functional Ecology 4: 399-407

Baird DJ, Barber I, Bradley M, Soares AMVM, Calow P (1991) A comparative study of genotype sensitivity to acute toxic stress using clones of *Daphnia magna* Straus. Ecotoxicology and Environmental Safety 21: 257-265

Calow, P (1992) The three Rs of ecotoxicology. Functional Ecology 6: 617-619

Finney DJ (1978) Statistical Method in Biological Assay, 3rd edn, Charles Griffen and Co, London

Forbes TL, Forbes VE (1993) A critique of the use of distribution-based extrapolation models in ecotoxicology. Functional Ecology 7: 249-254

Forbes VE, Depledge MH (1996) Environmental stress and the distribution of traits within populations. In: Baird DJ, Maltby L, Greig-Smith PW, Douben PET (eds) ECOtoxicology: Ecological Dimensions. Chapman and Hall, London, pp 71-86

Forbes VE, Forbes TL (1994) Ecotoxicology in Theory and Practice. Chapman and Hall, London

Forbes VE, Møller V, Depledge MH (1995) Intrapopulation variability in sublethal response to heavy metal stress in sexual and asexual gastropod populations. Functional Ecology 9: 477-484

Gaddum JH (1933) Reports on biological standards. III. Methods of biological assay depending on a quantal response. Medical Research Council, Special Report Series, no. 183, London

Holloway GJ, Crocker HJ, Callaghan A (1997) The effects of novel and stressful environments on trait distribution. Functional Ecology, in press

Koehn, RK, Bayne BL (1989) Towards a physiological and genetical understanding of the energetics of the stress response. Biological Journal of the Linnean Society 37: 157-171

Lam P (1996) Interpopulation differences in acute response of *Brotia hainanensis* (Gastropoda, Prosobranchia) to cadmium: genetic or environmental variance? Environmental Pollution 94: 1-7

Møller V, Forbes VE, Depledge MH (1994) Influence of acclimation and exposure temperature on the acute toxicity of cadmium to the freshwater snail *Potamopyrgus antipodarum* (Hydrobiidae). Environmental Toxicology and Chemistry 13: 1519-1524

Møller V, Forbes VE, Depledge MH (1996) Population responses to acute and chronic cadmium exposure in sexual and asexual estuarine gastropods. Ecotoxicology 5: 313-326

Slob W (1994) Uncertainty analysis in multiplicative models. Risk Analysis 14: 571-576

Soares AMVM, Baird DJ, Calow P (1992) Interclonal variation in the performance of *Daphnia magna* Straus in chronic bioassays. Environmental Toxicology and Chemistry 11: 1477-1483

Van Leeuwen C, Van der Zandt PTJ, Aldenberg T, Verhaar HJM, Hermens JLM (1992) Application to QSARs, extrapolation and equilibrium partitioning in aquatic effects assessment. I. Narcotic industrial pollutants. Environmental Toxicology and Chemistry 11: 267-282

Van Leeuwen CJ, Hermens JLM (1995) Risk Assessment of Chemicals. Kluwer Academic Publishers, Kordrecht, The Netherlands

Wagner C, Løkke H (1991) Estimation of ecotoxicological protection levels from NOEC toxicity data. Water Research 25: 1237-1242

Polymorphism in Glutathione S-Transferase Loci as a Risk Factor for Common Cancers

Richard C Strange, John T Lear and Anthony A Fryer.
Clinical Biochemistry Research Group, School of Postgraduate Medicine, Keele University, North Staffordshire Hospital, Stoke-on-Trent, Staffordshire, England.

Introduction

Allelism has been found in human glutathione S-transferase (GST) genes of the alpha, mu, theta and pi families with the best characterised examples being those in mu class GSTM1 and theta class GSTT1. Isoenzymes encoded by these genes catalyse the detoxification of various reactive toxic and mutagenic compounds including epoxides resulting from the cytochrome P450-mediated metabolism of polycyclic aromatic hydrocarbons as well as lipid and DNA products of oxidative stress (Hayes and Strange, 1995, Smith et al, 1995). Homozygosity for null alleles or those encoding low activity variants are likely therefore, to be associated with a biochemical consequence. However, while accumulating evidence suggests the importance of different GST, it remains unclear precisely which *in vivo* processes are influenced by these polymorphisms. In this chapter we discuss firstly, a new polymorphism in GSTM3 and secondly, the role of GST polymorphisms in determining cancer susceptibility with particular reference to allelism at GSTM1, GSTT1 and GSTM3 and their interactions with cytochrome P450 (CYP) genotypes in basal cell carcinoma of skin (BCC).

Identification of Polymorphism at GSTM3.

Five mu class genes (M1-M5) on chromosome 1p13 have been identified. Three GSTM1 alleles, GSTM1*0, GSTM1*A, GSTM1*B are identified. GSTM1*0 is deleted and homozygotes (GSTM1 null genotype) express no GSTM1 protein. GSTM1*A and GSTM1*B differ by one base in exon 7 and encode monomers that form active homo- and heterodimeric enzymes (Seidegard et al, 1988).

Until recently, allelism in other mu genes had not been reported though variation in the level of GSTM3 expression is reported in several human tissues (Nakajima et al, 1995, Hand et al, 1996). Inskip et al (1995) identified two alleles, GSTM3*A and GSTM3*B that differed in intron 6 by a 3 base pair deletion in GSTM3*B resulting in a recognition motif (-aagata-) for the YY1 transcription

---

Abbreviations: GST, glutathione S-transferase; CYP, cytochrome P450; BCC, basal cell carcinoma; UV, ultraviolet radiation; ROS, reactive oxygen species; OR, odds ratio; RR, rate ratio; HR, hazard ratio.

factor in GSTM3*B. We speculated that expression of GSTM3*B and GSTM3*A is differently regulated by YY1 (Inskip et al, 1995). We also found evidence that GSTM3*B and GSTM1*A are in linkage dysequilibrium (Inskip et al, 1995) possibly explaining why GSTM3 expression in lung is related to the GSTM1 genotype (Nakajima et al, 1995); individuals with GSTM1*A/GSTM3*B should express more GSTM3 than those with GSTM1*O/GSTM3*A or GSTM1*B/GSTM3*A because GSTM3*B is inducible by YY1. The finding of linkage dysequilibrium may also explain the discrepancy in results from studies of the influence of GSTM1 on lung cancer risk; thus, the GSTM1/GSTM3 haplotype rather than GSTM1 genotype alone, determines risk. This hypothesis is untested. Recent studies indicate that polymorphism in GSTM3 influences susceptibility to laryngeal cancer (Jahnke et al, 1996) and multiple BCC (see below).

GST Genotypes and Cancer Susceptibility

Influence of GST on Mutational Spectra: Study of the association between mutations in target genes and GST genotypes may better unravel the protective actions of these enzymes as particular genotypes and haplotypes should identify detoxification-deficient subjects who suffer increased carcinogen-DNA adducts (Kato et al, 1995) or mutations in genes such as p53 (Ryberg et al, 1994). We obtained evidence for a protective role for GSTM1 in epithelial ovarian cancer (Sarhanis et al, 1996); of 23 p53 immunohistochemically positive ovarian tumours, 20 (87.0%) were GSTM1 null. The frequency distributions of GSTM1 genotypes in the p53 immunopositive and negative subjects were different (exact p=0.002) and those for GSTT1 genotypes approached significant difference (exact p=0.057). In the 23 p53 immunopositive patients, sequencing identified p53 mutations in 10 patients; 9 were GSTM1 null (90.0%). In the 13 immunohistochemistry positive patients with no mutations, 11 patients were GSTM1 null (84.6%). We believe that these data reflect the role of GSTM1 in the detoxification of the products of oxidative stress (Strange, 1996). Thus, chronic failure to effectively detoxify these products of the repair process of the ovarian epithelium results, in some patients, in p53 being damaged causing persistent expression of mutant protein. Alternatively, oxidative stress effects damage to other genes not including p53 resulting in over-expression of wild type p53. It is emphasised that the apparent protective effect of GST is not universal (Strange, 1996).

Case-control studies of the influence of GST alleles on cancer risk: Many studies designed to test the hypothesis that GST influences the susceptibility to cancer have focused on cancers associated with exposure to tobacco-derived chemicals. Thus, an increased frequency of GSTM1 null in cigarette smokers with lung cancer compared with controls has been reported. However, while some studies have supported these results, much data is conflicting and presently the influence of GSTM1 null on risk is unclear (Smith et al, 1996, Strange, 1996). The reason for the discrepancies in GSTM1 data in lung and other cancers is unclear.

Each genotype has a relatively weak influence that may be masked by confounding factors (e.g. allelism at other loci). The importance of careful selection of controls has also been emphasised (Strange, 1996).

Influence of GST polymorphisms on susceptibility to BCC: The view that the GSTs are part of cellular antioxidant defences (Hayes and Strange, 1995) receives support from studies in cutaneous cancers, particularly BCC. Ultraviolet radiation (UV) is a major causative factor, though the relationship between exposure and risk is poorly understood. Indeed, compared with cutaneous squamous cell cancer, BCC are more common on generally less exposed sites especially the trunk, with lesions infrequently found on the forearms or backs of the hands. Also, while the incidence of BCC is increasing, it is the proportion of tumours on the trunk that demonstrates the greatest increase. UV has pleiotrophic effects on skin including the formation of reactive oxygen species (ROS) that can effect mutations in key target genes.

BCC is the commonest cancer in Caucasians with incidences in the United States reported to be as high as 300/100,000 people with annual increases of about 10% (Karagas et al, 1994). A further remarkable feature is the risk suffered by patients of developing further BCC at different sites. Importantly, the risk of further lesions depends on the number present; 27% of patients with 1 tumour suffer a further tumour within 5 years compared with 90% of those with 10 or more lesions (Karagas et al, 1994). These data suggest susceptibility to BCC is not merely dependent on UV exposure but also on host genetic factors. This view is supported by data showing no association between LOH at chromosome 9q and UV exposure, assessed by site or UV-related mutations in p53. A candidate gene (*patched*) for Gorlins syndrome, a familial condition associated with multiple BCC, maps to chromosome 9q22 and LOH in this region occurs in 60-70% of sporadic BCC (Gailani et al, 1996). The authors concluded that additional environmental agents act with UVB in the pathogenesis of BCC. Suggested agents included arsenicals, polycyclic aromatic hydrocarbons, ionizing radiation and/or UVA. The importance of host response is also demonstrated by studies showing that susceptibility to UVB-induced inhibition of contact hypersensitivity is a better indicator of non-melanoma skin cancer risk than cumulative UV exposure (Schmieder et al, 1992) and that GSTM1 A/B is protective against multiple BCC (Heagerty et al, 1994). Identifying which genes mediate susceptibility to BCC has important implications as these patients are also at increased risk of more serious cutaneous cancers such as squamous cell carcinoma and malignant melanoma and, perhaps haematological malignancies (Lear et al, 1996a).

The concept of susceptibility to BCC is complex, as genetic factors could influence tumour numbers and accrual (Heagerty et al, 1994, Lear et al, 1996b, 1997a). We have used a molecular epidemiological approach to identify GST genes that influence susceptibility to:

i. BCC regardless of tumour numbers.
ii. increased numbers of primary BCC.
iii. a faster accrual of tumours (number of lesions/year).

iv. lesions in less sun-exposed sites.
v. a faster time between 1st presentation and next tumour.

Accordingly, we recruited 856 Caucasians with BCC. In total, 566 patients suffered one tumour and 290 patients between 2-35 tumours.

i. Influence of GST genotypes on susceptibility to BCC: Table 1 shows the frequencies of GST and CYP genotypes in BCC patients, these were not different from those in controls indicating that these polymorphisms do not influence susceptibility to BCC (Lear et al, 1996b).

Table 1: GST and CYP genotypes in patients with single and multiple basal cell carcinoma

GSTM1		null	A	B	A/B
		392 (52.9%)	213 (28.7%)	114 (15.4%)	22 (3.0%)
GSTT1		null	expresser		
		122 (17.4%)	579 (82.6%)		
GSTM3		AA	AB	BB	
		208 (72.7%)	66 (23.1%)	12 (4.2%)	
CYP2D6	EM	HET	PM		
		429 (63.4%)	205 (30.3%)	43 (6.4%)	
CYP1A1	Ile/Ile	Ile/Val	Val/Val		
exon 7 mutation		572 (85.0%)	94 (14.0%)	7 (1.0%)	
		m1m1	m1m2	m2m2	
Msp1 3'-flanking		561 (82.0%)	119 (17.4%)	4 (0.6%)	

ii Influence of GST genotypes on tumour numbers: The data shown in Table 1 were further analysed using a Poisson regression approach to identify genotypes and characteristics associated with numbers of BCC. Table 2 shows median numbers of tumours in patients with and without particular characteristics. These data show that GSTM1 null and GSTM3 AA, in combination with skin type 1 (individuals who always suffer an inflammatory response with burning but without tanning on exposure to UV), are associated with increased susceptibility to larger numbers of BCC. A rate ratio was calculated allowing estimation of the quantitative importance of a characteristic (Table 3). Thus, the rate ratio for males (1.43) against females is equal to the mean number of BCC in males

Table 2: Factors influencing numbers of primary BCC (uncorrected)

	median numbers of BCC	
	with	without
Male gender	2.38	1.66
Blue or green eyes	2.39	1.92
Skin type 1	2.99	2.12
GSTM1 null and skin type 1	4.57	2.00
GSTM3 AA and skin type 1	3.57	2.08

(2.38)/mean number of BCC in females (1.66), when gender alone is considered.

Neither GSTM1 null nor GSTT1 null alone were significant risk factors. CYP2D6 EM was associated with increased numbers of BCC (Table 3). Male gender, skin type 1, and blue and green eyes were also significantly associated with increased numbers of BCC. Table 3 shows significant, age-corrected interactions. Thus, the rate ratio for the interaction between GSTM1 null and skin type 1 (2.702) was the highest identified (Yengi et al, 1996, Lear et al, 1996b).

iii Factors influencing accrual of new BCC: 169 patients were included in the study of genotype influences on BCC accrual (the median number of new tumours/year). Male gender appeared a particularly influential factor though both CYP2D6 EM and GSTT1 null were significantly associated with the rate at which new tumours appeared. The influence of GSTM1 null on the rate of accrual of tumours also approached significance. Comparison of the rate ratios indicated GSTT1 null was the most influential. No significant interactions between genes were identified. A significant interaction between CYP2D6 EM and male gender was observed (Table 3).

iv Factors influencing BCC site: Since the trunk is generally less exposed to UV, patients with truncal tumours may represent a high risk group because they are less able to detoxify the products of UV-induced damage. There is little data on genetic factors that influence formation of truncal rather than non-truncal tumours, though exposure to arsenic predisposes to truncal tumours, suggesting that factors other than UV alone are important. We studied the influence of genotypes and characteristics on tumour site in 345 Caucasians with BCC. 80.1% of first tumours were on the head/neck, 11.4% on the trunk, 6.5% on lower limbs and 2.0% were on upper limbs. Thus, 20% of BCC patients develop at least one tumour at sites believed to suffer relatively less exposure. 170 patients suffered one tumour and 175 patients 2-30 tumours. The mean number of primary tumours in patients whose first tumour was truncal (n=40; mean tumour number=3.96) was greater than those whose first tumour was not truncal (n=312; mean tumour number=2.58; p=0.0297).

Using logistic regression, the age-corrected proportion of GSTT1 null and CYP1A1 Ile/Ile genotypes were significantly greater in patients with at least one truncal tumour. The importance of GSTT1 and CYP1A1 genotypes was emphasised by the increased significance of the interaction (combination of GSTT1 null and CYP1A1 Ile/Ile) (Table 3). GSTM1 null, GSTM3 AA, CYP2D6 EM and CYP1A1 m1m1 were not associated with tumour site.

Patients whose first tumour is truncal represent a high risk group. Skin type 1 was not different between the two groups. GSTT1 and CYP1A1 were important, even after correction for BCC number as well as age at first presentation and gender. The influence of the highly significant interactive term (GSTT1 null and CYP1A1 Ile/Ile) suggests the effects are important. The data supports the view that GSTT1 null is associated with faster appearance of further tumours and the presence of truncal tumours. Thus, it appears that this genotype exerts its effect on the rate of BCC appearance because it predisposes to tumours on chronically and intermittently exposed sites. GSTT1 null individuals may be more susceptible to BCC following even relatively little UV exposure resulting in an increased number of tumours at a younger age and the development of lesions on intermittently exposed sites such as the trunk. No GSTM1 null effect was identified (Lear et al, 1997a).

v **Factors influencing time to presentation of next BCC:** This concept is important as an understanding of the factors that influence time to next lesion may allow clinicians to focus on susceptible individuals who are likely to present with further tumors quickly (i.e. within 2 years). Thus, lesions detected early are likely to be smaller and more easily removed with better cosmetic outcome.

Survival analysis was used to study factors that influence time from first presentation to presentation of a subsequent tumour. Cox's Proportional Hazards Regression was used to determine which factors influenced the time from first presentation to presentation of a subsequent tumour (Lear et al, 1997b). Data were adjusted for number of tumours at first presentation. The presence of more than one tumour at first presentation was significantly associated with a decreased time to next tumour ($p<0.0001$, hazard ratio 2.72, median time to next tumour: single tumour at first presentation 5.75 years, multiple tumours at first presentation 3.58 years). GSTT1 null was associated with a significantly reduced time to presentation of next tumour (Table 3) whereas GSTM1 null, CYP2D6 EM, CYP1A1 Ile/Ile and CYP1A1 m1m1 were not. Table 3 shows significant two factor interactions, corrected for number of tumours at first presentation. Thus, highly significant interactions between the presence of the first tumour on the trunk and each of GSTM1 null and CYP2D6 EM were identified (Fig. 1).

These interactions appeared particularly significant as all of such cases demonstrated further tumours within 5 years whereas only half of the cases without these combinations suffered a further lesion during this time. Other interactions between genotypes and patient characteristics, including skin type 1 and male gender were not significant. Our data indicate patients with a truncal tumour at presentation, especially males and those presenting with more than one lesion should receive meticulous and more frequent follow-up to expedite early diagnosis. Our data show that inter-individual differences in the efficiency of detoxification reactions also determine susceptibility. Thus, the highly significant influence of GSTM1 null or CYP2D6 EM on time to next tumour presentation suggests a possible use of genetic markers in a follow-up strategy.

Table 3: Association of CYP and GST genotypes with BCC numbers and accrual

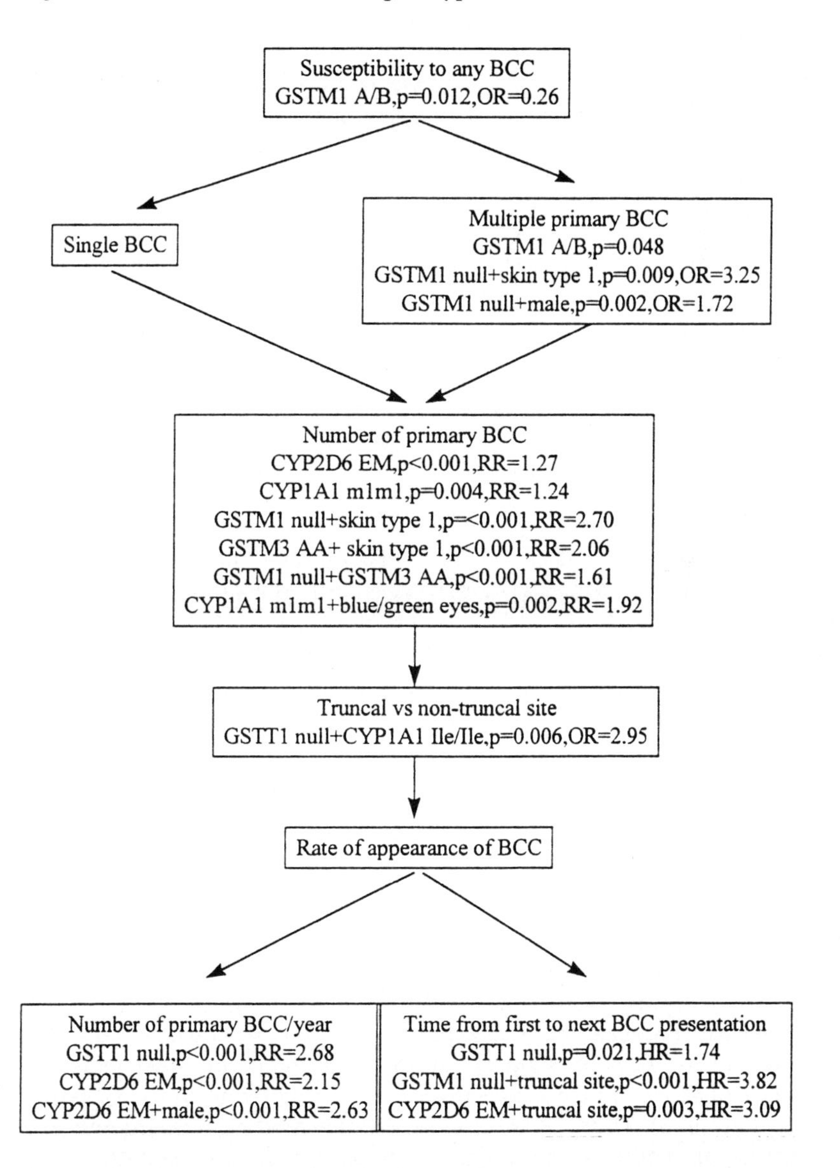

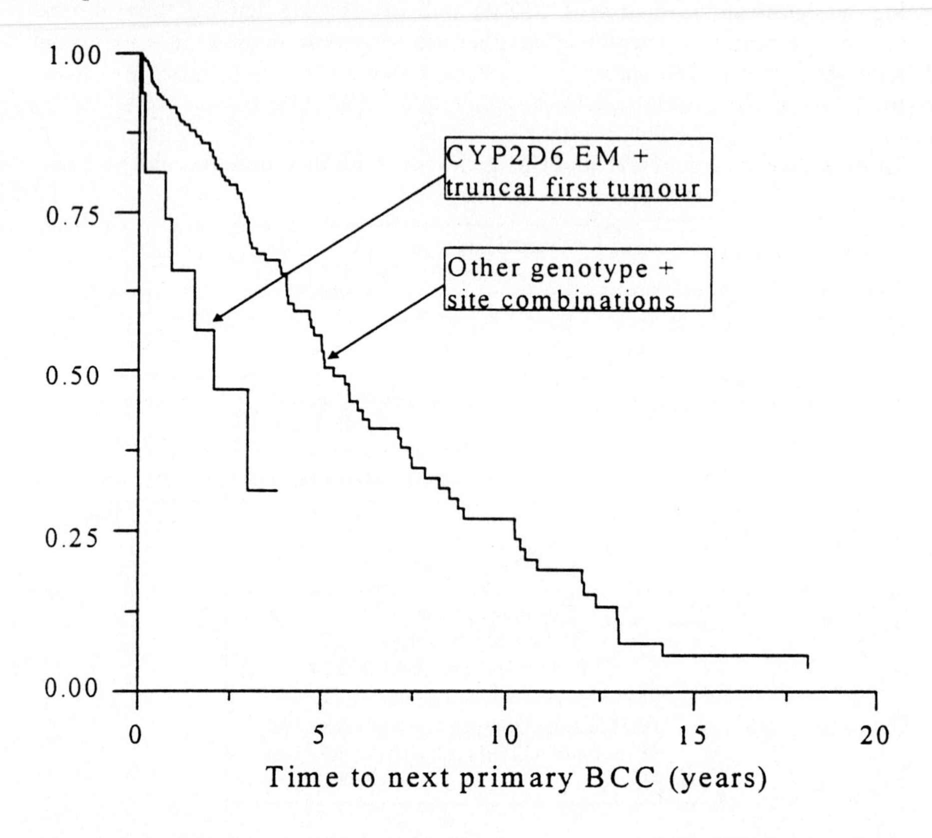

Fig. 1 Cumulative event-free probability curve for time to presentation of subsequent primary BCC in patients with the CYP2D6 EM genotype and whose first tumour was truncal compared with all other BCC patients.

Summary

Though a developing body of data indicates polymorphism at GST genes influences cancer susceptibility, it is unclear why a genotype is associated with one cancer but not another. We believe the GST exert a critical role in normal cell house-keeping activities. GSTM1, GSTM3 and GSTT1 influence tumorigenesis because these enzymes utilise the products of UV-induced oxidative stress. Further support for the importance of these genes in the protection of skin from UV comes from studies in systemic lupus erythematosus (Ollier et al, 1996). Thus, GSTM1 null is associated with increased anti-Ro (but not anti-La) antibodies, a phenotype associated with photosensitivity.

At present there is no basis for predicting which cancers will be influenced by GST polymorphisms though other studies do indicate that the GSTs are critical in the metabolism of environmental carcinogens. For example, GSTT1 null confers an increased risk of astrocytoma (Hand et al, 1996). While brain tumours are not clearly associated with environmental pollutants, N-methyl-N-nitrosourea, processed meats and occupation have been implicated. Why GSTT1 but not GSTM1 or GSTM3 influences the risk of astrocytoma is unclear. GSTM3 appears a good susceptibility candidate, as some astrocytes demonstrate strong expression (Hand et al, 1996). Susceptibility to squamous cell cancer of the larynx, a pathology associated with chronic consumption of tobacco and alcohol, is also influenced by allelism at GSTM3 (Jahnke et al, 1996). The roles of CYP2D6 and CYP1A1 are even more unclear, though the finding that systemic agents such as arsenic predispose to multiple BCC, suggests that CYP2D6-mediated hepatic detoxification of photosensitizing agents may be important.

Importantly, the extent of altered risk conferred by genotypes is generally 2-3 fold and it is necessary to identify which other genes interact with the GST so that haplotypes associated with 10-20 fold increases in risk can be defined.

Acknowledgements

We gratefully acknowledge the support of the Cancer Research Campaign, Arthritis and Rheumatism Council and Wellbeing.

References

Gailani MR, Leffell DJ, Zeigler AM, Gross EG, Brash DE, Bale AE (1996) Relationship between sunlight exposure and a key genetic alteration in basal cell carcinoma. J Natl Cancer Instit 88:349-353

Hand PA, Inskip A, Gilford J, Alldersea J, Elexpuru-Camiruaga J, Hayes J D, Jones PW, Strange RC, Fryer AA (1996) Allelism at the glutathione S-transferase GSTM3 locus: interactions with GSTM1 and GSTT1 as risk factors for astrocytoma. Carcinogenesis 17:1919-1922

Hayes, JD and Strange, RC (1995) Potential contribution of the glutathione S-transferase supergene family to resistance to oxidative stress. Free Rad Res Commun 22:193-207

Heagerty AHM, Fitzgerald D, Smith A, Bowers B, Jones P, Fryer AA, Zhao L, Alldersea J, Strange RC (1994) Glutathione S-transferase GSTM1 phenotypes and protection against cutaneous malignancy. Lancet 343:266-268

Inskip A, Elexperu-Camiruaga J, Buxton N, Dias PS, MacIntosh J, Campbell D, Jones PW, Yengi L, Talbot A, Strange RC, Fryer AA (1995) Identification of Polymorphism at the Glutathione-S-Transferase, GSTM3 Locus: Evidence for linkage with GSTM1*A. Biochem J 312:713-716

Jahnke V, Matthias C, Fryer AA, Strange RC (1996) Glutathione S-transferase and cytochrome P-450 polymorphism as risk factors for squamous cell carcinoma of the larynx. Amer J Surg 172:671-673

Karagas MR for the Skin Cancer Prevention Study Group (1994) Occurrence of cutaneous basal cell and squamous cell malignancies among those with a prior history of skin cancer. J Invest Dermat 102:10S-13S

Kato S, Bowman ED, Harrington AM, Blomeke B, Shields P (1995) Human lung carcinogen-DNA adduct levels mediated by genetic polymorphisms in vivo. J Natl Cancer Inst 87:902-907

Lear JT, Smith AG, Jones PW, Fryer AA, Strange RC (1996a) Multiple basal cell carcinoma and haematological malignancy. Brit Med J 313:298-299

Lear JT, Heagerty AHM, Smith A, Bowers B, Payne CR, Smith CAD, Jones PW, Gilford J, Yengi L, Alldersea J, Fryer AA, Strange RC (1996b) Multiple cutaneous basal cell carcinomas: glutathione S-transferase (GSTM1, GSTT1) and cytochrome P450 (CYP2D6, CYP1A1) polymorphisms influence tumour numbers and accrual. Carcinogenesis 17:1891-1896

Lear JT, Smith A, Bowers B, Heagerty AHM, Jones PW, Gilford J, Alldersea J, Strange RC, Fryer AA (1997a) Tumor site in cutaneous basal cell carcinoma: Influence of glutathione S-transferase, GSTT1 and cytochrome P450, CYP1A1 genotypes and their interactions. J Invest Dermatol in press

Lear JT, Strange RC and Fryer AA (1997b) Relationship between sunlight exposure and a key genetic alteration in basal cell carcinoma. J Natl Cancer Inst in press.

Nakajima T, Elovaara E, Anttila S, Hirvonen A, Camus A-M, Hayes JD, Ketterer B, Vainio H (1995) Expression and polymorphism of glutathione S-transferase in human lungs: risk factors in smoking-related lung cancer. Carcinogenesis 16:707-711

Ollier W, Davies E, Snowden N, Alldersea J, Fryer AA, Jones P, Strange RC (1996) Association of homozygosity for glutathione S-transferase GSTM1 null alleles with the Ro+/La- autoantibody profile in patients with systemic lupus erythematosus. Arth Rheumat in press.

Ryberg D, Kure E, Lystad S, Skaug V, Stangeland L, Mercy I, Børresen A-L, Haugen A (1994) p53 mutations in lung tumours: Relationship to putative susceptibility markers for cancer. Cancer Res 54:1551-1555

Sarhanis P, Redman C, Perrett C, Brannigan K, Clayton RN, Hand P, Musgrove C, Suarez V, Jones P, Fryer AA, Farrell WE, Strange RC (1996) Epithelial ovarian cancer: influence of polymorphism at the glutathione -transferase GSTM1 and GSTT1 loci on p53 expression. Brit J Cancer 74:1757-1761

Schmieder GJ, Yoshikawa T, Mata SM, Streilein JW, Taylor, JR (1992) Cumulative sunlight exposure and the risk of developing skin cancer in Florida. J Dermat Surg Oncol 18:517-522

Seidegard J, Vorachek WR, Pero RW, Pearson WR (1988) Hereditary differences in the expression of the human glutathione S-transferase activity on trans-stilbene oxide are due to a gene deletion. Proc Natl Acad Sciences USA 85:7293-7297

Smith G, Stanley LA, Sim E, Strange RC, Wolf CR (1995) Metabolic polymorphisms and cancer susceptibility. Cancer Surveys 25:27-67

Strange RC (1996) Glutathione S-transferases and cancer susceptibility, In: Proceedings 1995 International ISSX-Workshop on Glutathione S-transferases, Taylor and Francis

Yengi L, Inskip A, Gilford J, Alldersea J, Bailey L, Smith A, Lear JT, Heagerty AHM, Bowers B, Hand P, Hayes JD, Jones PW, Strange RC, Fryer AA (1996) Polymorphism at the glutathione S-transferase GSTM3 locus: Interactions with cytochrome P450 and glutathione S-transferase genotypes as risk factors for multiple basal cell carcinoma. Cancer Res 56:1974-1977

Population Responses to Contaminant Exposure in Marine Animals: Influences of Genetic Diversity Measured as Allozyme Polymorphism

Anthony J. S. Hawkins
Plymouth Marine Laboratory, Natural Environment Research Council, West Hoe, Plymouth PL1 3DH, Devon, United Kingdom

Abstract

Current understanding of the genetic and metabolic basis of relations between heterozygosity and animal performance under non-polluted conditions is relevent to interpreting apparently inconsistent findings concerning the relative advantage of allozyme polymorphism during contaminant exposure. Many interrelated factors which may influence and even compromise those relations include species (lifestyle, reproductive behaviour etc), lifestage, environmental influences and a variety of background genetic effects (limited parentage, null alleles, aneuploidy, genomic imprinting etc). Nevertheless, there is some promise that single-locus responses may be diagnostic for specific pollutants. In addition, limited evidence to date supports the *a priori* expectation that reduced energy requirements for maintenance metabolism may facilitate longer survival of multiple-locus heterozygotes during exposure to contaminants with toxic effects that result either in the reduced acquisition or availability of metabolizable energy, and/or a reduction in the efficiency with which metabolizable energy is used to fuel metabolic processes. More work is required to fully establish this trend in response to specific contaminant types, to assess any direct consequences of underlying differences in protein metabolism, and to resolve the interactive effects of contaminant mixtures. But the functional value of genetic variation within populations is confirmed. Reduced genetic diversity may not only compromise the capacity of an impacted population for genetic adaptation in the face of further environmental challenge, but may also result in increased energy requirements, lower production efficiency and reduced reproductive output. These metabolic consequences of reduced genetic polymorphism would further lower that population's potential for survival under lethal conditions of contaminant exposure, and also affect the genetic makeup of populations through differential reproduction under conditions of sublethal stress.

Introduction

It has often been suggested that the substantial genetic polymorphism which is evident both within and between animal populations may be used for detecting and monitoring the genetic effects of marine pollution (e.g. Nevo et al. 1986, Gillespie and Guttman 1993). Yet there are suprisingly few studies of the effects of pollutants on the genetics of populations. And although findings to date include convincing evidence for the natural selection of resistant genotypes in response to contaminant exposure (e.g. Nevo et al. 1977, Guttman 1994), the genetic and metabolic mechanisms that impart differential tolerance remain unclear. Recent discussions of findings to date have emphasised an apparent lack of consistent effects, especially concerning the relative advantage of enzyme polymorphism measured electrophoretically as heterozygosity or homozygosity for allozymes, with little or no indication of the likely basis of that variation during exposure to contaminants (Hummel and Paternello 1994, Beaumont and Toro 1996, Depledge 1996).

Much the greatest understanding of how genetic diversity may influence animal performance has emerged from the study of widespread associations whereby multiple-locus heterozygosity measured as enzyme polymorphism increases with growth, viability and other fitness-related traits in relations that underly "hybrid vigour" or "heterosis" under non-polluted conditions (Mitton and Grant 1984, Zouros and Foltz 1987, Hawkins and Day 1996). The main purpose of this paper is to consider how current understanding of the genetic and metabolic basis of those relations observed between multiple-locus heterozygosity and animal performance under non-polluted conditions may help (i) to explain the apparently variable effects of heterozygosity on survival during contaminant exposure, and (ii) to develop testable *a priori* expectations based upon the metabolic effects of different contaminant types.

The Genetic and Metabolic Basis of Relations between Multiple-locus Heterozygosity and Animal Performance under Non-polluted Conditions

Multiple-locus heterozygosity has been shown to vary in positive relations with growth, fecundity and/or viability in many plant and animal species (Mitton and Grant 1984, Zouros and Foltz 1987). The main physiological basis of faster production in more heterozygous individuals is reduced energy requirements for maintenance and growth, and which has been shown to result from lower intensities of protein turnover in mammals, shellfish and finfish (reviewed by Hawkins and Day 1996; see also Hedgecock et al. 1996, Bayne and Hawkins 1997). Protein turnover provides the metabolic flux that enables repair and cellular sanitation, regulation, development and adaptation (Hawkins, 1991). Because of the very obvious need to maintain the regulation and specificity of associated proteolysis, total energy costs associated with protein turnover are very significant. The result is that genetic differences in the intensity of protein turnover, and which are in some way linked with allozyme heterozygosity, affect energy costs of maintenance and growth, and hence growth efficiency (net energy gain/total energy absorption) among animals generally (Hawkins and Day 1996).

In addition to establishing higher growth efficiency and faster growth, experimental studies have shown that individuals with higher multiple-locus heterozygosity may be generally advantaged under conditions which, although non-polluted, are stressful in some other way. In each case, it seems likely that multiple-locus heterozygotes performed better by virtue of lower energy requirements for maintenance processes, which helped to minimise net energy losses under conditions when metabolizable energy was limited either in response to anoxia (Borsa et al. 1992), the inhibition of feeding processes (e.g. Scott and Koehn 1990) or reduced food availability (e.g. Gentili and Beaumont 1988).

Physiological consequences of multiple-locus heterozygotes are also evidenced under conditions of ecological heterogeneity in space and time (Hawkins 1996). Given fluctuations of environmental factors within the normal range for that species, multi-locus heterozygotes show an average phenotypic advantage, with reduced phenotypic variance (Hawkins 1996). Again, genetic influences on protein metabolism appear to underly these observations. Given fluctuating environmental conditions, the reduced metabolic intensity associated with slower protein turnover may result in greater potential scope for activity, decreased metabolic sensitivity, lower energy costs incurred during acclimation, and greater independence of net production from those environmental fluctuations (Hawkins 1995, 1996). As described above for experimental observations, these advantages may be most evident under conditions that result in the reduced availability of metabolizable energy.

These physiological interrelations provide an adaptive mechanistic explanation for heterozygote advantage, and support the long-standing theory that high protein polymorphism results from a balance of selective forces stemming from ecological and/or environmental heterogeneity. They also confirm the general influence of biochemical genetic polymorphism on metabolism and performance. However, the specific genetic basis of differences in protein metabolism and energy requirements remains unclear. Indeed, throughout the last decade, there has been considerable controversy concerning both the generality and the underlying genetic cause of heterozygosity associations. Many studies failed to find significant correlations, and no study has yet unambiguously established that the studied loci are directly responsible for heterozygosity-associations, or whether they interact with, or are neutral markers of, other responsible loci (Hedgecock et al. 1996). Nevertheless, there is now a general realisation that multiple-locus heterozygosity is typically associated with less than 25% of the variance measured between individuals, and that relations are not uniformly dependent on allozyme variability (Britten 1996). Instead, physiological consequences associated with multiple-locus heterozygosity may vary according to a variety of factors that include species (lifestyle/reproductive behaviour), age, environmental influences and background genetic effects (e.g. Gaffney 1994). In particular, it appears that associations vary according to the studied enzymes, rather than with heterozygosity *per se*. Enzymes that are involved in protein catabolism make among the greatest contributions to heterozygosity associations (e.g. Koehn et al. 1984), as is consistent with the established significance of differences in protein turnover (Hawkins and Day 1996).

Understanding the Population Genetics of Animal Survival during Contaminant Exposure

Studies of the genetic effects of marine pollution have reported conflicting effects of heterozygosity, without clear understanding of the mechanisms involved. For example, at the level of the single gene, numerous experimental studies of selection for different genotypes have reported longer survival of both heterozygotes and homozygotes (e.g. Lavie and Nevo 1986a, Hawkins et al. 1989b). Similarly, for multiple-locus effects, there has been no apparent consistency in either average heterozygote (Nevo et al. 1986, Beaumont and Toro 1996) or average homozygote (Battaglia et al. 1980, Patarnello et al. 1991, Guttman 1994) superiority during exposure to a variety of pollutants (Table 1).

To help explain the apparently variable effects of heterozygosity on survival during contaminant exposure, it is firstly important to distinguish between single-locus and multiple-locus effects. Evidence that selection may depend on the gene is conclusive, such as for different phosphoglucose isomerase (PGI) genotypes in separate species exposed to dissolved metals (e.g.. Lavie and Nevo 1986a, Hawkins et al. 1989b, Guttman 1994, Beaumont and Toro 1996). Yet at the level of the single locus, there is no general rule for genotypic fitness, with selection both for heterozygous and homozygous genotypes. One simply should not expect any single genotype to be most fit under all circumstances. Genetic components of fitness depend upon the gene, the species, sex, life history, stage and physiological condition, as well as upon separate interactions with the effects of independent environmental variables such as temperature, food availability and different contaminants (e.g. De Nichola et al. 1992, Beaumont and Toro 1996).

Interactions between gene and environment are known to be diverse, but there are few examples involving marine pollution. Working with a gastropod, Lavie and Nevo (1986b) reported differential survivorship of allozyme genotypes that were specific for each type of pollutant, as well as for their interaction. Alternatively, De Nichola et al. (1992) described opposite effects of Cd and Zn on survivorship of phosphoglucomutase (PGM) genotypes in isopods, but with a dominant influence of Zn. Under extreme concentrations of pollutants, collective evidence indicates that the relative functional differences associated with separate genes become more significant, such that homozygotes at specific enzymes may be better adapted (e.g. Lavie and Nevo 1986a, Guttman 1994).

Table 1. Summary of findings relating survival with multiple-locus heterozygosity (MLH) during contaminant exposure

Heterozygote Association	Observations
Survived longer	
Nevo et al. 1986	Gastropods: more heterozygous species, more resistant during experimental exposure to inorganic (metals and NaCl) and organic (detergents and crude oil) pollutants
Beaumont and Toro 1996	Mussels: heterozygotes (PGI, GSR, LAP, ODH) survived longer during experimental copper exposure (100 ppb)
Died faster	
Battaglia et al. 1980	Mussels: reduced MLH for 5 loci (AP, LAP, PGI, IDH, PGM) among survivors exposed to increasing pollution in the Lagoon of Venice
Patarnello et al. 1991	Barnacles: reduced MLH for 3 loci (MPI, PGI, PGM) among survivors in a polluted area in the Lagoon of Venice
Guttman 1994	Mosquitofish: reduced MLH genetic diversity at 3 (ADA, PGI, IDH) of 5 loci among survivors exposed in nature to coal plant effluent in South Carolina

Given that selection varies according to the gene, with a tendency towards homozygosity under extreme conditions that markedly narrow the normal niche-width, it should be clear that any consequences associated with higher average multiple-locus heterozygosity during contaminant exposure are not related to heterozygosity *per se*. Rather, heterozygosity effects are locus-specific. Significantly, locus-specific responses to contaminant exposure were inferred by Newman et al. (1989) who, studying the influences of mercury and arsenate on eight enzyme loci in mosquitofish, described how consequences associated with multiple-locus heterozygosity stemmed mostly from the summation of single-locus effects. Observed results may depend on the metabolic significances of measured allele products, such as for enzymes involved in protein catabolism, or they may stem from "associative overdominance", in which heterozygosity at the studied alleles acts as a marker of other linked loci that affect metabolism (Gaffney 1994). Whatever, correlations are influenced by the enzymes chosen for analysis.

Analyses of differential selection according to multiple-locus heterozygosity must take a variety of additional genetic effects into account. Most evidently, correlations may be compromised among animals of limited parentage, such as in the offspring from hatchery crosses, when relations may be obscured by linkage and epistatic effects. Other possible influences include spontaneous segmental aneuploidy and genomic imprinting, as well as possible mis-scoring of heterozygotes due to null alleles. All of these potential influences have contributed to past controversy concerning both the generality and underlying genetic cause of heterozygosity associations under non-polluted conditions (Gaffney 1994). As such, they may also contribute to the apparently variable effects of heterozygosity on survival during contaminant exposure.

To better predict the consequences of average multiple-locus heterozygosity for relative performance during exposure to pollutants, it is necessary to consider the metabolic and physiological basis of heterozygosity associations, and the likely implications for survival during exposure to different contaminant types. As explained above, the main and most consistent difference underlying faster growth in more heterozygous animals is slower protein turnover, representing the metabolic basis of lower energy requirements and higher growth efficiency. Protein turnover does not in itself effect the isolation, elimination or detoxification of contaminants. But there may be other functional consequences of protein turnover during the response to pollutants, which are undoubtedly complex. For example, faster protein turnover has been shown to enhance the incorporation of copper within proteins throughout the body, including metalloenzymes and other high molecular weight proteins, as well as smaller detoxifying metallothionein proteins (Hawkins et al. 1989, Hawkins unpublished data). Alternatively, a significant proportion of proteolysis may be effected by enzymes within the lysosomal vacuolar system (Hawkins and Day 1986), within which metals and other contaminants are sequestered by marine invertebrates prior to excretion (e.g.. Viarengo 1989). It is therefore conceivable that individuals with a greater facility for lysosomally-mediated turnover may benefit indirectly from a greater capacity to sequester contaminants. Other more certain benefits of faster protein turnover are in facilitating repair, acclimation and adaptation. Continuous replacement and renewal enables both sensitive and rapid adjustment of intracellular concentrations of proteins, including the synthesis of appropriate new enzymes, as well as the removal of redundant proteins (Hawkins 1991). This has been evidenced as faster metabolic acclimation of individual mussels with more intense protein turnover, in response to environmental temperature change (Hawkins et al. 1987). The faster acclimation afforded by more intense protein turnover may be particularly important in the face of small but frequent environmental changes, including sublethal fluctuations in the uptake or availability of contaminants.

There is, then, considerable uncertainty concerning the direct effects of faster protein turnover. More obvious are the indirect consequences of high associated energy costs, as have been clearly demonstrated under non-polluted conditions (refer above). Bearing in mind that energy requirements for movement, reproduction etc vary according to lifestyle and lifestage, it seems likely that lower energy requirements for maintenance may enable more heterozygous individuals to survive longer when the metabolic effects of contaminant exposure result either in the reduced availability of metabolizable energy, and/or a reduction in the efficiency with which metabolizable energy is used to fuel metabolic processes. It is to be expected that survival will be enhanced both as a result of there being more energy to support metabolic requirements that include detoxification mechanisms, and through the slower utilisation of energy reserves.

Any effects of reduced energy requirements in more heterozygous individuals will depend upon the mode of toxic action during contaminant exposure. Therefore, it is useful to consider which classes of environmental contaminants are known to reduce either the availability or efficiency of production of metabolizable energy in marine animals. Availability of metabolizable energy may be diminished following the inhibition of feeding, digestion and/or absorption as occurs following non-specific narcosis that is induced by hydrocarbons (e.g.. Donkin et al. 1991), neurotoxic pesticides (Donkin et al. 1997) and metals (e.g.. Viarengo 1989), or simply by the inhibition of oxidative metabolism as has been observed for the organic toxicant dibutyltin (Snoeij et al. 1987). Alternatively, efficiency of energy production may be reduced by uncouplers of oxidative metabolism in mitochondria, that inhibit the coupling between electron transport and phosphorylation reactions, thereby inhibiting ATP synthesis but without affecting respiration. Contaminant classes that are known to induce such metabolic uncoupling include tributyltin (Snoeij et al. 1987) and phenols (Buikema et al. 1979).

There is sufficient consistency in the impact of contaminant classes on the availability and/or utilization of metabolizable energy that one can justify *a priori* expectations for the superiority of average multiple-locus heterozygotes in the face of those contaminant classes alone, bearing in mind the other influencing factors (species, background genetic effects etc) discussed above. And in this respect, it may be significant that both studies reporting longer survival of multiple-locus heterozygotes to date have involved the controlled experimental exposure of animals to contaminants that affect the availability and/or utilization of metabolizable energy, including dissolved metals and/or organic pollutants (Nevo et al. 1986, Beaumont and Toro 1996) (Table 1). Alternatively, the three studies to date that have reported shorter survival of multiple-locus heterozygotes were all undertaken in natural populations exposed to unknown mixtures of chemicals (Battaglia et al. 1980, Patarnello et al. 1991, Guttman 1994) (Table 1). Under such uncontrolled natural conditions, the associated mechanisms of toxicity were presumably more diverse, both in terms of time scale and effect (biochemical, pathological, immunotoxic, cancers etc.), so that survival may not have been as dependent on the medium term availability of energy. Instead, the relative functional differences associated with separate genes appear to have become more significant, such that homozygotes at specific enzymes were better adapted.

Conclusions

Relations between heterozygosity and animal performance under non-polluted conditions indicate the importance of distinguishing between single-locus and multiple-locus effects. Considering single-locus effects, then selection during contaminant exposure may either be for heterozygous or homozygous genotypes, depending upon the studied gene itself, as well as the species, sex, life history stage, physiological status, environmental conditions, and a variety of background genetic effects. There is some promise of diagnostic responses, as indicated by differential survivorship of allelic isozyme genotypes that are specific for each type of pollutant and for their interaction, as well as parallel responses at the same loci to similar types of contaminant but in different species. However, more work is required to screen for and to confirm consistent single-locus responses under defined conditions.

Multiple-locus effects are also locus-specific, and may be confounded by the same variables that influence single-locus effects. There is no evidence to date that the faster protein turnover which underlies greater energy requirements in less heterozygous individuals may confer net survival advantage under specific circumstances or lifestages. Alternatively, limited evidence does support the *a priori* expectation that reduced energy requirements for maintenance metabolism may facilitate longer survival of multiple-locus heterozygotes during exposure to dissolved contaminants with metabolic effects that result either in the reduced availability of metabolizable energy, and/or a reduction in the efficiency with which available energy is used to fuel metabolic processes. More work is required to fully confirm this trend in response to specific contaminant types, bearing in mind that any such advantage may vary according to species/lifestyle, lifestage and environmental influences.

It has long been appreciated that any reduction in genetic diversity as results from natural selection during contaminant stress may compromise the ability of an impacted population for genetic adaptation in the face of further environmental challenge. But in addition, depending on the selected genes, this paper has described how reduced genetic polymorphism measured as allozyme heterozygosity may result in increased energy requirements, lower production efficiency and reduced reproductive output. Such metabolic consequences of reduced genetic diversity further lower that population's potential for survival, and alter performance in response to sublethal effects. Associated changes in relative reproductive effort in response to sublethal contaminant exposure may further affect the genetic makeup of populations, representing an additional indirect influence of genetic diversity.

Comparison of the effects and consequences of different contaminant types may usefully draw upon quantitative structure-activity relationships such as have been established for structurally related organic compounds, predicting a common mechanism of toxicity, together with effects that are additive when present in a complex mixture (e.g. Donkin et al. 1991). But contaminants may induce effects via more than one mechanism of toxicity, and the effects of different mixtures may either be additive or antagonistic (Widdows and Donkin

1991). Therefore, before one can reasonably expect to predict the effects of genetic variation on population responses to contaminant exposure in the natural environment, it will be necessary to develop a better understanding of the interactive effects of contaminant mixtures.

The potential rewards are substantial. When these questions are answered, we will better understand the functional value of genetic variation in populations and species subjected to contaminant impact, with the potential for greater benefits from monitoring population genetic structure within programmes of remediation, enhancement and conservation.

Acknowledgements

I wish to thank Amanda Day for helpful comments upon the manuscript.

References

Battaglia B, Bisol PM, Fossato VU, Rodinò E (1980) Studies on the genetic effects of pollution in the sea. Rapp P-v Réun Cons int Explor Mer 179:267-274

Bayne BL, Hawkins AJS (1997) Protein metabolism, the costs of growth and genomic heterozygosity: experiments with the mussel *Mytilus galloprovincialis* Lmk. Physiol Zool, in press

Beaumont AR, Toro JE (1996) Allozyme genetics of *Mytilus edulis* subjected to copper and nutritive stress. J Mar Biol Ass UK 76:1061-1071

Borsa P, Jousselin Y, Delay B (1992) Relationships between allozyme heterozygosity, body size, and survival to natural anoxic stress in the palourde *Ruditapes decussatus* L. (Bivalvia: Veneridae). J Exp Mar Biol Ecol 155:169-181

Britten HB (1996) Meta-analyses of the association between multilocus heterozygosity and fitness. Evolution 50:2158-2164

Buikema AL, McGinniss MJ, Cairns J (1979) Phenolics in aquatic ecosystem: a selected review of recent literature. Mar Environ Res 2:87-181

Depledge M (1996) Genetic ecotoxicology: an overview. J Exp Mar Biol Ecol 200:57-66

DeNichola M, Gambardella C, Guarino SM (1992) Interactive effects of cadmium and zinc pollution on PGI and PGM polymorphisms in *Idotea baltica*. Mar Poll Bull 24:619-621

Donkin P, Widdows J, Evans SV, Brinsley MD (1991) QSARs for the sublethal responses of marine mussels (*Mytilus edulis*). Sci Total Environ 109/110:461-476

Donkin P, Widdows J, Evans SV, Staff FJ, Yan T (1997) Effect of neurotoxic pesticides on the feeding rate of marine mussels (*Mytilus edulis*). Pestic Sci 49:196-209

Fevolden SE, Garner SP (1986) Population genetics of *Mytilus edulis* (L.) from Oslofjorden, Norway, in oil-polluted and non oil-polluted water. Sarsia 71:247-257

Gaffney PM (1994). Heterosis and heterozygote deficiencies in marine bivalves: more light? In: Beaumont AR (ed) Genetics and evolution of aquatic organisms. Chapman and Hall, London, pp 146-153

Gentili MR, Beaumont AR (1988) Environmental stress, heterozygosity and growth rate in *Mytilus edulis*. J Exp Mar Biol Ecol 120:145-153

Gillespie RB, Guttman SI (1993) Allozyme frequency analysis of aquatic populations as an indicator of contaminant-induced impacts. In: Gorsuch JM, Dwyer FJ, Ingersoll CG, LaPoint TW (eds) Environmental toxicology and risk assessment, Vol.2, ASTM STP 1173. American Society for Testing and Materials, Philadelphia, pp 134-145

Guttman SI (1994) Population genetic structure and ecotoxicology. Environmental Health Perspectives 102(12):97-100

Harrison FL, Lam JR, Novacek J (1988) Partitioning of metals among metal-binding proteins in the bay mussel, *Mytilus edulis*. Mar Environ Res 24:167-170

Hawkins AJS (1991) Protein turnover: a functional appraisal. Functional Ecology 5:222-233

Hawkins AJS, Day AJ (1996) The metabolic basis of genetic differences in growth efficiency among marine animals. Journal of Experimental Marine Biology and Ecology 203:93-115

Hawkins AJS (1996) Temperature adaptation and genetic polymorphism in aquatic animals. In: Johnston IA, Bennett AF (eds) Animals and Temperature: Phenotypic and Evolutionary Adaptation. Society for Experimental Biology Seminar Series 59, Cambridge University Press, Cambridge, pp 103-126

Hawkins AJS (1995) Effects of temperature change on ectotherm metabolism and evolution: metabolic and physiological interrelations underlying the superiority of multi-locus heterozygotes in heterogeneous environments. J Thermal Biol 20:23-33

Hawkins AJS, Bayne BL, Day AJ, Rusin J, Worrall CM (1989a) Genotype-dependent interrelations between energy metabolism, protein metabolism and fitness. In: Ryland JS, Tyler PA (eds) Reproduction, genetics and distributions of marine organisms. 23rd European Marine Biology Symposium, Olson and Olson, Fredensborg, pp 283-292

Hawkins AJS, Rusin J, Bayne BL, Day AJ (1989b) The metabolic/physiological basis of genotype-dependent mortality during copper exposure in *Mytilus edulis*. Mar Environ Res 28:253-257

Hawkins AJS, Wilson IA, Bayne BL (1987) Thermal responses reflect protein turnover in *Mytilus edulis*. Funct Ecol 1:339-351

Hedgecock D, McGoldrick DJ, Manahan DT, Vavra J, Appelmans N, Bayne BL (1996) Quantitative and molecular genetic analyses of heterosis in bivalve molluscs. J Exp Mar Biol Ecol 203:49-59

Holley ME, Foltz DW (1987) Effects of multiple-locus heterozygosity and salinity on clearance rate in a brackish water clam, *Rangia cuneata* (Sowerby). J Exp Mar Biol Ecol 111:121-131

Hummel H, Patarnello T (1994) Genetic effects of pollutants on marine and estuarine invertebrates. In: Beaumont AR (ed) Genetics and evolution of aquatic organisms. Chapman and Hall, London, pp 425-434

Koehn RK, Diehl WJ, Scott TJ (1988) The differential contribution by individual enzymes of glycolysis and protein catabolism to the relationship between heterozygosity and growth rate in the coot clam, *Mulinia lateralis*. Genetics 118:121-130

Lavie B, Nevo E (1986a) Genetic selection of homozygote allozyme genotypes in marine gastropods exposed to cadmium pollution. The Science of the Total Environment 57:91-98

Lavie B, Nevo E (1986b) The interactive effects of cadmium and mercury pollution on allozyme polymorphisms in the marine gastropod *Cerithium scabridum*. Mar Poll Bull 17:21-23

Lavie B, Nevo E (1987) Differential fitness of allelic isozymes in the marine gastropods *Littorina punctada* and *Littorina neritoides*, exposed to the environmental stress of the combined effects of cadmium and mercury pollution. Environ Management 11:345-349

Lavie B, Nevo E, Zoller U (1984) Differential viability of phosphoglucose isomerase allozyme genotypes of marine snails in nonionic detergent and crude oil-surfactant mixtures. Environ Res 35:270-276

Mitton JB, Grant MC (1984) Associations among protein heterozygosity, growth rate and developmental homeostasis. Ann Rev Ecol Syst 15:479-499

Nevo E (1988) Natural selection in action: the interface of ecology and genetics in adaptation and speciation at the molecular and organismal levels. In: Yom-Tov W, Tchernov E (eds) The zoogeography of Israel. Dr W Junk Publishers, Dordrecht, pp 411-438

Nevo E, Noy R, Lavie, B, Beiles A, Muchtar S (1986) Genetic diversity and resistance to marine pollution. Biol J Linn Soc 29:139-144

Nevo E, Shimony T, Libni M (1977) Thermal selection of allozyme polymorphism in barnacles. Experientia 43:1562-1564

Newman MC, Diamond SA, Mulvey M, Dixon P (1989) Allozyme genotype and time to death of mosquitofish, *Gambusia affinis* (Baird and Girard) during acute toxicant exposure: a comparison of arsenate and inorganic mercury. Aquatic Toxicology 15:141-156

Oakeshott JG (1976) Selection at the Adh locus in *Drosophila melanogaster* imposed by environmental ethanol. Genet Res 26:265-274

Patarnello T, Guiñez R, Battaglia B (1991) Effects of pollution on heterozygosity in the barnacle *Balanus amphitrite* (Cirripedia: Thoracica). Mar Ecol Prog Ser 70:237-243

Scott TM, Koehn RK (1990) The effect of environmental stress on the relationship of heterozygosity to growth rate in the coot clam *Mulinia lateralis* (Say). J Exp Mar Biol Ecol 135:109-116

Snoeij NJ, Pennicks AH, Seinen W (1987) Biological activity of organotin compounds – an overview. Environ Res 44:335-353

Viarengo A (1989) Heavy metals in marine invertebrates: mechanisms of regulation and toxicity at the cellular level. Aquatic Sciences 1:295-317

Widdows J, Donkin P (1991) Role of physiological energetics in ecotoxicology. Comp Biochem Physiol 100C:69-75

Zouros E, Foltz DW (1987) The use of allelic isozyme variation for the study of heterosis. In: Rattazzi MC, Scandalios JG, Whitt GS (eds) Isozymes: current topics in biological and medical research. Alan R. Liss, New York, pp 255-270

Extrahepatic Metabolism in Target Organ Toxicity

(Chair: C.R. Wolf, United Kingdom,
and C. Friis, Denmark)

The Use of Transgenic Animals to Assess the Role of Metabolism in Target Organ Toxicity

C. Roland Wolf, Sandra J. Campbell, A. John Clark[2], Austin Smith[3], John O. Bishop[3] and Colin J. Henderson
Imperial Cancer Research Fund Molecular Pharmacology Unit, Biomedical Research Centre, Ninewells Hospital and Medical School, Dundee, DD1 9SY, UK.
[2] Roslin Institute, Edinburgh, UK;
[3] Centre for Genome Research, University of Edinburgh, Edinburgh, UK

To date, there are few reports on the use of transgenic animals to establish the biological role and the regulation of drug metabolizing enzymes. *In vitro* models such as mammalian, yeast, insect and bacterial systems represent rapid and simple methods for assessing the role of cytochrome P450s in drug metabolism and chemical carcinogenesis, but may not necessarily reflect the *in vivo* situation, where other factors such as drug disposition, governed by absorption, distribution and excretion, are important. *In vivo* models provide the opportunity for the study of the biological effects of gene expression under more physiologically relevant conditions, and also allow the inclusion of pharmaco-toxicological endpoints.

Despite intensive study, the endogenous role(s) of the majority of drug metabolising enzymes, i.e. cytochrome P450 and glutathione-S-transferase (GST), remain to be elucidated. Transgenic animals constitute a powerful tool to unravel the physiological function(s) of these enzymes. Using this approach, for example, it is possible to replace an endogenous gene with the corresponding human sequence, allowing the study of a human drug metabolising enzyme in a heterologous system. Alternatively, replacement of an endogenous coding sequence with a mutant gene, or indeed removal or inactivation of the gene, as in 'knockout' mice, allows the investigation of the role of this gene and its products in normal homeostasis or in toxicological responses (Figure 1). In relation to transgenic animals, there are a number of approaches which can be adopted, some examples of which will be considered below (Figure 2), along with the major limitations and potential solutions (Figure 3).

The development of a regulatable promotor system would allow controlled expression of genes *in vitro*, for example in tissue culture, or *in vivo* in transgenic animals. Expression of a protein under the control of an inducible promotor allows the assessment of the effects of a single gene product in an otherwise normal animal at the appropriate developmental time. Previous efforts in this area have utilised the promotors of either heat shock proteins (Schweinfest et al, 1988), metallothionein (Hu and Davidson, 1990). or steroid-responsive genes (Ko et al, 1989), but have suffered from 'leakiness', i.e. detectable levels of expression in the uninduced state, and relatively poor-fold

Figure 1 Why use transgenic animals to study drug metabolising enzymes?

[1] function of normal vs mutant proteins
[2] study human enzyme
[3] mechanism of protein regulation
[4] establish role in normal homeostasis
[5] establish role in toxicological response
[6] establish pathways of chemical toxicity ie role of specific metabolites

Figure 2 Major approaches

[1] introduction of transgenes
[2] gene deletion
[3] substitution of genes
[4] introduction of promotor sequence
[5] enhancer traps

Figure 3 Limitations and Solutions

[1] Background activity, due to
(a) structural homology from same gene family
(b) overlapping substrate specificities between gene families
Solution: -delete background

[2] Species differences in other enzymes or toxicological response
Solution: -exploit sex/strain differences -eg Ah^R deficiency, female DA rat
-exploit differences in kinetic properties or susceptibility to inhibitors
-use different species, e.g. *Drosophila*

[3] Species differences in gene regulation
Solution: -express protein under homologous promotor

increases in expression after induction. We have utilised the promotor of the CYP1A1 gene, which is not only not constitutively expressed, but also highly inducible by compounds such as polycyclic aromatic hydrocarbons and polychlorinated hydrocarbons (Gelboin, 1980) to create a construct in which an 8.5 kb fragment of genomic DNA from the rat CYP1A1 promotor is cloned upstream of the *lacZ* reporter gene. Transgenic mice generated with this construct were studied to determine transgene expression, which was found to be completely absent in untreated animals. Treatment of these mice with 3-methylcholanthrene resulted in a huge increase (> 1000-fold) in transgene expression in a variety of tissues, including liver, adrenal, kidney and intestine, and at lower levels in other tissues such as spleen, lung, pancreas and reproductive organs (Campbell et al, 1996). The pattern of transgene expression closely mirrored the reported expression profile for CYP1A1 protein. These data support the idea that the CYP1A1 promotor can drive expression of heterologous genes in a tightly regulated manner, and provides a model system to establish the identity of the factor(s) which regulate expression of the CYP1A1 gene.

An *in vivo* drug metabolism model has been developed for *Drosophila melanogaster* (Jowett et al, 1991) and has been used as an extremely valuable genotoxicity model. Several biological endpoints are available for this purpose and the ease of *Drosophila* genetics makes this organism extremely amenable to genetic manipulation. In our studies the major phenobarbital inducible cytochrome P450 from rat (CYP2B1) was inserted into a vector containing a transposable P element. P elements are very mobile elements in the *Drosophila* genome, readily integrating into chromosomes. The vector also contained a phenotypic marker for red eye colour. The constructs were injected into fertilized eggs, and the transgenes were identified by eye colour and by Southern blotting. The larval *Drosophila* promotor (LSP1) directed the expression of CYP2B1 to the fat body of the larvae as verified by immunoblotting using anti-CYP2B1 antibodies. The catalytic properties of the expressed CYP2B1 toward various alkoxy-resorufin analogues were similar to the values found for the purified enzyme, showing that the insect P450-reductase coupled well with the mammalian P450. Using a Somatic Mutation And Recombination Test (SMART), it was found that CYP2B1 expression led to a more than ten-fold increase of genotoxicity of cyclophosphamide when compared to control. In a similar approach, canine CYP1A1 has been expressed in *Drosophila*, and its role in the toxication of 7,12 dimethylbenz(a)anthracene was studied in larvae (Komori et al, 1993), using premature mortality of the larvae as the biological endpoint. Expression of CYP1A1 in the larvae was under the control of a heat shock promotor, and it was found that DMBA toxicity was increased after induction of the heterologous CYP1A1. These examples demonstrate that *Drosophila* has a valuable role to play in the investigation of cytochrome P450 and chemical mutagenesis. However, drug metabolism in insects and mammals is rather distinct and this may limit comparisons between the two classes.

The physiological role(s) of an increasing number of mammalian proteins has been elucidated by the use of transgenic animals. Several techniques for gene transfer into animals have been developed (Kollias and Grosveld, 1992; Boyd and Samid, 1993). In one approach, the gene of interest is introduced by microinjection into the germ line of mice. However, as integration of the transgene is random, positional effects of expression and effects on other genes have to be taken into account. In another technique, pluripotent embryonal stem (ES) cells are transfected with the transgene together with a selection marker. Cells which have taken up the transgene are introduced into blastocysts which are implanted into foster mice. Chimerae can be readily identified by their mosaic hair colour pattern, and by breeding those chimerae, in which the transgene incorporates into the germ line, homozygotes for the transgene can be obtained. In yet another approach, a transgene may be introduced into somatic cells *in vitro* e.g. by retroviral transfer. Cells with the heterologous gene are then implanted into the adult organism where they produce the protein encoded by the foreign gene (Valerio, 1992).

We have recently established transgenic mice which specifically express the human glutathione S-transferase GSTA1 in the epidermal cells of the mouse skin (Simula et al, 1993). Expression was achieved by using the epidermal-specific keratin VI promotor, and was confirmed by Western blotting and enzymatically by masuring the alpha-class specific GST isomerase activity towards androsten 3, 17-dione as substrate. The skin was chosen as the site of expression of the enzyme for several reasons. Firstly, it is an easily accessible target tissue, both for the presentation of chemicals and for the assessment of toxicological endpoints. Secondly, GSTA1 is not expressed in mouse skin, providing an ideal background onto which to introduce the transgene. Finally, with respect to chemically induced skin-tumor formation, the mouse skin is a well characterized tissue. Skin painting experiments should allow the direct evaluation of the role of this and other transgenes in chemical carcinogenesis. Pilot skin painting experiments are now being conducted to investigate the protective role of the GST from carcinogen-induced DNA adduct and tumor formation.

The fact that some P450 enzymes are difficult to induce *in vitro* by xenobiotics means that transgenic animals are a very valuable tool to study the mechanisms underlying the inductive process. Ramsden et al.(1993), generated transgenic mice containing various lengths of the 5«flanking region of CYP2B2 as well as the entire coding and 3«non-coding region of this gene (Ramsden et al, 1993). They found that constructs containing 800 bp of the promotor region were constitutively expressed in mouse liver and kidney at high levels, but this construct was not inducible by phenobarbital. Only the inclusion of an additional 19 kb of 5«flanking region led to a very low basal, but phenobarbital-inducible, hepatic expression of the transgene, analogous to the expression pattern observed for CYP2B2 in rat liver. The authors concluded that sequence elements which had been previously identified 50-100 bp upstream from the transcription start site of CYP2B genes are not sufficient on their own to mediate phenobarbital induction. However, it cannot be ruled out that gene-positional effects may have influenced induction of the CYP2B transgenes in this study. The regulatory regions of CYP1A1 have also been studied in transgenic animals, 2.6 kb of the 5'flanking region of CYP1A1 having been combined with a chloramphenicol acetyltransferase (CAT) reporter construct (Jones et al, 1991). It was found that the expression pattern of the reporter was similar to the pattern of CYP1A1 expression *in vivo*.

Several groups have started to characterize the physiological and pharmacotoxicological function of drug metabolizing enzymes using gene deletion or 'knock outs'. Recently, homologous recombination techniques have been used to study the aryl hydrocarbon (Ah) receptor (AHR) mediated regulation of cytochromes P450 *in vivo*. The AHR has also been implicated in dioxin-mediated teratogenesis, apoptosis and immunosuppression. The AHR gene was inactivated in embryonic stem cells using a positive-negative selection strategy (Fernandez-Salguero et al, 1995), with the positive selection marker (neo) interrupting the AHR-gene in the targeting vector, and the negative selection marker (HSV-thymidine kinase) situated several kb upstream of the AHR-gene segment. Upon homologous recombination, the neo selection marker is retained and the HSV-TK marker lost. Thus, ES cells with the inactivated copy of the AHR will survive in a medium containing G-418 and ganciclovir, whereas cells in which non-homologous recombination had taken place should be killed by ganciclovir due to the expression of the HSV-TK. The inactivation of the AHR yielded pleomorphic symptoms. In the first instance, a high proportion of the animals died in the first few days after birth, the precise causes of death being undetermined. Treatment with dioxin induced CYP1A1 and UGT1*6 in AHR $^{+/+}$ animals but not in AHR $^{-/-}$ mice, demonstrating that a functional AHR is required for the induction in the liver. Histological examination of tissues isolated from the mice with the inactivated AHR revealed pronounced liver fibrosis and a highly reduced liver-to-body weight ratio. The study appeared to demonstrate that AHR plays also an important role in the development and the maintenance of the immune system, as evidenced by a decreased population of the spleen by lymphocytes, in particular in young AHR $^{-/-}$ mice as compared to their normal littermates. However, the mechanisms for this effect are not clear at present.

A second group has inactivated the AHR, this time by replacing exon 2, the region of the gene encoding the basic/helix-loop-helix domain essential for receptor dimerization and DNA binding (Schmidt et al, 1996), with a neomycin resistance gene under the control of an exogenous promotor. Homozygous null mice displayed no overt phenotypical changes from their wild-type counterparts, as litters were found in Mendelian proportions indicating no *in utero* lethality, although the null mice did exhibit a slower growth rate and thus decreased body weight in the post-natal period. CYP1A1 was not inducible in these homozygous null mice, and detailed analysis of the AHR null mice revealed that the livers of these animals were significantly smaller (~25%) than those of their wild-type littermates. The livers of the null mice were visibly different in the first two weeks of life, being pale in colour and of a spongy texture, although these features had disappeared by one month after birth. The authors suggested a metabolic defect in hepatocyte function as the underlying cause for these observations, indicating a temporal requirement for AHR expression. After the first few weeks, most of the AHR null mice began to suffer from portal fibrosis and hypercellularity, while approximately 50% of these animals also possessed enlarged spleens consistent with congestive splenomegaly, which the authors speculated may have been a result of decreased liver blood flow. This study could not distinguish whether the phenotypic changes noted in these AHR mice were a direct result of defective liver growth and maturation, or whether the mice were more susceptible to an unidentified environmental agent as a consequence of the gene deletion. No explanation is immediately apparent for the different phenotype observed from the two studies reported above: genetic background, previously cited in other cases where deletion of the same gene by two groups has led to differing phenotypic consequences, is unlikely to be the cause in this case, since both AHR deletions were on a 129xC57BL/6 background. Equally, the possibility of partial inactivation, a 'leaky' deletion, or formation of a new gene product at the targeted allele, seems unlikely, since both groups showed the absence of functional AHR protein by a lack of an inductive response of CYP1A1 or CYP1A2 to treatment with TCDD, and in the case of Schmidt et al (1996) by an absence of immunoreactive AHR protein on Western blots.

CYP1A2 is a highly conserved cytochrome P450 in mammals, which has been reported to be involved in the oxidative metabolism of xenobiotics, including the activation of polycyclic aromatic hydrocarbons, which are also capable of inducing the expression of CYP1A2. Two groups have recently established mouse lines in which CYP1A2 has been deleted by homologous recombination in ES cells. Pineau et al. (1995) inserted a neomycin resistance gene into the first coding exon of the CYP1A2 gene, and, although heterozygote animals appeared normal, homozygous mutant mice died within one hour of birth, displaying severe respiratory distress. The penetrance of this phenotype was incomplete, with 19/599 animals surviving to adulthood, appearing phenotypically normal and being able to reproduce. Those homozygous mutant mice which died exhibited decreased expression of surfactant apoprotein in type II alveolar cells and a reduced gallbladder size. Liang et al (1996) created their CYP1A2 knockout mouse by replacing part of exon 2, and all of exons 3 to 5, with the hypoxanthine phosphoribosyltransferase gene. In contrast to the previous study, homozygous null mutants were completely viable and appeared phenotypically indistinguishable from their heterozygote or wild-type littermates. Further study of these animals revealed a role for CYP1A2 in zoxazolamine metabolism, paralysis times with this muscle relaxant being increased more than nine-fold in the CYP1A2 null mouse. The difference in phenotypes observed in these two studies is of general interest in the creation of knockout mice. The authors of the second study (.Liang et al, 1996) suggest several possible reasons for the discrepancy, either the genetic background, which was C57BL/6 in their report, and CF-1 or Swiss Black in the first, or the presence of a viral or respiratory pathogen in a genetically susceptible host. Alternatively, the possibility remains that the phenotypic differences lie in the different approaches used to create the targeted allele.

CYP2E1, represented by a single gene in the mouse, is the major P450 responsible for the metabolism of ethanol, and it also carries out the metabolism of a wide range of xenobiotics, including acetaminophen, carbon tetrachloride, pyrazole and dimethylnitrosamine, as well as of endogenous chemicals such as acetone and arachidonic acid. Lee et al (1996) generated a CYP2E1 null mouse by replacing part of intron 1, exon 2 and part of intron 2 with the bacterial neomycin resistance gene under the control of the phosphoglycerate kinase-1 promotor. Mice homozygous null for CYP2E1 were found to be phenotypically indistinguishable from wild-type mice, but to exhibit increased resistance to acetaminophen toxicity, such that at 400 mg/kg acetaminophen, 50% of wild-type mice died compared to none in the null group. Histological examination of liver tissue and estimation of serum bilirubin, creatinine and alkaline phosphatase confirmed that acetaminophen hepatotoxicity was mediated by CYP2E1. Interestingly, CYP2E1 cannot be the only enzyme involved, since at higher doses of acetaminophen (600mg/kg) toxicity began to be manifest in the null mice also.

We have recently generated a mouse line which is nulled for both murine GST Pi genes, P1 and P2 (Henderson, manuscript in preparation). GST Pi expression was originally reported to be greatly elevated in rat pre-neoplastic hepatic foci (Morimura et al, 1993; Kora et al, 1996), and in drug-resistant cell lines (Wareing et al, 1993; Cowan et al, 1986; Black and Wolf, 1991) . In addition, GST Pi has been found to be altered in a variety of human tumours, such as ovarian (Green et al, 1993; Hamada et al, 1994), colon (Mulder et al, 1995), lung (Bai et al, 1996; Hida et al, 1994), bladder (Singh et al, 1994), gastric (Schipper et al, 1996) and testicular tumours (Katagiri et al, 1993), its expression being positively correlated with malignant transformation and resistance to chemotherapeutic drugs, and inversely correlated with patient survival. Interestingly, however, there are reports that survival is greater in those patients whose tumours stain positive for GST pi, for example in renal carcinoma (Grignon et al, 1994). In a similar vein, it has also been reported that in human prostatic carcinoma, expression of GST Pi is completely absent, whereas in normal prostatic and benign hyperplastic tissue, GST Pi expression is unaltered (Lee et al, 1994) . The same authors found that the absence of GST Pi expression in malignant prostatic tissue was due to hypermethylation of the promotor. These findings were confirmed in a subsequent study, although GST Pi expression was found to correlate with the basal cell phenotype, which is absent in prostatic carcinoma, leading to the suggestion that GST Pi may be more involved in the process of epithelial differentiation than in malignant transformation (Cookson et al, 1997).

Our laboratory reported the cloning and characterisation of the murine GST Pi genes, of which there are two (Bammler et al, 1994) , and we have used this knowledge to develop mouse lines in which either both genes have been deleted, or in which only GST P1 is missing (Henderson, manuscript in preparation). In the mouse, GST P1 is expressed at much higher levels than GST P2, and displays far greater catalytic activity towards a range of substrates. In addition, the expression of GST P1 is sexually differentiated in the liver, such that male mice have a level approximately one order of magnitude higher than that in females (Bammler et al, 1994).

Initial results with the double knockout (GSTP1/P2) mouse line suggest that GST pi is not essential for survival, since such mice displayed no phenotypic changes or reproductive difficulties, with litter sizes and sex distribution no different from their wild-type counterparts. Male homozygous null mice had a consistently greater body weight (~10%) than the corresponding wild-type males, although the significance of this finding, if any, is unclear. Characterisation of these animals is continuing, to discover differences in their ability to metabolise and excrete a number of xenobiotics associated with GST pi. Treatment of wild-type mice with acetaminophen resulted in the observation of a sexual differentiation in terms of the hepatotoxic effects of this compound, as evidenced by histological examination of the liver, and estimation of serum levels of enzymes considered diagnostic for hepatocellular damage, such that male mice were much more sensitive than females. When the GSTP1/P2 null male mice were similarly treated, they exhibited increased resistance to the hepatotoxic effects of the drug, relative to their wild-type male littermates. In humans and mice, acetaminophen is activated to its hepatotoxic metabolite, N-acetyl-*p*-benzoquinonimine (NABQI), by the cytochrome P450 monooxygenase system, principally CYP2E1, although CYP3A4 and CYP1A2 may also play a role (Patten et al, 1993; Thummel et al, 1993). The sexual dimorphism in renal toxicity observed in mice (Hu et al, 1993) has been attributed to the sexually differentiated expression of CYP2E1 in this organ (Henderson et al, (1990). However, in the liver, CYP2E1 expression is not sexually differentiated, and this situation is not altered following the deletion of GSTP1/P2. Detoxification and clearance of NABQI is believed to be achieved by conjugation with glutathione, either spontaneously, or by enzymatic means, with the major GST involved being GST Pi, in both rats and humans (153). On this basis, one might expect that a mouse nulled for GST P1 and P2 would exhibit increased sensitivity to the effects of acetaminophen, having a decreased ability to metabolise the hepatotoxic metabolite NABQI. However, the reverse appears to be the case, thus rendering unclear the mechanism of involvement of GST pi in acetaminophen metabolism. However, it is interesting to note that the hepatic level of GST pi may account for the sexual dimorphism in hepatotoxicity seen in the mouse following acetaminophen treatment, since male mice, with much higher levels of GST pi, are more sensitive to the effects of this drug than female mice, and a similar pattern is seen between male wild-type and GST P1/P2 null mice.

Peroxisome proliferators cause a pleiotropic response in the livers of both mice and rats, the animals exhibiting cellular hypertrophy and hyperplasia, an increased number and size of peroxisomes, and an induction in transcription of the enzymes involved in the β-oxidation of fatty acids. These coordinate events may represent an adaptive response in the process of lipid homeostasis; the mechanism is still unclear, but these events are indicative of the importance of peroxisome proliferation in the regulation of fatty acid metabolism. A peroxisome proliferator-activated receptor (PPARα), a member of the steroid receptor superfamily, was first cloned from the mouse, and a further five isoforms have been described, with PPAR homologues also having been identified in humans, rats and frogs. The identification of a peroxisome proliferator response element in the promotor region of target genes lends strength to the suggestion that PPARs act as transcription factors, although direct interaction of peroxisome proliferators with PPAR remains to be demonstrated. However, on the basis of structural similarities, it is assumed that a ligand-binding domain exists within PPAR. Lee et al (1993) produced a PPARα knockout mouse by replacing an 83bp segment of exon 8, within the putative ligand-binding domain, with a neomycin resistance gene. The resultant homozygous null mice were apparently phenotypically normal; however, when challenged with peroxisome proliferators, these mice failed to display the associated pleiotropic response, leading to the conclusion that PPARα is an essential requirement for the action of these chemicals and the mediation of the effects of peroxisome proliferators.

References

Bai F, Nakanishi Y, Kawasaki M, Takayama K, Yatsunami J, Pei XH, Tsuruta N, Wakamatsu K, and Hara N (1996) Immunohistochemical expression of glutathione S-transferase-Pi can predict chemotherapy response in patients with nonsmall cell lung carcinoma. Cancer 78:416-21.

Bammler TK, Smith CA, and Wolf CR (1994) Isolation and characterization of two mouse Pi-class glutathione S-transferase genes. Biochem J 298:385-90.

Black SM, and Wolf CR (1991) The role of glutathione-dependent enzymes in drug resistance. Pharmacol Ther 51:139-54.

Boyd AL, and Samid D (1993) Molecular biology of transgenic animals. J. Animal Sci. 71:1-9.

Campbell SJ, Carlotti F, Hall PA, Clark AJ, and Wolf CR (1996) Regulation of the CYP1A1 promotor in transgenic mice: an ezquisitively sensitive on-off system for cell-specific gene regulation. J. Cell Sci. 109:2619-2625.

Cookson MS, Reuter VE, Linkov I, and Fair WR (1997) Glutathione S-transferase PI (GST-pi) class expression by immunohistochemistry in benign and malignant prostate tissue. J Urol 157:673-6.

Cowan KH, Batist G, Tulpule A, Sinha BK, and Myers CE (1986) Similar biochemical changes associated with multidrug resistance in human breast cancer cells and carcinogen-induced resistance to xenobiotics in rats. Proc Natl Acad Sci USA 83:9328-32.

Fernandez-Salguero P, Pineau T, Hilbert DM, McPhail T, Lee SST, Kimura S, Nebert DW, Rudikoff S, Ward JM, and Gonzalez FJ (1995) Immune system impairment and hepatic fibrosis in mice lacking the dioxin binding Ah receptor. Science 268:722-726.

Gelboin HV (1980) Benzo[a]pyrene metabolism, activation and carcinogenesis: role and regulation of mixed function oxidases. Physiol. Rev. 60:1107-1166.

Green JA, Robertson LJ, and Clark AH (1993) Glutathione S-transferase expression in benign and malignant ovarian tumours. Br J Cancer 68:235-9.

Grignon DJ, Abdel Malak M, Mertens WC, Sakr WA, and Shepherd RR (1994) Glutathione S-transferase expression in renal cell carcinoma: a new marker of differentiation. Mod Pathol 7:186-9.

Hamada S, Kamada M, Furumoto H, Hirao T, and Aono T (1994) Expression of glutathione S-transferase-pi in human ovarian cancer as an indicator of resistance to chemotherapy. Gynecol Oncol 52:313-9.

Henderson CJ, Scott AR, Yang CS, and Wolf CR (1990) Testosterone-mediated regulation of mouse renal cytochrome P-450 isoenzymes. Biochem J 266:675-81.

Hida T, Kuwabara M, Ariyoshi Y, Takahashi T, Sugiura T, Hosoda K, Niitsu Y, and Ueda R (1994) Serum glutathione S-transferase-pi level as a tumor marker for non-small cell lung cancer. Potential predictive value in chemotherapeutic response. Cancer 73:1377-82.

Hu MCT, and Davidson N (1990). A combination of the derepression of the lac operator-repressor system with positive induction by glucocorticoid and metal ions provides a high-level-inducible gene expression system based on the human metallothionein IIA promotor. Mol. Cell. Biol. 10:6141-6151.

Hu JJ, Lee MJ, Vapiwala M, Reuhl K, Thomas PE, and Yang CS (1993) Sex-related differences in mouse renal metabolism and toxicity of acetaminophen. Toxicol Appl Pharmacol 122:16-26.

Jones SN, Jones PG, Ibarguen H, Caskey CT, and Craigen WJ (1991) Induction of the CYP1A1 dioxin-responsive enhancer in transgenic mice. Nucleic Acids. Res. 19:6547-6551.

Jowett T, Wajidi MFF, Oxtoby E, and Wolf CR (1991) Mammalian genes expressed in Drosophila: a transgenic model for the study of mechanisms of chemical mutagenesis and metabolism. EMBO J., 10:1075-1081.

Katagiri A, Tomita Y, Nishiyama T, Kimura M, and Sato S (1993) Immunohistochemical detection of P-glycoprotein and GSTP1-1 in testis cancer. Br J Cancer 68:125-9.

Ko MSH, Sugiyama N, and Takano T (1989) An auto-inducible vector conferring high glucocorticoid inducibility upon stable transformant cells. Gene 84:383-389.

Kollias G, and Grosveld F (1992) The study of gene regulation in transgenic mice. In: Transgenic animals (Kollias G, and Grosveld F, eds.) London: Academic Press, pp. 79-99.

Komori M, Kitamura R, Fukuta H, Inoue H, Baba H, Yoshikawa K, and Kamataki T (1993) Transgenic drosophila carrying mammalian cytochrome-P-4501A1 - an application to toxicology testing. Carcinogenesis 14:1683-1688.

Kora T, Sugimoto M, and Ito K (1996) Cellular distribution of glutathione S-transferase P gene expression during rat hepatocarcinogenesis by diethylnitrosamine. Int Hepatol Comm 5:266-273.

Lee SS, Pineau T, Drago J, Lee EJ, Owens JW, Kroetz DL, Fernandez Salguero PM, Westphal H, and Gonzalez FJ (1993) Targeted disruption of the alpha isoform of the peroxisome proliferator-activated receptor gene in mice results in abolishment of the pleiotropic effects of peroxisome proliferators. Mol Cell Biol 15:3012-22.

Lee WH, Morton RA, Epstein JI, Brooks JD, Campbell PA, Bova GS, Hsieh WS, Isaacs WB, and Nelson WG (1994) Cytidine methylation of regulatory sequences near the pi-class glutathione S-transferase gene accompanies human prostatic carcinogenesis. Proc Natl Acad Sci USA 91:11733-7.

Lee SS, Buters JT, Pineau T, Fernandez Salguero P, and Gonzalez FJ (1996) Role of CYP2E1 in the hepatotoxicity of acetaminophen. J Biol Chem 271:12063-7.

Liang HC, Li H, McKinnon RA, Duffy JJ, Potter SS, Puga A, and Nebert DW (1996) Cyp1a2(-/-) null mutant mice develop normally but show deficient drug metabolism. Proc Natl Acad Sci USA 93:1671-6.

Morimura S, Suzuki T, Hochi S, Yuki A, Nomura K, Kitagawa T, Nagatsu I, Imagawa M, and Muramatsu M (1993) Trans-activation of glutathione transferase P gene during chemical hepatocarcinogenesis of the rat. Proc Natl Acad Sci USA 90.

Mulder TP, Verspaget HW, Sier CF, Roelofs HM, Ganesh S, Griffioen G, and Peters WH (1995) Glutathione S-transferase pi in colorectal tumors is predictive for overall survival. Cancer Res 55:2696-702.

Patten CJ, Thomas PE, Guy RL, Lee M, Gonzalez FJ, Guengerich FP, and Yang CS (1993) Cytochrome P450 enzymes involved in acetaminophen activation by rat and human liver microsomes and their kinetics. Chem Res Toxicol 6:511-8.

Pineau T, Fernandez-Salguero P, Lee SST, McPhail T, Ward JM, and Gonzalez FJ (1995) Neonatal lethality associated with respiratory distress in mice lacking cytochrome P450 1A2. Proc. Natl. Acad. Sci. USA 92:5134-5138.
Ramsden R, Sommer KM, and Omiecinski CJ (1993) Phenobarbital induction and tissue-specific expression of the rat cyp2b2 gen

Extrahepatic Cytochrome P450: Role in *In Situ* Toxicity and Cell-Specific Hormone Sensitivity

Margaret Warner, Heike Hellmold, Malin Magnusson, Tove Rylander, Eva Hedlund, Jan-Åke Gustafsson.
Department of Bioscience, NOVUM, Karolinska Institute, S-141 86 Huddinge, Sweden

Summary

It is clear that members of the Cytochrome P450 supergene family are responsible for the majority of activations of procarcinogens to ultimate carcinogens in the body. These procarcinogens include the food mutagens (heterocyclic amines), pesticides, polycyclic aromatic hydrocarbons and nitrosamines. The Cyp P450 profile in a cell can indicate the capacity of that cell to form reactive metabolites. Furthermore, environmental factors, through their action on P450s, influence the fate of procarcinogens in a cell. This is because different isoforms of P450 are regulated differently by ethanol, diet and environmental inducers, have different substrate specificities and different propensity to be inhibited or activated by dietary components. Cyp P450 (through steroid inactivation), can also influence sensitivity of cells to hormones. Age and hormone related regulation of P450 isoforms such as 1A1, 2B1 and 2A3 in the breast suggest that *in situ* activation of carcinogens and hormone inactivation can occur in the breast. In the brain and endometrium most of the #P450 isoforms remain to be identified.

Introduction

It is now obvious that, in addition to the old classical view of hormones as chemicals which are synthesized by certain organs and carried via the circulation to remote target sites, there are very important hormones which act on cells adjacent to those in which they are synthesized (Knight, 1996; Ashwell et al., 1996; Luscher et al., 1996; Balthazart et al., 1996; Fleming et al., 1996; Navar et al., 1996; Zeldin et al., 1997). One characteristic of these "paracrine" systems is that the hormones need not be synthesized in large amounts because their targets are close to the site of synthesis.

In a similar fashion, it is clear that chemically-induced cytotoxicity or carcinogenesis does not necessarily require transport of reactive metabolites from the liver to the target tissue. Instead, reactive intermediates can be formed within tissues and can act on cells in which they are synthesized or on adjacent cells within a tissue (Utrecht, 1996; Zhou et al., 1996). As with the paracrine hormones, very small amounts of metabolites need to be synthesized and very little of the enzymes involved in formation of the metabolites are necessary. In fact, it can be argued that for such systems, to maintain specificity for their targets, the formation of these "paracrine" molecules must be limited to a very small number of cells.

As with hormones, activation and inactivation of cytotoxins and procarcinogens are very often catalysed by members of the cytochrome P450 supergene family (Nelson et al., 1993). Each tissue in the body has its own specific P450 profile and, therefore, a unique capacity to activate and /or inactivate procarcinogens and hormones. In tissues like the brain, breast and endometrium, where the P450 content is low, it is not uncommon to hear the view expressed that low levels can be equated with lack of physiological and toxicological significance. One has to reflect that many important P450 isoforms, which play crucial roles in the body, are present in tissues at very low concentrations. These include: aromatase in the brain and breast (Naftolin et al., 1975) ; prostacyclin synthase in the endothelium (Tanabe et al., 1995); and epoxygenases of arachidonic acid in the vascular system (Amruthesh et al., 1995; Harder et al., 1995) and pancreas (Zeldin et al., 1997).

With regard to the role of P450 in these tissues in the *in situ* formation of carcinogens it is perhaps necessary to recall that according to present understanding, mutation, at the right site on DNA in a single stem cell, is enough to initiate carcinogenesis. What is most important is not the number of cells which have adducts in their DNA or the total amount of DNA adducts formed, but the potential of the damaged cell with the critical mutation to propagate with the mutation, i.e. clonal expansion of an initiated cell.

The questions about P450 in the brain, breast and endometrium which we will address in this study are: How much P450 is there in these tissues? Which isoforms are expressed? In which cells are they located? How are they regulated? What are their physiological functions? What are the toxicological consequences of P450 in these tissues?

Materials and Methods

Preparation of microsomes and membrane fractions: For microsomal preparation, tissue from 10-20 rats were pooled to obtain sufficient material. The tissue was immersed in ice-cold homogenisation buffer composed of 100 mM TRIS, pH 7.4, 20% glycerol, 1.15% KCl, 0.2 μM dithiothreitol and 1mM EDTA. The tissue was minced with scissors and subsequently homogenised with a polytron homogeniser (PT, Kinematica, Lucerne Switzerland). PMSF, 0.2 mM, was added prior to homogenisation , and the homogenates were filtered through a piece of gauze. Microsomes were prepared by differential centrifugation (Warner et al., 1994) and resuspended and stored at -70° in 50 mM sodium phosphate buffer, pH 7,4, containing 1 mM EDTA and 20% glycerol. Protein content was determined according to Bradford using bovine serum albumin as a standard. Breast microsomes were prepared from pools of 20 rat breasts. For preparation of liver and lung microsomes, samples of approximately 0.5 g from each rat liver and lung were taken. The pooled samples were used for preparation of microsomes. For isolation of brain P450 a total membrane fraction was used. This was obtained by centrifugation of the brain homogenate at 100,000 x g for 1h.

Purification of P450 by hydrophobic chromatography: The microsomes were resuspended in solubilization buffer, 20% glycerol, 0.5%(w/v) sodium cholate, 0.2% emulgen 911 and 0.2 mM EDTA (pH 7.5), and P450 extracted by chromatography on p-chloroamphetamine columns as described previously (Warner et al., 1994). P450 was quantitated according to Omura and Sato (1964) using a Hitachi U-3200 spectrophotometer.

Western blotting: Proteins were precipitated with methanol and separated by SDS gel electrophoresis on 9% gels. In order to detect P450 signals it was necessary to load P450 from approximately 4 g wet weight breast tissue. This represents 0.6 pmol breast P450 per 0.3cm well. For the brain , it was necessary to load 30 pmol P450 /lane. For comparison, 1 pmol P450 of liver and 10 μg of microsomal lung protein was loaded per 0.3 cm well. Proteins were transfered electrophoretically onto PVDF membranes. Transfer buffer was tris-glycine pH 8.8. The membrane was blocked in PBS containing 0.2% Nonidet P-40 and 10% fat free milk. All filter washing and all antibody dilutions were made in this buffer. Secondary antibodies, horseradish peroxidase coupled, (Bio-Rad), were used and antigen antibody complexes were visualised by enhanced chemiluminescence (Amersham).

Antibodies : Generous gifts of antibodies were obtained as follows: Rabbit IgGs raised against peptides from human P450 1A1 and 1A2 from Dr. R.J. Edwards (Department of Clinical Pharmacology, Royal Postgraduate Medical School, London, England); rabbit IgG against rat P450 2A1 also recognizing human P450 2A6 from Dr. F.J. Gonzalez (National Cancer Institute, National Institutes of Health, Bethesda, NMD); sheep IgG against rat P450 4A isozymes from Dr. G. Gibson, (University of Surrey, Guildford); mouse IgG against human P450 19 (aromatase) from Dr. E.R. Simpson (University of Texas, Dallas) (Mendelson 85) and mouse IgG specific to human P450 2D6 from Dr U. Meyer (Biocenter of the University of Basle, Switzerland). Rabbit IgG against rat P450 2E1 was obtained from Oxygene Dallas (Dallas, TX) . Rabbit IgG's against rat P450 2B1/2 and P450 3A isozymes were obtained from Human Biologics, Inc. (Phoenix, Arizona). Microsomes from cell lines expressing human P450 1A1, 1A2, 2D6, 2B6, 2A6, 3A4, 2C8 and 2C9 were obtained from Gentest Corporation (MA, USA) and used as positive controls and to test the crossreactivity of the antibodies.

Human breast, endometrium: Normal human breast tissue from breast reduction surgery and endometrium was obtained at Huddinge hospital. These studies were approved by the Human Ethical Committee at Huddinge Hospital. Informed consent was obtained from all subejcts and the patients were also requested to complete a questionnaire concerning factors that may influence the expression of cytochrome P450, such as medication, smoking, alcohol consumption, parity and diet. Samples from mammary ductal carcinomas were a kind gift from Dr. Martin Bäckdahl (Department of Surgery, Karolinska Institute, Sweden) The excised tissue was frozen in liquid nitrogen and stored at 80°C until further use.

Animals: Sprague-Dawley rats were kept on hardwood bedding under standardized conditions of light (6 am-8 pm) temperature(21 ± 1 °C) and humidity. Food and water were available ad libitum. Pups were weaned on day 21. Tamoxifen was administered by subcutaneous implants in 2 cm -long silicon tubes. Animals were killed by decapitation after light CO_2 anaesthesia. BNF and PCN were administered intraperitoneally as corn oil suspensions. Neuroleptics and SSRB were administered orally by daily gavage.

Results

The results are summarized in table 1 and are not described further in the text.

Discussion

The discovery of multiple forms of P450 in the breast and endometrium has raised several questions about the role of these isoforms in carcinogenesis in these tissues. Cancers in these tissues are hormone dependent, and tissue specific P450, through metabolic inactivation, might alter the sensitivity of cells as well as the duration of action of steroid hormones. In addition, several classes of chemical carcinogens are activated by some of the P450s found in the breast (1A1, 2B1, 2A3) and this activation *in situ* might be responsible for initiation of carcinogenesis.

Since the P450 profile of these tissues can change with age and endocrine status of rats, their capacity to activate or inactivate hormones and procarcinogens may also be changing and there might be "windows" in time when these tissues are more sensitive to chemical carcinogens.

In the case of induction of Cyp 1A1 during pregnancy and lactation, the mechanism of regulation and significance of Cyp 1A1 induction remains to be elucidated. It is not yet established whether there is involvement of the aryl hydrocarbon receptor (AhR) (Hirose et al., 1996; Whitlock et al., 1996). No endogenous activator has been identified for this receptor system but several components of the diet and several environmental poisons, are powerful inducers of the system. Cyp 1A1 is one of a battery of genes whose transcription is increased when the AhR is activated. Although the AhR is essential for the transcriptional activity of the Cyp 1A1 gene, a ligand is not always necessary (Sadek and Hoffmann 1994). The mechanism of the non-ligand-dependent activation is not well understood but is thought to involve phosphorylation of the receptor and /or its partner proteins (Chen and Tukey, 1996).

Table 1. Identification and localization of P450 in the rat brain, and human and rat breast and endometrium.

	Brain	Breast	Endometrium	Liver
P450 content pmol/g	20-50	5-60	500-1,500	10,000
Constitutive isoforms detected	3β-diol hydroxylase P450 7B (DHEA hydroxylase) 4A 2D4, 2C22 Cyp 19	1A1, 1B1 2A, 2B, 2C, 2D, 2E1 3A, 4A Cyp 19	1B1	2A, 2C, 2D, 2E 3A, 4A
% of constitutive forms identified	10%	most	less than 1%	most
Inducible forms	1A1, 1A2 2B, 2C, 2D 3A, 4A	1A1 2E1 3A	?	1A1, 1A2, 1B1 2B, 2C 3A, 4A
Regulation	age hormones ethanol solvents neuroleptics Serotonin uptake blockers	age hormones tamoxifen PCN ethanol	?	age sex hormones diet solvents ethanol pharmaceuticals
Location	neurons glia endothelium	fat epithelium stroma	?	hepatocytes

Another potentially important function of P450 in these tissues could be the bolism of hormone antagonists like tamoxifen. If tamoxifen is metabolized in these tissues, this could explain, why in some situations, there is more agonist than antagonist activity of this antiestrogen and why resistance to tamoxifen develops. In addition, drugs like tamoxifen may be inducers of breast P450 and in this way may increase their own metabolism. Increased metabolism could lead to increased inactivation of the antihormone or increased production of cytotoxic metabolites.

In the liver, tamoxifen is one tenth as potent as phenobarbital as an inducer. Both Cyp 2B and 3A are induced by tamoxifen in the liver and these isoforms can metabolize steroids (Kimmett et al., 1996). Cyp 3A is one of the major pathways for the elimination of tamoxifen (Mani et al., 1993, 1994).

One physiological role for these P450 could be to alter the sensitivity of the cells which harbour them to certain hormones. After lactation, the breast undergoes involution. This involves apoptosis or programmed cell death. If pups are removed from the mothers for more than 72h, apoptosis is initiated. If pups are removed for shorter periods of time, lactation can be resumed when the pups are returned (Lin et al., 1997). Administration of glucocorticoids and progesterone will prevent involution of the breast (Werb et al., 1996; Feng et al., 1995). Enzymes like Cyp 2B and 3A, which can inactivate glucocorticoids and progesterone, may play a role in elimination of these hormones from specific cells in the breast during the postlactational period. Since Cyp 2B is induced in the breast in the postlactational period, the time course and mechanism of this induction is essential for the understanding of the role of Cyp 2B in the breast. The mechanisms of induction of 2B in the liver have not been satisfactorily elucidated. Several xenobiotics, including phenobarbital, pregnenolone-16α-carbonitrile and tamoxifen, induce Cyp 2B and 3A in the liver (Perieria and Lechner, 1995; Nuwaysir et al., 1995; White et al., 1993). In the case of Cyp 3A, glucocorticoids are involved but there are age related changes which suggests that factors other that the glucocorticoid receptor are involved (Perieria and Lechner, 1995). The mechanism of phenobarbital induction and the regions of the 5'-flank of the 2B genes which confer inducibility are still debated issues (Park et al., 1996).

One potentially beneficial intervention which results from knowing the P450 profiles in extrahepatic tissues is dietary manipulation of the catalytic activity of these P450s. The catalytic activity of certain forms of P450 can be inhibited by micronutrients found in fruits and vegetables (Wattenberg, 1990; Boone et al., 1990; Walker, 1996). This might be one of the ways by which diets rich in fruits and vegetables can protect against chemical carcinogenesis.

Much less is known about constitutive forms of P450 in the brain. The ones so far described 3β-diol-hydroxylase (Warner et al 1989), Cyp 7B (Stapleton et al., 1993), 2D4 (Wyss et al., 1995), 4A (Stromstedt et al., 1994) still account for a small fraction of the P450 protein in the brain. However, immunohistochemical studies reveal that the inducible forms are highly specifically located in neurons (Hedlund et al., 1996). It also appears that certain cells in the brain have a high sensitivity to CNS active drugs and solvents and respond by induction of P450.

These include neurons in the cerebellum, olfactory lobes, ventral tegmental area, substantia nigra pars compacta, supraoptic nucleus and the superior olive. The physiological and pharmacological roles and the relationship between expression of P450 in neurons and neuronal degeneration remains to be determined.

Acknowledgements

This work was supported by grants from the Swedish Society for Medical Research and the Swedish Cancer Society.

References

Amruthesh SC, FLack JR, Ellis EF(1992) Brain synthesis and cerebrovascular action of epoxygenase metabolites of arachidonic acid. J. Neurochem 58: 503-510

Ashwell JD, King LB, Vacchio MS (1996) Cross-talk between the T cell antigen receptor and the glucocorticoid receptor regulates thymocyte development. Stem Cells 14: 490-500

Balthazart J, Foidart A, Absil P, Harada N (1996) Effects of testosterone and its metabolites on aromatase-immunoreactive cells in the quail brain: relationship with the activation of male reproductive behavior. J Steroid Biochem. & Mol. Biol56: 185-200

Boone CW, Kelloff GJ, Malone WE (1990) Identification of candidate cancer genopreventive agents and their evaluation in animal models and human clinical trials: A review. Cancer Research 50: 2-9

Chen YH, Tukey RH (1996) Protein kinase C modulates regulation of the CYP1A1 gene by the aryl hydrocarbon receptor. J.Biol. Chem 271: 26261-26266

Feng Z, Marti A, Jehn B, Altermatt HJ, Chicaiza G, Jaggi R(1995) Glucocorticoid and progesterone inhibit involution and programmed cell death in the mouse mammary gland. J.Cell Biol. 131, 1095-1103

Fleming I, Bauersachs J Busse R (1996) Paracrine functions of the coronary vascular endothelium. Mol Cell. Biochem, 157: 137-145

Harder DR, Campbell WB Roman RJ (1995) Role of cytochrome P-450 enzymes and metabolites of arachidonic acid in the control of vascular tone. J Vascular Research 32:79-92

Hedlund E, Wyss A, Kainu T, Backlund M, KöhlerC, Pelto-Huikko M, Gustafsson J-Å, Warner M.(1996) Cytochrome P450 2D4 In The Brain: Specific Neuronal Regulation By Clozapine And Toluene. Molecular Pharmacology 50: 342-50

Hirose K, Morita M, Ema M, Mimura J, Hamada H, Fujii H, Saijo Y, Gotoh O, Sogawa K, Fujii-Kuriyama Y (1996) cDNA cloning and tissue-specific expression of a novel basic helix-loop-helix/PAS factor (Arnt2) with close sequence similarity to the aryl hydrocarbon receptor nuclear translocator (Arnt). Mol Cell Biol 16: 1706-1713

Kimmett SM, McNamee JP, Marks GS (1996) Chick embryo liver microsomal steroid hydroxylations: induction by dexamethasone, phenobarbital, and glutethimide and inactivation following the in ovo administration of porphyrinogenic compounds. Can J Physiol Pharmacol 74: 97-103

Knight PG (1996) Roles of inhibins, activins, and follistatin in the female reproductive system. Frontiers in Neuroendocrinology 17: 476-509

Li M, Liu X, Robinson G, Bar-Peled U, Wagner K-U, Toung WS, Henninghausen L and Furth PA (1997) Mammary-derived signals activate programmed cell death during the first stage of mammary gland involution Proc Natl Acad Sci USA 94: 3425-3430

Luscher TF, Tanner FC, Noll G (1996) Lipids and endothelial function: effects of lipid-lowering and other therapeutic interventions. Current Opinion In Lipidology 7: 234-240

Mani C, Pearce R, Parkinson A Kupfer D (1994) Involvement of cytochrome P4503A in catalysis of tamoxifen activation and covalent binding to rat and human liver microsomes. Carcinogenesis 15: 2715-2720

Mani C, Gelboin HV, Park SS, Pearce R, Parkinson A, Kupfer D (1993) Metabolism of the Antimammary Cancer Antiestrogenic Agent Tamoxifen .1. Cytochrome P-450-Catalyzed N-Demethylation and 4-Hydroxylation. Drug Metab Dispos, 21: 645-656

Naftolin F, Ryan KJ, Reddy VV, Flores F, Petro Z, Kuhn M, White RJ, Takaoka Y, Wolin L (1975) The formation of estrogen by central neurocrine tissue. Recent Progr. Horm Res 31:295-319

Navar LG, Inscho EW, Majid SA, Imig JD, Harrison-Bernard LM, Mitchell KD (1996) Paracrine regulation of the renal microcirculation. Physiological Reviews 6: 425-536

Nelson DR, Kamataki T, Waxman DJ, Guengerich FP, Estabrook RW, Feyereisen R, Gonzalez FJ, Coon MJ, Gunsalus IC, Goth , Okuda K, Nebert DW (1993) The P450 Super family: Update on New Sequences, Gene Mapping, Accession Number, Early Trivial Names of Enzymes, and Nomenclature. DNA and Cell Biol 12:1-51

Nuwaysir EF, Dragan YP, Jefcoate CR, Jordan VC, Pitot HC (1995) Effects of tamoxifen administration on the expression of xenobiotic metabolizing enzymes in rat liver. Cancer Res 55, 1780-1786

Omura T, Sato, R (1964) The carbon monoxide-binding pigment of liver microsomes. Evidence for its hemoprotein nature. J. Biol Chem 239: 2370-2378

Park Y, Li H, Kemper B (1996) Phenobarbital induction mediated by a distal CYP2B2 sequence in rat liver transiently transfected in situ. J Biol Chem 271: 23725-23728

Pereira TM, Lechner MC (1995) Differential regulation of the cytochrome P450 3A1 gene transcription by dexamethasone in immature and adult rat liver. Eur J Biochem 229, 171-177

Sadek CM, Allen-Hoffmann BL(1994) Suspension-mediated induction of Hepa 1c1c7 Cyp1A-1 expression is dependent on the Ah receptor signal transduction pathway. J Biol Chem 269: 31505-31509

Stapleton G, Steel M, Richardson M, Mason JO, Rose KA, Morris RGM, Lathe R (1993) A Novel cytochrome P450 expressed primarily in brain J Biol Chem 270:29739-29745

Strömstedt M, Warner M, Gustafsson J-Å (1994) Cytochrome P450 of the 4A subfamily in the brain. J. Neurochem 63: 671-676.

Strömstedt M, Warner M, Banner C, Gustafsson J-Å (1993) Role of brain P450 in regulation of the level of anaesthetic steroids in the brain. Mol. Pharmacol. 44: 1077-1083

Tanabe T, Miyata A, Nanayama T, Tone Y, Ihara H, Toh H, Takahashi E, Ullrich V (1995) Human genes for prostaglandin endoperoxide synthase-2, thromboxane synthase and prostacyclin synthase. Adv. in Prostaglandin, Thromboxane, & Leukotriene Research 23: 133-135

Uetrecht JP (1996) Reactive metabolites and agranulocytosis. Eur J Haematology 57: 83-88

Walker R (1996) Modulation of toxicity by dietary and environmental factors. Environmental Toxicol Pharmacol 2: 181-188

Warner M, Strömstedt M, Möller L. Gustafsson J-Å (1989) Distribution and regulation of 5α-androstane-3β,17β-diol hydroxylase in the rat central nervous system. Endocrinology124, 2699-2706.

Warner M, Wyss A, Yoshida S, Gustafsson, J-Å (1994) Cytochromes P450 in the brain. Methods in Neuroscience, 22: 51-66

Wattenberg LW. (1990) Inhibition of carcinogenesis by minor nutrient constituents of the diet. Proc Nutr Soc 49: 173-183

Werb Z., Sympson CJ, Alexander CM, Thomasset N, Lund LR, MacAuley A, Ashkenas J, Bissell MJ. (1996) Extracellular matrix remodeling and the regulation of epithelial-stromal interactions during differentiation and involution. Kidney International - Supplement 54:S68-74

White IN, Davies A, Smith LL, Dawson S, De Matteis F (1993) Induction of CYP2B1 and 3A1, and associated monoxygenase activities by tamoxifen and certain analogues in the livers of female rats and mice. Biochem. Pharmacol., 45, 21-30

Whitlock JP Jr, Okino ST, Dong L, Ko HP, Clarke-Katzenberg R., Ma Q, Li H (1996) Cytochromes P450: induction of cytochrome P4501A1: a model for analyzing mammalian gene transcription. FASEB Journal, 10, 809-818

Wyss A, Gustafsson JA Warner M (1995). Cytochromes P450 of the 2D subfamily in rat brain. Mol Pharmacol 47,1148-1155.

Zeldin DC, Foley J, Boyle JE, Moomaw CR, Tomer KB, Parker C, Steenbergen C, Wu S (1997) Predominant expression of an arachidonate epoxygenase in islets of langerhans cells in human and rat pancreas. Endocrinology. 138 1338-1346

Zhou LX, Pihlstrom B, Hardwick JP, Park SS, Wrighton SA Holtzman JL Metabolism of phenytoin by the gingiva of normal humans - the possible role of reactive metabolites of phenytoin in the initiation of gingival hyperplasia Clin Pharm Ther, 60: 191-198

Expression of Xenobiotic-Metabolizing Cytochrome P450s in Human Pulmonary Tissues

Hannu Raunio[1], Jukka Hakkola[1], Janne Hukkanen[1], Olavi Pelkonen[1], Robert Edwards[2], Alan Boobis[2] and Sisko Anttila[3]
[1] Department of Pharmacology and Toxicology, University of Oulu, Oulu, Finland
[2] Royal Postgraduate Medical School, London, UK
[3] Finnish Institute of Occupational Health, Helsinki, Finland

Abstract

The purpose of the study was to obtain a comprehensive picture of the expression of cytochrome P450s (CYP) in the human lung, broncho-alveolar macrophages (BAM), and peripheral blood lymphocytes. The methods used were reverse transcriptase-polymerase chain reaction (RT-PCR) with gene-specific primers and immunohistochemistry with specific anti-peptide antibodies. In RT-PCR, CYPs 1A1, 2B6/7, 2E1, 2F1, 3A5 and 4B1 were detected in cDNA prepared from whole lung tissue. BAMs expressed CYPs 1B1, 2B6/7, 2C, 2E1, 2F1, 3A5 and 4B1. These tissues lacked CYPs 1A2, 2A6, 2D6 and 3A7. In peripheral blood lymphocytes, only CYP1B1 and CYP2E1 mRNAs were consistently detected. In immunohistochemistry with anti-CYP3A antibodies, epithelial staining of CYP3A5 was observed in 100% of individuals, while only about 20% exhibited CYP3A4 staining. CYP3A5 protein was localized in the bronchial wall, bronchial glands, bronchiolar epithelium, alveolar epithelium, vascular endothelium and alveolar macrophages. The results indicate that several different xenobiotic-metabolizing CYPs are present in the human lung, possibly contributing to *in situ* activation of pulmonary procarcinogens.

Introduction

A key question concerning organ-specific chemical carcinogenesis is whether the actual target tissue has the capability to convert procarcinogens to reactive metabolites. One way to obtain such information is to screen for the presence of CYP enzymes that are known to mediate procarcinogen activation. The lung tissue is capable of catalyzing xenobiotic metabolism and reactive metabolites can be formed by several pulmonary cell types in experimental animals and humans (Raunio et al., 1995).

Peripheral blood lymphocytes have been widely used as a surrogate for lung tissue. Most studies on lymphocytes have concentrated on CYP1A1 because of its involvement in the activation of polycyclic aromatic hydrocarbons to DNA-binding adducts. In several of these studies, there has been no correlation between CYP1A1-associated parameters in lymphocytes and lung (Pelkonen, 1992).

The purpose of our studies was to first screen for the expression patterns of CYP genes in whole human lung, isolated BAMs, and peripheral blood lymphocytes at the mRNA level, and then, based on the information thus obtained, to assess the tissue distribution of some of the most interesting CYP proteins.

Materials and Methods

Whole lung samples were obtained from patients who received surgery for tumorous lung lesions. BAM samples were obtained during standard diagnostic bronchoalveolar lavage. Lymphocytes were isolated from blood samples taken from healthy volunteers using standard centrifugation methods. RT-PCR analysis of these tissues was done as described (Hukkanen et al, 1997) using gene-specific primers.

For immunohistochemistry, three anti-CYP3A antibodies were used: a polyclonal antibody detecting both CYP3A4 and CYP3A5, a polyclonal anti-peptide antibody specific for CYP3A4, and an anti-peptide antibody specific for CYP3A5. The characteristics of the antibodies and details of the immunohistochemical analysis are described in Anttila et al. (1997).

Results

CYP gene expression: Specimens from lung tissue were combined into three pools according to the smoking histories of the individual patients. In ethidium bromide-stained agarose gels, an amplification product of CYP1A1 mRNA was detected in the current smoker pool, but not in the non-smoker and ex-smoker pools. Products of CYP2B6/7, CYP2E1, CYP2F1, CYP3A5 and CYP4B1 were found in all three pools, whereas CYP1B1, CYP1A2, CYP2A6, CYP2C8-19 and CYP2D6 products were not detected (Table 1).

Amplification products of CYP2B6/7, CYP2C8-19, CYP2E1, CYP2F1, CYP3A5 and CYP4B1 were detected in each of the BAM samples, while CYP1A1, CYP1A2, CYP2A6, CYP3A4 and CYP3A7 were not detected in any of them. CYP1B1 mRNA was present in about 40% of the samples. Amplification with CYP2D6 primers yielded multiple-size products.

In peripheral blood lymphocyte samples, only CYP1B1 and CYP2E1 mRNAs were consistently detectable in all samples studied. Low levels of CYP2B6/7 mRNA were detected in most of the samples, and CYP2C8-19, CYP3A5 and CYP4B1 mRNAs were found only in a minority of the samples (Table 1).

Tissue distribution of CYP3A proteins: Since the lung samples appeared to express CYP3A5 but not CYP3A4, a more detailed analysis was done both at the mRNA and protein levels. Eight individual lung samples were assayed for the

Table 1. Summary of expression of CYPs in lung, macrophages and lymphocytes

CYP	Lung tissue	BAM	Lymphocytes
1A1	+[1]	-	-
1A2	-	-	-
1B1	-	+/-	+
2A6	-	-	-
2B6/7	+	+	+/-
2C8-19	-	+	+/-
2D6	-	-[2]	-[2]
2E1	+	+	+
2F1	+	+	-
3A4	-	-	-
3A5	+	+	+/-
3A7	-	-	-
4B1	+	+	+/-

+, mRNA detected in all samples; +/-, mRNA detected in some of the samples;
-, mRNA not detected in any samples. The detection criteria was correct-size amplification product visibility in ethidium-bromide stained agarose gels.
1CYP1A1 was detected only in current smoker pool in lung tissue
2CYP2D6 amplification produced multiple bands of different sizes in lymphocytes and macrophages

presence of members in the CYP3A subfamily, i.e. CYP3A4, CYP3A5, and CYP3A7. CYP3A5 mRNA was found in all samples, CYP3A4 mRNA was found in 1/8 sample, but CYP3A7 in none of the samples.

In immunohistochemistry, positive immunostaining in one or several cell types of the lung was observed in all patients examined with the non-specific anti-CYP3A antibody and with the anti-peptide CYP3A5 antibody. The intensity of staining varied considerably between individuals. With the specific anti-peptide CYP3A4 antibody, epithelial staining was observed in only 5/27 and staining of alveolar macrophages in 12/27 cases. CYP3A5 was localized in the ciliated and mucous cells of the bronchial wall, bronchial glands, bronchiolar ciliated and terminal cuboidal epithelium, type I and type II alveolar epithelium, vascular and capillary endothelium, and alveolar macrophages. CYP3A4 was present in bronchial glands, bronchiolar ciliated and terminal epithelium, type II alveolar epithelium and alveolar macrophages. There was no significant association between smoking habits of the patients and the staining intensity with any of the antibodies used, indicating a lack of induction of CYP3A forms by exposure to cigarette smoke.

Discussion

The present results on CYP expression in the whole lung agree well with earlier studies done with different methods, validating the RT-PCR approach used. The pattern of CYP expression in BAMs was similar to, but not identical with, that of the whole lung. An interesting observation is that CYP1A1 levels are very low or non-existent in BAMs and lymphocytes, and that CYP1B1 is expressed in both of these cell types. CYP1A2 appears to be lacking from all of the tissues studied.

CYP mRNA detection in lymphocytes was less consistent than detection of the other tissues studied here. Only CYP1B1 and CYP2E1 mRNAs was found in each individual sample. Some lymphocyte samples in this study contained low levels of CYP2B6/7, CYP2C, CYP3A5, and CYP4B1 mRNAs. This makes it unlikely that the corresponding proteins are present at significant levels in lymphocytes.

In conclusion, several individual *CYP* genes appear to be expressed in whole lung, BAMs, and peripheral blood lymphocytes. BAMs and whole lung have some similarities in CYP expression profiles, as exemplified by the uniform presence of CYP2B6/7, CYP2E1, CYP2F1 and CYP4B1 mRNAs. Lymphocytes differ markedly from both BAMs and whole lung, CYP2E1 being the only common mRNA found in all these tissues. This is reflected by the inconclusive results obtained in studies on the correlation in metabolic markers between peripheral blood lymphocytes and lung. Therefore, the usefulness of peripheral blood lymphocytes as a surrogate for lung tissue in epidemiological studies involving CYP parameters can be questioned.

Table 2. Some procarcinogens metabolised by CYP3A enzymes

- Aflatoxins
- Benzo(a)pyrene
- Benzo(a)pyrene-7,8-diol
- 1-nitropyrene
- 5-methylchrysene
- 6-aminochrysene
- 3,4-dihydroxy-3,4-dihydrobenz(a)anthracene
- 3,4-dihydroxy-3,4-dihydro-7,12-dimethylbenz(a)anthracene
- 9,10-dihydroxy-9,10-dihydrobenz(b)fluoranthene

These data also establish that CYP3A5 is the predominant CYP3A form in normal human lung, and that CYP3A4 is expressed in only about 20% of individuals. The presence of CYP3A proteins in the lung suggests that they can be of importance in the metabolism of inhaled xenobiotics including tobacco smoke-derived and other procarcinogens (Table 2). Interindividual differences

References

Anttila S, Hukkanen J, Hakkola J, Stjernvall T, Beaune P, Edwards RJ, Boobis AR, Pelkonen O, Raunio H (1997) Expression and localization of CYP3A4 and CYP3A5 in human lung. Am J Respir Cell Mol Biol 16:242-249

Hukkanen J, Hakkola J, Anttila S, Piipari R, Karjalainen A, Pelkonen O, Raunio H (1997) Detection of mRNA encoding xenobiotic-metabolizing cytochrome P450s in human broncho-alveolar macrophages and peripheral blood lymphocytes. Mol Carcinogen, in press

Pelkonen O (1992) Carcinogen metabolism and individual susceptibility. Scand J Work Environ Health 18 Suppl 1:17-21

Raunio H, Pasanen M, Mäenpää J, Hakkola J, Pelkonen O (1995) Expression of extrahepatic cytochrome P450 in humans. In: Pacifici GM, Fracchia GN (eds) Advanves in drug metabolism in man. European Comission, Office for Official Publications of the European Communities, Luxenbourg, pp 234-287

Comparison of GST Theta Activity in Liver and Kidney of Four Species

Ricarda Thier, Frederike A. Wiebel, Thomas G. Schulz[1], Andreas Hinke[1,2], Thomas Brüning and Hermann M. Bolt
Institut für Arbeitsphysiologie an der Universität Dortmund, Ardeystr. 67, 44139 Dortmund, FRG
[1] Institut für Arbeits- und Sozialmedizin, Universität Göttingen, Waldweg 37, 37073 Göttingen, FRG
[2] Urologische Universitätsklinik Marienhospital Herne, Widumer Str. 8, 44627 Herne, FRG

Glutathione transferases (GSTs) catalyzing the conjugation of glutathione with electrophilic substrates are important enzymes in the metabolism of xenobiotics. Several isozymes exhibit polymorphisms in humans. The two deletion polymorphisms of hGSTM1 and hGSTT1 result in total loss of enzyme activity in homozygous null genotype (GSTM1*0 and GSTT1*0 respectively) individuals (Seidegård et al. 1988; Pemble et al. 1994). Individuals that are heterozygous for hGSTT1 show distinctly lower enzyme activities than individuals carrying two functional alleles of hGSTT1 (Wiebel et al. 1996). A similar effect is conceivable for the hGSTM1 polymorphism but has not been verified so far.

Recently, a great number of studies have investigated the frequency of the GSTM1*0 or GSTT1*0 genotypes in certain groups of patients, e.g. suffering from specific carcinomas, compared to the general population of the area in question. While these epidemiologic data allow the identification of groups that might be on higher risk they are unsuitable to explain the mechanism leading to this higher sensitivity.

Due to lack of human data modern risk assessment is often based on extrapolation of toxicological data from animal studies. However, differences in toxic response or in target organs can result from differences in xenobiotic metabolism between species. Therefore adequate extrapolation requires information about these species differences. The presence of polymorphisms in humans further complicates the extrapolation from animal data.

We have compared glutathione transferase (GST) theta activity in liver and kidney of rats, mice, hamsters and humans differentiating the three distinct phenotypes seen in humans (Wiebel et al. 1996). Human tissue samples came either from the Urologische Universitätsklinik Marienhospital Herne (kidney tissue) or were gifts kindly granted by D. Davies, Royal Post Medical School,

---

*Abbreviations: GST, glutathione transferase; DCM, dichloromethane; EPNP, 1,2-epoxy-3-(4'-nitrophenoxy)propane; MC, methyl chloride; NC, non conjugator; LC, low conjugator; HC, high conjugator.

Table 1: Means ± standard deviations of GSTT1-1 enzyme activities towards MC and DCM in human, rat, mouse and hamster tissues.

	EPNP-Activity (nmol/min/mg cytosolicprotein)	MC-Activity (k/mg cytosolic protein)		DCM-Activity (nmol/min/mg cytosolic protein)	
Species	Liver (n)	Liver (n)	Kidney (n)	Liver (n)	Kidney (n)
Human, NC	n.d.[a] (2)	n.d. (2)	n.d. (2)	n.d. (2)	n.d. (2)
Human, LC	n.qu[a] (5)	0.053 ± 0.010 (11)	0.022 ± 0.009 (8)	0.62 ± 0.30 (11)	1.38 ± 0.52 (8)
Human, HC	n.qu[a] (5)	0.104 ± 0.009 (12)	0.049 ± 0.010 (4)	1.60 ± 0.48 (12)	3.05 ± 0.72 (4)
Rat	48.04[a] (8)	0.080 ± 0.013 (10)	0.045 ± 0.005 (8)	3.71 ± 0.28 (8)	1.71 ± 0.28 (8)
Mouse, male	215.70[a] (5)	0.558 ± 0.083 (5)	0.087 ± 0.023 (5)	18.2 ± 2.22 (5)	3.19 ± 0.46 (5)
Mouse, female	274.50[a] (5)	0.720 ± 0.107 (5)	0.102 ± 0.026 (5)	29.7 ± 6.31 (5)	3.88 ± 0.90 (5)
Hamster	51.47[a] (6)	n.d. (6)	n.d. (6)	0.27 ± 0.20 (6)	0.25 ± 0.21 (6)

n.d. = not detectable, n.qu. = not quantifiable.
[a] Pooled samples, mean of two measurements.

London, UK (liver cytosolic preparations). The known theta specific substrates methyl chloride (MC), dichloromethane (DCM) and 1,2-epoxy-3-(p-nitrophenoxy)-propane (EPNP) were used to assay the GST Theta activity in cytosolic preparations of the different organs. MC-activity was determined by the gas chromatographic method established by Peter et al. (1989). Formation of formaldehyde out of DCM was measured spectrophotometrically according to Nash (1953) and EPNP-conjugation according to Fjellstedt et al. (1973) and Habig et al. (1974).

No GSTT1-1 activity can be found in non-conjugator samples, whereas high conjugators show a twofold higher activity towards MC and DCM than low conjugators in all organs investigated. The activity of all human samples and kidney samples of all species towards EPNP is too close to the detection limit to differentiate the two human conjugator phenotypes or even to be quantified.

Thus, only the determination of MC- and DCM-activity in human tissue allows the distinction between the three phenotypes of hGSTT1 polymorphism while measurement of EPNP-activity is unsuitable for this aim. EPNP-activity in the mouse is 4-5 times higher than in rat and hamster (Table 1).

All species show 2-5 times higher activity towards MC in liver cytosol than in kidney cytosol. For both organs the relation between species is: Mouse>HC >Rat>LC>Hamster>NC. DCM-activity ratios differ: In rats, mice and hamsters GST activity in liver cytosol is also 2-5 times higher than in the kidney cytosol. Surprisingly, in humans DCM-activity in kidney cytosol is twice as high as in liver cytosol. Here the relation between species is Mouse>Rat>HC>LC>Hamster >NC in liver but Mouse>HC>LC/Rat>Hamster/NC in kidney cytosol (Table 1).

Comparison of hGSTT1-1-activity in human cytosols shows that MC activity is higher in liver than in kidney, while DCM-activity is higher in kidney (Figure 1). Hence, the relation of the two activities deviates, indicating that the substrate affinity of the enzyme differs in these organs, or that an additional enzyme activity is involved. Given the fact that the enzyme activity follows the polymorphism for GSTT1-1, however, it seems more likely that the differences are due to an alternative expression of hGSTT1-1 in different organs.

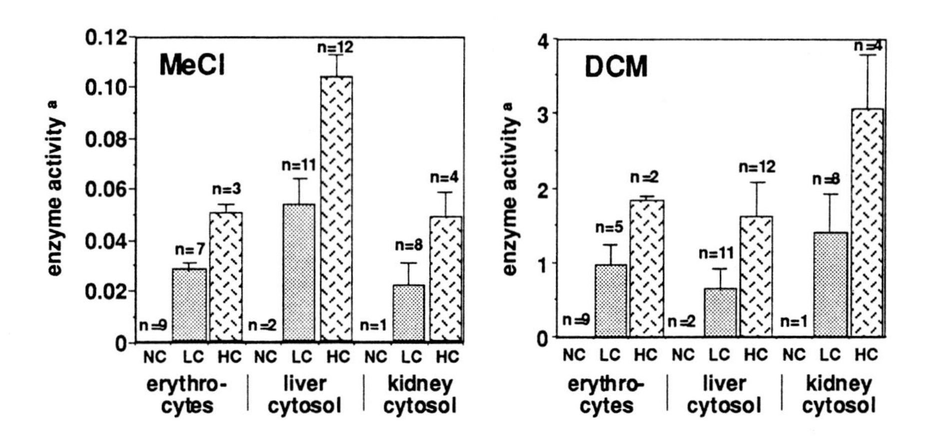

Figure 1: Activity of hGSTT1-1 in erythrocytes, liver and kidney.

[a] Enzyme activity given as metabolic constant (MC) and nmol/min (DCM), respectively, per 100 μl erythrocytes or per mg cytosolic protein

These results emphasize the importance to consider specific conditions at possible target sites when extrapolating from species to species in risk assessment.

Acknowledgements

We thank Ms. A. Dommermuth und S. Reich for their assistance. The studies were supported by the Bundesministerium für Arbeit und Sozialforschung.

References

Fjellstedt TA, Allen RH, Duncan BK, Jacobi WB (1973) Enzymatic conjugation of epoxides with glutathione. J Biol Chem 248:3207-3707

Habig WH, Pabst MJ, Jacobi WB (1974) Glutathione S-transferases - The first enzymatic step in mercapturic acid formation. J Biol Chem 249:7130-7139

Nash T (1953) The colorimetric estimation of formaldehyde by means of the Hantzsch reaction. Biochem J 55:416-421

Pemble SE, Schröder KR, Taylor JB, Spencer SR, Hallier E, Bolt HM (1994) Human glutathione S-transferase theta (GSTT1): cDNA cloning and the characterization of a genetic polymorphism. Biochem J 300 271-276

Peter H, Deutschmann S, Reichel C, Hallier E (1989) Metabolism of methyl chloride in human erythrocytes. Arch Toxicol 63:351-355

Seidegård J, Vorachek WR, Pero RW, Pearson WR (1988) Hereditary differences in the expression of the human glutathione transferase active on trans-stilbene oxide due to a gene deletion. Proc Natl Acad Sci USA 85:7293-7297

Wiebel FA, Vollmer D, Dommermuth A, Thier R, Bolt HM (1996) Inheritance of polymorphic GSTT1-1 and its effect on the formation of albumin adducts by endogenous ethylene oxide, Naunyn Schmiedeberg's Arch Pharmacol, Suppl 354:81

The Young Scientist Poster Award

The Young Scientist Poster Award is given to the first author of the best poster, who must be younger than 35 years.

The winner is announced at the EUROTOX Congress Dinner, and the abstract is published in the Congress Proceedings.

The winner 1997 was T.A. McGoldrick, Aberdeen

Haloalkene Conjugate Toxicity: The Role of Human Renal Cysteine Conjugate C-S Lyase

T.A. McGoldrick[1], E.A. Lock[2] and G.M. Hawksworth[1]

[1] Departments of Medicine and Therapeutics and Biomedical Sciences, University of Aberdeen, Aberdeen, UK

[2] Zeneca, Central Toxicology Laboratory, Alderley Park, Macclesfield, UK

Many halogenated alkenes undergo glutathione conjugation, followed by metabolism to the cysteine conjugates, which are nephrotoxic *in vivo* and *in vitro* in the rat (Lock EA (1988), CRC Crit. Rev. in Toxicol. 19, 1: 23-42). The toxicity is largely dependent on the activity of the renal C-S lyase.

Human proxinmal tubular (HPT) cells were isolated from histologically normal nephrectomies. Confluent cultures were exposed to S-(1,2-dichlorovinyl)-L-cysteine (DCVC), S-(1,1,2,2-tetrafluoroethyl)-L-cysteine, S-(1,2,2-trichlorovinyl)-L-cysteine, S-(2-chloro-1,1-difluoroethyl)-L-cysteine or S-(1,2-dichlorovinyl)glutathione (DCVG) at concentrations ranging from 100 µM to 1 mM for 2 to 48 hours. Toxicity was assessed using the MTT assay.

Of the cysteine conjugates only DCVC displayed toxicity, the TD_{50} at 24 hours being 10 µM. Addition of aminooxyacetic acid (AOAA, 250 µM) almost completely abrogated this toxicity. DCVG (500 µM) caused a 90% reduction in viability after 48 hours. This was prevented by the addition of AOAA or acivin (250 µM). The dichlorovinyl N-acetyl-cysteine conjugate (500 µM) caused a 60% decrease in viability after 48 hours, when applied to the basolateral surface of HPT cells grown on porous membranes.

These results suggest that both C-S lyase activity and the rate of formation of the cysteine conjugate are involved in the toxicity of these compounds.

Subject Index

Printing: Druckhaus Beltz, Hemsbach
Binding: Buchbinderei Schäffer, Grünstadt